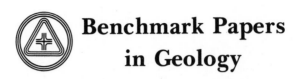 **Benchmark Papers in Geology**

Series Editor: Rhodes W. Fairbridge
Columbia University

Published Volumes

Additional volumes in preparation

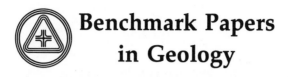

Benchmark Papers
in Geology

—— A *BENCHMARK* ® Books Series ——

METAMORPHISM AND
PLATE TECTONIC REGIMES

Edited by
W. G. ERNST
University of California, Los Angeles

Dowden, Hutchinson & Ross, Inc.
Stroudsburg, Pennsylvania

Distributed by
HALSTED PRESS *A Division of John Wiley & Sons, Inc.*

Copyright © 1975 by **Dowden, Hutchinson & Ross, Inc.**
Benchmark Papers in Geology, Volume 17
Library of Congress Catalog Card Number: 74-23373
ISBN: 0-470-24431-3

77 76 75   1 2 3 4 5
Manufactured in the United States of America.

Library of Congress, Cataloging in Publication Data

Ernst, Wallace Gary, 1931-   comp.
    Metamorphism and plate tectonics.

    (Benchmark papers in geology ; v. 17)
    Includes indexes.
    1. Metamorphism (Geology)--Addresses, essays,
lectures. 2. Plate tectonics--Addresses, essays,
lectures. I. Title.
QE475.A2E76   551.1'3   74-23373
ISBN 0-470-24431-3

Exclusive distributor: **Halsted Press**
a division of John Wiley & Sons, Inc.

Acknowledgments
and Permissions

ACKNOWLEDGMENTS

24th INTERNATIONAL GEOLOGICAL CONGRESS, MONTREAL—*24th International Geological Congress*
   Blueschist Metamorphism and Plate Tectonics
U.S. GEOLOGICAL SURVEY—*U.S. Geological Survey Professional Paper 700-C*
   On-Land Mesozoic Oceanic Crust in California Coast Ranges

PERMISSIONS

The following papers have been reprinted with the permission of the authors and copyright holders.

AMERICAN ASSOCIATION FOR THE ADVANCEMENT OF SCIENCE—*Science*
   Mantle-Derived Peridotites in Southwestern Oregon: Relation to Plate Tectonics

AMERICAN GEOPHYSICAL UNION—*Journal of Geophysical Research*
   Mid-ocean Ridges and Geotherm Distribution During Mantle Convection
   Plate Tectonic Emplacement of Upper Mantle Peridotites Along Continental Edges

AMERICAN JOURNAL OF SCIENCE (YALE UNIVERSITY)—*American Journal of Science*
   Do Mineral Parageneses Reflect Unusually High-Pressure Conditions of Franciscan Metamorphism?
   Occurrence and Mineralogic Evolution of Blueschist Belts with Time

THE CLARENDON PRESS, OXFORD—*Journal of Petrology*
   Evolution of Metamorphic Belts

ELSEVIER PUBLISHING COMPANY
   Physical Earth Planetary Interiors
      Classification of Eclogites in Terms of Physical Conditions of Their Origin
   Tectonophysics
      An Experimental Investigation of the Gabbro–Eclogite Transformation and Some Geophysical Implications
      Pressure and Temperature Conditions and Tectonic Significance of Regional and Ocean-Floor Metamorphism

GEOLOGICAL SOCIETY, NIEDERMENDIG—*Geologischen Rundschau*
   Sind die Alpinotypen Peridotitmassen vielleicht tektonisch verfrachtete Bruchstücke der Peridotitschale?

GEOLOGICAL SOCIETY OF AMERICA—*Geological Society of America Bulletin*
   Eclogites and Eclogites: Their Differences and Similarities
   Peridotite–Gabbro Complexes as Keys to Petrology of Mid-oceanic Ridges
   Thermal Structure of Island Arcs

ROYAL NETHERLANDS ACADEMY OF ARTS AND SCIENCES—*Koninklijke Nederlandse Akademie van Wetenschappen—Amsterdam, Overgenomen iet het Verslag van de gewone vergadering der Afd. Natuurkunde*
Overdruk van Tektonische Oorsprong of Diepe Metamorfose?

THE ROYAL NETHERLANDS GEOLOGICAL AND MINING SOCIETY—*Geologie en Mijnbouw*
The Duality of Orogenic Belts
Some Differences Between Post-Paleozoic and Older Regional Metamorphism

THE ROYAL SOCIETY, LONDON—*Philosophical Transactions of the Royal Society of London*
Metamorphism in the Mid-Atlantic Ridge near 24° and 30°N
The Troodos Massif, Cyprus and Other Ophiolites as Oceanic Crust: Evaluation and Implications

SPRINGER-VERLAG—*Contributions to Mineralogy and Petrology*
Blueschist Alteration During Serpentinization

Series Editor's Preface

The philosophy behind the "Benchmark Papers in Geology" is one of collection, sifting, and rediffusion. Scientific literature today is so vast, so dispersed, and, in the case of old papers, so inaccessible for readers not in the immediate neighborhood of major libraries that much valuable information has been ignored by default. It has become just so difficult, or so time consuming, to search out the key papers in any basic area of research that one can hardly blame a busy man for skimping on some of his "homework."

This series of volumes has been devised, therefore, to make a practical contribution to this critical problem. The geologist, perhaps even more than any other scientist, often suffers from twin difficulties—isolation from central library resources and immensely diffused sources of material. New colleges and industrial libraries simply cannot afford to purchase complete runs of all the world's earth science literature. Specialists simply cannot locate reprints or copies of all their principle reference materials. So it is that we are now making a concerted effort to gather into single volumes the critical material needed to reconstruct the background of any and every major topic of our discipline.

We are interpreting "geology" in its broadest sense: the fundamental science of the planet Earth, its materials, its history, and its dynamics. Because of training and experience in "earthy" materials, we also take in astrogeology, the corresponding aspect of the planetary sciences. Besides the classical core disciplines such as mineralogy, petrology, structure, geomorphology, paleontology, and stratigraphy, we embrace the newer fields of geophysics and geochemistry, applied also to oceanography, geochronology, and paleoecology. We recognize the work of the mining geologists, the petroleum geologists, the hydrologists, the engineering and environmental geologists. Each specialist needs his working library. We are endeavoring to make his task a little easier.

Each volume in the series contains an Introduction prepared by a specialist (the volume editor)—a "state of the art" opening or a summary of the object and content of the volume. The articles, usually some thirty to fifty reproduced either in their entirety or in significant extracts, are selected in an attempt to cover the field, from the key papers of the last century to fairly recent work. Where the original works are in foreign languages, we have endeavored to locate or commission translations. Geologists, because of their global subject, are often acutely aware of the oneness of our world. The selections cannot, therefore, be restricted to any one country, and whenever possible an attempt is made to scan the world literature.

To each article, or group of kindred articles, some sort of "highlight commentary" is usually supplied by the volume editor. This should serve to bring that article into historical perspective and to emphasize its particular role in the growth of the field. References, or citations, wherever possible, will be reproduced in their entirety—for by this means the observant reader can assess the background material available to that particular author, or, if he wishes, he too can double check the earlier sources.

A "benchmark," in surveyor's terminology, is an established point on the ground, recorded on our maps. It is usually anything that is a vantage point, from a modest hill to a mountain peak. From the historical viewpoint, these benchmarks are the bricks of our scientific edifice.

Rhodes W. Fairbridge

Contents

V. EVOLUTION OF METAMORPHIC FACIES TYPES WITH TIME

VI. SUMMARY

Contents by Author

Introduction

Metamorphism involves important changes in the texture, mineralogy, and/or bulk composition of a preexisting, largely solid rock; the processes result from the attendance of physical and chemical conditions different from those prevailing during the formation of the original rock. Excluded from metamorphism are those limiting pressure–temperature (P–T) conditions under which, at profound depths, partial melting ensures, or in the near-surface environment where diagenesis and weathering occur. The physicochemical principles governing metamorphism have been elucidated over the past 80 years and seem firmly established (see, e.g., Barrow, 1893, 1912; Grubenmann, 1904; Van Hise, 1904; Goldschmidt, 1911; Becke, 1913; Eskola, 1914, 1939; Tilley, 1925; Harker, 1932; Turner, 1948; Yoder, 1952; Thompson, 1955; Fyfe, Turner, and Verhoogen, 1958; Korzhinskii, 1959).

Two main types of metamorphism may be distinguished, depending on whether the changes are largely the result of mechanical or chemical processes. Generally both are involved to some extent. Inasmuch as the tectonics and the P–T regimes of an area are thought to reflect lithospheric plate motions, virtually all manifestations of metamorphic processes are at least distantly related to sea-floor spreading and global tectonics. However, many fewer researches have been addressed to the complex relationships between recrystallization and deformation—let alone lithospheric plate motions.

In general, mechanical effects of granulation and mylonitization are confined to, or at least are more apparent at, shallow depths, where rocks possess considerable strength and consequently behave brittlely; such rocks occur in shear zones and in terms of plate tectonics are generally inferred to be characteristic of a conservative, or transform, type of plate junction. Of course, mylonite zones are also developed well within continental crust, not necessarily associated with a plate junction. In contrast, relatively high temperature, relatively low pressure thermochemically promoted recrystallization and hydrothermal alteration are typical of the deeper parts of the

1

oceanic crust in the vicinity of an oceanic ridge—a divergent plate junction. The most widespread and structurally–petrologically complicated metamorphic events take place at and near convergent plate junctions, however. Here the thermal regime, hence metamorphism, will be a function of the rate of plate consumption, plate thickness, degree of uncoupling versus dissipative shear along the convergent junction, and, to a lesser extent, contrasts in radiogenic heat supplied by continental crust-capped lithosphere in contrast to oceanic crust-capped lithosphere. Crustal structures exhibit great variation, too, depending on whether continents collide, or oceanic crust is subducted or overrides the stable oceanic or continental crust-capped plate. Moreover, strong contrasts in tectonics and in recrystallization history are evident even along the extension of a single terrane, such as central and western portions of the Outer Metamorphic Belt of Japan and the high Tauern of the eastern Alps versus the western, Franco-Italian Alps.

The collection of papers presented here attempts to place the recrystallization process and metamorphic belts in a plate tectonic framework. We shall virtually ignore the predominantly mechanical processes and will concern ourselves with recrystallization occasioned by chiefly physicochemical changes. Because these changes occur mainly at nonconservative lithospheric boundaries, we will confine our attention to them. The concept of belts and the areal distribution of such terranes is first presented. We then examine the thermal regimes computed for divergent and convergent plate margins, and relate metamorphism to these environments. Eclogites, the ophiolite suite, and blueschists represent particularly significant problems in this regard, so are considered next. Finally, clues to a secular change in the earth's geothermal gradient with time are considered, based on an evolution of metamorphic facies types. The summary integrates the various principles of metamorphism with plate tectonic regimes.

A companion Benchmark collection, Subduction Zone Metamorphism, brings together significant papers dealing with specific blueschist terranes (New Zealand, New Caledonia, Japan, California, the Western Alps). Too little is yet known about oceanic ridge metamorphism to assemble an analogous volume; also, although relatively high temperature, relatively low pressure metamorphic belts occurring in island arcs and near continental margins have been intensively studied, the detailed integration with plate tectonics is not at all clear, so a collection of appropriate papers has not been attempted.

References

Barrow, G. (1893) On an intrusion of muscovite–biotite–gneiss in the southeastern Highlands of Scotland, and its accompanying metamorphism: *Quart. Jour. Geol. Soc. London,* **49,** 330–358.

Barrow, G. (1912) On the geology of Lower Dee-side and the southern Highland border: *Proc. Geol. Assoc. London,* **23,** 274–290.

Becke, F. (1913) Über Mineralabstand und Struktur der Kristallinischen Schiefer: *Akad. Wiss. Wien Denkschr., Math.–Nat. Kl.,* **75,** 1–53.

Eskola, P. (1914) On the petrology of the Orijarvi region in southwestern Finland: *Bull. Comm. Geol. Finlande,* No. 40, 279 p.

Eskola, P. (1939) Die metamorphen Gesteine, *in Die Entstehung der Gesteine* by T. F. W. Barth, C. W. Correns, and P. Eskola, Springer, Berlin, 422 p.

Fyfe, W. S., Turner, F. J., and Verhoogen, J. (1958) Metamorphic reactions and metamorphic facies: *Geol. Soc. America Mem.,* **73,** 260 p.

Goldschmidt, V. M. (1911) Die Kontaktmetamorphose im Kristianiagebiet: *Oslo Vidensk. Skr., v. I, Mat-Natur. Kl.,* **11,** 405 p.

Grubenmann, U. (1904) *Die kristallinen Schiefer:* Berntraeger, Berlin, 280 p.

Harker, A. (1932) *Metamorphism:* Methuen, London, 360 p.

Korzhinskii, D. S. (1959) *Physicochemical Basis of the Analysis of the Paragenesis of Minerals:* Consultants Bureau, New York, 142 p.

Thompson, J. B., Jr. (1955) The thermodynamic basis for the mineral facies concept: *Amer. Jour. Sci.,* **253,** 65–103.

Tilley, C. E. (1925) A preliminary study of metamorphic zones in the southern Highlands of Scotland: *Quart. Jour. Geol. Soc. London,* **81,** 100–112.

Turner, F. J. (1948) Mineralogical and structural evolution of the metamorphic rocks: *Geol. Soc. America Mem.,* **30,** 342 p.

Van Hise, C. R. (1904) A treatise on metamorphism: *U.S. Geol. Survey Monogr.,* **47,** 1286 p.

Yoder, H. S., Jr. (1952) The MgO–Al₂O₃ SiO₂–H₂O system and the related metamorphic facies: *Amer. Jour. Sci.,* Bowen Vol., 569–627.

I
Metamorphic Belts in Time and Space

Editor's Comments on Papers 1 and 2

1 **Miyashiro:** *Evolution of Metamorphic Belts*

2 **Zwart:** *The Duality of Orogenic Belts*

In 1961, Miyashiro published what has become a classic paper (Paper 1) which first called attention to the fact that contrasting metamorphic paragenetic sequences exist; prior to the appearance of this paper, most petrologists believed that the "normal" metamorphic terrane was characterized by mineral assemblages such as those prevalent in the Scottish Highlands and in New England. One of Miyashiro's major contributions was thus the recognition that three distinctive metamorphic facies series occur world-wide in rocks of different ages. These principal types are (1) the high-temperature, low-pressure (andalusite-sillimanite) type; (2) the moderate-temperature, moderate-pressure (kyanite-sillimanite) type; and (3) the low-temperature, high-pressure (jadeite-glaucophane) type. Two intermediate facies series were also distinguished. At present we regard each metamorphic terrane as a more-or-less unique product of its particular P-T-compositional history, and are aware that the mineral facies series may change to a modest degree even along the strike of a particular belt. Clearly variations in the metamorphic parageneses exist from terrane to terrane; each has developed in response to the specific physical and chemical conditions attending the local recrystallization.

Miyashiro was also the first to clearly demonstrate that in the circumpacific terranes at least, metamorphic belts of roughly the same age are paired. In all cases the relatively high pressure terrane is situated adjacent to the oceanic basin and contains portions of ophiolite suites as the principal igneous units, whereas the relatively low pressure metamorphic complex has developed within the confines of continental crust or a more-or-less sialic island arc, and is characterized by the abundance of both intrusive and extrusive calc-alkaline igneous rocks. With the advent of plate tectonics, the reasons for the spatial disposition of paired metamorphic belts and the petrologic contrasts have become more readily understandable (see Parts II and III). Prior to Miyashiro's 1961 paper, such parallelism was not even recognized.

Finally, Miyashiro pointed out that the relatively high pressure metamorphic belts seem to be developed most abundantly in more recent geologic times; as now exposed, very ancient metamorphic terranes apparently were characterized by relatively high temperatures and low pressures. This phenomenon suggests that either the earth's geothermal gradient has changed with the passage of time (i.e., the crust, at least, is cooling down), or the nature of orogenic activity has evolved with time. This subject will be taken up again in Part V.

The paired nature of orogenic belts, with special reference to metamorphic rocks, was explored further by Zwart in a perceptive article published in 1967 (Paper 2). He accepted Miyashiro's low-pressure and high-pressure metamorphic types, as signaled by index minerals such as andalusite and cordierite on the one hand, versus glaucophane, jadeitic pyroxene, lawsonite, and kyanite on the other. Zwart pointed out that low-pressure terranes are characterized by a marked telescoping of the metamorphic zones, indicating a high geothermal gradient relative to the high-pressure belts.

He also recognized the prevalence of migmatites and granitic rocks in the low-pressure terranes, the scarcity of ophiolites, and, perhaps most significantly, that these characteristic mineral assemblages were associated with a broad zone of orogeny + subsequent uplift and typical open folding. These features are to be compared with a lack of coeval calc-alkaline igneous rocks, an abundance of relatively cold alpine-type ultramafics, a narrow orogenic zone, and the prevalence of nappe structures in the high-pressure belt. Zwart postulated that the high-pressure type of metamorphism is favored by intense deformation; in contrast, a thermal regime characterized by very high heat flow seems required to account for high-temperature, low-pressure terranes.

1

Reprinted from *Jour. Petrology*, **2**(3), 277–311 (1961) with permission of
The Clarendon Press, Oxford

Evolution of Metamorphic Belts

by A K I H O   M I Y A S H I R O

Geological Institute, Faculty of Science, University of Tokyo, Japan

ABSTRACT

The metamorphic facies series in regional metamorphism may be classified into the following categories according to an order of increasing rock pressure: (1) andalusite–sillimanite type, (2) low-pressure intermediate group, (3) kyanite–sillimanite type, (4) high-pressure intermediate group, and (5) jadeite–glaucophane type.

In Japan and other parts of the circum-Pacific region, a metamorphic belt of the andalusite–sillimanite type and/or low-pressure intermediate group and another metamorphic belt of the jadeite–glaucophane type and/or high-pressure intermediate group run side by side, forming a pair. The latter belt is always on the Pacific Ocean side. They were probably formed in different phases of the same cycle of orogeny. Their origin is discussed.

Regional metamorphism under higher rock pressures appears to have taken place in later geological times.

The metamorphic facies series of contact metamorphism are briefly discussed.

INTRODUCTION

THIS paper discusses the classification, origin, and historical development of regional metamorphism.

Regional metamorphism takes place in the deeper parts of orogenic belts, while the regional metamorphic rocks are found in what are here called *metamorphic belts*.

Individual metamorphic facies correspond to a certain range of temperature, rock pressure, and other external conditions (Eskola, 1915, 1920, 1939; Korzhinskii, 1959). The range of temperature and pressure in a metamorphic terrain and metamorphic belt, however, is usually too wide to be definitive of a single metamorphic facies. Even in a single metamorphic terrain, the variation in temperature would be expressed usually by a series of metamorphic facies. Such a series will be called a *metamorphic facies series* or simply a *facies series*. Thus, from the viewpoint of external conditions, each metamorphic terrain is characterized by a certain metamorphic facies series.

Many authors consider that the mineral facies of regional metamorphism are the greenschist, epidote–amphibolite, amphibolite, and granulite facies. Indeed, this facies series appears to exist in Palaeozoic metamorphic belts of Europe and North America. In many other metamorphic belts, however, different kinds of metamorphic facies series are developed. The series of the greenschist, epidote–amphibolite, amphibolite, and granulite facies is only one of the various possible facies series of regional metamorphism.

A metamorphic facies series can be represented by a curve or a group of curves in a temperature–pressure diagram. According to the location of the curve or curves, the facies series may be classified into various types.

Belts of regional metamorphism are well exposed in Japan. Great progress has been made in the geological and petrological study of these belts in recent years, so that they now afford an important basis for the comparative study of the metamorphic belts of the world.

CLASSIFICATION OF FACIES SERIES OF REGIONAL METAMORPHISM

The classification of metamorphic facies series means a classification of metamorphism and of metamorphic terrains from the viewpoint of certain external conditions. After a survey of the petrographic descriptions of metamorphic rocks from different parts of the world, the writer reached the following conclusion. So far as such a classification is to be based upon easily observable mineralogical differences, the kinds or types of metamorphic facies series in the world are rather limited in number. At the present state of our knowledge, the facies series of regional metamorphism may be most conveniently classified into five categories: three standard types and two intermediate groups. Each of the intermediate groups involves various kinds of metamorphism that are intermediate in character between two of the standard types. As will be discussed later, the present classification is intended to be based on the operating rock pressure. Each standard type of metamorphism is accompanied by characteristic igneous activity in the geosynclinal as well as in the orogenic period.

The systematic classification and nomenclature of subfacies should be based on the veritable classification of facies series. Different parts of one facies belonging to different facies series may well be treated as different subfacies. In this paper, however, a new system of nomenclature of subfacies will not be proposed, the classification and nomenclature of metamorphic facies proposed by Eskola (1939) being adopted.

Kyanite–sillimanite type

This standard type of facies series is characterized by the stability of kyanite in a lower grade and that of sillimanite in a higher grade so far as chemical conditions permit. Andalusite and glaucophane are absent. The type metamorphic terrain is in the main part of the Grampian Highlands of Scotland, as studied by Barrow (1893, 1912), Tilley (1925), Harker (1932), Wiseman (1934), Snelling (1957), Chinner (1960), and others.

The facies series of this type is composed of the greenschist, epidote–amphibolite, and amphibolite facies, in order of increasing temperature. In this type, almandine garnet occurs commonly in the middle and high grades, not only in pelitic but also in basic metamorphic rocks. Staurolite is also common, while cordierite is absent. The metamorphic belt is usually divided into progressive metamorphic zones characterized by the following minerals: chlorite, biotite, almandine, kyanite, and sillimanite. The zone of staurolite may be delineated as an independent zone between those of almandine and kyanite. The progressive mineralogical variations are summarized in Fig. 1.

This type of regional metamorphism or mineral zoning was sometimes called the Dalradian or Barrovian one. The main part of the Caledonian metamorphic belt in Norway and that of the Appalachian metamorphic belt in North America appear to belong to this type. In the highest grade of this facies series, the

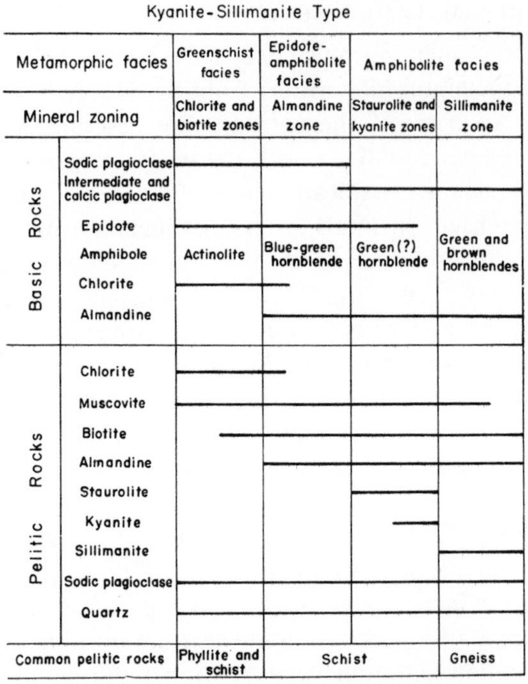

Kyanite-Sillimanite Type

Metamorphic facies	Greenschist facies	Epidote-amphibolite facies	Amphibolite facies	
Mineral zoning	Chlorite and biotite zones	Almandine zone	Staurolite and kyanite zones	Sillimanite zone
Basic Rocks				
Sodic plagioclase				
Intermediate and calcic plagioclase				
Epidote				
Amphibole	Actinolite	Blue-green hornblende	Green (?) hornblende	Green and brown hornblendes
Chlorite				
Almandine				
Pelitic Rocks				
Chlorite				
Muscovite				
Biotite				
Almandine				
Staurolite				
Kyanite				
Sillimanite				
Sodic plagioclase				
Quartz				
Common pelitic rocks	Phyllite and schist	Schist		Gneiss

FIG. 1. Progressive mineralogical variations in regional metamorphism in the main part of the Grampian Highlands of Scotland (Harker, 1932; Wiseman, 1934; &c.).

granulite facies may occur (Chapman, 1952). If all the world is considered, this type may not be more abundant than the other two standard types. It is not justified to regard this type as being 'normal' or ordinary in comparison with the other two, as was done by many authors.

This type of regional metamorphism appears to be always accompanied by the emplacement of a large amount of granitic rocks. Synkinematic granites are usually abundant in the high-grade parts of the metamorphic terrain. Much smaller amounts of gabbroic and ultrabasic rocks are also usually present.

Andalusite–sillimanite type

This standard type of facies series is characterized by the stability of andalusite in a lower grade and of sillimanite in a higher grade, so far as the chemical conditions permit. Kyanite, staurolite, and glaucophane are absent. The type metamorphic terrain is in the central part of the Abukuma Plateau in Japan, as studied by Miyashiro (1953, 1958), Shido (1958), and Shido & Miyashiro (1959).

The facies series of this type is composed of the greenschist and amphibolite facies. The epidote–amphibolite facies is practically absent. In this type, almandine occurs in some pelitic metamorphic rocks of high grades, and does not occur in ordinary basic metamorphic rocks. In the low and middle grades,

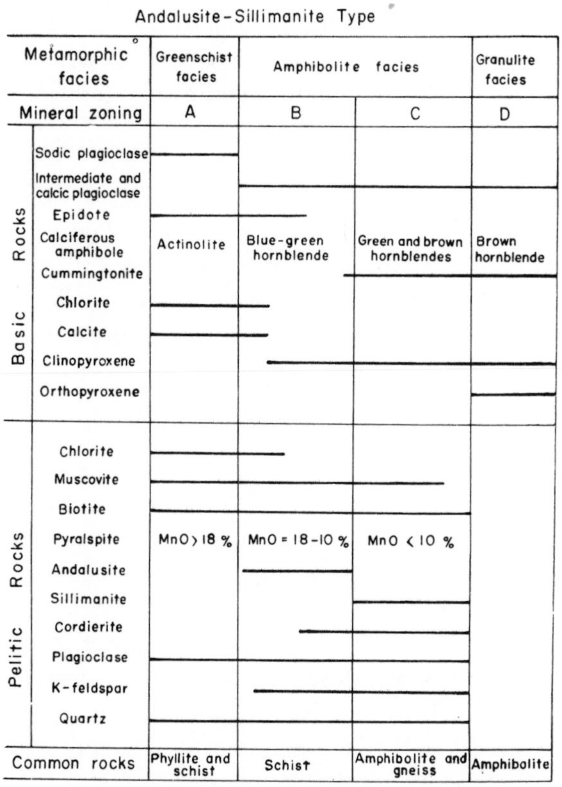

Andalusite-Sillimanite Type

Metamorphic facies	Greenschist facies	Amphibolite facies		Granulite facies
Mineral zoning	A	B	C	D

Basic Rocks:

	A	B	C	D
Sodic plagioclase				
Intermediate and calcic plagioclase				
Epidote				
Calciferous amphibole	Actinolite	Blue-green hornblende	Green and brown hornblendes	Brown hornblende
Cummingtonite				
Chlorite				
Calcite				
Clinopyroxene				
Orthopyroxene				

Pelitic Rocks:

	A	B	C	D
Chlorite				
Muscovite				
Biotite				
Pyralspite	MnO > 18 %	MnO = 18-10 %	MnO < 10 %	
Andalusite				
Sillimanite				
Cordierite				
Plagioclase				
K-feldspar				
Quartz				
Common rocks	Phyllite and schist	Schist	Amphibolite and gneiss	Amphibolite

Fig. 2. Progressive mineralogical variations in regional metamorphism in the central Abukuma Plateau, Japan (Miyashiro, 1953, 1958; Shido, 1958; Shido & Miyashiro, 1959).

pyralspite garnet is rare and, when it occurs, is manganiferous (Miyashiro, 1953). Cordierite occurs commonly. The progressive mineralogical variations are summarized in Fig. 2.

The highest grade of regional metamorphism in the type metamorphic terrain is in the amphibolite facies, as stated above. The contact metamorphism by synkinematic intrusions in the terrain produced the same type of metamorphic facies series, though the highest grade of the contact metamorphism reached what may be regarded as a part of the granulite facies, since orthopyroxene was formed there. This highest-grade part is also shown in Fig. 2.

Formerly, this type of regional metamorphism was called the central Abukuma type (Miyashiro, 1958). The Buchan type of Read (1952) is not the same as this, but belongs to the low-pressure intermediate group to be mentioned later. The Ryoke metamorphic belt of Japan and some metamorphic belts formed in Early Palaeozoic time in New South Wales, Australia, appear to belong to this type.

Many geologists regarded the regional metamorphism of this type as contact metamorphism only, because of the occurrence of andalusite and cordierite. The occurrence of such minerals, however, depends upon the operating temperature and pressure, and not upon the geological distinction between the contact and regional types. Metamorphism is to be regarded as regional providing the thermal structure of the resultant metamorphic terrain is independent of individual plutonic masses.

This type of regional metamorphism appears to be always accompanied by the emplacement of a large amount of granitic rocks. Synkinematic granites are usually abundant in the high-grade parts of the metamorphic terrain. Granite may occur abundantly, however, not only inside but also outside the regional metamorphic terrain, as in the case of the Ryoke belt. Much smaller amounts of gabbroic and ultrabasic rocks are also usually present.

Jadeite–glaucophane type

This standard type of facies series is characterized by the stability of the jadeite–quartz assemblage and glaucophane. The type metamorphic terrain is in the Kanto Mountains in Japan, as studied by Seki (1958, 1960 b, c, 1961b).

The facies series in the type metamorphic terrain is composed of the glaucophane–schist and greenschist facies. This terrain belongs to the Sanbagawa metamorphic belt. In the Bessi district of the same metamorphic belt, not only rocks of the glaucophane–schist facies but also rocks of the epidote–amphibolite facies and even eclogite are exposed, as will be described later (p. 295). These rocks may represent higher-grade parts of the present type of facies series. It is not certain, however, whether the metamorphic terrain of the Bessi district is of the jadeite–glaucophane type or of the high-pressure intermediate group to be mentioned later.

Lawsonite and pumpellyite usually occur in the metamorphic terrains of this type. The stable form of Al_2SiO_5 is probably kyanite, as was found in the Bessi district (Banno, 1957). Pyralspite garnet occurs commonly, and is almandine in some cases and more manganiferous in others. Andalusite, sillimanite, and cordierite do not occur. Note that jadeite in quartz-free rocks would not be characteristic of this type. The progressive mineralogical variations in the type metamorphic terrain are summarized in Fig. 3.

The regional metamorphism of this type and of similar nature, producing glaucophane, was called glaucophanitic metamorphism in some cases. At least some parts of the Kamuikotan metamorphic belt of Hokkaido and of the Franciscan metamorphic terrain of California appear to belong to this type.

This type of regional metamorphism is usually accompanied by the emplacement of a large volume of ultrabasic rocks (mostly serpentinite). Gabbroic rocks are also present, but granitic rocks are absent.

Formerly, many authors stated that the formation of glaucophane schists is due to a special kind of metasomatism, and Turner (1948) and others denied the existence of the glaucophane–schist facies. On the other hand, the glaucophane–schist facies was advocated by de Roever (1955a) and Miyashiro & Banno (1958).

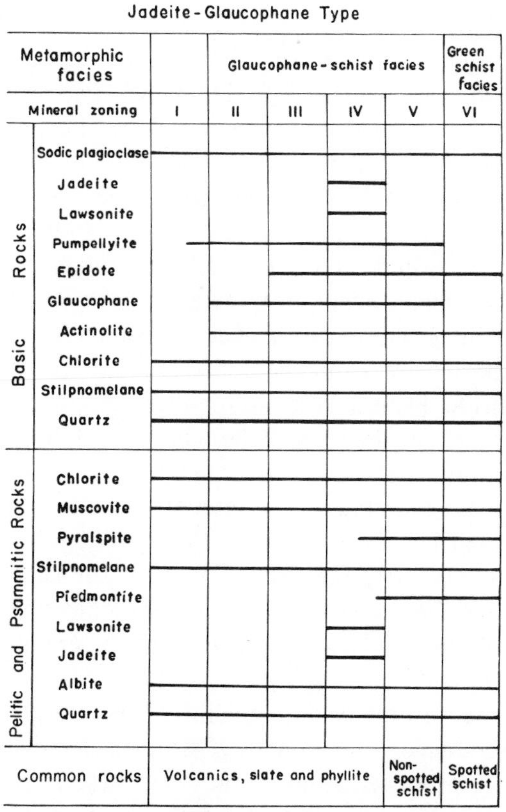

Jadeite - Glaucophane Type

Metamorphic facies		Glaucophane-schist facies					Green schist facies
Mineral zoning	I	II	III	IV	V	VI	

Basic Rocks

Mineral	I	II	III	IV	V	VI
Sodic plagioclase						
Jadeite						
Lawsonite						
Pumpellyite						
Epidote						
Glaucophane						
Actinolite						
Chlorite						
Stilpnomelane						
Quartz						

Pelitic and Psammitic Rocks

Mineral	I	II	III	IV	V	VI
Chlorite						
Muscovite						
Pyralspite						
Stilpnomelane						
Piedmontite						
Lawsonite						
Jadeite						
Albite						
Quartz						

Common rocks	Volcanics, slate and phyllite	Non-spotted schist	Spotted schist

FIG. 3. Progressive mineralogical variations in regional metamorphism in the Kanto Mountains, Japan (Seki, 1958, 1960 b, c, 1961b). Recrystallization is very incomplete in zone I.

It appears, however, that the existence of the glaucophane–schist facies has now become widely accepted and need not be advocated further.

It is surprising that no one had dealt with the glaucophanitic metamorphism as a progressive one having zones not only of the glaucophane–schist facies but also of some other facies, before Banno (1958a) and Seki (1958) succeeded in clarifying the mineral zoning of glaucophanitic terrains in Japan.

Low-pressure and high-pressure intermediate groups

As will be discussed later (p. 284), the prevailing rock pressure for the andalusite-sillimanite type is lower than that for the kyanite–sillimanite type, which latter is, in turn, probably lower than that for the jadeite–glaucophane type. There are many cases of regional metamorphism that may be regarded as being intermediate in character between the andalusite–sillimanite and kyanite–sillimanite types. Probably the prevailing rock pressure in these cases was intermediate between the pressures for the two standard types concerned. These cases of metamorphism will be said to belong to the *low-pressure intermediate group*.

As an example of the low-pressure intermediate group, the metamorphic terrain in the north-eastern Grampian Highlands (Aberdeenshire and Banffshire) of Scotland may be cited. In this area, andalusite and cordierite occur in pelitic metamorphic rocks, just as in the andalusite–sillimanite type, but staurolite also occurs just as in the kyanite-sillimanite type. Read (1952) called such a metamorphism the Buchan type. The association of andalusite and staurolite is not rare in the world. Many of the Precambrian metamorphic terrains in the Baltic and Canadian shields have this mineral association, and appear to belong to the low-pressure intermediate group. The coexistence of andalusite, staurolite, and kyanite as reported by Yamaguchi (1951) from Korea, as well as the coexistence of andalusite, kyanite, and sillimanite as reported by Hietanen (1956) from Idaho, indicates that the terrains belong to the low-pressure intermediate group.

There are other cases of regional metamorphism that may be regarded as being intermediate in character between the kyanite–sillimanite and jadeite–glaucophane types. Probably the prevailing rock pressure in these cases was intermediate between the pressures of the two standard types concerned. These cases of metamorphism will be said to belong to the *high-pressure intermediate group*.

For example, there exist many glaucophanitic metamorphic terrains where the jadeite–quartz assemblage is not found. If we consider that the jadeite–quartz assemblage was not produced in any grades of these terrains, then the metamorphism belongs to the high-pressure intermediate group. In this case, however, conclusive determination would be rather difficult, because jadeite is apt to be overlooked in many cases and the jadeite-bearing part may have been lost by some later events. Metamorphic terrains that resemble those of the kyanite–sillimanite type, except for the rare, sporadic occurrence of glaucophanic amphibole, probably belong to the high-pressure intermediate group.

*Discrimination between types of metamorphism with the aid of minerals of solid-
    solution series*

When some minerals characteristic of a type of metamorphism occur in a metamorphic terrain, one can readily identify the type. In many cases, however, appropriate characteristic minerals do not occur owing to the limited range in chemical composition of the metamorphic rocks developed. Even in such cases, one may be able to recognize the type through investigation of the solid-solution relations of some metamorphic minerals, as exemplified below.

Pyralspite garnet may occur in pelitic metamorphic rocks in any types of regional metamorphism, but the Mn content of pyralspite is much higher in the andalusite–sillimanite type than in the kyanite–sillimanite type (Miyashiro, 1953, 1958). The Na content of common hornblende in ordinary basic metamorphic rocks is high in the jadeite–glaucophane type and low in the kyanite–sillimanite type, and is still lower in the andalusite–sillimanite type (Shido, 1958; Shido & Miyashiro, 1959). Plagioclase becomes more calcic with rising temperature in any types of metamorphism, but the rate of increase of the Ca content is greater in the andalusite–sillimanite type than in the kyanite–sillimanite type (Miyashiro, 1958; Shido, 1958). In the jadeite–glaucophane type, intermediate and calcic plagioclases could not be formed.

Temperature, rock pressure, and other external conditions

The temperature and rock pressure corresponding to the three standard types of metamorphic facies series can be inferred from the experimentally determined stability relations of characteristic minerals.

The kyanite–sillimanite equilibrium curve was experimentally determined by Clark, Robertson, & Birch (1957) and by Clark (1960), as shown by line (*k*) in Fig. 4. Though the stability relations of andalusite have not been determined, they are probably approximately as shown in the same figure (Miyashiro, 1953; Thompson, 1955; Clark *et al.*, 1957; Miyashiro, 1960*a*).

The equilibrium curve for the reaction: jadeite+quartz = albite was experimentally determined by Birch & LeComte (1960), as shown in Fig. 4.

Typical temperature/rock-pressure curves corresponding to the three standard types of regional metamorphism are shown by arrows (1), (2), and (3) in Fig. 4. It is clear that the prevailing rock pressure in the andalusite–sillimanite type is lower than that in the kyanite–sillimanite type, which latter in turn is lower than that in the jadeite–glaucophane type. If it be assumed that the pressures shown in this figure were produced lithostatically by the weight of the overlying rocks, the depth of regional metamorphism is much greater than that so far inferred from geological studies. As our geological studies are not generally complete, such a great depth may have been actually reached in a locally and temporarily thickened crust during orogeny. It is also possible that orogenic compression played an important part in producing such pressures, at least in some cases.

The upper and lower limits of metamorphic temperature would differ in different types. It is only rarely that the metamorphism of the jadeite–glaucophane type reaches so high a temperature as to produce biotite, whereas that of the kyanite–sillimanite type commonly produces biotite in the middle- and high-grade parts. In the metamorphism of the andalusite–sillimanite type, most of the metamorphic rocks are higher in grade than the biotite isograd, and on the lower-temperature side of the biotite isograd recrystallization is usually very incomplete. Although the lowest temperature for the formation of biotite should vary to some extent with rock and water pressures, the above relations might well be considered as suggestive that the metamorphic temperature in the jadeite–glaucophane type is generally lower than that in the kyanite–sillimanite

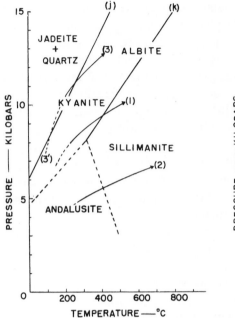

FIG. 4. Stability relations of andalusite, kyanite, and sillimanite and of the jadeite–quartz assemblage and albite. The solid lines (*k*) and (*j*) represent the experimentally determined equilibrium curves for the kyanite–sillimanite transformation, and for the reaction jadeite+quartz = albite, respectively. The arrows (1), (2), and (3) represent typical temperature/rock-pressure relations in regional metamorphism of the kyanite–sillimanite type, andalusite–sillimanite type, and jadeite–glaucophane type, respectively.

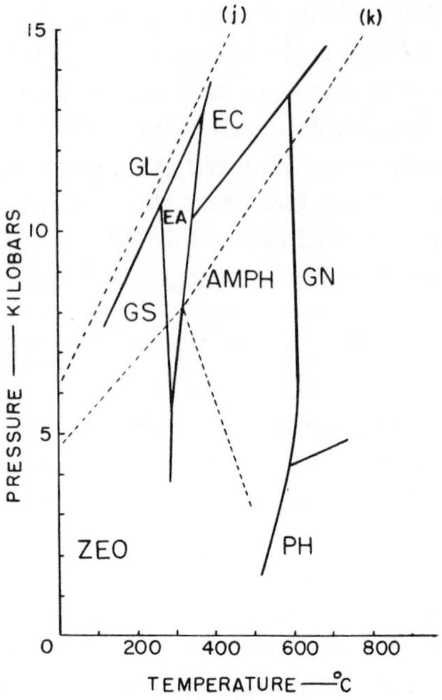

FIG. 5. Estimated temperatures and rock pressures of metamorphic facies. The solid lines (*k*) and (*j*) correspond to those of Fig. 4. GL, glaucophane–schist facies; EC, eclogite facies; GS, greenschist facies; EA, epidote–amphibolite facies; AMPH, amphibolite facies; GN, granulite facies; PH, pyroxene–hornfels facies; ZEO, zeolite facies.

type, which latter, in turn, tends to be lower than that in the andalusite–sillimanite type, as shown in Fig. 4.

In a high grade of the andalusite–sillimanite and kyanite–sillimanite types of regional metamorphism, where sillimanite is stable, muscovite decomposes by reaction with quartz to produce sillimanite, potassium feldspar, and water. The equilibrium temperature of the reaction was thermodynamically calculated for a wide range of rock and water pressures (Miyashiro, 1960*b*). The result is 450°–590° C. The decomposition of biotite by reaction with quartz takes place in the highest grade (granulite facies), probably of the andalusite–sillimanite and kyanite–sillimanite types. This temperature was estimated at 700°–800° C. (Miyashiro, 1960c).

The temperatures and rock pressures of metamorphic facies are roughly shown in Fig. 5. Note that there is a fundamental difference between this figure and fig. 2 of Francis (1956), who attached too great an importance to the kyanite–sillimanite type in comparison with the other standard types.

In recent years the behaviour of water in metamorphism has been discussed by Thompson (1955) and others. As the water content of metamorphic rocks decreases with increasing grade of metamorphism, water is evidently a mobile component. The fact that zonal mapping on a regional scale is usually possible in metamorphic terrains indicates that the prevailing water pressure is rather uniform or regular within the terrains. There are some indications that the prevailing water pressure was generally much lower than the rock pressure during regional metamorphism. Kennedy (1960) showed that pyrophyllite, instead of sillimanite+quartz and kyanite+quartz, is stable up to 600° C or higher temperatures under a pressure of more than 4 kilobars, providing the water pressure is equal to the rock pressure. On the other hand, if the prevailing pressure of water is equal to the rock pressure, quartzo-feldspathic rocks would begin to melt at temperatures near 600° C under pressures of more than 4 kilobars. Therefore, pyrophyllite, instead of sillimanite+quartz and kyanite+quartz, is stable nearly up to the minimum melting-point of quartzo-feldspathic rocks, if the water pressure is equal to the rock pressure. This is not, however, the case with ordinary metamorphic terrains. Actually there are wide areas of metamorphic rocks where sillimanite and kyanite, in association with quartz, are stable at temperatures much lower than the melting-point of quartzo-feldspathic rocks. For this reason the prevailing water pressure is considered to be generally much lower than the rock pressure during metamorphism.

Shaw (1956) showed that, generally, the iron in metamorphic rocks tends to be progressively reduced with increasing metamorphism. In the central Abukuma Plateau, not only the water content but also the oxygen content of metamorphic rocks decreases with increasing metamorphism, whereas in the Grampian Highlands the water content decreases but the oxygen content does not decrease (Miyashiro, 1958, pp. 267–8). These facts suggest that the mobility of oxygen would differ in different metamorphic terrains. In the central Abukuma Plateau, a recent investigation of oxide minerals in metamorphic rocks did not reveal the existence of local difference in the chemical potential of oxygen (Banno & Kanehira, 1961). Many authors showed from studies of mineral parageneses, however, that metamorphic rocks are closed with regard to oxygen. This conclusion was reached by Eugster (1959) in considering the metamorphosed iron formation of Michigan (low-pressure intermediate group), by Thompson (1957) in the northern Appalachians (kyanite–sillimanite type), by Chinner (1960) in the Grampian Highlands (kyanite–sillimanite type), and by Banno & Kanehira (1961) in glaucophanitic metamorphic rocks of the Bessi district,

Japan. It is now clear that the partial pressure of oxygen in metamorphic rocks is generally independent of that of water.

Variation of metamorphic facies series within a metamorphic belt

In many metamorphic belts, the character of the metamorphic facies series produced differs to some extent in different parts. So far as is known to the writer, the variation within a metamorphic belt is not so great as to result in the metamorphic belt including two of the above-mentioned standard types, but is such that the metamorphic belt involves one standard type and one adjacent intermediate group.

For example, the main part of the Grampian Highlands of Scotland is of the kyanite–sillimanite type, but a smaller part of the Highlands is of the low-pressure intermediate group.

PAIRED METAMORPHIC BELTS IN THE MAIN PART OF JAPAN

Two pairs of metamorphic belts

In the main part of Japan, i.e. Honsyu (Honshu), Sikoku (Shikoku), and Kyusyu (Kyushu), there are four belts of regional metamorphic rocks, running roughly parallel along the island arc of Japan. These four may be considered to belong to two pairs, here called the *older* and *younger*. As shown in Fig. 6, the older pair of metamorphic belts is composed of the *Hida metamorphic belt* on the northern side and the *Sangun metamorphic belt* on the southern. Large parts of the older pair are now covered by the Japan Sea, and the original structure is greatly obscured by later events. On the other hand, the younger pair of metamorphic belts is to the south of the older pair, and is composed of the *Ryoke–Abukuma metamorphic belt* on the northern, and the *Sanbagawa (Sambagawa) metamorphic belt* on the southern side. The original structure of this pair is relatively well preserved.

Each pair of metamorphic belts probably represents deeper parts of an orogenic belt in geological time. The ages of the orogenies and metamorphisms have been a subject of hot dispute amongst Japanese geologists in recent years. The geological evidence was not, in most cases, conclusive. For an historical review of this problem see Minato (1960), although the old radiometric dating which he quoted is quite unreliable. New isotopic dating is now giving important evidence relevant to this problem (Kuno *et al.*, 1960; Miller *et al.*, 1961; Banno & Miller, 1961). At the present state of our knowledge, it is most probable that the age of orogeny and regional metamorphism in the older pair is Late Palaeozoic and/or Early Mesozoic, whereas that in the younger pair is Late Mesozoic. The former would correspond to the Akiyosi (Akiyoshi) cycle of orogeny proposed by Kobayashi (1941), and the latter to his Sakawa cycle. The K–A ages so far obtained may be summarized as follows:

Hida metamorphic belt	180–90 million years (Triassic)
Ryoke–Abukuma metamorphic belt	91–102 million years (Cretaceous)
Sanbagawa metamorphic belt	84–93 million years (Cretaceous)

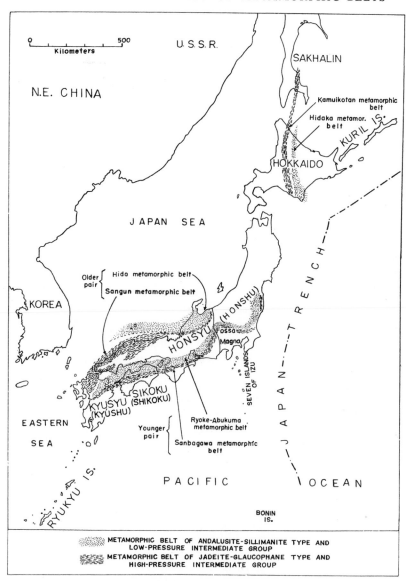

FIG. 6. Metamorphic belts of Japan, restored to the original state.

The regional metamorphism of the Sangun belts has not been dated, but some metamorphic terrains of the belt are covered by unmetamorphosed Middle Triassic sediments.

The tectonic development of Japan may be outlined as follows: From Early Palaeozoic time (at least Silurian), there was a large geosyncline roughly along the present site of the island arc of Japan. Thick sediments, especially of Late Palaeozoic time, were deposited there. The geosynclinal pile was subjected to orogeny and regional metamorphism first in Late Palaeozoic and/or Early

Mesozoic time, and secondly in Late Mesozoic time. Thus, two pairs of metamorphic belts were formed, and constituted the backbone of the Japan arc. Afterwards, in the Tertiary period, the main site of active tectonic movement shifted to the zone along the Japan Trench, which itself was formed probably by this tectonic movement. Thus, the Japan Trench and the adjacent zone to the west (Kuril Islands, north-eastern Honsyu, Seven Islands of Izu, and Bonin Islands) are a Cenozoic and present-day orogenic belt (Sugimura, 1960). The western part of Japan became a more stable region in Cenozoic time.

In each of the two pairs, the metamorphic belt on the northern side (i.e. continental side) is of the andalusite–sillimanite type and/or low-pressure intermediate group accompanied by abundant granitic rocks, whereas that on the southern side (i.e. Pacific Ocean side) is of the jadeite–glaucophane type and/or high-pressure intermediate group accompanied by abundant ultrabasic rocks. As will be shown on later pages, the occurrence of such paired metamorphic belts is a common feature not only in Japan but also in other parts of the circum-Pacific region. I have always called such paired metamorphic belts *metamorphic belts of the Japan type*, because they are most typically developed and most thoroughly investigated in Japan (Miyashiro, 1959).

There follows a rather detailed review of our knowledge of the metamorphic belts of the younger pair, which will help in understanding similar relations that are less completely preserved or less sufficiently investigated in other parts of the circum-Pacific region.

Ryoke–Abukuma metamorphic belt

As shown in Figs. 6 and 7, the Ryoke–Abukuma metamorphic belt is composed of two wings, each of which is convex towards the south-east (i.e. towards the Pacific Ocean). The western wing is called the *Ryoke metamorphic belt*, and runs from Takato in central Honsyu to central Kyusyu. On the other hand, the eastern wing is now divided into three areas: (1) a small area north of the Kanto Mountains, (2) Tukuba (Tsukuba) district, and (3) Abukuma Plateau. The metamorphic terrains of the Abukuma Plateau are divided into two main masses by granitic intrusions, i.e. a central part (Gosaisyo, Takanuki, and Nakoso districts) and a southern part (Hitati district).

The two wings meet in an area of later tectonic disturbance, usually called the *Fossa Magna*, in central Honsyu. This area is at the intersection of the Japan arc with the Izu–Marianas arc, which latter runs southward from the Seven Islands of Izu through the Bonin (Ogasawara) Islands to the Marianas Islands (Fig. 6). It is conceivable that the Ryoke–Abukuma metamorphic belt was originally in a single arcuate form and afterwards deformed into the present form, composed of two wings, by later events related to crustal disturbances of the Izu–Marianas arc.

Fig. 8 shows a provisional classification of granitic rocks (in a broad sense) in the main part of Japan according to the geological ages of their emplacement.

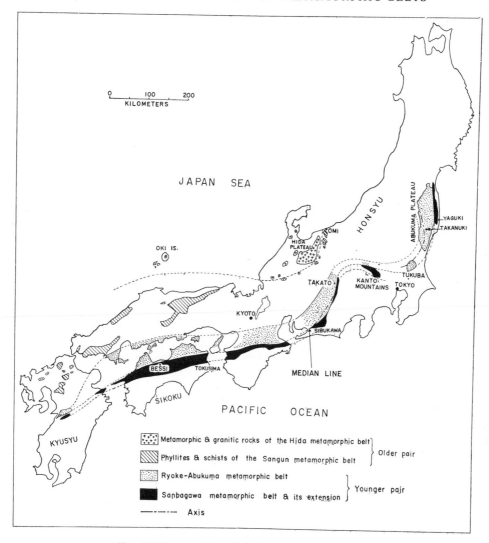

Fig. 7. Metamorphic belts in the main part of Japan.

It is clear that most of the granitic rocks now exposed widely in Japan, both inside and outside the Ryoke–Abukuma metamorphic belt, were formed by the Late Mesozoic orogeny that caused the Ryoke–Abukuma metamorphism.

The granitic rocks inside the Ryoke–Abukuma metamorphic belt may be classified into synkinematic and post-kinematic. The facies series of regional metamorphism in the belt is practically identical with that formed through contact metamorphism by the synkinematic granites, whereas it differs from that formed through contact metamorphism by the post-kinematic granites. A much smaller volume of gabbroic rocks is also present. Usually, the intrusion of gabbroic rocks was earlier than that of granitic rocks. The thermal structure of

21

this metamorphic belt was controlled mainly by the regional uprise of isogeo-
thermal surfaces through the geosynclinal pile and was partly modified by the
distribution of synkinematic intrusions. Synkinematic granites tend to be more
abundant in high-grade parts of the thermal structure.

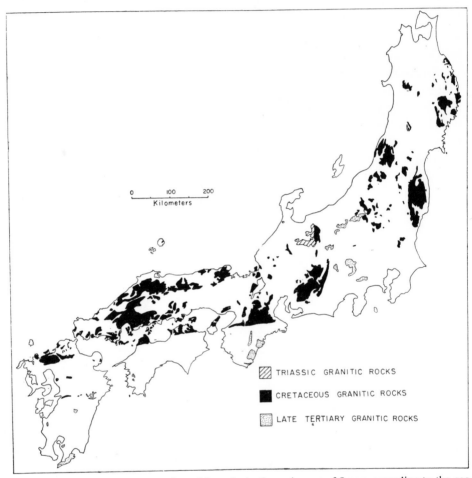

FIG. 8. Provisional classification of granitic rocks in the main part of Japan, according to the geological ages of their emplacement. Note that Cretaceous granitic rocks are widely distributed not only
inside, but also outside, the Ryoke–Abukuma metamorphic belt.

The metamorphic terrain of the Abukuma Plateau has an axis towards which
the metamorphic grade increases from both sides, as shown in Fig. 7. The axis
runs, near its western margin, roughly parallel to the elongation of the metamorphic belt. It appears that certain parts of the Ryoke metamorphic belt also
has an axis of the highest metamorphic grade, which in the eastern part is situated
near the south-eastern margin of the belt. In the Ryoke belt, the metamorphic
rocks were derived mostly from pelitic and psammitic sediments of Palaeozoic
formations and only rarely from basic volcanics, whereas in the Abukuma

Plateau, metamorphic rocks of basic as well as psammitic and pelitic compositions are abundant.

Pelitic metamorphic rocks in low grades are slates and phyllites, both with recrystallized biotite, and those in middle grades are schists, sometimes with andalusite or cordierite. Pelitic rocks in high grades are gneisses, sometimes with sillimanite or cordierite. Thus, this metamorphism is of the andalusite–sillimanite type. The reaction of muscovite with quartz to produce sillimanite, potassium feldspar, and water takes place in the higher-grade part of the sillimanite-bearing area. In the Ryoke belt most of the metamorphic rocks are higher in grade than the biotite isograd. In other words, the rocks on the lower-grade side of the biotite isograd are too poorly recrystallized to be classed as metamorphic.

There is some doubt about the type of metamorphism in the Hitati (Hitachi) district in the southern part of the Abukuma Plateau. There, andalusite and sillimanite occur just as in all the other parts of the Ryoke–Abukuma metamorphic belt, but chloritoid, which has not been found in other parts of the belt, occurs. Moreover, there are some data suggesting the presence of a zone of the epidote–amphibolite facies between the zones of the greenschist and amphibolite facies (Kuroda, 1959), and recrystallization of rocks on the lower-grade side of the biotite isograd appears to have advanced so as to produce chlorite–muscovite slates. All these data suggest that the metamorphism of the Hitati district may be of the low-pressure intermediate group, unlike all the other parts of the Ryoke–Abukuma metamorphic belt.

In the Ryoke–Abukuma belt, the terrain on the continental side of the axis is exposed mainly in the Ryoke wing, whereas the terrain on the Pacific Ocean side is exposed mainly in the Abukuma Plateau. The latter is richer in basic volcanic rocks in the geosynclinal pile, and is richer in gabbroic rocks of the orogenic phase, than the former. The latter also shows more conspicuous development of schistosity, as well as a more advanced degree of recrystallization in low grades, than the former. Such differences in character between the opposite sides of the axis of the metamorphic belt of the andalusite–sillimanite type, or low-pressure intermediate group, would be a commonly observed feature in paired metamorphic belts of the circum-Pacific region. Somewhat similar relations will be mentioned later (pp. 298 & 302) in the cases of the paired metamorphic belts of Hokkaido and California.

Detailed investigations of the metamorphic facies series in this belt were made in the central part of the Abukuma Plateau (Gosaisyo, Takanuki, and Nakoso districts) by Miyashiro (1953, 1958), Shido (1958), and Shido & Miyashiro (1959). Hence, this region was taken as the type metamorphic terrain of the andalusite–sillimanite type, as shown in Fig. 2. In this region pelitic rocks are in subordinate amount and basic rocks are much more abundant. The metamorphic terrain was divided into three progressive metamorphic zones, A, B, and C, on the basis of the progressive variation in calciferous amphiboles in the basic metamorphic rocks. Zone A is characterized by actinolite, and belongs to the greenschist facies. Zone B is characterized by common hornblende with $Z =$ blue-green, whereas zone C is characterized by common hornblende with $Z =$ green or brown (without a bluish tinge). Zones B and C belong to the amphibolite facies. As mentioned before, the contact metamorphism by synkinematic intrusions produced a metamorphic facies series practically identical to the facies series of the regional metamorphism. In the contact metamorphism by synkinematic gabbros, however, not only zones A, B, and C but also zone D, representing a higher grade than zone C, is produced. This zone is also shown in Fig. 2. Zone D is characterized by the appearance of orthopyroxene, and belongs to the lower-pressure part of the granulite facies. (Whether this zone belongs to the granulite facies or to the pyroxene–hornfels facies depends upon the definition of the two facies. In this paper, this zone is put into the granulite facies.)

The compositional variation from actinolite to common hornblende in the passage from zones A to B is abrupt and discontinuous. Throughout zones B and C, hornblendes in the basic metamorphic rocks have large contents of Al in four- and sixfold co-ordination (Shido, 1958; Shido & Miyashiro, 1959). Cummingtonite is widespread in amphibolites of zone C.

In zones A and B, pyralspite garnet does not occur in common pelitic rocks, but is confined to somewhat more manganiferous metasediments. When it occurs, it is highly manganiferous, and the Mn content decreases with increasing metamorphic grade. In zone C, pyralspite is common in pelitic rocks and is almandine in composition (Miyashiro, 1953). Biotite also shows a tendency towards a decrease in the Mn content as well as an increase in the Ti content with increasing metamorphic grade. The crystal structure of potassium feldspar changes gradually from the microcline structure to the orthoclase with increasing metamorphism (Shido, 1958). The high–low inversion point of quartz is lower in the higher-grade metamorphics than in the lower-grade ones (Iiyama, 1954).

The opaque minerals are mainly pyrite, pyrrhotite, magnetite, and ilmenite in zone A, and pyrrhotite, magnetite, and ilmenite in zones B and C (Banno & Kanehira, 1961).

Sanbagawa metamorphic belt

The Sanbagawa (Sambagawa) metamorphic belt was called the Sanbagawa–Mikabu or Nagatoro metamorphic belt by some authors. The main part of this belt runs along the eastern and southern sides (i.e. Pacific Ocean side) of the Ryoke metamorphic belt. Between the Ryoke and Sanbagawa metamorphic belts there is a long fault, called the *median line* or *median tectonic line*, as shown in Fig. 7. This fault cuts off granitic and metamorphic rocks of the Ryoke belt from metamorphic rocks of the Sanbagawa belt. Mylonite was formed along this fault. It would have originated at a later stage or immediately after the Late Mesozoic orogeny. The eastern extension of the Sanbagawa metamorphic belt outcrops in the Kanto Mountains and probably farther in the north-eastern

6233.3   X

margin of the Abukuma Plateau (Yaguku and Yamagami districts). Thus, the Sanbagawa metamorphic belt and its extension lie on the Pacific Ocean side of the full length of the Ryoke–Abukuma metamorphic belt. The tectonic features of this belt were analysed by Nakayama (1959) and others.

The Sanbagawa metamorphic belt is accompanied by abundant ultrabasic rocks (serpentinite, dunite, &c.) and gabbros. Granitic rocks are absent. A fault comparable to the median line is not clear in the Abukuma Plateau, where the easternmost extension of the Sanbagawa metamorphic belt is intruded by post-kinematic granites of the Ryoke–Abukuma belt.

In most of the main part of the Sanbagawa metamorphic belt as well as in the Kanto Mountains, the grade of metamorphism increases towards the median line. In the opposite direction the Sanbagawa metamorphic terrain grades into unmetamorphosed Palaeozoic formations. At some places in Sikoku a marked tectonic zone was found near the southern margin of the metamorphic terrain (Kojima & Suzuki, 1958). The grade of metamorphism appears to be related to the geographical position within the metamorphic belt, but not to the stratigraphical position in the geosynclinal pile. At some places the grade increases towards the lower stratigraphical horizon, and at others it increases towards the higher. Generally, the increase of metamorphic grade is not related to the distribution of igneous masses. An exceptional situation appears to exist in the Bessi district, where the metamorphic grade becomes highest in the vicinity of large basic and ultrabasic masses, which are situated about 6 km to the south of the northern margin of the Sanbagawa metamorphic belt.

The metamorphic rocks of this belt were derived mostly from pelitic, psammitic, and basic volcanic rocks of Middle and Upper Palaeozoic formations. The metamorphism is of the jadeite–glaucophane type at least in some parts, and may be of the high-pressure intermediate group in others. Glaucophane occurs in many parts of the belt, and the distribution of glaucophane is related to the metamorphic grades of the terrains but not to the distribution of ultrabasic and gabbroic masses. Jadeite associated with quartz was found in a certain part of the Kanto Mountains and Sibukawa district (Seki, 1960 *b*, *c*). The jadeite–quartz–albite assemblage was found, in which probably the solid solutions of jadeite and albite play an essential role. In most of the metamorphic belt, the metamorphic temperature was too low to produce biotite, and only in and near the Bessi district was biotite formed in pelitic rocks.

The lower-grade part of the metamorphic facies series was elaborately investigated in the Kanto Mountains by Seki (1958, 1960 *b*, *c*, 1961 *b*). Hence this region was taken as the type metamorphic terrain of the jadeite–glaucophane type. Here Seki showed that the area of almost unmetamorphosed Palaeozoic formations grades with increasing metamorphism into the zone of crystalline schists without albite porphyroblasts (called non-spotted schists), and further into the zone of crystalline schists with conspicuous albite porphyroblasts (called spotted schists). The zone of non-spotted schists and lower grades belongs

25

to the glaucophane–schist facies, whereas the zone of spotted schists belongs to the greenschist facies, as shown in Fig. 3. Pumpellyite tends to occur in lower grades, whereas piemontite tends to occur in higher grades. Probably, piemontite can have a large content of Mn only at relatively high temperatures (Miyashiro & Seki, 1958a). Seki divided the metamorphic region into six zones on the basis of progressive mineralogical changes (Fig. 3). It is interesting that lawsonite and jadeite (associated with quartz) are confined to a middle-grade part (zone IV) of the region. It is conceivable that the temperature–pressure curve corresponding to this metamorphism grazes the stability field of the jadeite–quartz assemblage, as is exemplified by a line (3)–(3′) in Fig. 4. It is also conceivable that the production of lawsonite in this part made the associated albite become more sodic, and hence promoted the formation of jadeite.

The metamorphic facies series in the higher-grade part of the Sanbagawa belt was investigated in the Bessi district by Banno (Banno, 1959; Miyashiro & Banno, 1958). Here, the metamorphic terrain was divided into four zones: Ia, Ib, II, and III, in the order of increasing metamorphism. Zone Ia is characterized by the occurrence of actinolite and pumpellyite, and zone Ib by the occurrence of actinolite (and not pumpellyite). Zone II is characterized by the occurrence of common hornblende with $Z =$ blue-green. Glaucophane occurs in zones Ia, Ib, and the lower-grade part of zone II. Zone III is characterized by the occurrence of biotite in pelitic metamorphic rocks. In the higher-grade part of zone III, diopside and kyanite were found in some basic rocks. Zones Ia and Ib belong to the glaucophane–schist facies, and zone III belongs to the epidote–amphibolite facies. Jadeite has not been found in the Bessi and surrounding districts, and it is conceivable that the metamorphism here belongs to the high-pressure intermediate group. The opaque minerals contained in the metamorphic rocks of this district are mainly pyrite, pyrrhotite, haematite, and magnetite, and only in zone III does ilmenite also occur (Banno & Kanehira, 1961).

The mineralogy of the Sanbagawa metamorphic belt is greatly diversified. Not only glaucophane (Seki, 1958; Iwasaki, 1960b), jadeite (Seki & Shido, 1959; Seki, 1960 a, b, c, 1961a), lawsonite (Seki, 1957, 1958), pumpellyite (Seki, 1958), piemontite (Miyashiro & Seki, 1958a; Hashimoto, 1959), kyanite (Banno, 1957), and stilpnomelane (Kozima, 1944), but also magnesio–riebeckite (Miyashiro & Iwasaki, 1957), magnesio–arfvedsonite (Banno, 1958b), aegirine–augite (Banno, 1959), aegirine–jadeite (Iwasaki, 1960a; Kanehira & Banno, 1960), paragonite (Banno, 1960), and ferriphengite (Kanehira & Banno, 1960) were found. Jadeite occurs not only in quartz-bearing metamorphic rocks but also in quartz-free veins associated with ultrabasic intrusives.

In the Bessi district a large mass of dunite occurs within the highest-grade metamorphic zone. This mass is probably an intrusive emplaced almost simultaneously with the regional metamorphism. In parts of the dunite mass, many sub-parallel bands of eclogite, up to 1 m thick, are interlayered with bands of dunite and harzburgite. Some parts of the eclogite are composed of diopside and pyrope with a small amount of common hornblende, whereas other parts are composed of almandine–pyrope, omphacite ($Na_2O = 2·84$ per cent), common hornblende, and epidote (Miyashiro & Seki, 1958b; Shido, 1959). This dunite–eclogite complex would represent the highest-temperature member belonging to the same pressure group as the metamorphic rocks of the Bessi district.

Older pair of metamorphic belts

As stated before, the older pair is composed of the Hida and Sangun metamorphic belts.

The Hida metamorphic belt is exposed mainly in the Hida Plateau of central Honsyu (Fig. 7). The western extension of the belt is mostly beneath the Japan Sea, and only a small part is exposed in the Oki Islands. The exposed terrain of the belt is mainly composed of quartzo-feldspathic gneisses, amphibolites, and crystalline limestones, closely associated with numerous small granitic masses. Large granitic masses, collectively called the Hunatu granite, intruded these rocks and probably represent late- or post-kinematic intrusion of this belt.

The petrographic study of the Hida metamorphic belt has been made, but not yet in sufficient detail. In the largest exposed metamorphic area, about 20–40 km south of Toyama, the grade of metamorphism appears to increase southward, judging from the southward increase in grain size of limestones and graphite deposits (Nozawa, 1959). Sillimanite occurs in many parts of the area, whereas andalusite was found only from near its northern margin. The northeastern end of the Hida belt (Kurobe-gawa district) shows somewhat different characters, where kyanite, andalusite, sillimanite, and staurolite were found by Ishioka and others (e.g. Ishioka & Suwa, 1956). These features suggest that the Hida metamorphic belt belongs to the andalusite–sillimanite type and low-pressure intermediate group.

The eastern part of the Sangun metamorphic belt is an arcuate belt lying on the eastern and southern sides of the Hida Plateau. In this part, metamorphic rocks are exposed in small separate areas, such as Omi, Gamata, and Naradani, all arranged within the belt. Between the Hida and Sangun metamorphic belts, there appears to exist a median fault, although its exact location has become obscured by later granitic, sedimentary, and volcanic rocks. The Sangun metamorphic belt is accompanied by abundant ultrabasic rocks (mostly serpentinite).

The metamorphic terrain of the Omi district at the north-eastern end of the Sangun belt was studied by Banno (Banno, 1958a; Miyashiro & Banno, 1958). The metamorphic rocks developed there were derived most from sedimentary and basic igneous rocks. The metamorphic terrain was divided into two progressive metamorphic zones, i.e. a zone of chlorite and one of biotite in order of increasing temperature. The metamorphic grade increases from both sides towards the axial zone, where a large mass of serpentinite is exposed. Glaucophane is confined to the chlorite zone. The jadeite–quartz assemblage was not found, although jadeite occurs abundantly in quartz-free albitite masses within serpentinites associated with the metamorphic rocks. Common hornblende occurs in the biotite zone. At least the lower-grade part of the chlorite zone belongs to the glaucophane–schist facies, whereas the biotite zone belongs to the epidote–amphibolite facies.

In the Gamata and Naradani districts of the Sangun belt, the metamorphic rocks belong to the greenschist facies with actinolite, and neither glaucophane nor common hornblende occurs (Seki, 1959).

In most of the Sangun metamorphic terrains in western Honsyu, the metamorphic grade is so low that biotite is not formed and recrystallization is incomplete. The metamorphic rocks are mostly slates and phyllites. Glaucophane was found extremely rarely. In northern Kyusyu, however, the Sangun metamorphic terrain is intruded and highly metamorphosed by many granitic masses. Probably these granites are not genetically related to the Sangun metamorphism but are the northern extension of granitic areas genetically related to the Ryoke metamorphism.

Probably the Sangun metamorphic belts are of the high-pressure intermediate group.

PAIRED METAMORPHIC BELTS IN THE CIRCUM-PACIFIC REGION

The metamorphic belts in various countries around the Pacific Ocean are now to be considered briefly. It will be shown that metamorphic belts of contrasted characters occur in pairs, just as in the main part of Japan.

Hokkaido

Paired metamorphic belts are exposed, running in a north–south direction, in the central part of Hokkaido, the northernmost island of Japan. They are shown in Figs. 6 and 9. The metamorphism of these belts took place during the orogeny in Late Mesozoic time.

The eastern metamorphic belt of the pair is called the *Hidaka metamorphic belt*, which was geologically studied in detail by Hunahashi and his collaborators (Hunahashi, 1957). Here, metamorphic rocks derived from psammitic and pelitic sediments are widely developed, and andalusite and sillimanite occur in rocks of appropriate compositions and grades. Cordierite also occurs, but neither kyanite nor staurolite is present. This metamorphic belt belongs to the andalusite–sillimanite type.

This metamorphic belt has an axis of the highest metamorphic grade, along which so-called migmatite is exposed. On both sides of the axis, rocks of decreasing grades (i.e. gneisses and then schists) are exposed successively in parallel zones. At the western margin of the belt, the zone of schists is in thrust-contact with unmetamorphosed sedimentary terrain, whereas on the eastern side, the zone of schists grades outward into that of hornfelses, which latter grades into unmetamorphosed sediments. The development of schistosity is more pronounced on the western side of the axis than on the eastern. Plutonic rocks, both granitic and gabbroic, are abundant in the belt, and gabbroic rocks are more abundant on the western side of the axis. Peridotite is also present in the westernmost part. These asymmetrical features on the opposite sides of the axis are similar to relations in the Ryoke–Abukuma metamorphic belt.

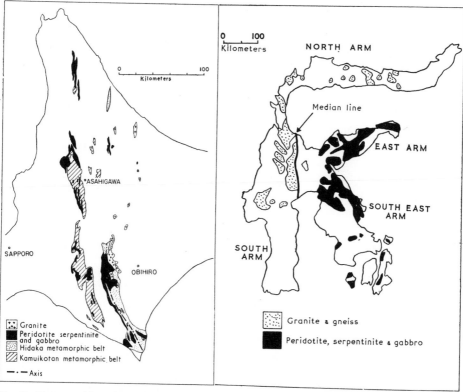

FIG. 9. Metamorphic and associated plutonic rocks of the Hidaka and Kamuikotan metamorphic belts in Hokkaido.

FIG. 10. Distribution of granite and gneiss, as contrasted with that of peridotite, serpentinite, and gabbro, in Celebes.

The western metamorphic belt of the pair is called the *Kamuikotan metamorphic belt*. The metamorphic rocks developed there were derived from basic volcanic and sedimentary rocks of Mesozoic age. Glaucophane is widespread, and jadeite occurs in association with quartz in a wide area (Shido & Seki, 1959). Most or all of the belt belongs to the jadeite–glaucophane type. Lawsonite and pumpellyite are also widespread.

The Kamuikotan metamorphic belt is accompanied by abundant ultrabasic rocks (mostly serpentinite), as described by Suzuki (1952). Formerly, many authors stated that glaucophane occurs in the contact zone around ultrabasic masses. Recent investigations, however, showed that the distribution of glaucophane has no definite relationship to ultrabasic masses, although in some contact zones recrystallization is so advanced that the metamorphic rocks are coarser-grained there than in other parts of the terrain, and the presence of coarse-grained glaucophane schists is very impressive.

A large part of the terrain belongs to the glaucophane–schist facies. In a certain part of the belt, however, Shido & Seki (1959) found epidote–hornblende schist, probably belonging to the epidote–amphibolite facies.

In an area to the north-west of Asahigawa, M. Hatano (personal communication) carried out zonal mapping of the Kamuikotan metamorphic terrain, and showed that there is an axis of the highest metamorphic grade that trends, roughly, in a north-north-east direction near the western margin of the exposed metamorphic terrain.

In geological structure the western part of Hokkaido is a northern extension of the main part of Japan, whereas the central and eastern parts of Hokkaido belong to a different geological province. The paired metamorphic belts in Hokkaido were formed probably along the western margin of the latter province. In south Sakhalin, a belt of metamorphic rocks similar to those of the Kamuikotan belt lies in a north–south direction. It is probably the northern extension of the Kamuikotan metamorphic belt, as shown in Fig. 6.

Celebes

The island of Celebes belongs to an orogenic belt of Late Mesozoic to Cenozoic age. The island shows a remarkable four-armed morphology, which is due to a connected double arc with its concave side towards the Pacific Ocean (Fig. 10). The north arm, the western part of central Celebes, and the south arm together constitute the inner arc, where Late Mesozoic and Tertiary granitic rocks are widespread, in association with biotite-rich schists (Bemmelen, 1949).

The granitic masses of the inner arc are cut abruptly on the eastern side by a great fault, called the *median line*, trending north–south. Mylonite was formed along the median line. The east arm, the central and eastern parts of central Celebes, and the south-east arm, which all lie to the east of the median line, together constitute the outer arc. In this arc basic and ultrabasic plutonic rocks of Mesozoic and Cenozoic ages are widely exposed, together with glaucophanitic metamorphic rocks. Basic volcanic rocks are abundant in the geosynclinal pile (Kündig, 1956). In some parts jadeite was found in association with quartz (de Roever, 1955b). It is claimed that metamorphic rocks of an older orogenic cycle remain (de Roever, 1947), but the evidence for it is not conclusive.

Thus, the inner and outer arcs constitute paired metamorphic belts. There are many pairs of double arcs in the circum-Pacific region. The structure of Celebes suggests that paired metamorphic belts may underlie the sedimentary and volcanic cover of the double arcs. In the double arc of Ryukyu (Riukiu), glaucophanitic metamorphic rocks are exposed in parts of the outer arc (Hanzawa, 1935; Yossii, 1935).

New Zealand

In New Zealand orogeny took place in Jurassic times, and geosynclinal sediments and volcanic rocks of Palaeozoic to Jurassic age were metamorphosed into schists. The schists are widely exposed in the South Island, and the southern, central, and northern parts of the schist belt are called the *Otago*, *Alpine*, and *Marlborough schists*, respectively (Grindley *et al.*, 1959), as shown in Fig. 11.

The axis of this metamorphic belt trends north-east in the northern part of

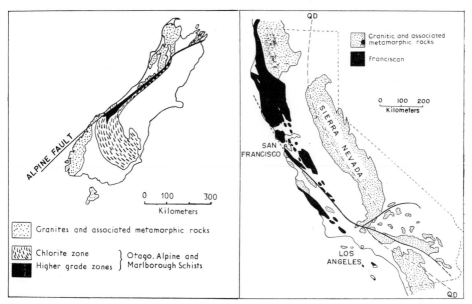

FIG. 11. Plutonic and metamorphic rocks of Palaeozoic and Mesozoic ages in the South Island of New Zealand.

FIG. 12. Distribution of basement rocks in California. The line *QD–QD* represents the 'quartz diorite boundary line' of Moore (1959).

the island, and turns to the south-east in the southern part. The belt is cut by a later fault, called the *Alpine fault*. In the southern and northern parts, the metamorphic grade increases from both sides towards the axis, whereas in the central part the grade increases westwards only, because there the western half of the metamorphic belt is lost by the Alpine fault.

The progressive metamorphism in this belt was studied by Turner (1938), Hutton (1940), Reed (1958), and others. Most of the metamorphic terrain is low in metamorphic grade, belonging to the chlorite zone (greenschist facies). Narrow zones of higher metamorphic grades, characterized by biotite, by almandine, and by oligoclase (greenschist to amphibolite facies), are present in the Alpine schists in the vicinity of the Alpine fault. Recently, Coombs *et al.* (1959) established the zeolite facies, which is lower in metamorphic grade than the greenschist facies.

Hutton (1940) found a crossite core in some actinolite crystals of this meta-

31

morphic belt, which probably belongs to the high-pressure intermediate group. Like the glaucophanitic metamorphic belts, it includes a wide, well-recrystallized, low-grade area. Pumpellyite, piedmontite, stilpnomelane, and kyanite were found. This metamorphic belt is accompanied neither by granitic nor by gabbroic and ultrabasic rocks.

To the west of this metamorphic belt, there is another belt of granitic and associated metamorphic rocks. The plutonism and metamorphism in this region appear to have taken place in Late Palaeozoic times. The metamorphic rocks are generally high in grade, and contain andalusite, sillimanite, and cordierite. Probably this metamorphism is of the andalusite–sillimanite type and/or low-pressure intermediate group. Granitic and gneissic rocks of probable Precambrian age occur at certain places within this metamorphic belt. Probably they represent the basement for Palaeozoic metamorphic rocks of this belt.

Thus, these two metamorphic belts in the South Island may be regarded as being in a pair, just as in Japan and Celebes. Although the ages of their metamorphism differ, the difference is not great.

California

The Sierra Nevada and the coast ranges of California belong to the Late Mesozoic to Early Tertiary orogenic belt along the west coast of North America. The metamorphic belt of the Sierra Nevada and that of the coast ranges appear to constitute a pair, as in many other parts of the circum-Pacific region (Fig. 12).

In the Sierra Nevada and adjacent areas, gigantic masses of granitic rocks are exposed in association with various metamorphic rocks. Much smaller amounts of gabbroic and ultrabasic rocks are also present. Generally the grade of regional metamorphism is not high, and only the contact aureoles around plutonic masses reach a high grade, producing andalusite. The granites are postkinematic. High-grade regional metamorphic rocks and synkinematic granites would be still hidden under the ground.

The coast ranges are underlain by the Franciscan formation. Recent investigations have revealed that metamorphism of the jadeite–glaucophane type took place, in the Franciscan, on a regional scale. Glaucophane, lawsonite, and jadeite (associated with quartz) were produced widely in poorly recrystallized greywackes (Bloxam, 1956; McKee, 1958). Well-recrystallized glaucophane schists occur in more limited areas. Basic volcanic rocks are abundant in the geosynclinal pile. Large amounts of gabbroic and ultrabasic rocks are associated with the Franciscan.

Two fault blocks, called *Salina* and *Anacapia*, composed of granitic and high-grade metamorphic rocks, occur within the area of the coast ranges. Probably they were originally in the southern extension of the Sierra Nevada, and were afterwards transported westward by movements along strike-slip faults (King, 1959). Radiometrically dated Precambrian plutonic rocks occur in Anacapia and areas to the south-east of the Sierra Nevada. The geosynclinal pile of the Sierra Nevada probably has a Precambrian crystalline basement.

Exposed plutonic rocks of the coast ranges, the Sierra Nevada, and adjacent regions, tend to become more acidic and more potassic towards the east. Moore (1959) showed that in the western part of the regions now under consideration quartz–diorite is the dominant 'granitic' rock, whereas in the eastern part granodiorite and quartz–monzonite are the dominant 'granitic' rocks, and that a boundary line, called the *quartz–diorite boundary line*, can be drawn between the two parts, as shown in Fig. 12. It is somewhat similar to relations in the Ryoke–Abukuma and Hidaka metamorphic belts.

ORIGIN AND DEVELOPMENT OF METAMORPHIC BELTS

Origin of paired metamorphic belts

Most of the younger orogenic belts in the circum-Pacific region have a pair of metamorphic belts, as described above. The characters of the paired metamorphic belts may be summarized as follows:

The metamorphic belt on the continental side will be called the *inner metamorphic belt*. It is of the andalusite–sillimanite type and/or low-pressure intermediate group. It has an axis of the highest metamorphic grade, if the original structure is preserved. The metamorphic rocks were usually derived mostly from sediments, which were probably deposited on the sialic basement. The basement may be of Precambrian age in some cases. Basic volcanic rocks are abundant in the geosynclinal pile in certain cases, but not so in many others. The inner metamorphic belt is always accompanied by the intrusion of a large amount of granitic rocks (in a wide sense), usually together with a much smaller volume of gabbroic types. A small volume of ultrabasic rocks may also be present.

The metamorphic belt on the Pacific Ocean side will be called the *outer metamorphic belt*. It is of the jadeite–glaucophane type and/or high-pressure intermediate group. It also has an axis of the highest metamorphic grade, if the original structure is preserved. The metamorphic rocks were usually derived largely from basic volcanic rocks as well as from sediments. The basement of their deposition is not known, and may be the basic ocean floor. The outer metamorphic belt is usually accompanied by the intrusion of a large volume of gabbroic and ultrabasic (mostly serpentinitic) rocks. Granitic rocks are absent.

In the younger pair of the main part of Japan and in Celebes, the two associated metamorphic belts are in direct contact with each other, having a fault between them. On the other hand, in Hokkaido and in New Zealand there is a zone of practically unmetamorphosed rocks between the two belts.

We may consider that the inner metamorphic belt was formed in the geosynclinal pile deposited on the sialic basement at the margin of the continental crust, whereas the outer metamorphic belt was formed in the geosynclinal pile deposited on the ocean floor outside the continental crust. Granitic rocks in the inner metamorphic belt would have formed by remelting or mobilization of the basement and geosynclinal sediments. On the other hand, in the outer metamorphic belt there would be no sialic basement to form granitic rocks by re-melting, and basic volcanic and plutonic rocks, together with ultrabasic rocks, would be formed in abundance by uprise from the basic and/or ultrabasic layer below. It is also possible that some outer metamorphic belts have a sialic base-ment that is too thin to produce granitic intrusions.

The regional metamorphism of the jadeite–glaucophane type and high-pressure intermediate group in the outer metamorphic belt represents higher rock pressures, and generally lower temperatures, than that of the andalusite–sillimanite type and low-pressure intermediate group in the inner metamorphic belt. On the lithostatic assumption, the higher rock pressure is due to the greater depth of down-buckling and/or underthrusting in the outer metamorphic belt than in the inner, during orogeny. The down-buckling of the outer metamorphic belt would not be generally simultaneous with that of the inner metamorphic belt. The two metamorphic belts were formed, probably, at different phases of the same cycle of orogeny. Probably, the outer metamorphic belt represents the site of the main down-buckle, and would be comparable with the present-day trench, whereas the inner metamorphic belt would represent a subordinate down-buckle of shorter duration. The commonly observed, very weak recrystallization in the lower-grade part of the inner metamorphic belt would be due mainly to the shorter duration of this episode.

The cause of lower temperature in the outer metamorphic belt is not clear. The absence of a sialic basement may be an important factor in this respect, because sialic rocks would produce a large amount of radiogenic heat and would transfer heat by intrusive movements. Recent measurements of heat flow on the ocean bottom have revealed that the amount of heat flow is much smaller in the trench than in other parts of the earth's surface. Whatever the cause of this low value may be, it would be related to the low temperature deduced for the outer metamorphic belt.

Historical development of metamorphic belts

The writer made a survey of petrographic descriptions of metamorphic rocks in the Canadian and Baltic shields, and was greatly impressed by the fact that andalusite, sillimanite, and cordierite are widespread, and staurolite and almandine are also rather common, whereas kyanite is rare, and jadeite and glaucophane are absent. This fact strongly suggests that the metamorphism in these Precambrian shields is mainly of the andalusite–sillimanite type and low-pressure intermediate group.

In the so-called Svecofennides of the Baltic shield, pelitic rocks of middle grades are characterized by the occurrence of andalusite and cordierite, whereas those of high grades are characterized by the occurrence of sillimanite, cordierite, and almandine (Eskola, 1914; Simonen, 1953; Magnusson *et al.*, 1960). Staurolite occurs in some parts (Seitsaari, 1951). In the so-called Karelides, pelitic rocks contain andalusite, cordierite, staurolite, and almandine (Eskola, 1927; Magnusson *et al.*, 1960). In the Baltic shield, kyanite occurs in a few localities in Sweden (Magnusson *et al.*, 1960; Bergström, 1960), but is associated with manganiferous andalusite in some of them. In a locality in Finnish Lapland, kyanite occurs in association with staurolite (Eskola, 1952). All these facts being taken into consideration, andalusite is much more widespread than kyanite in the metamorphic rocks of the Baltic shield.

In the Canadian shield, andalusite, sillimanite, and staurolite occur in the Grenville province as well as in northern Michigan of the Superior province (Engel & Engel, 1953; James, 1955). Cordierite also occurs in the Grenville province. In the vicinity of Great Slave Lake, andalusite, sillimanite, cordierite, and staurolite were described by a few authors. Heinrich and Corey (1959) found kyanite associated with manganiferous andalusite in some Precambrian rocks of New Mexico.

There is a wide area of radiometrically dated Precambrian metamorphic rocks in the southern part of the north-eastern provinces (Manchuria) in China and in the northern part of Korea. These rocks contain andalusite, sillimanite, and staurolite, and are devoid of kyanite, as in the above two shields. In central and southern Korea, kyanite, commonly in association with andalusite and staurolite, occurs in so-called Precambrian metamorphic rocks, although the evidence for a Precambrian age is not certain (e.g. Yamaguchi, 1951).

These Precambrian metamorphic terrains are in the northern region of the earth, where 'continental growth' with geological time is rather evident. The Precambrian shields in the southern region, such as those of India and Africa, where 'continental growth' is not so evident, appear to be somewhat different in character. In Precambrian metamorphic rocks in India, kyanite is widespread, and in those in Africa both kyanite and andalusite are widespread.

Some Palaeozoic metamorphic terrains belong to the andalusite–sillimanite type and low-pressure intermediate group, whereas others belong to the kyanite–sillimanite type. For example, Early Palaeozoic metamorphic belts of New South Wales in Australia (Joplin, 1942, 1943; Vallance, 1953 *a, b*) are of the andalusite–sillimanite type. The Palaeozoic metamorphic belt in the Scottish Highlands and Norway is mainly of the kyanite–sillimanite type, but contains some areas of the low-pressure intermediate group (Harker, 1932; Wiseman, 1934; Read, 1952; Clifford, 1958; Goldschmidt, 1915; Vogt, 1927). The Middle Palaeozoic metamorphic belt in the Appalachians is very similar (Barth, 1936; Billings *et al.*, 1952). The Late Palaeozoic metamorphic terrain of the Pyrenees appears to belong to the low-pressure intermediate group (Zwart, 1959; Allaart, 1959).

Most of the paired metamorphic belts in the circum-Pacific region were formed in Mesozoic time. They are composed of a belt of the andalusite–sillimanite type and/or low-pressure intermediate group and another belt of the jadeite–glaucophane type and/or high-pressure intermediate group, as stated before. The metamorphism in the Pennine nappes of the Alps would belong to the high-pressure intermediate group (Plas, 1959).

The great serpentine belt in eastern Australia is associated with the Brisbane metamorphic complex which contains some glaucophane-like amphiboles. If these rocks may be regarded as being in a pair with the granitic and metamorphic rocks to the west, they would represent the incipient formation of paired metamorphic belts in Palaeozoic time in the circum-Pacific region. The glaucophane schists in this case, however, are probably of the high-pressure intermediate group, and the metamorphism of the jadeite–glaucophane type appears to be confined to Mesozoic time. As regards the Cenozoic regional metamorphism, we have hardly any data with which to discuss it.

Thus, we may conclude that regional metamorphism of the andalusite–sillimanite type and low-pressure intermediate group took place widely from Precambrian through Mesozoic time, whereas that of the kyanite–sillimanite type was common in Palaeozoic time, and that of the jadeite–glaucophane type became common in Mesozoic time. The dominant types of regional metamorphism changed with geological age. It may be considered that regional metamorphism under lower rock pressures (that is, probably at shallower depths) appears to have taken place in all ages, whereas regional metamorphism under higher rock pressures (that is, probably at greater depths) appears to have taken place only in later geological times.

Working hypothesis

The rarity of glaucophane and lawsonite in early geological ages has already been noticed by Eskola (1939), de Roever (1956), and Plas (1959). de Roever claimed that it is a result of the secular decrease of geothermal gradient in the earth's crust through geological time. Metamorphic belts of the jadeite–glaucophane type, however, are commonly associated with those of the andalusite–sillimanite type. This fact is contradictory to de Roever's hypothesis.

For the cause of variation in types of regional metamorphism with geological ages, no definite answer can be given at present. A working hypothesis for it is, however, proposed below.

It is generally believed that the earth was formed about 4,500 million years ago, and the oldest rocks exposed on the earth are about 3,000 million years old. It may be assumed that the primitive sialic crust was formed during the period from 4,500 to 3,000 million years ago through partial melting and differentiation of the basic and/or ultrabasic layer of the earth. The outlines of the primitive crusts would have been somewhat similar to the shapes of the present-day continents, and the primitive crust would have been thicker in the central part of each mass than in the margins. The change of dominant types of regional metamorphism with geological time would be a result of a change in the spatial relationship between the sialic crust and orogenic belts.

In Precambrian time orogenic movements would have taken place mainly (but not exclusively) in the central part of the primitive sialic mass. The sialic crust there would have been relatively thick and strong and, consequently, would have resulted in relatively shallow down-buckles that caused metamorphism of the andalusite–sillimanite type and low-pressure intermediate group. In Palaeozoic time orogenic belts would have moved outward into the more marginal part of the sialic mass, where the crust was partly thick and partly thin. Orogenic movements would have produced down-buckles of greater depths in the thin parts of the crust, with the resultant formation of metamorphic belts of the kyanite–sillimanite type. In Mesozoic time orogenic belts would have moved still farther outward, to the margin of the sialic mass, and the main site of the orogenic down-buckle would have been in the geosynclinal pile that was deposited on a very thin sialic basement or directly on the basic ocean floor. Thus, the down-buckle in this case would have reached a great depth with resultant formation of metamorphic belts of the high-pressure intermediate group and jadeite–glaucophane type. On the other hand, on the nearby continental margin a smaller degree of down-buckling would have taken place to produce metamorphic belts of the andalusite–sillimanite type and low-pressure intermediate group.

In this hypothesis the 'continental growth' by orogenies during the period from Precambrian to Palaeozoic time was only apparent. Actually, it meant a gradual progress of recrystallization and hardening of the pre-existing primitive sialic crust from the centre towards the margin. The true growth of continents began in Mesozoic time, when the main site of orogeny passed over the margin of the primitive sialic masses into the oceanic regions.

FACIES SERIES OF CONTACT METAMORPHISM

The classification of metamorphism into regional and contact types is based on the geological relations. This classification, then, does not necessarily correspond to the classification of metamorphism based on the metamorphic facies series. Contact metamorphism appears to produce different facies series in different cases. In this respect, however, our knowledge is very incomplete, and only a cursory survey will be given below.

In the central Abukuma Plateau the metamorphic facies series developed in the contact aureoles of synkinematic granites is practically identical to that developed through regional metamorphism of the same region (i.e. of the andalusite–sillimanite type), whereas the facies series developed in the contact aureoles of the post-kinematic granites is different. Shido (1958) made a detailed investigation of the contact aureole around the post-kinematic Iritono granite in the central Abukuma Plateau.

In the Iritono contact aureole the low-grade zone is characterized by basic hornfelses with the actinolite–labradorite (or andesine)–quartz assemblage with, or without, chlorite. It is remarkable that epidote and zoisite are absent, and that the plagioclase is calcic despite the stable presence of actinolite. It may be that calcic plagioclase becomes stable at a lower grade than common hornblende in this aureole, owing to the very low rock pressure operating there. This assemblage does not belong to any metamorphic facies so far known. It is a new facies or subfacies, which might well be called the *actinolite–calcic plagioclase hornfels facies*. The associated hornfelses of sedimentary origin are biotite–plagioclase–quartz hornfelses, sometimes with muscovite or pyralspite.

The middle-grade zone of the Iritono contact aureole is characterized by basic hornfelses with the common hornblende–plagioclase assemblage with or without cummingtonite, clinopyroxene, and biotite. The colour (Z) of the common hornblende is brownish-green or brown. Blue-green common hornblende, such as is widespread in the middle grade of the associated regional metamorphism, is almost absent. The high-grade zone of the contact aureole is characterized by the appearance of orthopyroxene, producing the orthopyroxene–hornblende–plagioclase assemblage sometimes together with cummingtonite, clinopyroxene, and biotite. Almandine is stable in pelitic rocks of the middle grade, but the stability of almandine in the high-grade zone is not clearly established, because pelitic rocks are absent there. Conceivably, almandine is stable in the high-grade zone as well, should the chemical conditions permit.

Thus, the contact aureole of the Iritono district shows a metamorphic facies series that is different from any of the above-mentioned regional-metamorphic facies series. This facies series probably represents a rock pressure lower than that found in any type of regional metamorphism.

Some contact aureoles, such as those of the Kristiania district (Goldschmidt, 1911) and of the Comrie district (Tilley, 1924), contain andalusite and cordierite, and are devoid of almandine. Sillimanite may be present. Such aureoles would represent a still lower rock pressure than that in the Iritono contact aureole.

On the other hand, Compton (1960) described a contact aureole with staurolite together with andalusite, sillimanite, and cordierite. It would be of the low-pressure intermediate group.

The sanidinite facies is probably the highest-temperature member in some contact metamorphic facies series. This facies may well be divided into certain subfacies representing different temperatures and pressures. For example, pigeonite is stable only in a high-temperature part of the facies. Some acidic volcanic rocks of Japan contain pyralspite garnet and sillimanite (not mullite, according to an investigation by S. Aramaki). A lower-temperature part of the sanidinite facies is probably characterized by the stability of almandine, spessartine, sillimanite, cordierite with a large distortion index (Miyashiro *et al.*, 1955; Miyashiro, 1957), and osumilite (Miyashiro, 1956). On the other hand, a higher-temperature part of the facies is probably characterized by the stability of pigeonite, mullite, and cordierite with smaller distortion indices. Though the petrographic evidence is not available, spessartine is probably stable in the higher-temperature part as well, should the chemical conditions permit.

ACKNOWLEDGEMENTS

This paper is the result of my collaboration with some colleagues and students in the University of Tokyo during the last six years. I am greatly indebted to my collaborators, especially to Yotaro Seki, Fumiko Shido, and Shohei Banno, who inspired me with their many important discoveries. We enjoyed lively discussions with, and helpful advice from, one another. Naturally, I am solely responsible for possible mistakes in this paper.

Professor Hisashi Kuno, of the same university, read the manuscript with friendly criticism. I am also grateful to M. Gorai, R. S. Fiske, and many other friends for much advice.

REFERENCES

ALLAART, J. H., 1959. The geology and petrology of the Trois Seigneurs Massif, Pyrenees, France. *Leid. geol. Meded.* **22**, 97–214.

BANNO, S., 1957. Kyanite from the Bessi mining district (in Japanese). *J. geol. Soc. Japan*, **63**, 598.

—— 1958*a*. Glaucophane schists and associated rocks in the Omi district, Japan. *Jap. J. Geol. Geog.* **29**, 29–44.

—— 1958*b*. Notes on rock-forming minerals. (1) Magnesioarfvedsonite from the Bessi district. *J. geol. Soc. Japan*, **64**, 386–7.

—— 1959. Aegirinaugites from crystalline schists in Sikoku. Ibid. **65**, 652–7.

—— 1960. Notes on rock-forming minerals. (12) Finding of paragonite from the Bessi district, Sikoku, and its paragenesis. Ibid. **66**, 123–30.

—— & KANEHIRA, K., 1961. Sulfides and oxides in schists of the Sanbagawa and central Abukuma metamorphic terranes. *Jap. J. Geol. Geog.* **32**, 331–48.

—— & MILLER, J., 1961. New data on the age of the Ryoke and Sanbagawa metamorphism (in Japanese). *Kagaku*, **31**, 144.

BARROW, G., 1893. On an intrusion of muscovite–biotite–gneiss in the south-eastern Highlands of Scotland, and its accompanying metamorphism. *Quart. J. geol. Soc. Lond.* **49**, 330–58.

—— 1912. On the geology of Lower Dee-side and the southern Highland border. *Proc. geol. Ass. Lond.* **23**, 274–90.

BARTH, TOM F. W., 1936. Structural and petrologic studies in Dutchess County, New York. Part II. Petrology and metamorphism of the Paleozoic rocks. *Bull. geol. Soc. Amer.* **47**, 775–850.

BEMMELEN, R. W. VAN, 1949. *The geology of Indonesia.* The Hague: Government Printing Office.

BERGSTRÖM, L., 1960. An occurrence of kyanite in a pegmatite in western Sweden. *Geol. Fören. Stockh. Förh.* **82**, 270–2.

BILLINGS, M. P., THOMPSON, J. B., JR., & ROGERS, J., 1952. Geology of the Appalachian Highlands of east-central New York, southern Vermont, and southern New Hampshire. *Guidebook for field trips in New England*, 1–71. Geologists of Greater Boston, Boston.

BIRCH, F., & LECOMTE, P., 1960. Temperature-pressure plane for albite composition. *Amer. J. Sci.* **258**, 209–17.

BLOXAM, T. W., 1956. Jadeite-bearing metagraywackes in California. *Amer. Min.* **41**, 488–96.

CHAPMAN, C. A., 1952. Structure and petrology of the Sunapee quadrangle, New Hampshire. *Bull. geol. Soc. Amer.* **63**, 381–425.

CHINNER, G. A., 1960. Pelitic gneisses with varying ferrous/ferric ratios from Glen Clova, Angus, Scotland. *J. Petrol.* **1**, 178–217.

CLARK, S. P., JR., 1960. Kyanite revised. *J. geophys. Res.* **65**, 2482.

—— ROBERTSON, E. C., & BIRCH, F., 1957. Experimental determination of kyanite–sillimanite equilibrium relations at high temperatures and pressures. *Amer. J. Sci.* **255**, 628–40.

CLIFFORD, T. N., 1958. A note on kyanite in the Moine series of southern Ross-shire, and a review of related rocks in the Northern Highlands of Scotland. *Geol. Mag. Lond.* **95**, 333–46.

COMPTON, R. R., 1960. Contact metamorphism in Santa Rosa range, Nevada. *Bull. geol. Soc. Amer.* **71**, 1383–1416.

COOMBS, D. S., ELLIS, A. J., FYFE, W. S., & TAYLOR, A. M., 1959. The zeolite facies, with comments on the interpretation of hydrothermal synthesis. *Geochim. et Cosmoch. Acta*, **17**, 53–107.

ENGEL, A. E. J., & ENGEL, C. G., 1953. Grenville series in the northwest Adirondack Mountains, New York. Part I. General features of the Grenville series. *Bull. geol. Soc. Amer.* **64**, 1013–47.

ESKOLA, P., 1914. On the petrology of the Orijärvi region in south-western Finland. *Bull. Comm. géol. Finl.* **40**.

—— 1915. On the relations between the chemical and mineralogical composition in the metamorphic rocks of the Orijärvi region. Ibid. **44**.

—— 1920. The mineral facies of rocks. *Norsk. geol. Tidsskr.* **6**, 143–94.

—— 1927. Petrographische Charakteristik der kristallinen Gesteine von Finland. *Fortschr. Min.* **11**, 57–112.

—— 1939. Die metamorphen Gesteine, in *Die Entstehung der Gesteine*, by Tom. F. W. Barth, C. W. Correns, & P. Eskola, 263–407. Berlin: Julius Springer.

—— 1952. On the granulites of Lapland. *Amer. J. Sci.*, Bowen vol., 133–71.

EUGSTER, H. P., 1959. Reduction and oxidation in metamorphism, in *Researches in geochemistry*, 397–426. New York: Wiley & Sons.

FRANCIS, G. H., 1956. Facies boundaries in pelites at the middle grades of regional metamorphism. *Geol. Mag. Lond.* **93**, 353–68.

GOLDSCHMIDT, V. M., 1911. Die Kontaktmetamorphose im Kristianiagebiet. *Videnskaps Skrifter.* I, Mat.-Naturv. Kl. **11**.

—— 1915. Geologisch-petrographische Studien im Hochgebirge des südlichen Norwegens: III. Die Kalksilikatgneise und Kalksilikatglimmerschiefer im Trondhjem-Gebiete. Ibid. **10**.

GRINDLEY, G. W., HARRINGTON, H. J., & WOOD, B. L., 1959. The geological map of New Zealand, 1:2,000,000. *Bull. geol. Surv. N.Z.*, N.S., **66**.

HANZAWA, S., 1935. Topography and geology of the Riukiu Islands. *Sci. Rep. Tôhoku Univ.*, ser. II, **17**, 1–61.

HARKER, A., 1932. *Metamorphism.* London: Methuen & Co.

HASHIMOTO, M., 1959. Piedmontite from piedmontite-bearing muscovite–quartz schist from Inasato, Nagano Prefecture. *Bull. nat. Sci. Mus. Tokyo*, **4**, 183–7.

HEINRICH, E. WM., & COREY, A. F., 1959. Manganian andalusite from Kiawa Mountain, Rio Arriba County, New Mexico. *Amer. Min.* **44**, 1261–71.

HIETANEN, A., 1956. Kyanite, andalusite, and sillimanite in the schist in Boehls Butte quadrangle, Idaho. Ibid. **41**, 1–27.

HUNAHASHI, M., 1957. Alpine orogenic movement in Hokkaido, Japan. *J. Fac. Sci. Hokkaido Univ.*, ser. IV, **9**, 415–69.

HUTTON, C. O., 1940. Metamorphism in the Lake Wakatipu region, western Otago, New Zealand. *Mem. Dep. sci. industr. Res. N.Z.* **5**.

IIYAMA, T., 1954. High–low inversion point of quartz in metamorphic rocks. *J. Fac. Sci. Tokyo Univ.*, sec. II, **9**, 193–200.

ISHIOKA, K., & SUWA, K., 1956. Metasomatic development of staurolite schist from rhyolite in the Kurobe-gawa area, central Japan (a preliminary report). *J. Earth Sci. Nagoya Univ.* **4**, 123–40.

IWASAKI, M., 1960a. Clinopyroxene intermediate between jadeite and aegirine from Suberi-dani, Tokusima Prefecture. *J. geol. Soc. Japan*, **66**, 334–40.

—— 1960b. Colorless glaucophane and associated minerals in quartzose schists from eastern Sikoku, Japan. Ibid. 566–74.

JAMES, H. L., 1955. Zones of regional metamorphism in the Precambrian of northern Michigan. *Bull. geol. Soc. Amer.* **66**, 1455–88.

JOPLIN, G. A., 1942. Petrological studies in the Ordovician of New South Wales. I. The Cooma complex. *Proc. Linn. Soc. N.S.W.* **67**, 156–96.

—— 1943. Petrological studies in the Ordovician of New South Wales. II. The northern extension of the Cooma complex. Ibid. **68**, 159–83.

KANEHIRA, K., & BANNO, S., 1960. Ferriphengite and aegirinjadeite in a crystalline schist of the Iimori district, Kii Peninsula. *J. geol. Soc. Japan*, **66**, 654–9.

KENNEDY, G. C., 1960. Phase relations of some rocks and minerals at high temperatures and high pressures. *Trans. Amer. geophys. Un.*, **41**, 283–6.

KING, P. B., 1959. *The evolution of North America.* Princeton: Princeton Univ. Press.

KOBAYASHI, T., 1941. The Sakawa orogenic cycle and its bearing on the origin of the Japanese Islands. *J. Fac. Sci. Tokyo Univ.*, sec. II, **5**, 219–578.

KOJIMA, G., & SUZUKI, T., 1958. Rock structure and quartz fabric in a thrusting shear zone: the Kiyomizu tectonic zone in Shikoku, Japan. *J. Sci. Hiroshima Univ.*, ser. C, **2**, 173–93.

KORZHINSKII, D. S., 1959. *Physico-chemical basis of the analysis of the paragenesis of minerals.* New York: Consultant Bureau.

KOZIMA, Z., 1944. On stilpnomelane in green-schists in Japan. *Proc. imp. Acad. Japan*, **20**, 322–8.

KÜNDIG, E., 1956. Geology and ophiolite problems of East Celebes. *Kon. Ned. Geol. Mijnb. Genoot. Geol. Ser.* **16** (Gedenkboek H. A. Brouwer), 210–35.

KUNO, H., BAADSGAARD, H., GOLDICH, S., & SHIOBARA, K., 1960. Potassium–argon dating of the Hida metamorphic complex, Japan. *Jap. J. Geol. Geog.* **31**, 273–8.

KURODA, Y., 1959. Petrological study on the metamorphic rocks of the Hitachi district, north-eastern Japan. *Sci. Rep. Tokyo Kyoiku Daig.*, sec. C, **58**.

MCKEE, B., 1958. Jadeite alteration of sedimentary and igneous rocks. *Program, 1958 Annual Meetings, Geol. Soc. Amer.*, 108.

MAGNUSSON, N. H., THORSLUND, P., BROTZEN, F., ASKLUND, B., & KULLING, O., 1960. Description to accompany the map of the pre-Quaternary rocks of Sweden. *Sverig. geol. Unders. Afh.*, ser. Ba, **16**.

MILLER, J., SHIDO, F., BANNO, S., & UYEDA, S., 1961. New data on the age of orogeny and metamorphism in Japan. *Jap. J. Geol. Geog.* **32**, 145–51.

MINATO, M., 1960. On the age of metamorphism in the Japanese Islands. *Int. geol. Rev.* **2**, 901–11.

MIYASHIRO, A., 1953. Calcium-poor garnet in relation to metamorphism. *Geochim. et Cosmoch. Act*, **4**, 179–208.

—— 1956. Osumilite, a new silicate mineral, and its crystal structure. *Amer. Min.* **41**, 104–16.

—— 1957. Cordierite–indialite relations. *Amer. J. Sci.* **255**, 43–62.

—— 1958. Regional metamorphism of the Gosaisyo–Takanuki district in the central Abukuma Plateau. *J. Fac. Sci. Toyko Univ.*, sec. II, **11**, 219–72.

—— 1959. Abukuma, Ryoke, and Sanbagawa metamorphic belts (in Japanese). *J. geol. Soc. Japan*, **65**, 624–37.

—— 1960a. Thermodynamics of reactions of rock-forming minerals with silica. Part III. Andalusite, kyanite, sillimanite, and corundum. *Jap. J. Geol. Geog.* **31**, 107–11.

—— 1960b. Thermodynamics of reactions of rock-forming minerals with silica. Part IV. Decomposition reactions of muscovite. Ibid. 113–20.

—— 1960c. Thermodynamics of reactions of rock-forming minerals with silica. Part V. Decomposition reactions of phlogopite. Ibid. 241–6.

—— & BANNO, S., 1958. Nature of glaucophanitic metamorphism. *Amer. J. Sci.* **256**, 97–110.

—— IIYAMA, T., YAMASAKI, M., & MIYASHIRO, T., 1955. The polymorphism of cordierite and indialite. Ibid. **253**, 185–208.

—— & IWASAKI, M., 1957. Magnesioriebeckite in crystalline schists of Bizan in Sikoku, Japan. *J. geol. Soc. Japan*, **63**, 698–703.

—— & SEKI, Y., 1958a. Enlargement of the composition field of epidote and piedmontite with rising temperature. *Amer. J. Sci.* **256**, 423–30.

—— —— 1958b. Mineral assemblages and subfacies of the glaucophane schist facies. *Jap. J. Geol. Geog.* **29**, 199–208.

MOORE, J. G., 1959. The quartz–diorite boundary line in the western United States. *J. Geol.* **67**, 198–210.

NAKAYAMA, I., 1959. Tectonic features of the Sambagawa metamorphic zone, Japan. *Mem. Coll. Sci. Kyoto*, ser. B, **26**, 103–10.

NOZAWA, T., 1959. On the age of Hida metamorphic rocks (a preliminary note) (in Japanese). *J. geol. Soc. Japan*, **65**, 463–9.

PLAS, L. VAN DER, 1959. Petrology of the northern Adula region, Switzerland. *Leid. géol. Meded.* **24**, 411–602.

READ, H. H., 1952. Metamorphism and migmatization in the Ythan Valley, Aberdeenshire. *Trans. Edinb. geol. Soc.* **15**, 265–79.

REED, J. J., 1958. Regional metamorphism in southeast Nelson. *Bull. geol. Surv. N.Z.*, N.S., **60**.

ROEVER, W. P. DE, 1947. Igneous and metamorphic rocks in eastern Central Celebes, in *Geological explorations in the island of Celebes under the leadership of H. A. Brouwer*, 65–173. Amsterdam: North-Holland Publishing Co.

—— 1955a. Some remarks concerning the origin of glaucophane in the North Berkeley Hills, California. *Amer. J. Sci.* **253**, 240–4.

—— 1955b. Genesis of jadeite by low-grade metamorphism. Ibid. **253**, 283–98.

—— 1956. Some differences between post-Paleozoic and older regional metamorphism. *Geol. en Mijnb.* (NW ser.), **18**, 123–7.

SEITSAARI, J., 1951. The schist belt northeast of Tampere in Finland. *Bull. Comm. géol. Finl.* **153**.

SEKI, Y., 1957. Lawsonite from the eastern part of the Kanto Mountains. *Sci. Rep. Saitama Univ.*, ser. B, **2**, 363–73.

—— 1958. Glaucophanitic regional metamorphism in the Kanto Mountains, central Japan. *Jap. J. Geol. Geog.* **29**, 233–58.

SEKI, Y., 1959. Petrological studies on the circum-Hida crystalline schists. I. *Sci. Rep. Saitama Univ.*, ser. B, **3**, 209–20.

—— 1960a. Jadeite and associated minerals of meta-gabbroic rocks in the Sibukawa district, central Japan. *Amer. Min.* **45**, 668–79.

—— 1960b. Distribution and mineral assemblages of jadeite-bearing metamorphic rocks in Sanbagawa metamorphic terrains of central Japan. *Sci. Rep. Saitama Univ.*, ser. B, **3**, 313–20.

—— 1960c. Jadeite in Sanbagawa crystalline schists of central Japan. *Amer. J. Sci.* **258**, 705–15.

—— 1961a. Notes on rock-forming minerals. (17) Jadeite from Kanasaki of the Kanto Mountains, central Japan. *J. geol. Soc. Japan*, **67**, 101–4.

—— 1961b. Pumpellyite in low-grade regional metamorphism. *J. Petrol.* **2**, 407–23.

—— & SHIDO, F., 1959. Finding of jadeite from the Sanbagawa and Kamuikotan metamorphic belts, Japan. *Proc. Japan Acad.* **35**, 137–8.

SHAW, D. M., 1956. Geochemistry of pelitic rocks. Part III. Major elements and general geochemistry. *Bull. geol. Soc. Amer.* **67**, 919–34.

SHIDO, F., 1958. Plutonic and metamorphic rocks of the Nakoso and Iritono districts in the central Abukuma Plateau. *J. Fac. Sci. Tokyo Univ.*, sec. II, **11**, 131–217.

—— 1959. Notes on rock-forming minerals. (9) Hornblende-bearing eclogite from Gongen-yama of Higasi-Akaisi in the Bessi district, Sikoku. *J. geol. Soc. Japan*, **65**, 701–3.

—— & MIYASHIRO, A., 1959. Hornblendes of basic metamorphic rocks. *J. Fac. Sci. Tokyo Univ.*, sec. II, **12**, 85–102.

—— & SEKI, Y., 1959. Notes on rock-forming minerals. (11) Jadeite and hornblende from the Kamuikotan metamorphic belt. *J. geol. Soc. Japan*, **65**, 673–7.

SIMONEN, A., 1953. Stratigraphy and sedimentation of the Svecofennidic, early Archean supracrustal rocks in southwestern Finland. *Bull. Comm. géol. Finl.* **160**.

SNELLING, N. J., 1957. Notes on the petrology and mineralogy of the Barrovian metamorphic zones. *Geol. Mag. Lond.* **94**, 297–304.

SUGIMURA, A., 1960. Zonal arrangement of some geophysical and petrological features in Japan and its environs. *J. Fac. Sci. Tokyo Univ.*, sec. II, **12**, 133–53.

SUZUKI, J., 1952. Ultrabasic rocks and associated ore deposits of Hokkaido, Japan. *J. Fac. Sci. Hokkaido Univ.*, ser. IV, **8**, 175–210.

THOMPSON, J. B., JR., 1955. The thermodynamic basis for the mineral facies concept. *Amer. J. Sci.* **253**, 65–103.

—— 1957. The graphical analysis of mineral assemblages in pelitic schists. *Amer. Min.* **42**, 842–58.

TILLEY, C. E., 1924. Contact metamorphism in the Comrie area of the Perthshire Highlands. *Quart. J. geol. Soc. Lond.* **80**, 22–70.

—— 1925. A preliminary survey of metamorphic zones in the southern Highlands of Scotland. Ibid. **81**, 100–12.

TURNER, F. J., 1938. Progressive regional metamorphism in southern New Zealand. *Geol. Mag. Lond.* **75**, 160–74.

—— 1948. Mineralogical and structural evolution of the metamorphic rocks. *Mem. geol. Soc. Amer.* **30**.

VALLANCE, T. G., 1953a. Studies in the metamorphic and plutonic geology of the Wantabadgery–Adelong-Tumbarumba district, N.S.W. I. Introduction and metamorphism of the sedimentary rocks. *Proc. Linn. Soc. N.S.W.* **78**, 90–121.

—— 1953b. Studies in the metamorphic and plutonic geology of the Wantabadgery–Adelong–Tumbarumba district, N.S.W. II. Intermediate-basic rocks. Ibid. **78**, 181–96.

VOGT, TH., 1927. Sulitelmafeltets geologi og petrografi. *Norg. geol. Unders.* **121**.

WISEMAN, J. D. H., 1934. The central and south-west Highland epidiorites: a study in progressive metamorphism. *Quart. J. geol. Soc. Lond.* **90**, 354–417.

YAMAGUCHI, T., 1951. On the so-called Yonchon system and its regional metamorphism (in Japanese). *J. geol. Soc. Japan*, **57**, 419–38.

YOSSII, M., 1935. On some glaucophane-rocks from the Ryukyu Archipelago. *Sci. Rep. Tôhoku Univ.*, ser. II, **16**, 225–48.

ZWART, H. J., 1959. Metamorphic history of the central Pyrenees. Part I. *Leid. geol. Meded.* **22**, 419–90.

2

Reprinted from *Geologie en Mijnbouw,* **46e**(8), 283–309 (1967)

THE DUALITY OF OROGENIC BELTS

H. J. ZWART [1]

ABSTRACT

A twofold classification of the orogenic belts of the world is proposed, based in first instance on the facies series of the metamorphic rocks. The Hercynian and Alpine orogens of Europe are reviewed in detail and are found to differ respectively in several ways: (1) low pressure metamorphism with andalusite and cordierite vs. high pressure metamorphism with glaucophane, sodium pyroxene, lawsonite and kyanite, (2) thin vs. thick metamorphic zones, (3) abundant vs. few granites and migmatites, (4) few vs. abundant ophiolites and ultrabasites, (5) broad vs. narrow orogen, (6) small vs. large uplift, (7) scarce vs. dominant nappe structures.

The Caledonian orogen of Europe is examined and found to be intermediate in nature between the "Hercynotype" and "Alpinotype" orogens. The Svecofennian-Karelian belt of the Baltic shield is found to be a typical Hercynotype orogen.

The paired metamorphic belts of the Circumpacific region appear to be a new element in the earth's history and tend to emphasize the peculiar character of the Pacific Ocean.

It seems probable that the thermal history of the earth has not appreciably changed during the last 2500 million years. The pressure-temperature fields of the various facies and facies series based on field and experimental work are discussed. Finally the causes of regional metamorphism are considered.

I. INTRODUCTION

There seems to be general agreement among geologists that all orthotectonic mountain chains are the result of strong folding, overthrusting, and metamorphism of orthogeosynclines. These geosynclines are narrow, very long sedimentary troughs consisting usually of a eugeosynclinal part with initial magmatic ophiolites and a miogeosynclinal part lacking such magmatic features. Metamorphism and intrusion of granites are mostly confined to the eugeosyncline, whereas the miogeosyncline may be strongly folded but is devoid of metamorphism and magmatism. The deformation in such geosynclines is said to be "Alpinotype", a term coined by S t i l l e (1924) to indicate that strong folding and overthrusting took place. Obviously the Alps are taken as an example for this type of deformation. Further attempts to classify orogenic belts into certain types have thus far not been very successful, although recently Russian geologists have identified two kinds of mobile belts by the nature of their volcanism and metallogeny (B i l i b i n, 1959).

It is the intent of the present author to propose a classification, which is based not on the characteristics of the original geosyncline, but on the final products of the orogenesis, i.e. deformed and metamorphosed rocks. The criteria for this classification may seem somewhat unorthodox, namely the metamorphic mineral associations occurring in the regional metamorphic belts of the orogens. On this basis it can be shown that at least two fundamentally different types of mountain chains exist. As an exemple the two most recent orogens in Europa, the Alpine and the Hercynian, will be considered.

It is a well known fact that regional metamorphism is mainly due to a rise in temperature in the earth's crust and this is clearly shown by the progressive metamorphic zoning, present in

[1] Author's adress: Geologisk Institut, Aarhus Universitet, Århus C, Denmark.

almost all regional metamorphic terrains. Less importance has been attached to the pressures operating during metamorphism, although it is here that the possibility exists for the distinction of different orogenies. During the last decade research in the field of experimental petrology has made it possible to obtain data about the temperatures and pressures which are necessary for the formation of certain minerals. From this research it has become clear that pressure is an important factor in metamorphism.

In an important paper M i y a s h i r o (1961) described so-called "facies series", groups of metamorphic facies which belong together and whose formation depends on the pressure. He distinguished three such facies series with two intermediate series. These are a low pressure, an intermediate pressure and a high pressure series, characterized respectively by 1) andalusite and cordierite, 2) kyanite and sillimanite, and 3) glaucophane and jadeite. In each of these facies series a whole range of temperature dependent facies and subfacies can be distinguished. As the pressures under which these minerals form are relatively well known from experiments, and because it is assumed that pressures are directly dependent upon the weight of the overlying rocks, the corresponding depth at which the rocks were metamorphosed and deformed can be calculated. Any such calculations must however be checked with the available field data. It is possible that some correction has to be made, due to so-called tectonic overpressures, but it is doubtful whether they have any influence, and if so their influence is supposed to be relatively small.

Facies series and their temperature and pressure ranges have recently been reviewed by D e n T e x (1965).

M i y a s h i r o (1961) has called attention to the occurrence of paired metamorphic belts in the circumpacific region, the pairs being characterized by different facies series. D e R o e v e r (1956) has shown that the Mesozoic and Tertiary mobile belts of the world are typified by the occurrence of rocks in the high pressure glaucophane-schist facies, a facies which is rare in pre-Mesozoic belts. When the facies series of the Alpine and Hercynian orogenies in Europe are compared, important differences come to light, and this of course signifies large differences in depth of folding and metamorphism.

II. THE HERCYNIAN OROGENIC BELT OF EUROPE

1. General remarks

Many publications about the Hercynian fold belt of Europe have appeared during the last 50

years, among which those of S t i l l e (1924, 1951), K o s s m a t t (1927), L o t z e (1945), and A u b o u i n (1965) are outstanding. In most of these publications an attempt has been made to distinguish certain characteristics of the original geosynclines from which the fold belt has arisen. It is clear that the northern segment of this belt is very well known, with its division from north to south into a marginal trough, a Rheno-Hercynian zone, a Saxo-Thuringian zone and a Moldanubian zone. The latter zone – extending from the Bohemian massif through the Black Forest, the Vosges, and the Central Massif in France to southern Brittany – is according to most views a "Zwischengebirge" consisting largely of a pre-Hercynian massif. North of the Moldanubian zone in the Saxo-Thuringian and Rheno-Hercynian zones there is present a rather complete, strongly folded eu- and miogeosyncline. The vergence of the folds is towards the north and both the intensity and age of the folding decrease in the same direction. The Moldanubian Zwischengebirge or "Alemanische Scheitel" of S t i l l e (1951) is the locus of strong Hercynian metamorphism and granite intrusion. Many authors, the latest being A u b o u i n (1965), have tried to show that a similar succession of geosynclines and folds also exists to the south of the Moldanubian zone.

The Moldanubian zone is thus considered to represent a symmetry axis (symmétrie centrifugale, A u b o u i n) in the Hercynian orogen. This interpretation is, however, not necessarily valid, although rocks which have been folded and metamorphosed in the Hercynian foldbelt are indeed widespread. The difficulty is partly due to the presence of the Alps in which a large part of the Hercynian basement has been reactivated during the Alpine orogenesis. To the west, in the Montagne Noire and the Pyrenees, one might think of a southern geosyncline, but after closer consideration the geosynclinal character is not very evident. Moreover the metamorphism there is as strong as in the Moldanubian zone but the folding took place rather late (Sudetic). The vergence of the folds is only to the south in the Southern Pyrenees.

Another zone of centrifugal symmetry is claimed by A u b o u i n to exist in the highly metamorphic zone in the western part of the Iberian peninsula, the Ibero-Hesperian zone. Unfortunately, little is known about the exact age of the folding and the vergence of the folds. It might be possible that a centripetal axis exists between the two large divisions of the Hercynian chain in the Cantabrian mountains. Here the Leon line separates two quite different parts of a mountain chain with northerly overthrusts in the southern

44

part and southerly overthrusts in the northern part.

Apart from this typically miogeosynclinal Cantabrian basin, most of the rocks affected by the Hercynian orogeny occur in a regional metamorphic state, as for example in the Pyrenees, in the Hercynian cores of the Alps, in the Montagne Noire, in large parts of Spain and Portugal, in small but significant outcrops of Sardinia, and in Calabria (southern Italy). Only relatively small basins like the Cantabrian basin are completely devoid of Hercynian metamorphism. Most of the above mentioned regions are characterized not only by Hercynian folding and metamorphism of a Paleozoic sedimentary sequence, but also by the involvement and reworking of parts of an older basement. This is for example, the case in the whole Moldanubian zone, in the Ibero-Hesperian zone, and possibly in other areas as well. Because the metamorphism in this whole region is of the low pressure type – with andalusite and cordierite as typical minerals, and kyanite, glaucophane, lawsonite, and jadeite almost completely lacking – it is of importance to treat these areas individually, to see what evidence exists to assign the metamorphic mineral assemblages to the Hercynian orogeny.

Figure 1. is a map showing the distribution of the Hercynian regional metamorphic rocks and the Hercynian instrusive granites. The wide extent of this orogenic belt in Europe is obvious.

2. *The Moldanubian zone*

a. Bohemian massif.

When reviewing the literature about the Bohemian massif it becomes evident that at least two facies series are present. On the one hand there are rock types such as eclogites, garnet-peridotites, kyanite-granulites and kyanite-gneisses which belong to a high pressure series and on the other hand there are schists and migmatites with andalusite and cordierite that belong to the low pressure facies series. That many rocks of the Bohemian massif are polymetamorphic is assumed by most authors. Furthermore there is general agreement that the eclogites, granulites and kyanite gneisses belong to an older pre-Hercynian metamorphic complex whose exact age is still unknown. No generally accepted opinion exists about the age of metamorphism of the andalusite-cordierite bearing rocks. Dating of metamorphism can be done either by dating the youngest rocks which participated in the metamorphism, and their unconformable cover, or

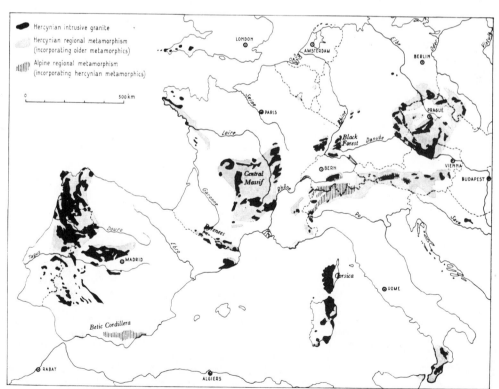

Fig. 1. — Distribution of Hercynian regional metamorphism and Hercynian intrusive granites.

by radiometric dating.

For the Bohemian massif it is sufficient to find rocks of Silurian, Devonian or Carboniferous age in a metamorphic state, because it is known that post-Hercynian metamorphism has not occurred. Two such occurences are known. One is mentioned by Bederke (1935) from the Altvatergebirge in Silesia where Devonian rocks are metamorphosed to phyllites and micaschists with staurolite and andalusite. The latter mineral occurs also in the underlying Kepernik gneisses which in addition contain kyanite. The second example is described by Schreyer (1966) from the Oberpfalz where sediments of the Saxo-Thuringian zone are transformed into biotite-muscovite-andalusite schists. The original sediments are of lower Paleozoic age; metamorphism is necessarily younger and must be Hercynian because the Caledonian orogeny is only of local occurrence in the Bohemian massif.

Age determinations from metamorphic rocks are as yet not very numerous. Rubidium-Strontium determinations from a perlgneiss with cordierite from the Bavarian Forest yielded an age of 335 m.y. Sillimanite and biotitegneisses, some of which are intercalated with cordierite-sillimanite-gneisses in the Moldanubian, gave ages from 330-345 m.y. (Davis & Schreyer, 1962). Potassium-Argon dates from biotite in a cordierite-gneiss from the southwestern part of the Bohemian massif gave an apparent age of 340 m.y. (Smejkal, 1965). From these data it appears that metamorphism producing andalusite and cordierite, both typical of low pressure metamorphism, occurred during the Hercynian orogeny.

Besides these cases, andalusite and cordierite are very widespread at many places in the Bohemian massif (fig. 2). Often a definite proof of the Hercynian age of the metamorphism cannot be given and must await further radiometric dating. A number of authors have, however, recently put forward strong arguments attributing the low pressure minerals to the Hercynian orogeny. Schreyer (1966), for example, considers the cordierite-sillimanite gneisses of the Oberpfalz near the Saxo-Thuringian boundary as due to the Hercynian metamorphism. The same applies to the perlgneisses of the Austrian Mühlviertel. Scharbert (1965) describes Hercynian cordierite bearing migmatites overprinting older rocks. Köhler (1950) mentions Hercynian cordierite-gneisses from the Austrian part of the Bohemian massif. Extensive Hercynian migmatization, more or less connected with the granites in the Moldanubian in southern Czechoslovakia, has recently been described by Suk (1964) and Dudek & Suk (1965, a and b). Many of

these rocks contain cordierite and sillimanite.

In contrast herewith many authors agree that the high pressure metamorphics, such as the eclogites and kyanite-gneisses, belong to an older orogeny. Schuman (1930), Schüller (1949), Misar, (1958), Bederke (1956), Scharbert (1962 and 1965), Zoubek (1948 and 1965), and Scheumann & Behr (1963) and several others consider that the eclogites and granulites were formed either during a Precambrian orogeny or during the Assyntian orogeny. Only the Münchberger gneiss massif with its granulite facies rocks is considered to be of Hercynian age (Würm, 1960). Again opinions are divided on this point (Watznauer, 1955).

Summarizing it can be stated that metamorphism generating andalusite and cordierite bearing rocks occurred during the Hercynian orogeny. It is not excluded that older metamorphics also carry these minerals, but there is no proof of it. High pressure metamorphism producing kyanite-gneisses, eclogites and kyanite-granulites is almost exclusively of pre-Hercynian age.

b. Black Forest, Odenwald, Spessart.

As has been mentioned, the Black Forest, the Vosges, the Central Massif and part of Brittany belong to the Moldanubian zone. These areas have in fact many features in common with the Bohemian massif. There occur crystalline rocks, ortho- and paragneisses of pre-Hercynian age, in which a number of undeformed intrusive granites and granodiorites were emplaced during the Hercynian orogeny.

Cordierite-sillimanite bearing migmatites are widespread among the crystalline rocks of the Black Forest (fig. 2; Wimmenauer, 1950; Mehnert, 1962). According to Hoenes (1949) and Wimmenauer (1950) two stages of regional anatexis can be distinguished, preceded and separated by stages of synkinematic metamorphism. The migmatites of the second stage of anatexis occur abundantly and are undeformed. These rocks carry andalusite, cordierite and sillimanite. They predate the intrusive granites, but the opinion that these migmatites are of Hercynian age is gaining favour (Wimmenauer, 1962; Mehnert, pers. comm.). It seems rather logical that large scale anatexis has lead to the abundant intrusive granites, a hypothesis also proposed by Langerfeldt (1961). If this is the case then the migmatites belong to the Hercynian cycle. Potassium-Argon dates from various metamorphic rocks yielded ages roughly between 260 and 360 m.y. and, according to Mehnert (1958), metamorphism has taken place near the end of the Devonian or the beginning of the Carboniferous, which would be early Hercynian.

46

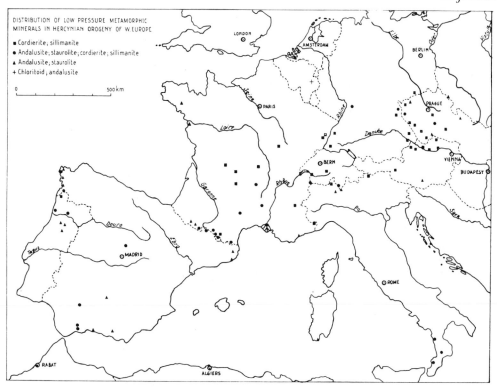

Fig. 2 — Distribution of low pressure mineral assemblages in West Europe. Each dot, triangle or square indicates a region where these minerals are widespread.

North of the Black Forest two smaller parts of the Moldanubian zone form the massifs of Odenwald and Spessart. In the Odenwald B o s s d o r f (1961) described schists between Hercynian granites. These schists were certainly metamorphosed during the Hercynian orogeny and contain andalusite, cordierite, sillimanite, and garnet; other schists carry plagioclase, andalusite and cordierite.

In the Spessart massif the only aluminium silicate is staurolite; kyanite, andalusite and cordierite have not been found there (M a t t h e s, 1954). Thus the metamorphics of the Odenwald clearly belong to the low pressure facies series. The lack of index minerals makes it impossible to assign the Spessart rocks to any facies series.

c. Vosges.

The geology of the Vosges is rather similar to that of the Black Forest; a pre-Hercynian basement is intruded by Hercynian granites and cordierite bearing migmatites have been formed in several places (V o n E l l e r, 1961). Their age of metamorphism has not been established with certainty; earlier opinions favoured a pre-Hercynian age, but later opinions mention the possibility of a Hercynian age. Andalusite and cordierite

which are certainly Hercynian occur of course in the contact metamorphic rocks around the granite plutons, e.g. in the well known Barr-Andlau contact aureole.

d. Central Massif and Montagne Noire.

The northern part of the Central Massif is the continuation of the Vosges and is referred to as the Arverno-Vosgien zone. According to R o q u e s (1941) at least two metamorphic events belonging to two orogenies have occurred; the first one is Precambrian, the second one post-Cambrian and by Roques referred to as probably Caledonian. Andalusite, cordierite, sillimanite, and staurolite are mentioned from several places (fig. 2.); kyanite is also encountered. J u n g (1946) stated that unmetamorphosed fossiliferous Lower Carboniferous rests unconformably on gneisses and granites. Consequently the metamorphism must be older than Lower Carboniferous. Some granites cut through the Carboniferous and hence are younger. According to D e m a y (1948) metamorphism, migmatization and granite intrusion are of Hercynian age, but pre-Stephanian.

47

Cordierite-sillimanite gneisses are widespread for instance in the Dordogne valley southwest of Clermont-Ferrand, and in the Sioule valley west of this town. C h e v e n o y (1958), who studied the crystalline schists in the northwestern part of the Central Massif, also distinguished two metamorphic episodes. The first one produced so-called ectinites − schists and gneisses − and "fundamental migmatites" with cordierite and sillimanite. The younger metamorphism gave rise to stratiform migmatites. The age of the original sediments in this region is mainly Cambro-Ordovician. According to C h e v e n o y the first metamorphism is Caledonian, the second one Hercynian. Lead α radiometric age determinations from zircons give ages for ectinites of 410-481 m.y., for the fundamental migmatites 340-385 m.y. and for the stratiform migmatites 235-298 m.y. The fundamental migmatites with cordierite could therefore have been formed in the Bretonnic phase of the Hercynian orogeny. F o r e s t i e r (1963), who worked between Clermont-Ferrand and Le Puy, also distinguished two metamorphisms; an older one probably Assyntian with micaschists carrying staurolite, andalusite, sporadic kyanite, sillimanite, and cordierite, and a younger one of Hercynian age with anatectic rocks carrying cordierite and sillimanite. The possibility that the andalusite in the first group belongs to the second metamorphism is suggested by Forestier. Generally the second metamorphism is lowgrade and the migmatitic rocks occur only locally. Some granites of definite Hercynian age carry large cordierites.

In the southern part of the Central Massif the relations are less complicated. In the Cévennes, P a l m (1957) described micaschists with andalusite, cordierite, sillimanite and garnet as Hercynian metamorphics. The micaschist zone is very thin in this area and amounts to only 500 m.

In the Montagne Noire the metamorphism is also of Hercynian age according to G è z e (1949). This is confirmed by S c h u i l i n g (1960) who described micaschists with andalusite, staurolite and garnet, and gneisses with sillimanite, cordierite and garnet. According to S c h u i l i n g the geothermal gradient must have been very high, namely 200°C/km.

Summarizing the literature it seems certain that in the Central Massif two orogenies have occurred. Each one was accompanied by regional metamorphism. The younger orogeny is most probably Hercynian and produced at several places schists and migmatites with andalusite, cordierite, staurolite and sillimanite. The older metamorphism is probably Assyntian but might be older; the evidence for a Caledonian orogeny is rather weak. The older metamorphics may also carry low pressure minerals but there is no direct evidence thus far. In the southern part of the Central Massif exclusively Hercynian metamorphics with andalusite and cordierite occur.

e.   Brittany.

In Brittany and Normandy at least three orogenies have occurred. The oldest one, of limited outcrop extent, is called the Pentevrian and is Precambrian in age. The next one, the widespread Cadomian orogeny, is correlated with the Assyntian orogeny elsewhere. The youngest one is the Hercynian orogeny. According to C o g n é (1957) the Hercynian orogeny produced only epizonal schists and migmatites. It occurred mainly in the southern part of Brittany. Staurolite-andalusite-schists occur in the Baie de Bourganeuf but are considered to be pre-Hercynian by C o g n é. Schists with andalusite, staurolite, kyanite and sillimanite could be partly due to Hercynian metamorphism. In any case low pressure minerals are found locally in Brittany. Minerals of the high pressure suite like kyanite, and eclogitic rocks occur at several places and they may belong to the Assyntian (Cadomian) orogeny. The glaucophane schists of Ile de Groix on the other hand are of Hercynian age. They may be indicative of high pressure metamorphism, but this occurrence is certainly exceptional.

f.   Pyrenees.

Although the Pyrenees do not belong to the Moldanubian zone proper, this chain will be treated here, as it shows many similarities to the Montagne Noire. Micaschists and migmatites in the Pyrenees are characterized by the abundant occurrence of andalusite, staurolite, cordierite and sillimanite (fig. 2.). There is no doubt about the Hercynian age of the metamorphism, first because the metasediments range in age from Cambro-Ordovician to Devonian and second because there is no Caledonian folding. Kyanite has been found in the St. Barthélemy massif (F o n t e i l-l e s *et al.*, 1964) in granulite facies rocks, which have been described by the present author as a Precambrian basement (Z w a r t, 1954). For these reasons the Hercynian metamorphics belong to the low pressure facies series.

g.   Summary of the Moldanubian zone.

From this review of the whole Moldanubian zone some general features seem to be well established.
1)   Almost the whole zone is polymetamorphic; at least two orogenies accompanied by regional metamorphism are superposed. The older orogeny is at least partly Assyntian, but still older orogenies may be present; the younger orogeny

48

is Hercynian.

2) Metamorphism of the Assyntian or older rocks belongs at least partly to a high or intermediate pressure facies series, with kyanite-schists and gneisses, kyanite-granulites and eclogites. Whether rocks in the low pressure facies series were also produced during this orogeny has not been established with certainty.

3) Hercynian metamorphism corresponds to the low pressure facies series, with andalusite and cordierite as typical minerals (see fig. 2.). Hercynian rocks attributable to high pressure metamorphism are rare (Münchberger gneiss massif).

4) Many of the Hercynian metamorphics occur as migmatitic and anatectic rocks.

5) Intrusive granites and granodiorites of undoubted Hercynian age occur abundantly throughout the whole Moldanubian zone (see fig. 1.).

3. The Ibero-Hesperian zone

In the Ibero-Hesperian zone, which comprises a large part of Spain and Portugal, many similarities to the Moldanubian zone can be distinguished. Hercynian metamorphics of the low pressure facies series and Hercynian granites are widespread. Furthermore, parts of an older basement occur in this zone, especially in Galicia. Eclogites and granulites are described as Precambrian by Parga-Pondal (1956) in the so-called "Complejo antiguo". More recently Den Tex & Vogel (1962) described a pre-Hercynian eclogite-granulite body with kyanite-gneisses. Floor (1966) investigated a pre-Hercynian riebeckite-gneiss and showed in the same publication that andalusite and cordierite are of Hercynian age. Farther east in the Cantabrian Mountains De Sitter (1961) mentioned unmetamorphosed Cambrian resting unconformably upon Precambrian phyllites and schists. This unconformity can be correlated with the Assyntian unconformity.

The Hercynian orogeny has put a marked stamp on the geology of Galicia, producing metamorphism and migmatization on a large scale. Because part of the Cambro-Silurian is also incorporated in the metamorphism and because no Caledonian orogeny exists here, it is certain that folding and metamorphism is of Hercynian age. Andalusite, cordierite, staurolite and sillimanite are reported from many localities in northwestern Galicia (fig. 2.; Floor, 1966).

In the continuation of this zone in northern Portugal several theses of the Amsterdam Geology Department have shown that andalusite and cordierite bearing schists and migmatites are to be attributed to the Hercynian metamorphism (Schermerhorn, 1959; Priem, 1962). More to the west near Oporto, Torre de Assunçao

(1962) described Hercynian cordierite-sillimanite bearing migmatites. As in Galicia, granites intruded during the Carboniferous are plentiful. Farther south in Portugal the Hercynian metamorphic zone continues, but the grade decreases and moreover there is little information in the literature about the mineral assemblages found there. Where this zone again enters Spain, metamorphism with andalusite, cordierite, staurolite, and sillimanite of Hercynian age has been described by Fabriès (1963) in the Sierra Morena. Another example in the same region is mentioned by Henke (1927), but here the andalusite and cordierite seem to be of contact metamorphic origin around Hercynian granites. Another example of low pressure metamorphism which also belong to the Hercynian chain is reported from the Sierra de Guadarrama north of Madrid (Heim, 1952). Here andalusite and cordierite occur in schists and migmatites. Kyanite is also found in that region, but whether it is Hercynian or older is uncertain.

Although the Ibero Hesperian zone of the Hercynian orogeny is far less well known than the Moldanubian zone, there are remarkable similarities between these two zones of this large orogenic belt. Both zones are very broad, i.e. several hundreds of kilometres. Low pressure metamorphism is very widespread and has affected Cambrian, Ordovician, Silurian, and Devonian sediments. In remnants of older orogenies high pressure metamorphism is found. Finally the occurrence of large bodies of intrusive granites of Carboniferous age is typical.

4. Various other occurrences

Besides in the Moldanubian and Ibero-Hesperian zone, other occurrences of Hercynian metamorphics and granites are to be found at various places in southwestern Europe, namely on some islands in the Mediterranean, in Calabria, in the Alps, and in the Betic Cordillera (see fig. 2.). In the latter two areas the Alpine orogeny has reworked some of the Hercynian metamorphics.

The islands of Corsica and Sardinia consist for an important part of Hercynian granite with some areas of regional metamorphic rocks. On the southern tip of Sardinia, I found in these schists abundant andalusite and a strongly altered mineral, probably cordierite. In Calabria Hercynian metamorphics with granites again occur. Kinzigites and sillimanite-cordierite-gneisses have been described from some localities by Teichmüller & Quitzow (1935), Bucca (1884) and Novarese (1931).

In the Alps part of the Hercynian basement has been spared from the Alpine metamorphism, for instance in the Aar and the Aiguilles Rouges massifs. Besides metamorphic rocks, Hercynian granites also occur abundantly in these massifs, e.g. the Central Aar granite. Cordierite has been mentioned from the latter massif, and from the Aiguilles Rouges massif by B e l l i è r e (1958). In the Monte Rosa nappe, which partly conists of Hercynian basement that has not been reworked, cordierite-sillimanite gneisses have been found by B e a r t h (1952). In the Tauern massif similar Hercynian cordierite bearing rocks have been reported by F r a s l (1955). In the Austroalpine and South-Alpine zones, where Alpine metamorphism has not taken place, andalusite and cordierite bearing rocks are known from several localities, for instance near Lago Maggiore (S u z u k i, 1930; N o v a r e s e, 1931) and from the Ötztaler Alpen (H a m m e r, 1925). It should be added that in the same zone kyanite is found in several places. Eclogites have been reported from the Koralpen and Saualpen. Therefore in this region at least two facies series seem to be present. Little is as yet known about polymetamorphism in this region, but it seems possible that pre-Hercynian material is also involved.

The andalusite-cordierite bearing rocks in the Hercynian massifs of the Alps seem to be connected to those of the Pyrenees by the Massif des Maures near Toulon on the Mediterranean coast. Here andalusite-schists are common, but from the same area kyanite is also reported (S c h o e l l e r & L u t a u d, 1931).

Low pressure assemblages have further been reported from parts of the Betic Cordillera, namely the Betic of Malaga and Sierra Nevada. In the Sierra Nevada, this mineral occurs together with staurolite in micaschists (B a r r o i s & O f f r e t, 1889). It is certain that these rocks belong to a pre-Alpine basement, but their attribution to the Hercynian orogeny is not certain. According to E g e l e r (1963) the metamorphism is older. Locally the rocks are overprinted by the Alpine metamorphism with kyanite and glaucophane as typical minerals (N i j h u i s, 1964). In the Betic of Malaga, andalusite, cordierite, and sillimanite occur at several places, e.g. in the massif of Velez Malaga and the Serrania de Ronda (M i c h e l L é v y & B e r g e r o n, 1889). In this region a Hercynian age of the metamorphism is assumed by H o e p p e n e r (1964). Alpine metamorphism is lacking here.

Finally the metamorphism in the Ardennes has to be mentioned. Although the occurrence of chloritoid (ottrelite) is well known from this area, andalusite from the same locality has been mentioned by C o r i n (1963). Besides these minerals garnet also occurs. As the grade of metamorphism is rather low, no other aluminium silicates occur there. The presence of andalusite indicates, however, that here metamorphism is again of the low pressure type, which is in accordance with the relatively small overburden consisting of Devonian and Carboniferous only.

5. Conclusions

From the preceding survey of the Hercynian orogenic belt of Europe the following conclusions can be drawn.

1) Regional metamorphism and the folding connected with it are extremely widespread and have affected a large part of Europe south of the Caledonian foreland in Northern Europe. The area involved is approximately 1500 x 2000 km in size.

2) Low pressure metamorphism is characteristic of the whole Hercynian belt. There are only scattered occurrences of intermediate or high pressure metamorphism, and in a number of cases the Hercynian age of these occurrences is in doubt.

3) Migmatites and intrusive granites are abundant throughout the whole Hercynian belt. Contact metamorphism around these granites is also characterized by andalusite and cordierite, indicating the shallow position of the granites.

4) Many of the regional metamorphic rocks are the product of the reactivation of older basement rocks.

III. THE ALPINE OROGENIC BELT OF EUROPE

The Alpine mountain chain in Europe is a very long, but narrow and sinuous range. At several places in this chain regional metamorphic rocks of Alpine age are exposed. The main occurrences are in the French, Italian and Swiss Alps, on the Island of Corsica, in the Betic Cordillera in southern Spain, and in the Dinaric ranges in Yugoslavia and Greece. From the Dinarides the Alpine orogen can be traced into Turkey.

The metamorphic rocks of the Alps proper are well known and the zonal or facies character of the metamorphism has been described by E. N i g g l i (1960), N i g g l i & N i g g l i (1965), W e n k (1962), B e a r t h (1966), V a n d e r P l a s (1959) and E l l e n b e r g e r (1960). The last three authors considered in particular metamorphic rocks with glaucophane and other sodium amphiboles. These rocks occur for instance in the French Alps (Vanoise) with lawsonite and sodium pyroxene; in the canton of Valais in Switzerland (Val de Bagnes, Zermatt region) with chloritoid, sodium pyroxene, garnet and kyanite; in the Italian Alps (Val d'Aosta) with the same paragenesis, and in the Upper Rhine Valley with phengite, sodium pyroxene, garnet, chloritoid and kyanite. It is obvious that the lawsonite-glauco-

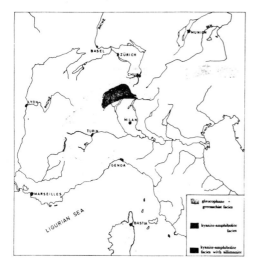

Fig. 3 — Distribution of metamorphic facies in the Alps.

Metamorphic rocks of Alpine age occur also in the Tauern window. Glaucophane has been found at two localities. Further stilpnomelane, chloritoid and garnet are common, and kyanite has been reported from a few places.

The minerals glaucophane, sodium pyroxene, lawsonite, stilpnomelane and phengite emphasize the difference between the Alpine and Hercynian metamorphism. Chloritoid, staurolite and sillimanite occur abundantly in both orogens and cannot be used for differentiating between both belts.

On the Island of Corsica which presumably is the continuation of the Penninic zone of the Alps, glaucophane, lawsonite and sodium-pyroxene have been reported by B r o u w e r & E g e-l e r (1951) and E g e l e r (1956).

Alpine metamorphic rocks are also exposed in the Betic Cordillera in southern Spain. Our knowledge about the mineral assemblages is mainly due to D e R o e v e r & N i j h u i s (1964) and N i j h u i s (1964). The following characteristic minerals are found in this region: glaucophane, sodium pyroxene, garnet, kyanite and stilpnomelane.

The distribution of sodium amphiboles in the continuation of the Alpine chain in Yugoslavia and Greese is shown on maps published by V a n d e r P l a s (1959). It is evident from them that blue amphiboles are rather common in this part of the Alpine orogen.

It is clear therefore that the whole Alpine chain of Europe is characterized by metamorphism in the lawsonite-glaucophane or the glaucophanitic greenschist facies of W i n k l e r (1965) with only locally (Tessin, Betic Cordillera) higher grade metamorphism in the kyanite- or sillimanite zone. High pressures and relatively low temperatures were responsible for this kind of metamorphism as has been postulated by D e R o e v e r (1956), M i y a s h i r o (1961) and W i n k l e r (1965). The widespread occurrence of the lawsonite-glaucophane facies not only in the European Alpine orogen, but also elsewhere in the world was originally noticed by E s k o l a (1939), but it has been especially emphasized by D e R o e v e r (1956), who considered glaucophane and lawsonite to be guide minerals for post Paleozoic metamorphism. This latter author stated that the formation of these minerals depended on a secular decrease of the geothermal gradient during the course of the earth's history. I will return to this important aspect later on.

Besides the presence of a number of typical minerals, the Alpine metamorphism shows in its evolution certain trends which seem to be widespread. Like many Hercynian metamorphics, the Alpine rocks are commonly plurifacial; different mineral assemblages follow one another in time.

phane or the glaucophane-schist facies occupies large portions of the Alps (fig. 3.). Eclogites have been described from the Zermatt region by B e a r t h (1965). These rocks constitute in this area the oldest metamorphic rocks of the Alpine cycle and have been subsequently changed into glaucophane schists and then into greenschists. Locally the eclogite character is still well preserved.

Besides sodium amphiboles and chloritoid, N i g g l i & N i g g l i (1965) have shown the distribution of stilpnomelane, staurolite, kyanite and sillimanite in the Swiss Alps. Stilpnomelane is found mainly in a region outside the glaucophane occurrences and apparently is of a lower grade. Staurolite, kyanite and sillimanite occur inside the arc where glaucophane is found in an area occupied by the canton Tessin situated between the Upper Rhine Valley and Valais. Staurolite is found in the same area as kyanite, but sillimanite occurs only in a small area near the root zone of the Pennine nappes (fig. 3.). The distribution of the aluminium silicates shows the same pattern as the map with anorthite percentages of plagioclase, published by W e n k (1964). The highest An-percentages can be found in the kyanite-sillimanite area, and towards the north, west and east the An-percentage gradually decreases, indicating the lower metamorphic grade. It is evident, as has been mentioned by N i g g l i & N i g g l i, that the area of highest grade metamorphism coincides with the Tessin culmination, the deepest exposed part of the pennine Alps.

The nature of the successive mineral facies is strikingly similar in many Alpine metamorphic rocks. Van der Plas (1959) described a succession from glaucophane-schists to schists with blue-green amphibole to greenschists with actinolite. Bearth (1965, 1966) found in the Zermatt region a succession from eclogite to glaucophane schist to greenschist. A similar succession is described from the ophiolite series of Grand Paradis in the Italian Alps by Nicolas (1966). Ellenberger (1960) investigated in the French Alps garnet-glaucophane-lawsonite-schists, in which these primary minerals are replaced by epidote, actinolite, chlorite and stilpnomelane, indicating greenschist facies conditions. A similar succession is known from Corsica (Egeler, 1956) and recently it has been described by Nijhuis (1964) from the Betic Cordillera, where glaucophane schists are followed by rocks in the greenschist facies and finally in the almandine-amphibolite facies. Outside the European Alpine zone, a similar plurifacial succession has been discovered by De Roever (1950) at Celebes in Indonesia.

Thus there is a general tendency in these metamorphic suites to show the succession glaucophane-schist to greenschist. Occasionally the succession is eclogite to glaucophane-schist to greenschist. If this evolution is to be translated in terms of temperature-pressure conditions it seems quite probable that it is an indication of a decrease in pressure, because most of the high pressure minerals, like glaucophane, lawsonite and sodium pyroxene are replaced by minerals of the greenschist facies. In some cases like the Betic Cordillera an increase in temperature also seems to have occurred. Bearth (1966) and Nicolas (1966) arrived at the same conclusions based on their work in the Alps.

Finally, there is one more striking difference between the Hercynian and Alpine orogenies, namely the amount of granite and migmatite present. In the Hercynian belt granite and migmatite are much more common than in the Alpine chain. A few granites of Alpine age occur in the Alps (Bergell, Adamello). There are a number of small bodies in the Dinaric chain, but there are none in Corsica and in the Betic Cordillera.

Alpine migmatites have been described by Wenk (1956) from Tessin, but it is quite possible that they are Hercynian migmatites overprinted by Alpine structures. The latter hypothesis was expressed by Niggli & Niggli (1965). The present author, after inspecting the rocks in the field, is also convinced that the migmatites are pre-Alpine and probably Hercynian. In the remainder of the Alpine chain the grade of metamorphism is too low to permit migmatization.

In fig. 4. the distribution of Alpine granites is represented. When compared with the distribution of the Hercynian granites (fig. 1.),

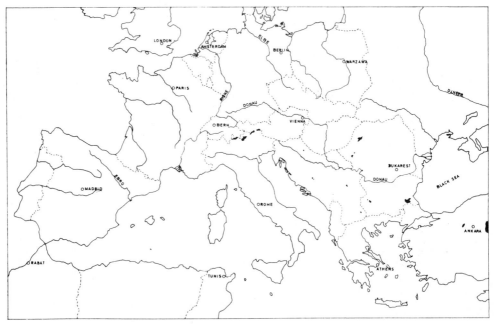

Fig. 4 — Distribution of Alpine granites in Europe.

the difference is striking. The scarcity of Alpine granites and migmatites cannot be due to the level of erosion, for as will be seen, the amount of uplift and erosion of the Alpine chain is much larger than that of the Hercynian orogen. Therefore, the divergence must be due to the thermal history of both orogenies.

In conclusion it can be stated that the Alpine orogen of Europe is characterized by (1) the occurrence of high pressure metamorphism, mainly in the lawsonite- glaucophane or glaucophanitic greenschist facies, (2) a general succession from glaucophane-schists to greenschists, and (3) a scarcity of granitic rocks.

IV. THE DUALITY OF THE HERYNIAN AND ALPINE OROGENIC BELTS

1. *Temperature Gradients*

It is obvious that low pressure metamorphism is directly dependent on the overburden and hence on the depth of the rocks during metamorphism. It is not excluded, however, that tectonic overpressures have to be taken into account, but these are mainly used to explain high pressure metamorphism. It is generally agreed that their effect does not exceed 10-15% of the pressure due to the weight of the overburden, although higher estimates have been made by S o b o l e v (1960) and C o l e m a n & L e e (1962). Nevertheless, in the present author's opinion the depth of metamorphism is still the dominant controlling factor.

In the Hercynian orogen there are a few cases where the stratigraphic control is sufficiently good to wårrant conclusions as to the depths of metamorphism and the temperature gradients that existed during the Hercynian metamorphism.

In the P y r e n e e s Zwart (1963) arrived at a depth of 3500-4000 m for the upper limit of the andalusite zone and a depth of about 5000 m for the cordierite-sillimanite zone. These are probably minimum depths for regional metamorphism and involve temperature gradients of about 150° C/km. A direct consequence of this high gradient is the thinness of the progressive metamorphic zones when going from the chlorite zone to the cordierite-sillimanite zone. In the Pyrenees the thickness of the individual metamorphic zones often amounts to not more than a few hundreds of metres, and from the chlorite to the cordierite-sillimanite zone not more than 1500 m. This confirms the high temperature gradient. There are mountains in the Pyrenees, like in the Bosost area, where the summit is in the chlorite zone and the base is in the cordierite-sillimanite zone.

Similar observations have been made by B e - d e r k e (1947) in the Altvatergebirge in Silesia,

also a part of the Hercynian belt. He arrived at a temperature gradient of 200°C/km. It should be added that at that time much less was known about the temperatures at which metamorphic minerals are produced. In spite of this, the order of magnitude is comparable with the present author's estimate.

F r i t s c h (1962) calculated a gradient of 60° C/km in the Austrian Alps where andalusite occurs in the highest grade gneisses.

Figures similar to these have been suggested by P a l m (1958) in the Cevennes (Central Massif) and S c h u i l i n g (1960) in the Montagne Noire. The latter author, also using the thickness of the metamorphic zones, postulated a geothermal gradient of 200°C/km, but this figure may in actuality be somewhat exaggerated as S c h u i - l i n g does not take into account the possible "telescoping" effect. Telescoping is the result of an increase in temperature during metamorphism whereby higher temperature minerals occur along with relics of lower temperature minerals; in such a situation the rocks are called plurifacial (D e R o e v e r & N i j h u i s, 1964). The result of telescoping of mineral facies is an apparent reduction in the thickness of individual mineral zones, because lower grade rocks do not often respond to an increase in temperature. This effect, especially when rather large temperature differences are involved, is quite characteristic of high temperature – low pressure metamorphism and is a consequence of the thinness of the metamorphic zones; the ascent of a heat front has only to be relatively small to cause the telescoping effect. By a detailed microscopic investigation the various mineral zones can be identified and their order of formation unraveled. Good examples of telescoping facies occur in almost all Pyrenean metamorphics (Z w a r t, 1963). In the majority of cases examined the metamorphism is of a progressive nature. Cases of retrograde metamorphism are rare.

From other areas in the Hercynian belt similar features are shown in photographs or figures. S c h e r m e r h o r n (1959) for instance depicts late biotite porphyroblasts in a phyllite matrix. Metamorphism has increased from the chlorite into the biotite zone. Unfortunately, in many cases the study of metamorphism as a chronological sequence of events is neglected.

If it is assumed that high pressure metamorphism is directly related to deep burial, then the temperature gradient during the Alpine metamorphism must have been very small, many times smaller than during the Hercynian metamorphism. Assuming a burial of about 30 km and a temperature of 300-450° for the lawsonite glaucophane facies one arrives at gradients between

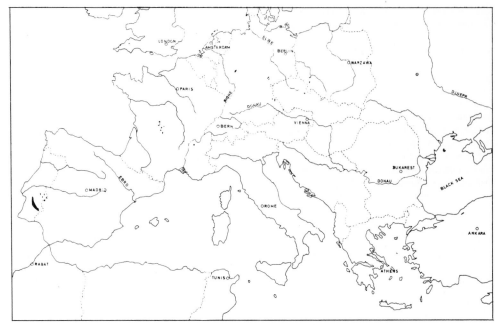

Fig. 5 — Distribution of Hercynian ophiolites in Europe.

10° and 15°C/km, which are considerably smaller than the gradients that presently exists at many places on the continents. The slow drop in temperature towards the surface resulted in very thick metamorphic zones and over large tracts the grade of metamorphism is equal. This is particularly well illustrated in the Pennine nappes where with a relief of more than 3000 m the grade of metamorphism is the same at the summits of the mountains and in the bottoms of the valleys, clearly showing that each individual zone has a thickness of several thousands of metres. This is in strong contrast to the thin metamorphic zones in the Hercynian chain.

2. Depth and Pressure Ranges

In the case of the low pressure metamorphism characteristic of the Hercynian orogen it is important to know the maximum depth at wich this type of metamorphism took place. This will make it possible to calculate the pressure range. Again this can be done in the Pyrenees, where a total thickness of 10-12 km of metamorphic rocks is exposed in the Canigou and Aston massifs. Together with the upper, unmetamorphosed strata of 4000 m thickness this gives a depth of approximately 15 km. The deepest exposed rocks are cordierite-sillimanite gneisses and lack kyanite, indicating that the rocks were still metamorphosed in the low pressure facies series.

Other possibly more deeply exposed parts of the Hercynian chain e.g. the Central Massif, are always characterized by the same low pressure minerals. Therefore it seems safe to conclude that low pressure metamorphism with andalusite and cordierite can take place at levels at least 15-20 km. deep. The range of these minerals might even extend deeper. The total thickness of the whole sequence of metamorphic rocks in the Central Massif has been estimated at 20 km by J u n g & R o q e s (1952). However these figures are probably exaggerated.

Our figures are in good agreement with experimental studies of the Al_2SiO_5 system and of cordierite.

The position of the triple point of the three aluminium silicates is of great importance. According to B e l l (1963) and K h i t a r o v *et al.* (1963) it lies at pressures of 8-9 kb. S c h u i l i n g (1962) and W i n k l e r (1965) placed it at a pressure of 7,5-8 kb and at a temperature of 560°C. Recent work of N e w t o n (1966) on the andalusite-kyanite reaction indicates even lower pressures for the triple point, about 4 kb with a temperature of 520° C. A l t h a u s (1966) has recently determined the triple point in experiments and arrived at a pressure of 6,5 kb and a temperature of 600°C. These last figures, especially for the pressures are in rather good agreement with field evidence. According to S c h r e y e r & Y o d e r (1964) cordierite is also a low pressure mineral and it could be stable to depths of about 25 km.

There is a general agreement that the lawsonite-glaucophane facies has been formed under high pressures of at least 8-10 kb (D e R o e v e r, 1956; M i y a s h i r o, 1961; C o l e m a n & L e e, 1962; W i n k l e r, 1965) .That means that, if tectonic overpressures are not considered, a depth of about 30 or more kilometres was necessary for the Alpine metamorphism. These figures are in so far in agreement with the conclusions reached from the study of the Hercynian chain, that high pressure metamorphism cannot be expected at depths of less than 20 kilometres.

3. *Plurifacial metamorphism*

The nature of plurifacial metamorphism in the Alpine and Hercynian orogenic belts is quite different. In the Hercynian metamorphics where greenschist facies rocks are often followed by andalusite-staurolite schists and these in turn by cordierite-sillimanite rocks, the sequence of mineral assemblages indicates a raise of temperature during the whole metamorphic period.

In the Alpine metamorphics the general trend is from glaucophane-schists to greenschist, a succession which presumably is due to a decrease in pressure during the metamorphism.

4. *Granites and migmatites*

A direct consequence of the steep geothermal gradients present during the Hercynian metamorphism is the melting of rocks. This began to take place in the cordierite-sillimanite zone and most rocks in this zone are migmatitic, usually containing quartzo-feldspathic mobilisates. At many places the rocks were completely mobilized in situ and gave rise to autochthonous granodiorites or granites, which often still carry cordierite or sillimanite as evidence of their original sedimentary nature. Continued mobilization and melting undoubtedly lead to the formation of real magmas which intruded surrounding or higher strata. The vast amount of Hercynian granites has been noticed by F a u l (1962) who used for these granites the comprehensive term "Hercynian complex". The granites are surrounded by contact aureoles also with low pressure mineral assemblages. Therefore it is not unexpected that many of the typical contact metamorphic aureoles, like the Barr-Andlau and many others, surround Hercynian granites.

We can conclude that Hercynian low pressure metamorphism and the formation of abundant granite and migmatite are directly linked. In many cases the granite is the product of the meta-

morphism and not its cause. This is principally evident from the time relations. Generally the regional metamorphism precedes the intrusion of the granites; this can be concluded from detailed structural analysis in regional metamorphic rocks where intrusive granites occur (Z w a r t, 1963). On a large scale it has also been proved that many Hercynian granites are relatively late Carboniferous (F a u l, 1962), whereas the metamorphism often is early Carboniferous or even late Devonian. It is difficult to say whether or not there are other, older granites, which might have been the source of the vast amount of heat required for the regional metamorphism at so shallow a level in the earth's crust. It is evident, however, that heat must have been introduced either by rising magmatic bodies, which in this case can only be granite, or by hot fluids or solutions coming from beneath.

For the Pyrenees the first explanation seems not unlikely because in most regional metamorphic areas of this mountain chain large bodies or sheets of orthogneiss occur, which might be the original heat-bringing granite responsible for the regional metamorphism. These gneisses are probably of early Hercynian age. The same situation exists in the Montagne Noire and the Cevennes.

The amount of granite and migmatite in the Alpine orogeny is very small when compared to the Hercynian belt. This feature is apparently related to the low geothermal gradient which prevented any appreciable melting of the rocks and the formation of granite.

Therefore the hypothesis that the higher grade metamorphism in the Tessin culmination is related to a thermal dome or a granite in the underground, as suggested by W e n k (1962) and C h a t t e r j e e (1961) is at least doubtful. It is more likely that the higher temperatures in this part of the Alps were the result of deeper burial (see N i g g l i & N i g g l i, 1965).

Introduction of heat in the Hercynian orogen was responsible for large scale migmatization and granitization. Lack of heat prevented these features in the Alpine chain.

5. *Initial Geosynclinal Magmatism*

One more noteworthy feature of the Hercynian orogeny of Europe has to be mentioned, namely the scarcity of ophiolites. On fig. 5. the ophiolites of the Hercynian belt are shown. It is evident that they are not very abundant. Among these ophiolites there are hardly any ultrabasites. Therefore it is difficult to distinguish a eu- and a miogeosynclinal zone in the Hercynian belt, except for the region north of the Moldanubian

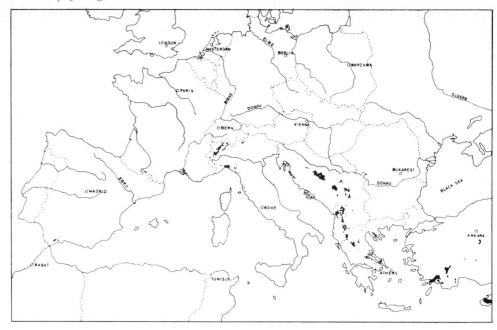

Fig. 6 — Distribution of Alpine ophiolites in Europe.

zone where such a division is possible. This is also clear from the tectonic map of Europe, where most of the Hercynian belt is shown to consist of miogeosynclinal rocks. Despite its wide extent a clear picture of the Hercynian geosyncline of a larger part of Europe cannot be given, and in my opinion the typical "Textbook" geosyncline of Stille or Aubouin may never have existed.

On the other hand the initial geosynclinal magmatism in the Alpine chain is much more important than in the Hercynides. Fig. 6. shows the distribution of the Alpine ophiolites. When compared with fig. 5. of the Hercynian ophiolites the more common occurrence in the Alpine belt is obvious. Among these Alpine ophiolites ultrabasites are rather important as for example in the ophiolite zone of Zermatt-Saas Fee in the Pennine nappes (Bearth, 1959). These ophiolites lead us to the Alpine geosyncline which certainly is much better defined than the Hercynian one. The division into a miogeosyncline without ophiolites and with important limestone development giving rise to the Helvetides, and a eugeosyncline with ophiolites and a much more monotonous limy shale facies evolving into the Pennine nappes is evident. Also the much narrower Alpine chain is more in accordance with the "geosynclinal theory" than the very broad Hercynian **zone.**

6. *Uplift and Erosion*

Because the whole Hercynian belt is characterized by the same low pressure metamorphism now exposed at the surface it is evident that the depth of erosion is not very large. At many localities where unmetamorphosed Upper Paleozoic is exposed erosion has not reached deeper than a few kilometres. In the metamorphic areas the depth of erosion usually lies between 5 and 15 km, and probably exceeds 20 km only rarely. Therefore the uplift of the Hercynian chain since the Carboniferous has been rather small and it seems quite obvious that this belt never possessed such lofty peaks as the present Alpine range. This has already been expressed by R i c h-t e r (1935) who claimed that the Hercynian chain of Europe never was a high mountainous one. This, of course, applies only to the overall picture. Locally there has occurred some considerable uplift in a rather short time, for example in the Pyrenees where an uplift of several kilometres has taken place since the late Tertiary. But even in this chain the total uplift since the Carboniferous has not exceeded 15 km. In general the uplift in the Hercynian orogeny amounts to an average of 1 mm per 10-25 years. The rate of uplift must locally have been greater immediately after the orogeny, because at some places in the Hercynian belt the Permian rests unconformably on the metamorphics.

The amount and speed of uplift after folding and metamorphism is far greater in the Alpine chain. Rocks undoubtedly formed at depths of more than 20 km and probably more than 30 km are exposed now in high mountainous regions. As the metamorphism in the Alps took place about 20-30 m.y. ago, uplift has amounted to about 1 mm per year, which is rapid compared to the average of 1 mm per 10-25 years in the Hercynides.

7. *Discussion*

From the different character of the orginal geosynclines – narrow in the Alps with extensive ophiolites, broad in the Hercynides without appreciable ophiolites – it is evident that by this time the future character of the orogeny has already been determined, at least if both mountain chains can be considered as typical. Because the ultrabasites in the Alpine chain probably originated from the mantle or at least from a very deep part of the crust, it may be assumed that the crust under the Alpine geosyncline was strongly thinned or perhaps even faulted so as to make introduction of mantle or lower crustal material possible. It is evident that the geosyncline did not have an oceanic floor, as the Hercynian basement is exposed at many places in the Alps and constitutes an integral part of the Pennine nappes. The crust under the Hercynian geosyncline apparently was sufficiently thick so as to impede the introduction of material from the lower crust or mantle.

The question arises how the metamorphics in the Alpine chain could have been buried so deeply as to result in such very high pressures. Several solutions have been brought forward. One is that tectonic overpressures add considerably to the pressure caused by the weight of the overburden. It is generally accepted, however, that these overpressures if they do exist, can have only a small influence amounting to about 1 or 2 kb. Far greater overpressures have been claimed by S o-b o l e v (1960). Tectonic overpressures have been recently discussed by R u t l a n d (1965). Several objections can be raised against tectonic overpressures being a powerful mechanism, (1) the same mineral assemblages, and not different ones, are often produced both during and after deformation, (2) zones of very strong movement in low pressure orogens fail to produce high pressure minerals, (3) the occurrence of apparently undeformed, non-schistose glaucophane rocks can hardly be due to tectonic overpressures, and (4) when stresses are applied to rocks they usually cause movements which immediately result in a reduction in the stresses. Therefore tectonic overpressures can hardly be the main cause of high pressure metamorphism. The main cause must be rather deep burial.

Deep burial in subsiding geosynclines has also been proposed by some geologists as the cause of high pressure metamorphism (e.g. W i n k l e r, 1965). Although this may hold true for certain cases, it does not apply to all occurrences of high pressure facies series.

In my opinion the origin of high pressure metamorphism must indeed be sought in the deep burial of the rocks; most likely it is a tectonic burial due to the superposition of nappes. It can be no accident that nappes are universally present in the European Alpine chain, both in the Alps and in the Betic Cordillera. The relation between high pressure metamorphism and nappe structures has already been suggested by D e R o e v e r (1953) from his studies of the Celebes glaucophane-schists and by E l l e n b e r g e r (1958) from the French Alps. The latter author proved that metamorphism postdates the nappe movements and that deep burial is due to the so-called "geosynclinal des nappes". N i g g l i & N i g g l i (1965) also believe that tectonic burial was the cause of the high pressure metamorphism in the Alps. This fits quite well the time relations of the metamorphism. W e n k (1962) and C h a t t e r j e e (1961) observed that metamorphic zones cut through the tectonic boundaries. They concluded from this that the distribution of metamorphic minerals is not related to the major structures. It is obvious, however, that when metamorphism is due to burial by the piling up of nappes, it could only start when a sufficient number of nappes have been stacked one on top of another. This means that high pressure metamorphism can take place only in a late stage of nappe movements, when the rocks are buried deeply enough and are warmed up sufficiently. Metamorphism will continue after the movements have ceased. There is no reason then why metamorphic boundaries should not cut through tectonic units.

Although it is not certain that the upper Pennine nappes west of the Tessin culmination can be connected with those to the east, or that the East Alpine nappes have covered the Pennine nappes, such assumptions can explain a total thickness of several tens of kilometres sufficient to account for the high pressures.

In the Betic Cordillera, the presence of Pennic style nappes has been proved beyond doubt and here the necessary depth could also have been achieved by the piling up of nappes.

Although it would be going too far to suppose that all areas with glaucophane and kyanite are characterized by nappe structures, it certainly applies to the Alpine European occurrences of this facies series.

Another consequence of the superposition of several nappes is the accumulation of a thick pile of crustal material, causing a downbuckle in the mantle. This would cause uplift by isostatic compensation, as soon as the compressive forces responsible for the nappe movements had disappeared. This downbuckle must have been rather deep in view of the rapid and large uplift of the Alps. These features may in turn be related to the narrowness of the Alps, amounting to not more than 200 km. Only in a relatively narrow zone could a thick pile of crustal material accumulate.

In the Hercynian belt – the width of which is thousands instead of hundreds of kilometres – the rocks have also been compressed and folded, but less strongly than in the Alpine chain. A deep downbuckle in the mantle could not possibly have occurred over so large an area. Consequently a large isostatic uplift cannot be expected in this orogenic belt.

In this way strong compression, thickening of the crust, high pressure metamorphism, narrowness of the mountain chain and rapid and large uplift in the Alpine belt are seen to be closely related. Likewise the less strong compression, low pressure metamorphism, broadness of the chain and small uplift in the Hercynian orogen, are also related.

The nature of the Alpine plurifacial metamorphism, which is due to a decrease in pressure, is most probably the result of the isostatic uplift. This implies that the formation of the greenschist facies must postdate the nappe movements. Although dating of the metamorphic events with regard to tectonic phases has only locally been done in the Alps, observations of the present author in the Zermatt region confirm that the greenschist facies metamorphism is later than the nappe movements.

If we try to describe the difference between the Alpine and Hercynian orogenies in terms of energy, it seems obvious that in the former the energy was mainly released in tectonic movements and in the latter in the form of heat. This explains both the contrast in amount of granite in both orogenies and also the the different rates of uplift.

The difference in amount of granite in the Hercynian and Alpine orogenic belts cannot be due to insufficient depth of erosion in the Alps, because this mountain chain is more deeply exposed than the Hercynian chain.

On the other hand, regions with low pressure metamorphism like the Hercynian chain, apparently do not contain Penninic style nappes. Some nappe structures have been postulated in the Central Massif, (Demay, 1934) and in the Pyrenees (Guitard, 1964) but in both cases any real stratigraphic proof is lacking and the evidence cannot be compared with that from the Alps. Of course I do not want to say that nappes cannot occur in the Hercynian chain. They have convincingly been described for instance by De Sitter (1959) from the Cantabrian mountains, but here they are more of the Helvetic type in unmetamorphosed rocks. The absence of Pennine-type nappes in the Hercynian orogen is also exemplified by the common occurrence of slate belts, strongly compressed regions with more or less vertical folds and cleavage. They occur for instance in Devon in England, in the Rhenic and Thuringer slate mountains in Germany, and in the Pyrenees. In the latter region the slates are underlain by metamorphics with mostly flatlying structures which can hardly be reconciled with nappe structures.

8. Definition of Hercynotype and Alpinotype Orogenies

From the previous description, based primarily on the mineral assemblages in the metamorphic rocks, it is clear that the Hercynian and Alpine orogenies of Europe are almost diametrically opposed. Their main differences warrant the distinction of a Hercynotype and an Alpinotype orogeny, with the following characteristics:

Hercynotype:
1) shallow, low pressure metamorphism; thin metamorphic zones
2) plurifacial metamorphism dependent on increase in temperature
3) abundant granite and migmatite
4) few ophiolites, ultrabasites practically absent
5) very wide orogen
6) small and slow uplift
7) nappe structures missing or rare.

Alpinotype:
1) deep, high pressure metamorphism; thick metamorphic zones
2) plurifacial metamorphism dependent on decrease in pressure
3) few granites and migmatites
4) abundant ophiolite with considerable amount of ultrabasite
5) relatively narrow orogen
6) large and rapid uplift
7) nappe structures predominant.

The main point that both orogenies have in common are that strong folding and metamorphism took place and that both processes were to a large extent simultaneous.

V. NATURE OF THE CALEDONIAN AND SVECOFENNIAN OROGENIC BELTS

If the Hercynian and Alpine orogenies are essentially different in character, it should be possible to examine older orogens and find ones similar to, or transitional between the Hercyno-type and Alpinotype orogenies.

1. *Caledonides*

The Caledonian mountain chain of Europe appears to be transional in nature between the Hercynian and Alpine chain. Starting with the mineral assemblages of the metamorphic rocks we encounter the typical Barrovian zones in Scotland: chlorite, biotite, garnet, staurolite, kyanite and sillimanite zone. These zones correspond to the greenschist, epidote-amphibolite and amphibolite facies. There are no lawsonite-glaucophane schists, either in the Scottish[2] or in the Scandinavian Caledonides, but the amphibolite facies rocks contain kyanite at many places. In Scotland the low pressure association with andalusite-cordierite occurs in a small region in northeastern Scotland, the so-called Buchan metamorphism, where kyanite-schists grade into andalusite-schists. This transition probably corresponds to the phase boundary between andalusite and kyanite, as suggested by Johnson (1963) and Chinner (1966).

Kyanite has been described from several places in the Norwegian Caledonides (fig. 7.); it is often associated with staurolite (Holtedahl, 1960). Stilpnomelane, probably also a high pressure mineral, was found by Murris (1957) at Västerbotten, Sweden. Andalusite and cordierite have not been mentioned from the Scandinavian Caledonides, but it is not unlikely that occurrences comparable to the Buchan area of Scotland do exist and will be found in the future.

The lack of eclogites[3] and lawsonite-glaucophane rocks on the one hand, and the scarcity of andalusite-cordierite rocks on the other, shows that the mineral assemblages in the Caledonian chain occupy a position intermediate between those of the Alps and of the Hercynides. Consequently, the pressures and depth of metamorphism must also have been intermediate. This intermediate facies series has hardly any typical minerals of its own but is characterized by the presence of kyanite and the absence of lawsonite-glaucophane.

---

[2] Glaucophane-schist and eclogites are known from Ayrshire (Bloxam & Allen, 1960), but the age of the metamorphism is supposed to be pre-Caledonian.
[3] Eclogites occur in the Caledonian chain in Sunnmore (Eskola, 1921; Gjelsvik, 1952). They may belong to the Caledonian metamorphism or they may be older. The garnet-peridotite from the same region is almost certainly much older.

It would be of interest to know the thickness of the individual metamorphic zones to see whether they also are intermediate, but reliable figures are not available. Several other properties of the Caledonian fold belt also confirm the intermediate position of the orogeny. On fig. 7. the Caledonian granites are shown. It is evident that they are more abundant than the Alpine granites (fig. 4.), but less abundant than Hercynian ones (fig. 1.). It is not unexpected that a certain amount of granite will have been formed, because at several places the sillimanite isograd is passed and anatexis becomes a possibility. Apparently the amount of heat introduced was not adequate to cause such large scale granitization as in the Hercynian belt.

Considering the ophiolites, a glance at fig. 7. indicates that they also occupy an intermediate position between those in the Hercynian and Alpine belts. Among these ophiolites ultrabasic rocks are quite common.

Other characteristics of the Caledonian orogeny are more like the Alpine belt. It is a rather narrow but very long mountain chain, like the Alps, but less sinuous. In Scandinavia the eastern foreland is exposed and in Scotland the western foreland is found west of the Moine thrust. The width of this orogeny can be estimated at about 400 km. with a length of at least 3000 km. The structure of the Caledonides is also more Alpinotype because of its large nappes, present both in Scandinavia and in Scotland. The presence of kyanite indicates that the depth of metamorphism

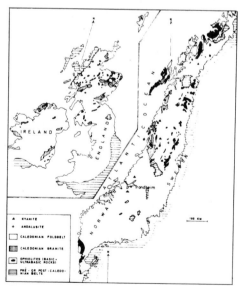

Fig. 7 — Map of the Caledonian foldbelt with distribution of kyanite, andalusite, Caledonian granites and ophiolites.

must have been rather great. This depth could not have been caused by geosynclinal subsidence but was achieved rather by the stacking of nappes which, like the Pennine nappes in the Alps, contained a large proportion of basement rocks.

The uplift of the Caledonides has been considerable and probably rapid. In Scotland the Devonian Old Red sandstone unconformably over-

lies high grade metamorphics indicating that the rocks were uncovered in a relatively short time.

The original geosyncline of the Caledonides is much more in accordance with textbook examples than the Hercynian one. According to S t r a n d (1961) the Caledonian geosyncline consists of two parts, (1) an eastern miogeosynclinal zone comprised of generally unmetamorphosed Eocam-

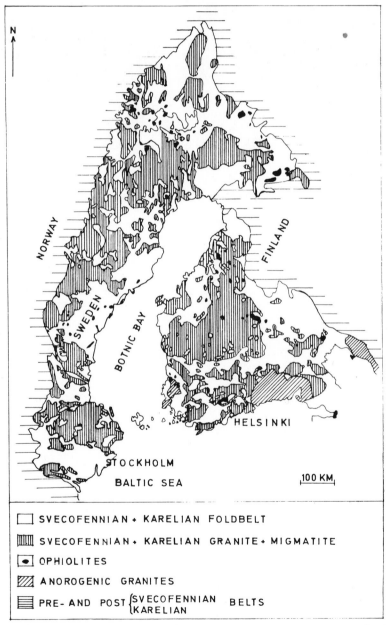

N

NORWAY

SWEDEN

BOTNIC BAY

FINLAND

HELSINKI

STOCKHOLM

BALTIC SEA

100 KM

☐ SVECOFENNIAN + KARELIAN FOLDBELT

▥ SVECOFENNIAN + KARELIAN GRANITE + MIGMATITE

⊡ OPHIOLITES

▨ ANOROGENIC GRANITES

▤ PRE- AND POST {SVECOFENNIAN} BELTS {KARELIAN}

Fig. 8 — Map of the Svecofennian and Karelian foldbelts with distribution of granites and migmatites, and ophiolites.

brian sparagmites and Cambro-Silurian sediments, and (2) a western eugeosynclinal zone comprised of extensively metamorphosed sediments with abundant ophiolites.

Summarizing the characteristics of the Caledonian orogeny of Europe it can be stated that it is intermediate between the Hercynian and Alpine belts of Europe. This is indicated by the intermediate character of its mineral assemblages, by the amount of granite and migmatite and by the occurrence of ophiolites. From the point of view of structures, the nature of the geosyncline, and the relative narrowness of the mountain chain, it resembles more the Alpine type orogeny.

2. *Svecofennian and Karelian orogenic belts*

Going back in the history of the earth, several other orogenies, each accompanied by regional metamorphism, are known to have occured. From Europe the following orogenies are known: (1) The Assyntian at the end of the Precambrian, (2) an unnamed orogeny with strongly metamorphosed and deformed rocks in southwest Norway having an age of about 900-1000 m.y. and probably corresponding to the Grenville orogeny in North America, (3) the Gothian orogeny, mainly in Sweden and about 1550 m.y. old, (4) the Svecofennian and Karelian about 1800 m.y. old, (5) the Belomorian in the USSR with an age of about 2000 m.y. and (6) the Saamian which is still older. The characteristics of many of these orogens are not always evident, and the older the orogen the less that has been preserved. Thus it is often difficult to assess their original properties. There is one clear exception, however; it is the Svecofennian and Karelian orogenies which together form an important part of the Baltic shield. These two orogenies, which are both of the same age, possess all the characteristics of a Hercynotype mountain chain.

M i y a s h i r o (1961) noticed already that the mineral assemblages found in this belt are clearly of the low pressure type because both andalusite and cordierite are rather widespread in the Finnish and Swedish parts of this mountain chain (S i m o n e n, 1960; M a g n u s s o n *et al.,* 1960).

Kyanite, on the contrary, has not been mentioned from the Finnish Svecofennian and Karelian orogenies and is rare in the Swedish Svecofennides. Eclogites have not been encountered, but charnockites are described from southwestern Finland by P a r r a s (1958). These rocks, however, do not contain kyanite but rather cordierite. They formed possibly under relatively low pressures.

Of particular importance are the rather thin metamorphic zones which can be observed in the

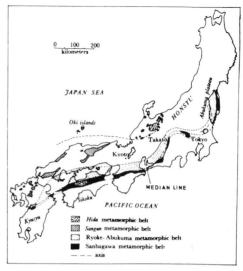

Fig. 9 — Map of main island of Japan showing paired belts;

Hida and Ryoke-Abukuma are low pressure belts; Sangun and Sanbagawa are high pressure, glaucophanitic belts.

Tampere schist belt. In the same area the telescoping of mineral facies is shown by high temperature minerals occurring in a low temperature matrix. This indicates a rapid increase in temperature during metamorphism, a situation similar to that in the Pyrenees. The present author has found for example, near Hameenlinna, epizonal phyllites with late andalusite and staurolite porphyroblasts.

One of the characteristics of the Hercynotype orogen, namely the widespread development of granite and migmatite, is even more evident in the Svecofennian-Karelian belt (fig. 8.). Intrusive granite and migmatite have not been separated on this map, but their extreme abundancy is beyond doubt. Ophiolites on the contrary are rather scarce as is shown on the same map, and among these ophiolites, ultrabasic rocks are rare.

Another feature, which the Svecofennian-Karelian orogen has in common with the Hercynian belt of Europe is its large size. It covers an area of 700 x 1500 km, but was originally larger because the western border is now covered by the Caledonian fold belt. The structures in this Precambrian mountain chain are much more like those in the Hercynian orogen. Nappes seem to be absent and folds with steep to vertical axial planes and steep schistosities are the predominant structure. The original geosyncline was also of the rather indistinct Hercynian type and without a clear distinction between a eu- and a miogeosynclinal zone. According to S i m o n e n (1960) the Karelian belt should constitute the miogeosyncline, and the Svecofennian the eugeosyncline. The main criterion for this division is the difference in the nature of the sediments, many quartzites and limestones in the Karelian belt, greywackes and greenstones in the Svecofennian belt. Ophiolites are largely confined to the Svecofennides. Because neither the presence or absence of metamorphism, nor the occurrence of granites can in the case of this orogen be used to distinguish between a eu- and miogeosyncline, the division of this geosyncline is less evident than the division of typical Alpinotype geosynclines.

It is obvious that in all respects the Svecofennian-Karelian belt is a typical Hercynotype orogen. The presence of andalusite-cordierite schists indicates that the rocks which are now exposed at the surface were not formed so deep in the earth's crust as has been thought previously. The amount of uplift of this part of the Baltic shield is apparently rather small, generally not more than 15 km and at many places even less. The often heard statement that the Precambrian orogens are very deeply eroded roots of mountain chains can no longer be upheld. The depth of erosion of the Svecofennian-Karelian orogen is hardly greater than that of the Hercynian, and far less than the Alpine or Caledonian chains.

Other parts of the Baltic shield, for instance the pre-Karelian Belomorides, contain much kyanite and may belong to a more Alpinotype orogeny. Here a much deeper part of a mountain chain is now visible.

VI. THE CIRCUMPACIFIC BELT

From our survey of the European orogenic belts it has become clear that each orogeny is characterized by only one (or exceptionally two) facies series in its regional metamorphic rocks. Consequently metamorphism has taken place within a restricted range of pressure and depth. If it is assumed that this is the normal behaviour for orogenic regions, then the circumpacific region with its paired metamorphic belts stands out as unique. The first description of these belts was by A. Miyashiro (1961). The paired belt in New Zealand has recently been described in more detail by Landis & Coombs (1966).

According to Miyashiro one such a paired belt consists of (1) an outer belt, on the Pacific side, with high pressure metamorphism in lawsonite-glaucophane facies, and (2) an inner belt, on the continental side with low pressure metamorphism and andalusite-cordierite. High and low pressure metamorphic belts occur side by side (fig. 9.).

On the Island of Honsyu (fig. 9.) two paired belts occur. In the case of the outer Sanbagawa – Abukuma belt metamorphism in both the high and low pressure belts was simultaneous and Cretaceous in age. In the case of the second paired belt, which is located more toward the continent, metamorphism occured earlier, during the early Mesozoic.

It is important to note that several features of the individual belts in Japan are similar to the European orogens. The distribution of granites and migmatites, for example, shows that they are widespread in the low pressure Abukuma belt, but lacking in the high pressure Sanbagawa belt (fig. 10.).

On the other hand the ophiolites and ultrabasic rocks are widely distributed in the Sanbagawa belt, but are quite scarce in the Abukuma belt (fig. 11.). The boundary between the Sanbagawa and the Abukuma belts is a large fundamental fault, called the median tectonic line. The difference in amount and rate of uplift in both belts is also significant. Uplift was small in the low pressure belt and large in the high pressure zone. A dissimilarity with the European orogens is that the low pressure belt is rather narrow, often not broader than 100 km, and does not cover such vast regions as the Hercynotype orogenies. It would be interesting to know the difference in structure between both belts and the cause of the deep burial in the high pressure belt. It may partly be due to deep geosynclinal subsidence, followed in turn by tectonic thickening.

A similar paired metamorphic belt occurs on Hokkaido, north of Honsyu, but here the low pressure belt lies on the oceanic side instead of on the continental side.

According to Miyashiro the outer belt with lawsonite-glaucophane facies metamorphism is part of a geosyncline, the sediments of which may have been deposited directly on the simatic oceanfloor, whereas the low pressure belt with its granites undoubtedly has a sialic substratum.

From Celebes another paired metamorphic belt is known with lawsonite-glaucophane schists and ultrabasites on the eastern oceanic side and andalusite-cordierite bearing rocks with many granites on the western side (Egeler, 1946; De Roever, 1950). Both belts are separated by the median tectonic line.

In a recent publication Landis & Coombs (1967) described a paired metamorphic belt from New Zealand. On the western, continental side metamorphism took place during the Cretaceous in the Buller geosyncline, producing the Tasman belt with andalusite, staurolite, cordierite and sillimanite, clearly low pressure metamorphic minerals. East of the median tectonic line low grade but high pressure metamorphism formed rocks of the zeolite, prehnite-pumpellyite and the lawsonite-glaucophane facies in the Wakatipu belt. The well known Dun Mountain ultrabasic belt of Permian age occurs in the latter zone, where granites are absent. Metamorphism in this belt also took place in the Cretaceous and was probably simultaneous with metamorphism in the Tasman belt. According to Landis and Coombs the eastern geosyncline is situated on the continental margin or possibly on the oceanic crust, whereas the Tasman belt rests on a sialic crust. Geothermal gradients in the low pressure belt must have amounted to 30-60°C/km; in the high pressure belt as low as 14°C/km.

Other occurrences of the lawsonite-glaucophane facies in the circumpacific region are in the well known Franciscan formation in California and in the northern Cascades. In the latter area two metamorphic belts have been described by M i s c h (1966). The narrow Shuksan belt consists mainly of glaucophane-schists and several bodies of ultrabasic rocks. The much wider Skagit belt is characterized by Barrovian, intermediate pressure metamorphism up to fairly high grades. Many large granite bodies are present in this belt. According to Misch the pressure-temperature ratio in the Shuksan belt was larger than in the Skagit belt. Both belts are separated by a large fault. These characteristics strongly remind one of the paired metamorphic belts elsewhere in the circumpacific region, although the belt on the continental side is in the intermediate rather than the low pressure facies series. Further inland Hietanen (1956) has described andalusite-cordierite bearing rocks near the Idaho batholith. In the same region kyanite is also encountered and in fact this could be the region where the high or intermediate pressure metamorphism on the oceanic side is grading into the low pressure metamorphism on the continental side. The large granites like the Sierra Nevada batholith, the Idaho batholith and the Canadian Coast Range batholith suggest that they belong to a low or intermediate pressure, and high temperature metamorphic belt.

VII. GENERAL DISCUSSION AND CONCLUSIONS

A number of points have to be discussed after the review of the properties of orogenic belts, namely: (1) the alternation of Hercynotype and Alpinotype orogenies, (2) the significance of the Circumpacific orogeny, (3) the pressure-temperature conditions of regional metamorphism, and (4) the causes of regional metamorphism.

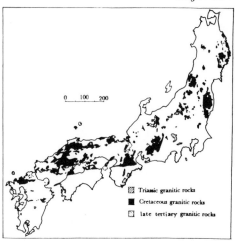

Fig. 10 — Distribution of granites on main island of Japan.

1. Alternation of Hercynotype and Alpinotype Orogenies

The succession in Europe of the Caledonian, Hercynian and Alpine orogenies with intermediate, low and high pressure metamorphism respectively, suggest that some cyclicity might exist in the successive development of orogenic belts. In this respect the hypothesis of D e R o e v e r (1956, 1964) concerning the distribution of glaucophane and lawsonite, almost exclusively restricted to post-Paleozoic orogens, is of great interest. According to D e R o e v e r this feature is due to

Fig. 11 — Distribution of ultrabasic rocks in Japan, M M = median tectonic line.

63

a secular decrease in the geothermal gradients during the earth's history and indicates a gradual cooling of the earth. If this were the case no such cyclicity of different types of orogens would exist. D e R o e v e r's statements have been discussed by M i y a s h i r o (1961) who, in the light of his discovery of paired belts, believes that De R o e v e r's hypothesis is unjustified. The discovery of glaucophane, and particularly lawsonite in late Precambrian or Paleozoic metamorphic rocks in the Ural mountains (D o b r e t-s o v *et al.*, 1966), definitely proves that D e R o e v e r's hypothesis, however interesting it may be, can no longer be upheld. This means that Alpinotype orogenies have occurred before the Mesozoic, and that in the course of the earth's history Alpinotype and Hercynotype orogenies have occurred more than once. Whether there is any regular alternation of the two types is a question which has to await the availability of more worldwide data about metamorphic belts.

One of the oldest orogenies of Hercynotype

character in Europe is the Svecofennian-Karelian orogeny. It was preceded by the Belomorian orogeny in which kyanite is rather common and consequently is of the high pressure or intermediate pressure type. It is clear therefore that the thermal properties of the earth have not changed considerably during the last 2200 m.y., a time span roughly equal to one half of the life of our planet.

Of course D e R o e v e r's observations that glaucophane and lawsonite are very rare in pre-Mesozoic orogens is quite correct. In the present author's opinion this is due to at least two circumstances. We have already seen that glaucophane-lawsonite rocks are unlikely to occur in the Hercynian and Caledonian chains of Europe, and possibly of the rest of the world also. This means that only early Paleozoic or Precambrian orogenic belts of the Alpinotype can be expected to contain these minerals. The rapid and large uplift, however, which is characteristic for this type of belt, results in the removal of all the lower grade

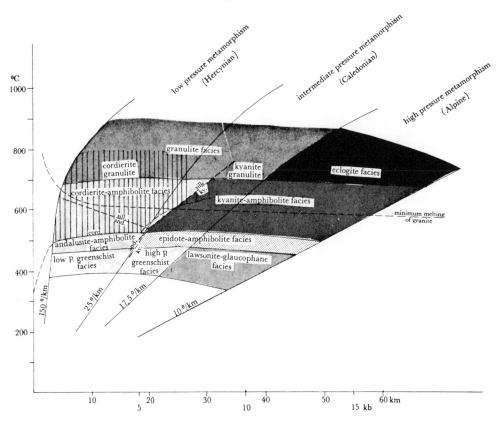

Fig. 12 — Temperature-pressure diagram showing metamorphic facies and the three facies series. Curves in the lines limiting the facies series are drawn to express that towards the depth the gradients in high temperature metamorphism probably are less steep. Gradients indicated in °C/km apply only to lower, straight part of these curves. Widely spaced vertical ruling is stability field of Mg-cordierite.

rocks by erosion. Only the high grade roots of such mountain chains – probably in kyanite-granulite or eclogite facies – are preserved. Another reason for the scarcity of glaucophane and lawsonite may be that these minerals have been replaced by minerals of the greenschist facies. As remarked before, this kind of plurifacial metamorphism is common in high pressure metamorphics. The chance of survival of lawsonite-glaucophane rocks is in pre-Mesozoic orogenic belts therefore rather small.

In the Hercynotype mountain chains, on the contrary, the preservation of low grade rocks in greenschist facies is quite common, because the amount of uplift is much smaller. Therefore these lowgrade schists are quite common, not only in the Hercynian belt, but also in the much older Svecofennian-Karelian orogens.

2. *The Circumpacific Orogenic Belt*

The only apparently new feature in the development of the earth seems to be the Circumpacific orogens with their paired metamorphic belts, low and high pressure metamorphism occurring side by side. From our description of the European mountain chains it is evident that such paired metamorphic belts do not occur there. This appears to be the normal situation for most orogenic belts. Moreover there is an important difference between the Alpinotype orogens and the high pressure belts of the Circumpacific orogens. According to M i y a s h i r o (1961) and L a n d i s & C o o m b s (1967) it can be assumed that the geosynclines of these zones are situated directly on oceanic crust. The alpinotype mountain chains on the other hand have their geosyncline situated on a sialic crust, as clearly indicated by the granitic or gneissose basements involved in the nappe structures of these mountain chains. Furthermore the Alpine metamorphics reach a grade as high as the amphibolite facies, whereas in the Circumpacific region the rocks only rarely reach the epidote-amphibolite facies (M i y a s h i r o, (1961). It is likely therefore that the Alpine belt and high pressure part of the circumpacific region are markedly different. It is possible that in the latter region deep burial is the result of subsidence, and not due to the stacking of nappes.

As long as no "fossil" paired metamorphic belts are discovered in the intercontinental orogens – and it is doubtful that this will ever happen – the Circumpacific belts are unique and probably represent a new development in the history of the earth. Any hypothesis about the origin and development of the Pacific ocean must take this into consideration.

3. *Pressure-temperature conditions of regional metamorphism*

A point which merits some further consideration is the pressure-temperature fields of the different regional metamorphic facies and facies series as based on natural occurrences. We have already concluded that the andalusite-cordierite metamorphism without kyanite definitely occurs at depths up to about 20 km, implying temperature° gradients varying from 150-25 °C/km and pressures ranging from 1-5.5 kb. The intermediate facies series also has a certain depth range, which is estimated at approximately 20-30 km, a pressure range of 5.5-8 kb and temperature gradients of 25-17°C/km. Finally the high pressure metamorphism should occur at depths between 30 and 60 km (involving crustal material), with pressures from 8-16 kb and temperature gradients of 17-10°C/km.

The above data have been represented in fig. 12. The supposed phase boundaries in the Al_2-SiO_5 system and the stability field of Mg-cordierite according to S c h r e y e r & Y o d e r (1964) are also shown. This diagram appears reasonable as far as natural mineral assemblages are concerned and is in good agreement with recent experimental work.

It should be emphasized once more that the occurrence of low pressure – high temperature metamorphics in large tracts of continents, like in the Hercynides and the Svecofennides must involve fairly large thicknesses of metamorphic rocks. The conclusion that intermediate and high pressure metamorphism occurs at greater depth is inescapable. Interpretation of major structures in such regions has to take this into serious consideration. The hypothesis that kyanite or glaucophane rocks occurring in belts of regional extent may have formed under an overburden of only a few kilometres undoubtedly needs serious reconsideration.

In fig. 12. it is assumed that granulite facies rocks are formed at moderate pressures, because they often contain cordierite and lack kyanite. This is the case in the Finnish granulites, in Saxony and in the southwestern Norwegian granulite terrain. In some cases these granulites are transitional to amphibolites, like in Norway and India, where the amphibolites belong to a low or intermediate pressure facies series. For these reasons it is probably that in the Hercynotype orogens the amphibolite facies rocks are underlain by cordierite-granulites.

In Paleozoic or younger Hercynotype chains slow uplift has exposed only the amphibolite facies, and granulite facies rocks are mainly restricted to Precambrian orogens. Even in the rather old Svecofennian orogen erosion has exposed underlying granulites only in a few rare cases.

4. Causes of regional metamorphism

There is no doubt that two mechanisms are required for regional metamorphism, one is deformation – almost all regionally metamorphosed rocks are deformed – the other is introduction of heat.

As far as the heat is concerned we have to distinguish between Hercynotype and Alpinotype mountain chains. In the first case the geothermal gradient is many times larger than the present day gradients on continents and in ocean basins. In the Alpinotype metamorphism however the gradients were considerably lower than the present day gradients. In the intermediate type mountain chains they must have been about equal to present day gradients.

It will be evident that a very large amount of heat has to be introduced in the Hercynotype orogen, especially because of the large regions involved. Whether this heat emanates from rising magmatic bodies or is due to some other mechanism is difficult to tell. The fundamental causes of orogenic metamorphism lie deep in the crust or in the mantle, and as yet remain unknown. As deformation and metamorphism usually occur simultaneously, the two processes must be related. The nature of their relationship is not clear, nor is there any evidence that one process has initiated the other.

In the Alpinotype belts no extra heat is required other than that resulting from the normal heat flow from the interior of the earth. As soon as the rocks are buried deeply enough, either by the stacking of nappes or by deep subsidence, metamorphism starts, deformation acting as a catalyst. Therefore, burial due to some tectonic action – nappe formation or subsidence – is the main cause of metamorphism in the Alpinotype and high pressure circumpacific belts. This kind of metamorphism can then be called "tectonic burial metamorphism", whereas in the Hercynotype orogens one can speak of "dynamo-thermal metamorphism". These terms express the profound differences between the two extreme types of regional metamorphism.

Acknowledgment. I am much indebted to Dr. J. L. Weiner for his aid in editing and correcting the final draft of the manuscript.

REFERENCES

Althaus, E. (1966) — Experimentelle Bestimmung des Stabilitätsbereichs von Disthen. Die Naturwissenschaften **2**, 42-43.

Aubouin, J. (1965) — Geosynclines. Elsevier, Amsterdam, 335 p.

Barrois, Ch. & A. Offret (1889) — Mémoire sur la Constitution géologique du sud de l'Andalousie. Mém. Acad. Sc. 79-169.

Bearth, P. (1952) — Geologie und Petrographie des Monte Rosa. Beiträge geol. Karte Schweiz 96.

—— (1959) — Über Eklogite, Glaucophanschiefer und metamorphe Pillowlaven. Schweiz. Min. Petr. Mitt. **39**, 267-286.

—— (1965) — Zur Entstehung alpinotyper Eklogite. Schweiz. Min. Petr. Mitt. **45**, 1, 179-188.

—— (1966) — Zur Mineralfaziellen Stellung der Glaucophangesteine der Westalpen. Schweiz. Min. Petr. Mitt. **46**, 1, 13-23.

Bederke, E. (1935) — Die Regionalmetamorphose im Altvatergebirge. Geologische Rundschau **26**, 108-124.

—— (1947) — Über der Wärmehaushalt der Regionalmetamorphose. Geologische Rundschau **35**, 26-32.

Bell, P. M. (1963) — Aluminium silicate system: Experimental determinations of the triple point. Science **139**, 1055-1056.

Bellière, J. (1958) — Contribution à l'étude pétrographique des schistes cristallins du massif des Aiguilles Rouges (Haute-Savoie). Thesis, Liège, 1-190.

Bilibin, Yu. A. (1959) — Geochemical types of orogenic zones (in Russian), Izbr. trudy, II, Izd. AN USSR.

Bloxam, T. W. & J. B. Allen (1960) — Glaucophaneschist, eclogite, and associated rocks from Knocknormal in the Girvan-Ballantrae Complex, south Ayrshire. Trans. Royal Soc. Edinb. **64**, 1, 1-28.

Bossdorf, R. H. H. (1961) — Das Kristallin von Gadernheim und Landenau im Odenwald. N. Jb. Miner. Abh. **95**, 370-419.

Brouwer, H. A. & C. G. Egeler (1951) — Sur le métamorphisme à glaucophane dans la nappe des schistes lustrés de la Corse. Kon. Ned. Akad. Wet. Proc. B, **54**, 130-139.

Bucca, L. (1884) — Sopra alcune roccie della serie cristallina di Calabria. Boll. Com. Geol. Italia V, 240-249.

Chevenoy, M. (1958) — Contribution à l'étude des schistes cristallins de la partie NW du Massif Central francais. Mém. Serv. Géol. France, 1-428.

Chatterjee, N. D. (1961) — The alpine metamorphism in the Simplon Area, Switzerland and Italy. Geol. Rundschau **51**, 1-72.

Chinner, G. A. (1966) — The distribution of pressure and temperature during Dalradian metamorphism. Quart. Journ. Geol. Soc. London **122**, 159-168.

Cogné, J. (1957) — Schistes cristallins et granites en Bretagne méridionale: Le domaine de l'anticlinal de Cornouaille. Mém. Expl. Carte géol. France, 1-382.

Coleman, R. G. & D. E. Lee (1962) — Metamorphic aragonite in the glaucophane schists of Cazadero, California. Am. Journ. Science **260**, 577-595.

Corin, F. (1963) — Über Knoten in Phylliten des Salmien von Recht, Vielsalm und Ottré (Ost-Ardennen). Geologische Mitteilungen Aachen **3**, 179-184.

Davis, G. L. & W. Schreyer (1963) — Altersbestimmungen an Gesteinen des Ostbayerischen Grundgebirges und ihre geologische Deutung. Geol. Rundschau **52**, 146-169.

Demay, A. (1934) — Contribution à la synthèse de la chaîne hercynienne d'Europe. Revue de Géographie physique et de Géologie dynamique **7**, 199-221.

------ (1948) — Tectonique antéstéphanienne du Massif Central. Mém. Carte géol. France, 1-246.

Dobretsov, N. L., V. V. Reverdatto, V. S. Sobolev, N. V. Sobolev, Ye. N. Ustakova & V. V. Khlestov (1965) — Distribution of regional metamorphic facies in USSR. International Geology Review 8, 11, 1335-1346.

Dudek, A. & M. Suk (1965a) — Zur geologische Entwicklung des Moldanubikums. Bericht Geol. Gesellschaft **10**, 147-161.

------ (1965b) — The depth relief of the granitoid plutons of the Moldanubicum. Neues Jb. Geol. u. Paläont. **123**, 1-19.

Egeler, C. G. (1946) — Contributions to the petrology of the metamorphic rocks of western Celebes. Thesis Amsterdam, 1-165.

------ (1956) — The Alpine metamorphism in Corsica. Geol. en Mijnb. **18**, 115-118.

------ (1963) — On the tectonics of the eastern Betic Cordillera. Geol. Rundschau **53**, 260-269.

Ellenberger, F. (1958) — Etude géologique du Pays de Vanoise. Mém. Expl. Carte géol. France, 1-561.

------ (1960) — Sur une paragénèse éphémère à lawsonite et glaucophane dans le métamorphisme alpin en Haute-Maurienne (Savoie). Bull. Soc. Géol. France **7**, 2, 190-194.

Eskola, P., T. F. W. Barth & C. W. Correns (1939) — Die metamorphe Gesteine, in: Die Entstehung der Gesteine. Berlin, Springer Verlag.

Eller, J. P. von (1961) — Les gneiss de Sainte-Marie-aux-Mines et les séries voisines des Vosges moyennes. Mém. Serv. Carte géol. Als. Lorr. 19.

Fabriès, J. (1963) — Les formations cristallines et métamorphiques du Nord Est de la Province de Séville (Espagne). Thèse Nancy, 1-267.

Faul, H. (1962) — Age and extent of the Hercynian complex. Geol. Rundschau **52**, 767-781.

Floor, P. (1966) — Petrology of an aegirine-riebeckite gneiss bearing part of the Hesperian massif: the Galiñeiro and surrounding areas, Vigo, Spain. Leidse Geol. Mededel. **36**, 1-203.

Fonteilles, M., G. Guitard & E. Raguin (1964) — Sur la présence de gneiss à disthène et cordiérite dans le massif du Saint-Barthélemy. C.R. Ac. Sc. **258**, 3524-3525.

Forestier, F. M. (1963) — Métamorphisme Hercynien et Antéhercynien dans le bassin du Haut-Allier (Massif Central francais). Bull. Serv. Carte Géol. France **271**, 59, 521-813.

Frasl, G. (1965) — Exkursionsführer Azopro-Exkursion Sept. 1965.

Fritsch, W. (1962) — Von der "Anchi" zur Katazone im Kristallinen Grundgebirge Ostkärntens. Geol. Rundschau **52**, 202-210.

Gèze, B. (1949) — Etude géologique de la Montagne Noire et des Cevennes méridionales. Mém. Soc. Géol. France **19**, 62, 1-215.

Gjelsvik, T. (1952) — Metamorphosed dolerites in the gneiss area of Sunnmøre on the west coast of southern Norway. Norsk Geol. Tidsskrift **30**, 33-134.

Guitard, G. (1964) — Un example de structure en nappe de style pennique dans la chaîne hercynienne: Les gneiss stratoïdes du Canigou (Pyrénées Orientales). C.R. Acad. Sc. Paris **258**, 4597-4599.

Hammer, W. (1925) — Cordieritführende metamorphe Granite aus den Ötztaler Alpen. Tschermaks Min. Petr. Mitt. **38**, 67-87.

Heim, R. C. (1952) — Metamorphism in the Sierra de Guadarrama. Thesis, Utrecht.

Henke, W. (1927) — Beitrag zur Geologie der Sierra

Morena nördlich von la Carolina (Jaén). Abh. Senckenberg. Naturforsch. Gesellschaft **39**, 2, 185-197.

Hietanen, A. (1956) — Kyanite, andalusite and sillimanite in schists in Boehls Butte Quadrangle, Idaho. Amer. Mineral. **41**, 1-27.

Hoenes, D. (1949) — Petrogenese im Grundgebirge des Südschwarzwaldes. Heidelberger Beiträge zur Mineralogie und Petrographie **1**, 121-202.

Hoeppener, R., St. Dürr & M. Mollet (1964) — Ein Querschnitt durch die Betischen Cordilleren bei Ronda (S.W. Spanien). Geol. en Mijnb. **43**, 282-298.

Holtedahl, O. (1960) — Geology of Norway. Norges Geologiske Undersøkelse **208**, 1-540.

Johnson, M. R. W. (1963) — Some time relations of movement and metamorphism in the Scottish Highlands. Geol. en Mijnb. **42**, 121-142.

Jung, J. (1946) — Géologie de l'Auvergne. Mém. Serv. Carte Géol. France.

Jung, J. & M. Roques (1952) — Introduction à l'étude zonéographique des formations cristallophylliennes. Bull. Serv. Carte Géol. France **235**, 1-61.

Khitarov, N. I., V. A. Pugin, Chzao Bin & A. B. Slutzky (1963) — Relations between andalusite, kyanite and sillimanite at moderate temperatures and pressures. Geokhimia **3**, 219-228.

Köhler, A. (1950) — Zur Entstehung der Granite der Südböhmischen Masse. Tschermaks Min. Petr. Mitt. **301**, 175-286.

Kossmatt, F. (1927) — Gliederung des varistischen Gebirgsbaues. Abh. Sächs. Geol. Landesanstalt **1**, 1-39.

Landis, C. A. & D. S. Coombs (1967) — Metamorphic belts and orogenesis in Southern New Zealand. 11th Pacific Science Congress, Tokyo, in press.

Langerfeldt, H. (1961) — Über Syenitbildung durch Palingenese und Kalifeldspat-Metablastesis im mittleren Schwarzwald. Jahrb. Geol. Landesamt Baden-Württemberg **5**, 19-51.

Lotze, F. (1945) — Zur Gliederung des Varisziden der Iberischen Meseta. Geotekt. Forsch. **6**, 79-92.

Magnusson, N. H., P. Thorslund, F. Brotzen, B. Asklund & O. Kulling (1960) — Description to accompany the map of the pre-Quaternary rocks of Sweden. Sveriges Geologiska Undersökning. Afh. Ser. Ba. 16.

Matthes, S. (1954) — Die Para-Gneise im mittleren Kristallinen Vor-Spessart und ihre Metamorphose. Abh. Hessischen Landesamt für Bodenforschung **8**, 1-86.

Mehnert, K. R. (1958) — Argon Bestimmungen an Kaliumminerallien VI. Geochimica et Cosmochimica Acta **14**, 105-113.

------ (1962) — Petrographie und Abfolge der Granitisation im Schwarzwald III. N. Jb. Miner. Abh. **98**, 2, 208-249.

Michel-Lévy, M. & A. Bergeron (1889) — Etude géologique de la Serrania de Ronda. Mém. Acad. Sc. 172-375.

Misar, Z. (1958) — Zur Altersfrage der kristallinen Serien und ihrer Metamorphose in Keprnik Gewölbe im Hohen Gesenke (Mähren). Neues Jb. Geol. u. Paläont. **106**, 277-292.

Misch, P. (1966) — Tectonic evolution of the northern cascades of Washington State. Sympos. on tect. Hist. and Min. Dep. of W. Cordillera in B.C. **8**, 101-148.

Miyashiro, A. (1961) — Evolution of metamorphic belts. Journal of Petr. **2**, 277-311.

------ (1967) — Aspects of metamorphism in the circum-Pacific region. Tectonophysics, in press.

Murris, R. J. (1957) — Geology and petrology of the Gieravarto-Jofjället region, Västerbotten, Sweden. Thesis Amsterdam, 1-117.

Newton, R. C. (1966) — Kyanite-andalusite equilibrium from 700° to 800° C. Science **153**, 170-172.

Nicolas, A. (1966) — Le complexe Ophiolites-Schistes lustrés entre Dora Maira et Grand Paradis (Alpes piémontaises). Thesis, Nantes, II, 183-299.

Niggli, E. (1960) — Mineral-Zonen der alpinen Metamorphose in den Schweizer Alpen. Int. Geol. Congr. 21st Session, Norden, **13**, 132-138.

Niggli, E. & C. R. Niggli (1965) — Karten der Verbreitung einiger Mineralien der alpidischen Metamorphose in den Schweizer Alpen (Stilpnomelan, Alkali-Amphibol, Chloritoid, Staurolith, Disthen, Sillimanit). Eclogae Geol. Helv. **58**, 335-368.

Nijhuis, H. J. (1964) — Plurifacial alpine metamorphism in the south-eastern Sierra de los Filabres south of Lubrin, SE Spain. Thesis Amsterdam, 1-151.

Novarese, V. (1931) — La Formazione Diorito-kinzigitica in Italia. Boll. uff. Geol. Italia **56**, 1-62.

Palm, Q. A. (1957) — Les roches cristallines des Cévennes médianes à hauteur de Largentière. Thèse Utrecht, 1-121.

Parga-Pondal, I. (1956) — Nota explicativa del mapa geológico de la parte N.O. de la provincia de la Coruna. Leidse Geol. Mededel. **21**, 468-484.

Parras, K. (1958) — On the charnockites in the light of a highly metamorphic rock complex in south western Finland. Academical dissertation, Helsinki, 1-137.

Plas, L. van der (1959) — Petrology of the northern Adula region, Switzerland. Leidse Geol. Mededel. **24**, 418-602.

Priem, K. N. A. (1962) — Geological, petrological and mineralogical investigations in the Serra do Marao region, Northern Portugal. Thesis Amsterdam, 1-160.

Richter, M. (1935) — War das Variscische Gebirge ein Hochgebirge? Geol. Rundschau **26**, 149-150.

Roever, W. P. de (1950) — Preliminary notes on glaucophane-bearing and other crystalline schists from S.E. Celebes, and on the origin of glaucophane-bearing rocks. Kon. Ned. Akad. Wet. Proc. B, **53**, 1455-1465.

—— (1953) — Tectonic conclusions from the distribution of the metamorphic facies in the island of Kabaena, near Celebes. 7th Pacific Sc. Congress (New Zealand) **2**, 71-81.

—— (1956) — Some differences between post-Paleozoic and older regional metamorphism. Geol. en Mijnb. **18**, 123-127.

Roever, W. P. & H. J. Nijhuis (1964) — Plurifacial alpine metamorphism in the eastern Betic Cordilleras (SE Spain) with special reference to the genesis of the glaucophane. Geol. Rundschau **35**, 324-335.

Roques, M. (1941) — Les schistes cristallins de la partie sud-ouest du Massif Central. Mém. Serv. Carte Géol. France, 1-530.

Rutland, R. W. R. (1965) — Tectonic overpressures, in: Controls of Metamorphism, 118-139, edited by W. S. Pitcher and G. W. Flinn; Oliver & Boyd, London.

Scharbert, H. G. (1962) — Die Granulite der südlichen Böhmische Masse. Geol. Rundschau **52**, 112-123.

—— (1965) — The Bohemian massif in Austria. Report for the Azopro excursion, Austria.

Schermerhorn, L. J. G. (1959) — Igneous, metamorphic and ore geology of the Castro Daire – Sao Pedro do Sul – Satao region (N. Portugal). Thesis Amsterdam, 1-617.

Scheumann, K. M. & H. J. Behr (1963) — Konvergenzerscheinungen am Rande des sächsischen Granulits. Abh. Sächs. Akad. Wiss. Math. Naturw. kl. **47**, 4, 1-18.

Schreyer, W. (1957) — Über das Alter der Metamorphose im Moldanubikum des südlichen Bayerischen Waldes. Geol. Rundschau **46**, 306-317.

—— (1966) — Metamorpher Übergang Saxothuringikum-Moldanubikum östlich Tirschenreuth/Opf., nachgewiesen durch Phasenpetrologische Analyse. Geol. Rundschau **55**, 491-508.

Schreyer, W. & H. S. Yoder (1964) — The system Mg-cordierite —H_2O and related rocks. Neues Jahrb. Miner. **101**, 271-342.

Schoeller, H. & L. Lutaud (1931) — Feuille d'Hyères. Carte Géol. de France, notice explicative, 1-24.

Schuiling, R. D. (1960) — Le dome gneissique de l'Agout. Mém. Soc. Géol. France **91**, 1-58.

—— (1962) — Die petrogenetische Bedeutung der drei Modifikationen von Al_2SiO_5. Neues Jahrb. Miner. **9**, 200-214.

Schüller, A. (1949) — Petrogenetische Studien zum Granulitproblem an Gesteinen der Münchberger Masse. Heidelberger Beiträge zur Mineralogie und Petrographie **1**, 279-340.

Schumann, H. (1930) — Über Moldanubische Paraschiefer aus dem niederösterreichischen Waldviertel zwischen Gföhler Gneiss und Bittescher Gneiss. Tschermaks Min. Petr. Mitt. **40**, 73-187.

Simonen, A. (1960) — Pre- quarternary rocks in Finland. Bull. geol. Finlande **191**, 1-49.

Sitter, L. U. de (1959) — The Rio Esla Nappe in the zone of Leon of the Asturian Cantabric mountain chain. Notas y Comunicaciones del Instituto Geológico y Minero de España **56**, 3-24.

—— (1961) — Le précambrien dans la chaîne cantabrique. C.R. Somm. Soc. Géol. France **9**, 253-254.

Smejkal, V. (1965) — Anomalous potassium-argon absolute ages of the migmatite cordierite gneisses from the SW part of the Czech massif. Krystalinikum **3**, 157-162.

Sobolev, V. S. (1960) — Role of high pressure in metamorphism. Intern. Geol. Congress. 21st session, Norden **14**, 72-82.

Stille, H. (1924) — Grundfragen der vergleichende Tektonik. Borntraeger, Berlin, 443 p.

—— (1951) — Das Mitteleuropäische variszische Grundgebirge im Bilde des Gesamteuropäischen. Geol. Jahrb. Beitr. **2**, 1-138.

Strand, T. (1961) — The Scandinavian Caledonides, a review. American Journal of Science **259**, 161-172.

Suk, M. (1964) — Material characteristics of the metamorphism and migmatization of Moldanubian paragneisses in Central Bohemia. Krystalinikum **2**, 71-105.

Suzuki, J. (1930) — Über die Staurolit-Andalusit-Paragenesis im Glimmergneis von Piodina bei Brissago (Tessin). Schweiz. Min. Petr. Mitt. **10**, 117-132.

Teichmüller, R. & H. W. Quitzow (1935) — Deckenbau im Appenninbogen. Abh. Gesellsch. Wiss. Göttingen **13**, 1-186.

Tex, E. den (1965) — Metamorphic lineages of orogenic plutonism. Geol. Mijnb. **44**, 105-132.

Tex, E. den & D. E. Vogel (1962) — A "Granulitgebirge" at Cabo Ortegal (NW Spain). Geol. Rundschau **52**, 95-112.

Torre de Assunçao, C. F. (1962) — Rochas graniticas do Minho e Douro. Servicos Geológicos de Portugal Mem. **10**. 1-70.

Watznauer, A. (1955) — Saxothuringicum-Lugicum, ein regional-tektonischer Vergleich. Freiberger Forschungshefte **C17**, 30-53.

Wenk, E. (1956) — Alpines und ostgrönländisch-kaledonisches Kristallin, ein tektonisch-petrogenetischer Vergleich. Verh. Naturf. Ges. Basel **67**, 2, 72-102.

—— (1962) — Plagioklas als Indexmineral in den Zentralalpen. Die Paragenese Calcit-Plagioklas. Schweiz. Min. Petr. Mitt. **42**, 1, 139-152.

Winkler, H. G. F. (1965) — Petrogenesis of metamorphic rocks. Springer Verlag, Berlin, 220 p.

Wimmenauer, W. (1950) — Cordieritführende Gesteine im Grundgebirge des Schauinsland gebietes. Neues Jb. Min. Geol. Paläont. **80**, 375-436.

—— (1962) — La métallogénie de la Forêt Noire.

Chronique des Mines et de la Recherche Minière **313**, 303-308.

Würm, A. (1960) — Über einige grundlegende Fragestellungen in der Münchberger Gneismasse. Geol. Rundschau **49**, 343-349.

Zoubek, V. (1948) — Remarks on the geology of the Crystallinicum of the Bohemian mass. Sbornik Stat. geol. Ust. Cesk. Republ. **15**, 381-398.

—— (1965) — Moldanubikum and seine Stellung im geologischen Bau Europas. Freiberger Forschungshefte **C190**, 129-140.

Zwart, H. J. (1954) — La géologie du Massif du Saint-Barthélemy (Pyrénées, France). Leidse Geol. Mededel. **18**, 1-229.

—— (1962) — On the determination of polymetamorphic mineral associations and its application to the Bosost area (Central Pyrenees). Geol. Rundschau **52**, 38-65.

—— (1963) — The structural evolution of the Paleozoic of the Pyrenees. Geol. Rundschau **53**, 170-205.

Erratum

On p. 67, left-hand column, the fourth line from the top should read: "V. Sobolev, E. N. Ushakova, and V. V. Khlestov."

II

Thermal Regimes near Lithospheric Plate Margins

Editor's Comments on Papers 3 and 4

3   **Oxburgh and Turcotte:** *Mid-ocean Ridges and Geotherm Distribution During Mantle Convection*

4   **Oxburgh and Turcotte:** *Thermal Structure of Island Arcs*

Metamorphic recrystallization takes place due to pronounced changes in the temperature, pressure, and chemical environment of the protoliths. Of these factors, thermal state is judged to be the most important in causing mineralogic reactions to occur. The earth's geothermal gradient—that is, the increase in temperature with depth of burial—is a measure of the thermal regime. Geothermal gradients have been calculated by numerous authors. Perhaps the best known and currently most widely accepted values are those computed for both Precambrian shield and oceanic areas by Clark and Ringwood (1964). These geotherms were calculated prior to the enunciation of the concept of plate tectonics, but remain valid today because they were based on the observed surface heat flow and the distribution of rock types of known thermal conductivities and heat-producing radioactive constituents as a function of depth. Of course, the geothermal gradients of Clark and Ringwood are only appropriate for plate interiors. Near plate margins the isotherms are markedly perturbed, owing to the dynamics of lithospheric plate motion, which allows the close approach to the surface of relatively hot asthenosphere at an accreting margin, and the deep underflow of relatively cold lithosphere at a consumptive plate junction.

The thermal structure (i.e., the disposition of the isotherms) at nonconservative plate boundaries has been calculated by many workers, including: McKenzie, 1967, 1969; Oxburgh and Turcotte, Paper 3, Paper 4, 1971; Turcotte and Oxburgh, 1972; Hasebe, Fujii, and Uyeda, 1970; Minear and Toksöz, 1970; Toksöz, Minear, and Julian, 1971; Toksöz, Sleep, and Smith, 1973; and Griggs, 1972; see also Matsuda and Uyeda, 1970.

The 1968 paper by Oxburgh and Turcotte is presented herein (Paper 3) as a plausible thermal model for a rising mantle plume or accreting plate boundary; steady-state convection cells were postulated. Assuming reasonable mantle properties, appropriate velocity and temperature distributions have been obtained. In cross section normal to the ridge, a mushroom-shaped mantle area nearly 400 km across appears to undergo partial fusion; the top of this zone approaches to within about 10 km of the oceanic crust–seawater interface. This thermal structure appears to account approximately for the breadth of the high-heat-flow region in the vicinity of the spreading center and for the increasing thickness of the lithosphere proceeding away from the accreting plate junction.

A computed model for the thermal regime near consumptive plate margins is represented in Paper 4 by Oxburgh and Turcotte; similar relationships have been presented by Hasebe, Fujii, and Uyeda (1970). Although their emphases were somewhat different, both calculated thermal structures have demonstrated the remarkable downwarping of the isotherms—consequently low heat flow (e.g., $<1.0\ \mu$cal(cm²/sec)— in the descending lithospheric slab. In contrast, the stable, nonsubducted plate exhibits an upward bowing of the isotherms—hence high heat flow (e.g., $>2.5\ \mu$cal(cm²/sec)—

as a result of frictional dissipation along the plate juction (the Benioff zone), and by bodily transfer of melts from both the asthenosphere and the downgoing partially melted, transformed (and in part hydrated?) oceanic crust. The growth and evolution in time and space of subduction zone mélanges, volcanic arcs and marginal seas, and the related metamorphism were considered at some length in this paper.

References

Clark, S. P., and Ringwood, A. E. (1964) Density distribution and constitution of the mantle: *Rev. Geophys.*, **2**, 35–88.

Griggs, D. T. (1972) The sinking lithosphere and the focal mechanism of deep earthquakes: p. 361–384, in *The Nature of the Solid Earth*, E. C. Robertson, J. F. Hays, and L. Knopoff, eds., McGraw-Hill, New York, 677 p.

Hasebe, K., Fujii, N., and Uyeda, S. (1970) Thermal processes under island arcs: *Tectonophysics*, **10**, 335–355.

McKenzie, D. P. (1967) Some remarks on heat flow and gravity anomalies: *Jour. Geophys. Res.*, **72**, 6261–6274.

McKenzie, D. P. (1969) Speculations on the consequences and causes of plate motions: *Geophys. Jour. Royal Astron. Soc.*, **18**, 1–32.

Matsuda, T., and Uyeda, S. (1970) On the Pacific-type orogeny and its model-extension of the paired belts concept and possible origin of marginal seas: *Tectonophysics*, **11**, 5–27.

Minear, J. W., and Toksöz, M. N. (1970) Thermal regime of a downgoing slab and global tectonics: *Jour. Geophys. Res.*, **75**, 1397–1419.

Oxburgh, E. R., and Turcotte, D. L. (1971) Origin of paired metamorphic belts and crustal dilation in island arc regions: *Jour. Geophys. Res.*, **76**, 1315–1327.

Toksöz, M. N., Minear, J. W., and Julian, B. R. (1971) Temperature field and geophysical effects of a downgoing slab: *Jour. Geophys. Res.*, **76**, 1113–1138.

Toksöz, M. N., Sleep, N. H., and Smith, A. T. (1973) Evolution of the downgoing lithosphere and the mechanisms of deep focus earthquakes: *Geophys. Jour., Royal Astron. Soc.*, **35**, 285–310.

Turcotte, D. L., and Oxburgh, E. R. (1972) Mantle convection and the new global tectonics: *Ann. Rev. Fluid Mech.*, **4**, 33–68.

3

Copyright © 1968 by the American Geophysical Union

Reprinted from *Jour. Geophys. Res.*, **73**(8), 2643–2661 (1968)

Mid-Ocean Ridges and Geotherm Distribution during Mantle Convection

E. R. Oxburgh[1] and D. L. Turcotte

Graduate School of Aerospace Engineering, Cornell University, Ithaca, New York 14850

A boundary-layer solution for steady cellular convection is applied to thermal convection within the earth's mantle. For the large Prandtl and Rayleigh numbers applicable for the mantle, each cell contains a highly viscous core, on the horizontal boundaries of the cell are thin thermal boundary layers, and on the vertical boundaries between cells thin thermal plumes drive the viscous flow. Temperature and velocity distributions applicable to the mantle are obtained. With accepted values of the Rayleigh number the theory predicts a reasonable velocity for continental drift. Predicted values for the surface heat flux are in good agreement with measurements made on the ocean floors. The thickness of the thermal boundary layer adjacent to the earth's surface varies from about 8 km in the vicinity of the vertical rise to about 60 km near the descending plume. The vertical rise has a width of about 300 km with a similar width for the descending plume. In deriving the conditions for incipient upper mantle melting from experimental data on the melting of basalt, it is shown that in the upper part of an ascending plume the mantle is likely to undergo partial melting over an extensive mushroom-shaped area having a maximum width of about 400 km and extending to within 10 km of the surface. It appears that in the ascending plume mantle material may begin to melt over a range of temperature and pressure conditions that permit the development of a range of basaltic magma types. Thermal arguments are presented against serpentinite and in favor of basalt as the main constituent of layer 3 of the oceanic crust. The shape and characteristics of the zone of fusion suggest an explanation for the anomalous gravity and seismic observations associated with mid-oceanic ridges.

INTRODUCTION

The geological evidence in favor of some kind of cellular movement in the mantle is strong but circumstantial [e.g., *Runcorn*, 1962; *Blackett*, 1965], but it has been considerably strengthened by the concept of ocean-floor spreading [*Hess*, 1962, 1965] and the associated interpretations of the linear magnetic anomaly patterns that characterize many parts of the ocean floors [*Vine and Matthews*, 1963; *Vine*, 1966]. On the other hand, objections or severe physical constraints on convective motion within the crystalline rock of the mantle have been advanced by others [*MacDonald*, 1965a; *McKenzie*, 1966], and the cellular character of any such motion has been questioned [e.g., *Elsasser*, 1963]. It is clearly desirable to seek further ways of testing the convection hypothesis.

Progress in testing the convection hypothesis has been hindered by the absence of any complete and internally consistent theoretical model for mantle convection relating the directions of movement and velocities of mantle material to its thermal behavior. The obstacles to the development of a rigorous model are formidable [*Knopoff*, 1964; *Tozer*, 1965], and in any case there is considerable uncertainty about many of the constants describing the physical behavior of the mantle at depth. In particular, it is not clear whether the mantle is better regarded as a plastic or a viscous body. *Gordon* [1965] has given strong arguments favoring a viscous model. The plastic model has been examined by *Orowan* [1965] and *Elsasser* [1963].

It is well known that cellular convection occurs when a confined fluid is heated from below. Solutions of the linearized stability problem that determine the onset of the convective motion were first obtained by *Rayleigh* [1916]. Convective motion is predicted if the dimensionless parameter $R = \alpha \, (T_{w1} - T_{w2}) \, d^3 \, g/\kappa \nu$ exceeds a critical value where α is the coefficient of thermal expansion, T_{w1} is the temperature

[1] Now at Division of Geological Sciences, California Institute of Technology, Pasadena, California. On leave from the Department of Geology and Mineralogy, University of Oxford, Oxford, England.

of the hot lower surface, T_{w2} is the temperature of the cold upper surface, d is the thickness of the layer of fluid, g is the acceleration of gravity, κ is the thermal diffusivity, and ν is the kinematic viscosity. The value of the critical Rayleigh number for free-surface boundary conditions is 657. The wavelength of the convective motion is $\lambda = 2^{3/2} d$. The extensive literature on the stability problem is summarized by *Chandrasekhar* [1961]. The predicted onset of convection is in good agreement with laboratory measurements.

For Rayleigh numbers that are large in comparison to the critical Rayleigh number, the linear theory is not applicable and it is necessary to solve a set of nonlinear partial differential equations. Extensions of the linear theory into the nonlinear regime have been given by *Malkus and Veronis* [1958], *Kuo* [1961], and *Platzman* [1965]. Numerical computations of cellular convection for Prandtl numbers of the order of 1 have been given by *Fromm* [1965].

For finite amplitude convection cells the Prandtl number, $Pr = \nu/\kappa$, enters the analysis as well as the Rayleigh number. Clearly the Prandtl number for mantle convection is large. A theoretical model for laminar cellular convection applicable for large values of the Prandtl and Rayleigh numbers has been given by *Turcotte and Oxburgh* [1967]. In this model temperature gradients are restricted to thin thermal boundary layers adjacent to the horizontal cell boundaries and to thin thermal plumes on the vertical boundaries between cells. The core of each convection cell is assumed to be highly viscous and isothermal. It has been shown by *Turcotte* [1967] that the boundary-layer theory for finite amplitude convection is in excellent agreement with laboratory measurements of heat flux. It is the purpose of this paper to examine the implications of the boundary-layer model for possible mantle convection.

The model adopted assumes that the mantle convection cells have the form of two-dimensional elongated rolls with a horizontal dimension in the direction of movement of thousands of kilometers and a persistence along the axis of the rolls of many times this distance. Oceanic ridges, such as the East Pacific rise and the mid-Atlantic ridge, are regarded as the sites of ascending limbs, and the ocean floor is regarded as the upper surface of the cells, as in the ocean-floor spreading hypothesis of continental drift [*Hess*, 1962]. A possible distribution of ascending and descending limbs is illustrated by Figure 1 [after *Girdler*, 1965]. That convection cells can be expected to have this form has been verified in the laboratory. A photograph of finite amplitude laminar convection cells, as obtained by *Somerscales and Dropkin* [1966], is given in Figure 2. It is seen that the convection cells have the form of elongated two-dimensional rolls.

THEORETICAL MODEL

To determine the steady structure of Benard cells, the conservation equations for mass, momentum, and energy are required. It has been shown by *Jeffreys* [1928] that the influence of the earth's rotation on convection in the mantle is negligible. *Jeffreys* [1930] has also shown that the Boussinesq approximation may be used if the temperature in the Boussinesq equations is taken to be the difference between the actual temperature and the adiabatic temperature at that point, provided that this temperature difference is not too large. The temperature T is defined to be the difference between the actual temperature and the adiabatic temperature at that point. The viscosity, thermal diffusivity, and coefficient of thermal expansion are taken to be constant; in the body force term of the momentum equation a linear relation is assumed between the variations of temperature and density:

$$\rho - \rho_0 = -\rho_0 \alpha (T - T_0) \qquad (1)$$

where T_0 is the temperature at which the density ρ is equal to the reference density ρ_0. Introducing $\Theta = T - T_0$ and $P = p + \rho_0 py$ we can write the equations for conservation of mass, momentum, and energy for a steady, laminar flow as

$$\nabla \cdot \mathbf{u} = 0 \qquad (2)$$

$$(\mathbf{u} \cdot \nabla)\mathbf{u} = -\frac{1}{\rho_0} \nabla P + \nu \nabla^2 \mathbf{u} + \alpha \Theta g \mathbf{j} \qquad (3)$$

$$(\mathbf{u} \cdot \nabla)\Theta = \kappa \nabla^2 \Theta \qquad (4)$$

Chandrasekhar [1961] considered the effect of a spherical geometry on the stability and did not find a strong geometrical influence. Therefore, we will consider only the two-dimen-

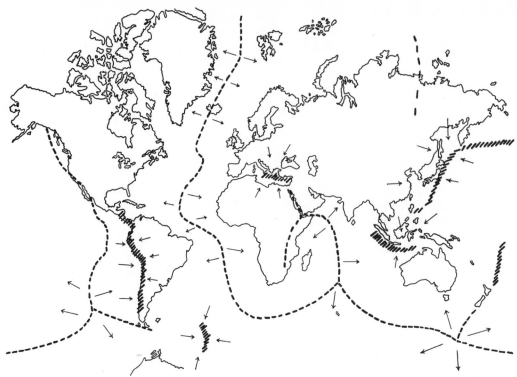

Fig. 1. The distribution of ascending and descending limbs over the earth's surface. The zones of shallow seismicity associated with mid-oceanic ridges and attributed to ascending limbs are denoted by a dashed line. The zones of deep-focus earthquakes associated with ocean trenches and attributed to descending limbs are denoted by slanted lines [after *Girdler*, 1965].

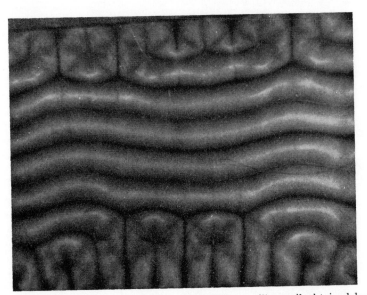

Fig. 2. Top view of laboratory cellular convection of a silicon oil obtained by *Sommerscales and Dropkin* [1966] in an 8-inch by 6-inch container, $d = 0.551$ inch; $R = 3040$, $Pr = 522$. The dark lines correspond to the surface traces of ascending and descending limbs.

sional planar problem. We neglect the production of heat by radioactivity in the convecting layer and assume that the gravitational acceleration is a constant. We solve for the flow in convection cells confined between two horizontal boundaries. It is convenient to let $T_0 = \frac{1}{2}(T_{w1} + T_{w2})$, so that $\Theta_{w1} = \frac{1}{2}(T_{w1} - T_{w2})$ on the lower boundary at $y = 0$ and $\Theta_{w2} = \frac{1}{2}(T_{w2} - T_{w1}) = -\Theta_{w1}$ on the upper boundary at $y = d$; T_{w1} and T_{w2} are constants. The condition that there be no flow through the boundaries requires that $v = 0$ at $y = 0$ and d. A second velocity boundary condition is required on the horizontal boundaries. The standard alternatives are the fixed wall boundary condition, $u = 0$ at $y = 0$ and d, or the free surface boundary condition that the tangential component of the shear stress vanish at the horizontal boundaries. The second condition would be applicable at the ocean floor if the oceanic crust is being transported by the cellular flow. Under a continental mass the fixed wall boundary condition might be applicable. The appropriate boundary condition on the lower cell boundary is not well defined since the boundary itself is not well defined. In this paper free surface boundary conditions are applied to both boundaries. Since our analysis is restricted to two-dimensional flows, the free surface boundary conditions require that $\partial u/\partial y = 0$ at $y = 0$ and d.

For a fluid with large Prandtl and Rayleigh numbers it is appropriate to use a boundary-layer theory to solve for the steady, finite amplitude cellular convection. The fluid is divided into two-dimensional rolls with rectangular cross sections; alternate rolls rotate in the clockwise and counterclockwise directions. Each cell has a height d and a width w. It is assumed that the aspect ratio of each cell is that given by the linear stability theory; that is, $w = 2^{1/2}d$. Since thermal gradients are assumed to be restricted to thin layers on the sides of each cell, it is appropriate to solve for the isothermal core flow in a rectangle with dimensions d and $2^{1/2}d$. For constant fluid properties an analytic solution for the flow field is easily obtained.

Thin thermal boundary layers are formed on the horizontal boundaries of each cell. With the prescribed boundary conditions the hot and cold boundary layers have similar temperature profiles. A solution for the boundary-layer temperature profiles is obtained by using the mean horizontal velocity at the horizontal boundaries and by setting the vertical velocity equal to zero in the boundary layers. The solution is a good approximation except in the vicinity of the vertical boundaries between cells. The local heat flux to the horizontal boundaries is also found.

When the two hot thermal boundary layers from adjacent cells meet, they separate from the lower horizontal boundary and form a hot thermal plume that rises to the upper surface. Similarly, when the two cold boundary layers from adjacent cells meet, they separate from the upper horizontal boundary and form a cold thermal plume that descends to the lower surface. The center line of each thermal plume is the dividing line between adjacent cells. A solution for the temperature distribution in the vertical plumes is obtained by using the mean vertical velocity at the vertical boundaries between cells and by setting the horizontal velocity equal to zero in the plumes. It is assumed that the initial temperature distribution at the base of the plume is the same as the temperature distribution in the thermal boundary layers adjacent to the base of the plume. The decrease in the maximum (minimum) temperature and the growth of the plume are obtained. The buoyancy force in the plumes drives the viscous flow. The magnitude of the velocities associated with the cellular convection is obtained by matching the integral of the temperature excess (deficit) within the plume to the velocity gradient at the edge of the plume.

When the convective plumes impinge on the horizontal surfaces, stagnation-point thermal boundary layers are formed. Using the velocities obtained from the core flow, we can obtain temperature profiles for the stagnation-point boundary layers. The stagnation-point thermal-boundary layers merge smoothly into the thermal-boundary layers on the horizontal boundaries. The stagnation-point heat flux is also determined.

The boundary-layer model for finite amplitude, steady cellular convection is illustrated in Figure 3. The analysis that leads to a solution has been given by *Turcotte and Oxburgh* [1967]. The results that will be applied to mantle convection are reproduced here. The horizontal

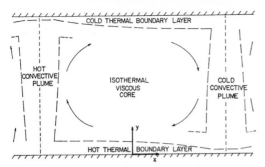

Fig. 3.   Illustration of the boundary-layer model
for cellular convection.

velocity component within each cell is given by

$$\frac{ud}{\kappa} = \sum_{n=1,3,\cdots} \left[\frac{3.56R^{2/3}\,\cos\,(n\pi y/d)}{n^2\pi^2\,\cosh\,(n\pi/2^{1/2})} \right]$$

$$\cdot \left[\frac{n\pi}{2^{1/2}} \tanh \frac{n\pi}{2^{1/2}} \cosh \frac{n\pi x}{d} - \frac{n\pi x}{d} \sinh \frac{n\pi x}{d} \right] \quad (5)$$

and the vertical velocity component is given by

$$\frac{vd}{\kappa} = \sum_{n=1,3,\cdots} \left[\frac{3.56R^{2/3}\,\sin\,(n\pi y/d)}{n^2\pi^2\,\cosh\,(n\pi/2^{1/2})} \right]$$

$$\cdot \left[\left(1 - \frac{n\pi}{2^{1/2}} \tanh \frac{n\pi}{2^{1/2}} \right) \sinh \frac{n\pi x}{d} \right.$$

$$\left. + \frac{n\pi x}{d} \cosh \frac{n\pi x}{d} \right] \quad (6)$$

where R is the Rayleigh number. The coordi-. nates x and y are measured from the center of the lower boundary as shown in Figure 3. The mean value of the horizontal velocity on the horizontal boundaries is given by

$$u_m d/\kappa = 0.142 R^{2/3} \quad (7)$$

and the mean value of the vertical velocity on the boundaries between cells is given by

$$v_m d/\kappa = 0.250 R^{2/3} \quad (8)$$

The temperature profiles in the horizontal boundary layers are given by

$$\Theta = \Theta_w \left[1 - \int_0^{0.188(y_1/d)(d/x_1)^{1/2}R^{1/3}} \right.$$

$$\left. \cdot \exp\,(-\eta)\,d\eta \right] \quad (9)$$

where x_1 and y_1 are the distances from the corners of the cell, as shown in Figure 3. This solution is valid for $x_1 > 0.0266d$. The local

heat flux to the horizontal boundaries due to the horizontal boundary layers is

$$\frac{q_w d}{2\Theta_{w1} k} = 0.106 \left(\frac{d}{x_1} \right)^{1/2} R^{1/3} \quad (10)$$

This relation is valid for $x_1 > 0.0266d$. The thickness of the thermal boundary layer, $y_{1\delta}$, is defined as the value of y_1 where $\Theta/\Theta_w = 0.1$. The maximum boundary layer thickness is at $x_1 = 2^{1/2}d$ and is given by

$$y_{1\delta}/d = 7.38 R^{-1/3} \quad (11)$$

This result verifies the boundary layer approximation. As long as the Rayleigh number is sufficiently large, the horizontal boundary layers are thin in comparison to the cell dimensions. Since the thickness of the plumes is of the same order as the thickness of the boundary layers, the plumes are also thin for large Rayleigh numbers. The temperature distribution in the thermal plumes is given by

$$\Theta = \Theta_w \left(\frac{d}{\pi y_1} \right)^{1/2}$$

$$\cdot \int_{-\infty}^{\infty} \left[1 - \frac{2}{\pi^{1/2}} \int_0^{0.635|\xi|} \exp\,(-\eta^2)\,d\eta \right]$$

$$\cdot \exp \left[-\left(0.50\,\frac{x_1}{d} R^{1/3} - \xi \right)^2 \frac{d}{y_1} \right] d\xi \quad (12)$$

The temperature distribution in the stagnation-point thermal-boundary layer is given by

$$\Theta = \Theta_w \left[1 - \frac{2}{\pi^{1/2}} \right.$$

$$\left. \cdot \int_0^{1.155(y_1/d)R^{1/3}} \exp\,(-\eta^2)\,d\eta \right] \quad (13)$$

This solution is valid for $x_1 < 0.0266d$. Comparing (13) with (9) shows that the structure of the stagnation-point thermal-boundary layer is identical to the structure of the horizontal thermal-boundary layer at $x_1 = 0.0266d$. The local heat flux to the horizontal boundaries due to the stagnation-point thermal-boundary layers is

$$q_w d/2\Theta_{w1} k = 0.652 R^{1/3} \quad (14)$$

This relation is valid for $x_1 < 0.0266d$. The thickness of the stagnation point thermal boundary layer is given by

$$y_{1\delta}/d = 1.01 R^{-1/3} \quad (15)$$

The mean heat flux to the horizontal boundaries Q_w is obtained by averaging q_w over the width of a cell using (10) and (15) with the result

$$Q_w d/2\Theta_w k = 0.165 R^{1/3} \qquad (16)$$

APPLICATION TO MANTLE CONVECTION

Before the boundary-layer theory for finite amplitude cellular convection can be applied to mantle convection, the properties of the mantle must be specified. Some of the properties are not well known, and some vary with depth in an unknown manner. Obviously, a theory that assumes constant values for the properties can be only approximately valid. For these reasons and others, close agreement between observed phenomena and theoretical predictions must be regarded as somewhat fortuitous.

The physical properties of the mantle have been studied by many authors including *Verhoogen* [1958], *Knopoff* [1964], *Clark and Ringwood* [1964], *McConnell* [1965], *Jaeger* [1965], *Tozer* [1965], and *MacDonald* [1965b]. Probably the most serious uncertainty relates to the kinematic viscosity ν. When the theoretical analysis of diffusion creep given by *Gordon* [1965] is compared with the analysis of the unlift of Fennoscandia by *Haskell* [1935] and the uplift of Lake Bonneville by *Crittenden* [1963], however, we conclude that for the upper mantle a mean value of $\nu = 10^{22}$ cm^2/sec should have an uncertainty of less than an order of magnitude [see also *McKenzie*, 1966].

The thermal state of the upper mantle is subject to considerable controversy. Some authors believe that the earth is cooling [see *Jeffreys*, 1929] and others believe that the earth is heating up because of radioactivity [see *Lubimova*, 1958]. There appears, however, to be no earth thermal history model in the literature that takes into account the possible effect of continued transfer of heat through the mantle by convection. The heat fluxes associated with a convecting mantle are so high that a steady-state thermal condition is not impossible. To explain the measured values of oceanic heat flux, it has been proposed that under oceanic areas the bulk of the radioactive components of the mantle are concentrated in the upper few hundred kilometers. With mantle convection such an assumption is no longer

necessary, and it is appropriate to assume a uniformly low distribution of radioactivity over the mantle. For a uniform distribution of radioactivity and a steady thermal state, the average thermal gradient in excess of the adiabatic across the convecting layer is taken to be $\beta = 1.5$ °K/km, and this value should have an uncertainty of less than a factor of 2. The thermal conductivity is taken to be $k = 2 \times 10^{-2}$ cal/cm sec °K, which is a factor of 3 higher than the thermal conductivity at the upper edge of the mantle but should be representative of an average value of the thermal conductivity through the upper mantle. The coefficient of thermal expansion is taken to be $\alpha = 2 \times 10^{-5}$ °K^{-1}; the thermal diffusivity, $\kappa = 3 \times 10^{-2}$ cm^2/sec; the acceleration of gravity, $g = 10^3$ cm/sec^2. The depth of the convecting layer d is assumed to be 1500 km. This value is uncertain by about a factor of 2. Since the width of each cell is $2^{1/2}d$ the corresponding cell width is 2100 km. Using the above values, the Rayleigh number is found to be $R = 5 \times 10^5$, which is near the center of the range of uncertainty. A reasonable estimate for the possible error in the Rayleigh number would be two orders of magnitude. The Prandtl number is $Pr = 3 \times 10^{23}$.

The mean horizontal velocity on the surface is found from (7) to be 1.8×10^{-7} cm/sec (5.68 cm/yr). Considering the range of uncertainty, this value is in good agreement with the velocities of continental drift obtained from paleomagnetic and other evidence; 10^{-7} cm/sec [*Tozer*, 1965], 1.25 cm/yr [*Orowan*, 1965], 3–5 cm/yr [*Allen*, 1965], and 1–4.4 cm/yr [*Vine*, 1966]. The mean vertical velocity on the centerline of the convective plumes from (8) is 3.2×10^{-7} cm/sec. The dependence of the horizontal component of velocity on the distance from the vertical rise is obtained from (5) and is given in Figure 4 for several depths. The dependence of the vertical component of velocity on the distance from the vertical rise is obtained from (6) and is given in Figure 5 for several depths. It is seen that vertical motion is concentrated in the regions adjacent to the boundaries between cells. The dependence of the temperature on the depth in the thermal boundary layer is obtained from (9) and is given in Figure 6 for several distances from the vertical rise. The dependence of the temper-

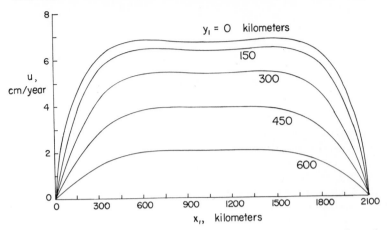

Fig. 4. Dependence of the horizontal component of velocity on the distance from the vertical rise at several depths.

ature on depth in the vicinity of the vertical rise is given in Figure 7. Included are profiles obtained from the stagnation-point thermal-boundary layer solution, (13), which are matched to the plume temperature outside the boundary layer. The stagnation-point solution is valid at distances of less than 40 km from the vertical rise. The temperature distribution in the rising convective plume, as obtained from (12), is given in Figure 8. The temperatures plotted are the difference between the local temperature and the adiabatic temperature at that depth. The maximum thickness of the thermal-boundary layer obtained from (11) is 139 km. The thickness of the stagnation-point boundary layer as obtained from (15) is 19.1 km.

From (10) and (14) the local heat flux to the surface can be obtained. The predicted values are given in Figure 9. Included for comparison are the measured values obtained on the East Pacific rise. The experimental points are taken directly from a figure published by *Lee and Uyeda* [1965]. The mean value of the heat flux as obtained from (16) is 3.93×10^{-6} cal/cm² sec. This value compares with the value 1.58×10^{-6} cal/cm² sec given by *Lee and Uyeda* [1965]. The more recent analysis of heat flow measurements by *Langseth et al.* [1966] gives similar values. Once again the agreement between theory and experiment is well within the range of possible error in the choice of mantle properties. The agreement between predicted and observed heat fluxes could be made closer

by the adoption of a more realistic thermal conductivity for the upper boundary layer, in place of the mean upper mantle conductivity used at present. To adopt this more realistic representation, however, would scarcely be justified in view of the other simplifying assumptions made.

ISOTHERM DISTRIBUTION

The steady-state distribution of isotherms predicted by the model is shown in Figure 10; the ascending and descending limbs are symmetrically related, as are the upper and lower boundary layers. Strong horizontal thermal

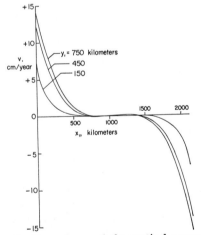

Fig. 5. Dependence of the vertical component of velocity on the distance from the vertical rise at several depths.

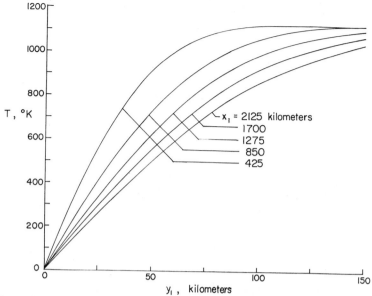

Fig. 6. Dependence of the temperature excess on depth at various distances from the vertical rise.

gradients exist only near the vertical plumes. The isotherm distribution shown for the uppermost part of the ascending limb is, however, unlikely to be realized in nature. As discussed in more detail below, partial fusion of the mantle takes place in this zone, and temperatures are lowered by latent heats of fusion and by the rapid transfer of heat to the surface by magmas.

It is of interest to note the extent of isotherm depression within the descending plume. The tectonic hypotheses that depend on the assumption that the base of the crust above the de-

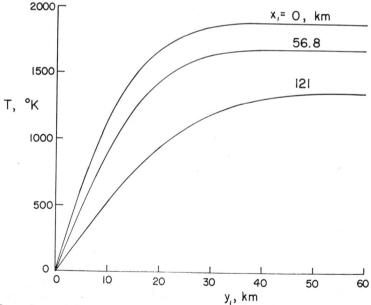

Fig. 7. Dependence of the temperature excess on depth at various distances from the vertical rise.

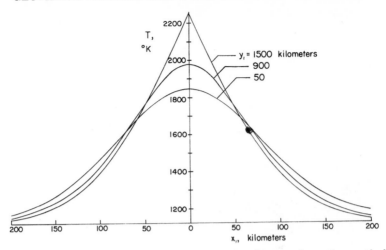

Fig. 8. Dependence of the temperature excess on the distance from the vertical rise at several depths.

scending limb of a mantle convection current is dragged downward and heated may be seen to be somewhat improbable or, at any rate, dependent on convective theory very different from that presented here. It is, however, difficult to visualize any convection model in which the depression of the isotherms in the descending plume is not greater than any reasonable amount of depression of the base of the crust.

UPPER PART OF ASCENDING PLUME

Mantle material traveling upward in the ascending plume undergoes a reduction in pressure that is large in comparison to its reduction in temperature (Figure 10). Whether this material melts depends on whether the geotherms displaced by convection intersect the isotherms for the beginning of mantle melting; in mantle material that does not have large-scale lateral compositional inhomogeneities, these isotherms will be approximately horizontal and parallel to the earth's surface.

It will be accepted here that the upper mantle is of broadly peridotitic composition and that basalt represents a low melting fraction derived from it. The evidence for these assumptions will not be reviewed here, but it is fully treated in the literature [e.g., *Yoder and Tilley*, 1962; *Oxburgh*, 1964; *Green and Ringwood*, 1966]. The melting behavior of basalts at high pressures should, therefore, provide a useful guide to the early melting behavior of the mantle. There are, however, considerable differences in melting

behavior of basalts having different chemical compositions, and there is no general agreement as to which basaltic types represent the unmodified first melting products of the mantle and which may have been contaminated by the solution of extraneous material or may have undergone differentiation. *Engel et al.* [1965] have found that olivine tholeiite basalt of extremely uniform composition is the most abundant volcanic rock of the ocean ridges. Its abundance, uniformity of composition, and occurrence in the deep oceans, where there is little opportunity for contamination with silicous crustal material,

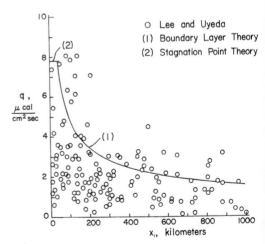

Fig. 9. Comparison of measured values of the heat flux as a function of the distance from the East Pacific rise with the boundary layer theory.

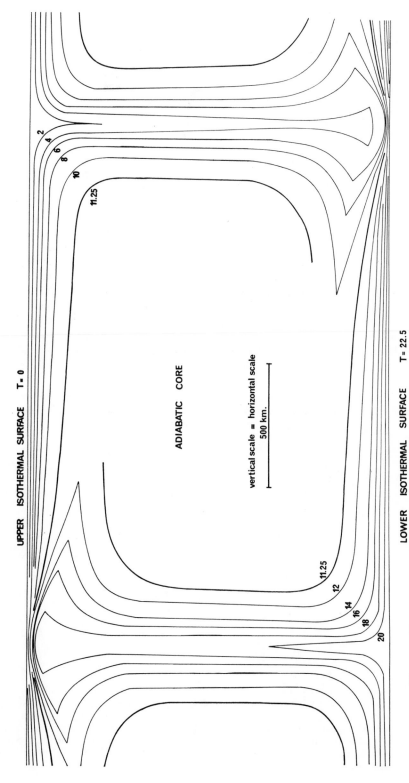

Fig. 10. The steady-state isotherm distribution predicted by the boundary-layer model; the values of the isotherms are given as superadiabatic temperatures $\times 10^{-2}$ °C.

83

all suggest that it may well be a relatively unmodified first melting fraction of the mantle.

Figure 11 shows the relationship between the predicted geotherm distribution within the upper part of the ascending plume in the absence of partial melting and the isotherms for the melting of olivine tholeiite. A zone of partial melting is defined by the intersection of the geotherms with melting isotherms of equal or lesser value. (The cross-sectional shape of this zone is shown in Figure 12.) The zone is somewhat more than 400 km wide and 200 km deep.

The amount of fusion occurring within this zone and the actual temperature distribution within it cannot be established without more precise information on the composition of the upper mantle than is available at present. These problems may be approached qualitatively, however, by considering the passage of a small reference volume of mantle through the zone of fusion; on the one hand the liquidus for the reference volume tends to be lowered by the progressive reduction in confining pressure during ascent, while on the other it is relatively elevated by the melting of the lower tempera-

ture fractions. The situation is represented diagramatically in Figure 13, where a family of liquidus curves corresponding to different confining pressures are plotted on a temperature-composition diagram. Consider the case of a particle that is following a streamline such that it enters the zone of fusion at a point where the conditions correspond to those at A. The diagram shows various possibilities for the subsequent changes in temperature, pressure, and liquid composition in the system represented by the reference particle. The limiting cases are paths A–X and A–W. The path A–X represents the situation in which the relative elevation of the liquidus through progressive melting is negligible in comparison with the depression through pressure reduction; this could happen only in a system with a bulk composition close to that of olivine tholeiite and would result in a temperature distribution within the zone of fusion that, for every depth, was close to the tholeiite liquidus. The other limiting case, path A–W, represents the converse situation in which a very small amount of melting is accompanied by a rapid relative

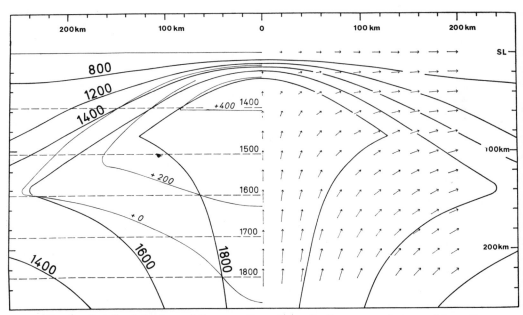

Fig. 11. Detail of the top of the ascending plume. Isotherms are shown as heavy lines (values in degrees Celsius in heavy numerals); liquidus temperatures for olivine tholeiite indicated by horizontal dashed lines, temperatures in degrees Celsius. The outline of the left half of the zone of fusion is indicated by the light line +0, and the zone within which the predicted temperature exceeds the liquidus temperature by 200°C or more by the light line +200. Small arrows are velocity vectors for particle motion.

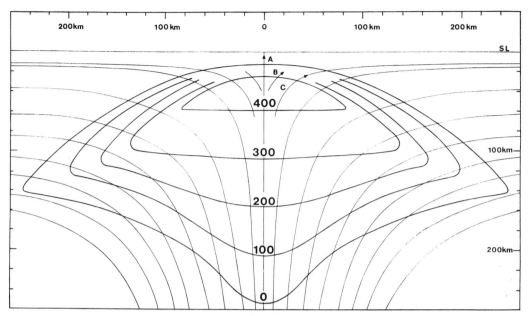

Fig. 12. Top of the ascending limb; heavy lines indicate zone of fusion and zones within it where the predicted temperature exceeds the olivine tholeiite fusion temperature by 100, 200°C, etc. Lighter lines indicate particle paths. For A, B, and C see Figure 14.

elevation in the liquidus temperature, so that the total amount of liquid produced in the fusion zone is very small and the temperatures within it correspond closely to the temperatures expected in the absence of any melting. The actual temperature distribution must lie between these extremes, e.g. path A–Y. Path B–Z would be followed by a particle that is on a path farther from the center of the rising column than A and that consequently enters the fusion zone at a lower temperature and pressure.

The compositional variations displayed by the melting of a peridotite-basalt system clearly cannot be properly represented in terms of two components, as is done in Figure 13, and the successive liquid compositions developed during passage through the fusion zone would vary in a complex fashion. In addition, the shape of the liquidus curves would probably vary with pressure; i.e., the P_8 curve would not have the same shape as the P_1 curve (Figure 13). This would reflect changes in the number and composition of the phases present in the system at different pressures.

Figure 12 shows a series of 'excess temperature' contours within the zone of fusion; the

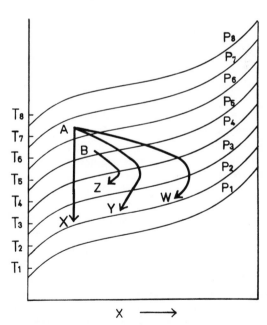

Fig. 13. Schematic relations between temperature, $T_1 < T_8$, pressure, $P_1 < P_8$, and liquid composition, x, within the zone of fusion. For discussion see text.

contours are drawn for every hundred degrees by which the predicted isotherms exceed the basalt fusion temperature. The 'excess heat' within the fusion zone would be sufficient to fuse olivine tholeiite equivalent to about 60% of the zone. This would correspond to fusion along path A–X in Figure 13. This path is the upper limiting case for the degree of fusion and, as discussed above, the degree of fusion that actually occurs is much less.

Figure 11 shows the velocity vectors for particle motion near the top of the hot plume computed for points on a 20-km square grid (see also Figures 4 and 5). From the velocity vectors streamlines may be derived, as shown in Figure 12. The streamlines diverge as the velocity decreases in the area of the stagnation point at the top of the hot plume, but they converge again after the corner is rounded and the flow becomes horizontal. The streamline separation is greater on the horizontal limb than in the rising plume, corresponding to horizontal velocities that are lower than the velocities in the plume. It can be seen from Figure 12 that only mantle material ascending along streamlines 100 km or less from the plume center line will pass through the fusion zone; mantle that has passed through the fusion zone forms the upper 130 km of the horizontal limbs. This would result in a compositional layering within the upper mantle. The uppermost mantle would be depleted in basalt and would pass downward into mantle material that had not passed through the fusion zone and was therefore unaffected [*Oxburgh*, 1965]. The depleted layer could be the source of some olivine-nodule xenoliths in basalts.

The model described above has several other petrogenetic implications. Within each half of the zone of fusion the combination of temperature and pressure conditions at every point is unique. This means that the liquids produced by the partial melting of a complex silicate system should vary continuously in composition throughout the fusion zone. In the parts of the fusion zone where the degree of melting is sufficiently great to allow their escape, these liquids should migrate upward to the surface to form hyperbyssal intrusions and surface eruptions. During their upward passage such liquids could mix with other liquids produced elsewhere within the fusion zone and having different compositions. It is evident that magmas erupted at the surface comprise a mixture of different liquids and that gradational variations in magma compositions are to be expected. It may be that such gradational variation in composition as has commonly been observed in volcanic sequences is partly to be explained in terms of the processes outlined above, rather than as entirely due to differentiation. If mantle material were of a composition such that its melting behavior followed a path similar to A–W in Figure 13, i.e. in the lower parts of the olivine tholeiite zone of fusion, little or no melting occurred, temperatures in the upper part of the zone should be sufficiently high to allow the formation of ultrabasic magmas within a narrow zone in the center of the ascending plume.

It should also be pointed out that, if ascending plumes are located under mid-ocean ridges, the main volcanic activity of the ridge should be concentrated within 200 km of the ridge crest, the lateral extent of the zone of fusion. Although this may be generally true, oceanic volcanic activity is certainly not confined to such a narrow belt. Active volcanos are found 500 km or more from the ridge; this may mean that the zone of fusion is wider than predicted by the present model.

Nature of Layer 3 in the Oceanic Crust

Seismic refraction measurements at sea have shown that in oceanic areas the M discontinuity is generally at little more than 10 km below sea level. Taking M as the base of the oceanic crust, the upper mantle is characterized by compressional wave velocities of 8.0–8.2 km/sec, whereas the oceanic crust comprises three horizontal layers, in descending order, layers 1, 2, and 3. Typical thicknesses and velocities are as follows [*Hill*, 1957; *Raitt*, 1963]:

	V_p, km/sec	Thickness, km
Layer 1	1.45 to 2.0	0.45, but variable
Layer 2	5.07 ± 0.63	1.71 ± 0.75
Layer 3	6.69 ± 0.26	4.86 ± 1.42

Layer 1 makes the floor of the oceans and may be sampled directly; it is known to be predominantly unconsolidated sedimentary material. Layers 2 and 3 present more of a problem; their character has been discussed by a number of authors, notably *Hess* [1962, 1965]. Briefly, however, layer 2 has a velocity compatible with a composition of indurated sediment, possibly containing intercalated volcanics. The velocity of layer 3 is compatible with basalt or serpentinite; a few other known crustal rocks have similar velocities but basalt and serpentinite are the only compositions that have been seriously proposed for layer 3. The work of *Vine and Matthews* [1963] and *Vine* [1966] has shown that the steep local gradients in the magnetic field on the flanks of mid-ocean ridges require that the magnetic material responsible for the anomalies lies predominantly in layer 2. This substantiates the idea that layer 2 has a significant basaltic component. The anomalies also suggest that layer 3 is either nonmagnetic or weakly so. This might appear to eliminate the possibility of a basaltic layer 3, in that basalt is generally rich in the magnetic oxides. Serpentinite, however, as proposed by Hess, both has a suitable velocity and is very weakly magnetic.

The model for mantle convection presented in this paper suggests some difficulties in the formation of a serpentinite crust [see also *Menard*, 1965]. In the system MgO-SiO$_2$-H$_2$O the upper temperature limit for the stability of serpentinite is known to be about 450°C at a pressure of 1.5 kb [*Scarfe and Wyllie*, 1967], i.e. the pressure at the base of the oceanic crust. It has been proposed [*Hess*, 1962, 1965] that as the peridotite of the mantle ascends in the rising limb of a convection cell and changes direction to flow away laterally, the olivine and enstatite of mantle material that is outermost cool sufficiently to react with interstitial water to produce serpentinite. Thus, layer 3 of what is recognized seismically as the crust would in fact have been produced by a change of phase within material that had previously been part of the upper mantle.

The principal difficulty with this model is that it is not clear how a seismic discontinuity sharp to a few tens of meters, as it appears to be in oceanic areas away from the ridges, could develop between the serpentinized mantle and the mantle proper; a more gradational contact would be expected. In addition, surface heat flow measurements indicate that, away from oceanic ridges the 450°C isotherm at which the serpentinization reaction proceeds, must be

considerably deeper than the M discontinuity.

Hess recognized this and proposed that the oceanic M discontinuity represents the 'fossilized' isotherm for the temperature of the serpentinization reaction [*Hess*, 1962]; this requires that at some temperature less than 450°C the serpentinization reaction rate drops sharply, effectively restricting the reaction to a rather narrow temperature interval, here called the reaction interval. The peridotite immediately below M and now at temperatures of less than 450°C must be assumed to have cooled through the reaction interval too rapidly for serpentinization to have taken place, whereas the peridotite immediately adjacent to it and, on the horizontal limb, above it, cooled sufficiently slowly for the reaction to proceed.

The model of mantle convection presented in this paper allows the cooling history of particles of mantle material following various flow paths to be established. This was done for the particles moving along three selected paths shown in Figure 12 as A, B, and C; the path A corresponds to an ascent along the center line of the plume ($x = 0$) and subsequent lateral motion along the upper surface ($y = 0$), and B and C are slightly deeper paths. The results are plotted in Figure 14. Although velocities are very low near the stagnation point ($x = 0$, $y = 0$), this does not prevent very rapid cooling on path A as it nears the upper surface. In fact path A gives more rapid cooling than either B or C; streamlines farther away from the surface pass more slowly through the reaction interval the deeper they are. The reaction interval is here arbitrarily designated

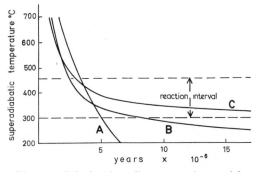

Fig. 14. Calculated cooling curves for particles following paths A, B, and C shown in Figure 12. For discussion see text.

as 450°–300°C, but the result would be similar for any other reasonable lower limit.

It is necessary to conclude that, if layer 3 is composed of serpentinite produced by hydration of the upper mantle, the M discontinuity is unlikely to have developed because the material below it did not have time to react; the deeper the streamline is, the more time there is available for the reaction. It is possible that a serpentinite layer 3 could develop if only the outer few kilometers of the mantle contained sufficient water for serpentinization to occur, but there does not seem to be any obvious way in which this could come about. It is proposed below that layer 3 comprises predominantly rocks of basaltic composition but that the magnetic properties of their oxide minerals may have been destroyed during low-grade metamorphism near the mid-oceanic ridge [*Cann and Vine*, 1966].

CRUSTAL STRUCTURE OF MID-OCEANIC RIDGES

The crustal structure of mid-oceanic ridges has been reviewed in great detail by workers from the Lamont Geological Observatory [*Le Pichon et al.*, 1965; *Talwani et al.*, 1965; *Heirtzler and Le Pichon*, 1965; *Ewing et al.*, 1966; *Langseth et al.*, 1966]. These workers were primarily concerned with the East Pacific rise and part of the mid-Atlantic ridge; although these ridges differ in a number of ways, gravity and seismic observations indicate that they have several significant features in common:

1. Although on the flanks of both ridges typically layered oceanic crust is found, beneath a zone extending several hundreds of kilometers away from the ridge crest on either side, no clear M discontinuity has been observed, and under a layer that is continuous with layer 2 on the flanks, a variety of P-wave velocities ranging from 6.4 to 8.2 km/sec are measured; particularly notable are velocities in the range 7.2 to 7.6 km/sec, which are absent on the flanks.

2. Both ridges are characterized by negative Bouguer gravity anomalies centered over their axial zones. These anomalies require the presence of a substantial mass deficiency immediately below the axial zone and somewhat smaller deficiencies on the flanks.

Talwani et al. [1965] have proposed solutions for the East Pacific rise and mid-Atlantic ridge structures that satisfy both gravity and seismic observations, and a regional crust and upper mantle structure has been proposed for the North Atlantic by *Tryggvason* [1961].

The model for mantle convection proposed in this paper has a bearing on the interpretation of both gravity and seismic observations. It is known in a general way that the effect of increase in temperature is to reduce seismic velocities in rocks, whereas increase in confining pressure is associated with an increase in velocity. A considerable amount of experimental work has been done in this field at moderate temperatures and pressures (reviewed by *Press* [1966]); relatively little is known about the temperature and pressure dependence of velocity at pressures up to 100 kb and at the high temperatures proposed here for the top of the ascending plume. In particular, the effects of partial melting on velocities in crystalline aggregates do not appear to have been investigated. It is to be expected, however, that the physical parameter most drastically affected by partial melting would be the rigidity; by comparison, the decrease in density could be relatively small. Taking the usual equations for the velocities of propagation of compressional waves V_p and shear waves V_s,

$$V_p = [(\tfrac{4}{3}\mu + \kappa)/\rho]^{1/2} \qquad (17)$$

$$V_s = (\mu/\rho)^{1/2} \qquad (18)$$

where k is the bulk modulus, μ is the rigidity, and ρ is the density of the transmitting medium, it is evident that a large reduction in rigidity accompanied by a small reduction in density might bring about a 10% reduction in V_p but a more marked reduction in V_s. A similar conclusion is reached by *Shimozuru* [1963].

In addition to direct effects on elastic wave velocity, the temperature distribution proposed here should have strong effects on Q, the anelasticity of the upper mantle [*Anderson et al.*, 1965; *Anderson*, 1967]. Although lateral variations in Q for the upper mantle have been observed in some places [*Oliver and Isacks*, 1967], there appear to be no studies bearing directly on mid-oceanic ridge structure.

In the absence of supporting experimental evidence it is proposed that compressional wave velocities in the range 7–8 km/sec correspond to peridotite ($V_p = 8.0$–8.2 km/sec) of the upper mantle that has begun to melt and that consists of an aggregate of weakly connected crystals the interstices between which are occupied by basaltic liquid. It should be possible to verify this suggestion by measurements of shear wave velocities and wave attenuation phenomena under oceanic ridges. It is seen from Figure 12 that tongues of completely crystalline upper mantle with normal mantle velocities are to be expected above the zone of partial fusion on either side. Partially serpentinized peridotite is almost the only other rock known to have compressional wave velocities in the range 7–8 km/sec (Birch in *Hess* [1962]), and it has been shown earlier that temperatures at the top of the plume are too high for the existence of serpentinite at the depths required. An alternative suggestion has been made by *Hess* [1962] that these anomalous velocities are the result of microfracturing in the peridotite where the direction of flow changes rapidly. This proposal is difficult to evaluate.

A detailed discussion of gravity profiles computed for the convection model proposed here will be presented elsewhere. It may be noted, however, that the geometry of the isotherms and thus roughly the densities, as proposed, is consistent with the solutions of *Talwani et al.* [1965] and *Tryggvason* [1961]. The present model, however, would require the anomalous masses to be rather larger than the values proposed by *Talwani et al.* [1965] and to be situated at somewhat greater depth. It is also possible that the density contrasts employed by the authors cited are too great. The authors employed the Nafe-Drake empirical relationship between seismic velocity and density. Whereas this relationship is known in many circumstances to give reasonable solutions, it is possible that in situations such as the one described here, where rocks are close to or above their temperatures for the beginning of fusion, the relationship breaks down as rigidity decreases more rapidly than density.

KINETIC IMPLICATIONS OF THE PROPOSED MODEL

It remains to consider some of the larger-scale and long-term implications of the convection model proposed here. For the cell dimensions and rates of movement assumed here, one complete convective overturn would require approximately 10^9 years, i.e. nearly a quarter of the age of the earth. Isotherms would, however, have reached an approximately steady-state configuration after 5×10^8 years. All these values would be somewhat reduced, however, if the aspect ratio of the model cell were modified as discussed below. Thus, the earliest stages of ocean spreading or continental drift by the spreading mechanism might be expected to be devoid of volcanic activity.

As briefly mentioned in an earlier section, the mantle material that passes through the zone of fusion subsequently changes direction and moves horizontally away from the ridge, forming the upper 130 km of mantle on the ridge flanks. The removal of the low melting fraction during passage through the fusion zone renders the uppermost mantle different in composition from the material that it overlies. The basaltic lava formed in the fusion zone should rise to the surface above and there be erupted upon earlier lavas, or, near the stagnation point, upon upper mantle material itself; in either case, the basalt would be expected to form a passive sheet resting on the upper mantle which is continuously transported away laterally upon it, in this way forming layer 3 of the oceanic crust. Thus, after the early stages, as long as the convective motion persists, new layer 3 should be generated at or near the mid-oceanic ridges.

If layer 3 is formed in the way suggested, its thickness gives some idea of the degree of fusion in the fusion zone. For a materials balance, a 4-km thickness of layer 3 (possibly with some of layer 2) added to a 130-km thickness of depleted upper mantle immediately below M, should have the same bulk composition as the upper mantle below 130 km which has not been melted. This suggests that the average degree of fusion within the fusion zone may be as little as 3%, although locally it would be rather higher, particularly if not all the fused material finds its way to the surface.

Figures 4, 5, and 11 show that particles on the upper surface of the convection cell undergo a rapid acceleration for about 300 km away from the stagnation point at the beginning of their horizontal paths. The widths of the magnetic anomaly bands [*Vine*, 1966] at different distances from the ridge indicate, however, that spreading rates at the surface do not vary in this fashion. It is evident that here the assumption of uniform viscosity has failed and that the cold upper surface layer of the convecting medium has such a high viscosity that it effectively moves as a rigid plate. The thickness of the layer behaving in this fashion would be related to the depth of the cold thermal boundary layer and would thus increase away from the ridge. A rigid surface layer of this kind should have a marked effect on the geometry of the descending flow, possibly along the lines suggested by *Oliver and Isacks* [1967].

CONCLUSIONS

A quantitative model for upper mantle convection has been proposed that satisfies reasonably well the requirements of heat transfer, velocity, and scale imposed by the hypothesis of ocean-floor spreading. It appears also that the model may be consistent with seismic observations on mid-oceanic ridges, although the experimental data necessary to demonstrate this are not available. Geotherm distributions predicted by the model have been used to explain mid-oceanic volcanic activity and the generation of layer 3 of the oceanic crust. A serpentinite layer 3 is not compatible with the thermal model used in this paper unless the position of the discontinuity is governed primarily by the availability of water for the serpentinization reaction.

A large number of simplifying assumptions have been made in setting up the model. The effects of radioactivity have been ignored; although many of the arguments for the strong concentration of radioactive constituents in the upper part of the mantle are invalid if mantle convection occurs, the currents proposed here take so long to complete a cycle that relatively low concentrations of radioactivity could have significant effects. One such effect could be the modification of the cell aspect ratio [*Tritten and Zarraga*, 1967; *Roberts*, 1967]. This could explain the existence of convection currents with the horizontal dimensions required for large continental displacements but at the same time sufficiently shallow to be unaffected by the high viscosities that appear to characterize the deeper parts of the mantle [*McKenzie*, 1966; *Gordon*, 1965]. The effects of more realistic boundary conditions on mantle convection must also be investigated.

It has also been assumed that the mantle behaves in a viscous fashion and that viscosity does not vary with temperature; the former assumption has been seriously questioned [e.g., *Orowan*, 1965] and the latter assumption is certainly wrong, but it is not yet known how this might affect convective circulation. The provisional agreement, however, between the model proposed here and the natural phenomena that it purports to describe is sufficiently good as to suggest that the simplifying assumptions either have effects that cancel each other or are less important than they first appear.

Acknowledgments. The research was partially supported by the Air Force Office of Scientific Research of the Office of Aerospace Research under contract AF 49(638)-1346. The research was undertaken when one of the authors (D. L. Turcotte) held a National Science postdoctoral research fellowship at the Department of Engineering Sciences, University of Oxford.

REFERENCES

Allen, C. R., Transcurrent faults in continental areas, *Phil. Trans. Roy. Soc. London, A, 258,* 82, 1965.

Anderson, D. L., Latest information from seismic observations, in *The Earth's Mantle,* edited by T. F. Gaskell, pp. 355–420, Academic Press, New York, 1967.

Anderson, D. L., A. Ben-Menahem, and C. B. Archambeau, Attenuation of seismic energy in the upper mantle, *J. Geophys. Res., 70,* 1441, 1965.

Blackett, P. M. S., Introduction; A symposium on continental drift, *Phil. Trans. Roy. Soc. London, A, 258,* vii, 1965.

Cann, J. R., and F. J. Vine, An area on the crest of the Carlsberg ridge: Petrology and magnetic survey, *Phil. Trans. Roy. Soc. London, A, 259,* 198, 1966.

Chandrasekhar, S., *Hydrodynamic and Hydromagnetic Stability,* Oxford University Press, Oxford, 1961.

Clark, S. P., and A. E. Ringwood, Density distribution and constitution of the mantle, *Rev. Geophys., 2,* 35, 1964.

Crittenden, M. D., Effective viscosity of the earth derived from isostatic loading of Pleistocene Lake Bonneville, *J. Geophys. Res., 68,* 5517, 1963.

Elsasser, W. M., Early history of the earth, in *Earth Science and Meteoritics,* pp. 1–30, North Holland, Amsterdam, 1963.

Engel, A. E. J., C. G. Engel, and R. G. Havens, Chemical characteristics of oceanic basalts and the upper mantle, *Bull. Geol. Soc. Am., 76,* 719, 1965.

Ewing, M., X. Le Pichon, and J. Ewing, Crustal

structure of the mid-ocean ridges, 4, Sediment distribution in the South Atlantic Ocean and Cenozoic history of the mid-Atlantic ridge, *J. Geophys. Res.*, *71*, 1611, 1966.

Fromm, J. E., Numerical solutions of the nonlinear equations for a heated fluid layer, *Phys. Fluids*, *8*, 1757, 1965.

Girdler, R. W., The formulation of new oceanic crust, *Phil. Trans. Roy. Soc. London, A, 258*, 252, 1965.

Gordon, R. B., Diffusion creep in the earth's mantle, *J. Geophys. Res., 70*, 2413, 1965.

Green, D. H., and A. E. Ringwood, The genesis of basaltic magmas, in *Petrology of the Upper Mantle, Publ. 444*, Department of Geophysics and Geochemistry, Australian National University, 1966.

Haskell, N. A., The motion of a viscous fluid under a surface load, *Physics, 6*, 265, 1935.

Heirtzler, J., and X. Le Pichon, Crustal structure of the mid-ocean ridges, 3, Magnetic anomalies over the mid-Atlantic ridge, *J. Geophys. Res., 70*, 4013, 1965.

Hess, H. H., History of ocean basins, in *Petrologic Studies*, pp. 599–620, Geological Society of America, New York, 1962.

Hess, H. H., Mid-oceanic ridges and tectonics of the sea floor, in *Submarine Geology and Geophysics, Colston Papers*, vol. 17, pp. 317–334, Butterworths, London, 1965.

Hill, M. N., Recent geophysical exploration of the ocean floor, *Phys. Chem. Earth, 2*, 129, 1957.

Jaeger, J. C., Application of the theory of heat conduction to geothermal measurements, in *Terrestrial Heat Flow Geophys. Monograph 18*, pp. 7–23, American Geophysical Union, Washington, D. C., 1965.

Jeffreys, H., Some cases of instability in fluid motion, *Proc. Roy. Soc. London, A, 118*, 195, 1928.

Jeffreys, H., *The Earth*, 2nd ed., Cambridge University, Cambridge, 1929.

Jeffreys, H., The instability of a compressible fluid heated below, *Proc. Cambridge Phil. Soc., 26*, 170, 1930.

Knopoff, L., The convection current hypothesis, *Rev. Geophys., 2*, 89, 1964.

Kuo, H. L., Solution of the non-linear equations of cellular convection and heat transport, *J. Fluid Mech., 10*, 611, 1961.

Langseth, M. G., X. Le Pichon, and M. Ewing, Crustal structure of the mid-ocean ridges, 5, Heat flow through the Atlantic ocean floor and convection currents, *J. Geophys. Res., 71*, 5321, 1966.

Le Pichon, X., R. E. Houtz, C. L. Drake, and J. E. Nafe, Crustal structure of the mid-ocean ridges, 1, Seismic refraction measurements, *J. Geophys. Res., 70*, 319, 1965.

Lee, W. H. K., and S. Uyeda, Review of heat flow data, in *Terrestrial Heat Flow, Geophys. Monograph 8*, pp. 87–190, American Geophysical Union, Washington, D. C., 1965.

Lubimova, H. A., Thermal history of the earth with consideration of the variable thermal conductivity of the mantle, *Geophys. J., 1*, 115, 1958.

MacDonald, G. J. F., Continental structure and drift, *Phil. Trans. Roy. Soc. London, A, 258*, 215, 1965a.

MacDonald, G. J. F., Geophysical deductions from observations of heat flow, in *Terrestrial Heat Flow, Geophys. Monograph 8*, pp. 191–210, American Geophysical Union, Washington, D. C., 1965b.

Malkus, W. V. R., and G. Veronis, Finite amplitude cellular convection, *J. Fluid Mech., 4*, 225, 1958.

McConnell, R. K., Isostatic adjustment in a layered earth, *J. Geophys. Res., 70*, 5171, 1965.

McKenzie, D. P., The viscosity of the lower mantle, *J. Geophys. Res., 71*, 3995, 1966.

Menard, H. W., Sea floor relief and mantle convection, in *Physics and Chemistry of the Earth, 6*, pp. 315–364, Pergamon, London, 1965.

Oliver, J., and B. Isacks, Deep earthquake zones, anomalous structures in the upper mantle and the lithosphere, *J. Geophys. Res., 72*, 4259, 1967.

Orowan, E., Convection in a non-Newtonian mantle, continental drift and mountain building, *Phil. Trans. Roy. Soc. London, A, 258*, 284, 1965.

Oxburgh, E. R., Petrological evidence for the presence of amphibole in the upper mantle and its petrogenetic and geophysical implications, *Geol. Mag., 101*, 1, 1964.

Oxburgh, E. R., Volcanism and mantle convection, *Phil. Trans. Roy. Soc. London, A, 258*, 142, 1965.

Platzman, G. W., The spectral dynamics of laminar convection, *J. Fluid Mech., 23*, 481, 1965.

Press, F., Seismic velocities, in *Handbook of Physical Constants, Mem. 97*, pp. 195–218, Geological Society of America, New York, 1966.

Raitt, R. W., The crustal rocks, *The Sea*, vol. 3, 85, Interscience, New York, 1963.

Rayleigh, J. W. S., On convective currents in a horizontal layer of fluid when the higher temperature is on the under side, *Phil. Mag., 32*, 529, 1916.

Roberts, P. H., Convection in horizontal layers with internal heat generation: Theory, *J. Fluid Mech., 30*, 33, 1967.

Runcorn, S. K., Palaeomagnetic evidence for continental drift and its geophysical cause, in *Continental Drift*, pp. 1–40, Academic Press, New York, 1962.

Scarfe, C. M., and P. J. Wyllie, Experimental redetermination of the upper stability limit of Serpentine up to 3-kb pressure (abstract), *Trans. Am. Geophys. Union, 48*, 225, 1967.

Shimozuru, D., Geophysical evidence for suggesting the existence of molten pockets in the earth's upper mantle, *Bull. Volcanol., 26*, 181, 1963.

Somerscales, E. F. C., and D. Dropkin, Experimental investigation of the temperature distribution in a horizontal layer of fluid heated from below, *Intern. J. Heat Mass Transfer, 9,* 1189, 1966.

Talwani, M., X. Le Pichon, and M. Ewing Crustal structure of the mid-ocean ridges, 2, Computed model from gravity and seismic refraction data, *J. Geophys. Res., 70,* 341, 1965.

Tozer, D. C., Heat transfer and convection currents, *Phil. Trans. Roy. Soc. London, A, 258,* 252, 1965.

Tritton, D. J., and M. N. Zarraga, Convection in horizontal layers with internal heat generation: Experiments, *J. Fluid Mech., 30,* 21, 1967.

Tryggvason, E., Wave velocity in the upper mantle below the Arctic-Atlantic Ocean and northwest Europe, *Ann. Geofis., 14,* 380, 1961.

Turcotte, D. L., A boundary-layer theory for cellular convection, *Intern. J. Heat Mass Transfer, 10,* 1065, 1967.

Turcotte, D. L., and E. R. Oxburgh, Finite amplitude convection cells and continental drift, *J. Fluid Mech., 28,* 29, 1967.

Verhoogen, J., Temperatures within the earth, *Phys. Chem. Earth, 1,* 17, 1958.

Vine, F. J., Spreading of the ocean floor: New evidence, *Science, 154,* 1405, 1966.

Vine, F. J., and D. H. Matthews, Magnetic anomalies over oceanic ridges, *Nature, 199,* 947, 1963.

Yoder, H. S., and C. E. Tilley, Origin of basalt magmas: An experimental study of natural and synthetic rock systems, *J. Petrol., 3,* 342, 1962.

(Received June 30, 1967;
revised December 1, 1967.)

4

Copyright © 1970 by the Geological Society of America

Reprinted from *Geol. Soc. America Bull.*, **81**, 1665–1688 (June 1970)

E. R. OXBURGH
D. L. TURCOTTE

Thermal Structure of Island Arcs

ABSTRACT

The ocean-floor spreading hypothesis requires that island arc areas be loci of descending mass transport and destruction of oceanic crust. All island arc areas are characterized by volcanic activity and in some, unusually high heat flow values have been measured for hundreds of kilometers behind the arc. Downward mass motion is associated presumably with the sinking of cool and thus relatively denser matter, and high temperature phenomena normally would not be expected where this was occurring. The only feasible means of generating heat in such areas appears to be by frictional dissipation along the seismically active zone of movement (the Benioff zone) between the moving oceanic crust and upper mantle, and the relatively stationary material that it underrides. It has been shown elsewhere that dissipative heating is of the right order of magnitude to explain the observations, but the required shear stresses are rather high; in this paper a detailed thermal structure for the Benioff zone and the regions on either side is presented. The cold descending material heats very slowly and should form a thin cold slab extending to depths of more than 500 km. Within the zone of movement, temperatures are buffered by the melting temperatures of the various components of the oceanic crust that partially fuse to give the main members of the calc-alkaline igneous suite. The thermal structure of the region above the Benioff zone is subject to uncertainty because processes other than lattice conduction (for instance, magmatic activity) are probably involved in the transfer of heat to the surface.

Observed rates of magmatic activity in island arcs require a relatively small degree of fusion of the descending material and probably only the oceanic crust is involved. The siliceous and presumably water-bearing sedimentary material of the oceanic crust is the most important source of material added to island arcs; growth occurs either by physical addition of sedimentary material, which, for mechanical reasons, may not be carried down, or by partial fusion of sedimentary material along the fault zone and by rise of the resultant magmas to the surface. Rates of growth suggested by these processes are compatible with the known ages of arcs. Older arcs should have larger crustal cross-sectional areas than younger ones, if both have been continuously active at similar rates.

INTRODUCTION

The importance of island arc structures in problems of large-scale tectonics was not really appreciated until the pioneering marine gravity work of Vening Meinesz (1932, 1937) and Vening Meinesz and others (1934), showed that island arcs and the deep-ocean trenches, which commonly lie on their oceanward side, were associated with strong perturbations of the Earth's gravitational field. Vening Meinesz's careful studies in the East Indies showed that strongly negative Bouguer and isostatic anomalies run parallel to the trench-arc system centered between the trench and the arc. He interpreted these observations and the occurrence of widespread seismicity along these zones as evidence that in these areas the crust was buckled downward in the trench and upward in the adjacent arc through the operation of subcrustal convection currents. As a result, the crust was in a dynamic rather than an isostatic equilibrium.

These ideas were developed by a number of authors (for example, Hales, 1936; Kuenen, 1936; Hess, 1937a, 1937b). In 1948, Hess published a new, detailed bathymetric chart of the western north Pacific extending from southern Japan to New Guinea. This area (Figs. 1, 2, and 3) contains a number of well-developed island arcs. In his discussion of the chart, Hess noted (1) the previously unrecognized continuity of the deep trenches along the island arcs, (2) the close spatial association of seismic activity, volcanism, and gravity anomalies along the arc-trench lineaments, and (3) the distribution of peridotitic bodies and glaucophane schists along the arc system. Some of these features are also found in continental orogenic belts, and thus the possibility arose that island arc systems in oceanic areas were the surface expression of the same subcrustal processes that gave

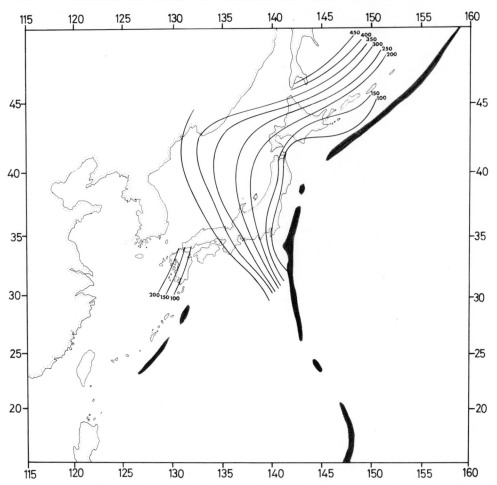

Figure 1.   Seismicity and ocean trenches for Japan and part of the northwestern Pacific. Depth to the Benioff zone is shown by 50-km contours, and trends of trenches are indicated by the dark tone. (In part, *after* the Association for Geological Collaboration of Japan, 1965.)

rise to Alpine-type orogenic belts in continental areas.

In 1962, Hess introduced the concept of ocean-floor spreading: that new oceanic crust is continuously created at the mid-oceanic ridges and that the whole of the oceanic crust is in motion away from active ridges. If large changes in the radius of the Earth are discounted (Birch, 1968), conservation of volume and surface area require that oceanic crust be destroyed at the same rate at which it is created. Destruction of oceanic crust appears to occur at oceanic trenches where the crust and subjacent upper mantle is sliding down an inclined fault into the mantle beneath the island arc region. During the last five years the interpretation of linear magnetic anomalies aligned with oceanic ridges (Vine and Mathews, 1963; Vine, 1966; Heirzler and others, 1968; Morgan, 1968; Le Pichon, 1968) has provided quantitative evidence in support of the ocean-floor spreading hypothesis.

On mid-oceanic ridges, heat flow measurements have been available for a number of years, and the general pattern of high heat flow associated with the ridge crest is consistent with the sea-floor spreading hypothesis (Langseth and others, 1966; Turcotte and Oxburgh, 1967; McKenzie, 1967; Oxburgh and Turcotte, 1968a). Heat-flow measurements in island-arc regions, however, are less abundant, but an interesting and somewhat surprising pattern has emerged from recent studies in the western Pacific (*see* Vaquier and

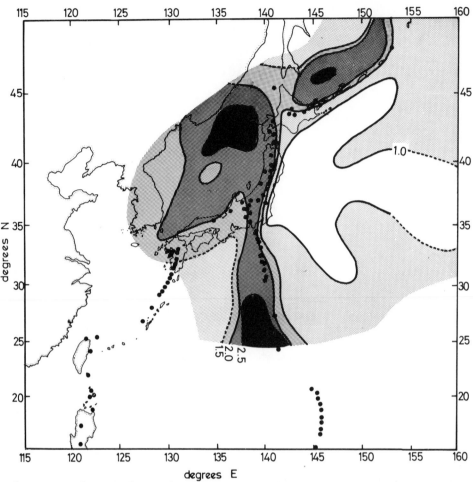

Figure 2. Heat flow and volcanicity for Japan and part of the northwestern Pacific. Volcanoes are shown as black dots; heat flow (*after* Vaquier and others, 1966) contours, in μ cal cm^{-2} sec^{-1}.

others, 1966; Sclater and Menard, 1967). It appears that on the trench side of the island arc system the heat flow is typical of deep-ocean basins, whereas on the island arc side, heat flow may be substantially higher. In view of the abundant volcanic activity characteristic of island arcs the high heat flow is hardly surprising. However, an explanation is required for the presence of these thermal phenomena in a region of descending mass transport. One might expect that the descending mass transport in trench areas would result in a strong depression of the isotherms and, therefore, a lower than average surface heat flux. The body force required to drive the descending flow is certainly due to the temperature deficiency and, thus, a mass excess in the descending

material. Several authors (McKenzie and Sclater, 1968; Oxburgh and Turcotte, 1968b; Turcotte and Oxburgh, 1968) have concluded that frictional dissipation along the inclined fault zone is the probable explanation for the thermal phenomena observed at the surface.

In this paper we present a more thorough analysis of this problem. The characteristics of island arc systems that relate to the problem are presented, the alternative sources for heat are examined, and a two-dimensional thermal structure for island-arc regions is derived. We examine the implications of this structure for the production of magmas and the contribution that they and other processes may make to the growth of the island arc. As noted above, island arcs have many features in common with

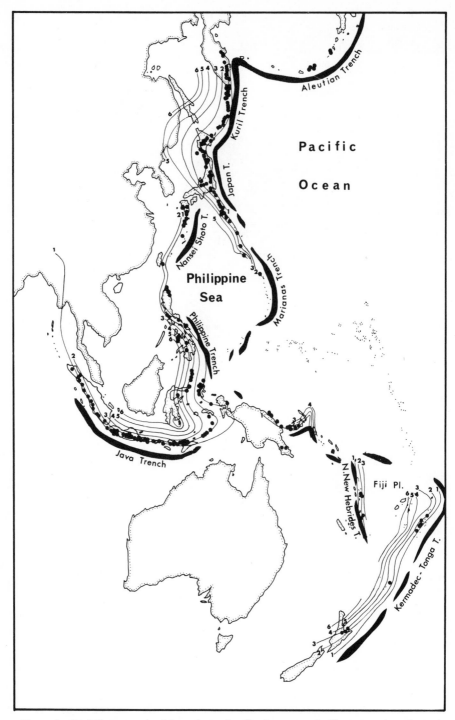

Figure 3. Benioff zones, volcanicity and trenches for the western Pacific; conventions shown in Figures 1 and 2. Data from Gutenberg and Richter (1954), Association for Geological Collaboration of Japan (1965); Hamilton and Gale (1968) and Dr. T. Hamada (1968) private commun.

continental orogenic belts so that the results given here may also be applicable to continental regions.

THERMAL OBSERVATIONS

Surface evidence for thermal phenomena occurring at depth come from heat flow measurements and observed volcanism. Surface measurements of heat flow in island arc regions prior to 1964 were summarized by Lee and Uyeda (1965). These authors gave a mean heat flow for oceanic trenches of 0.99 μ cal/cm^2 sec (21 measurements), somewhat lower than the mean value for ocean basins (1.28 μ cal/cm^2 sec). No special attention was given, however, to variations in the heat flux across island arc-trench systems. Vaquier and others (1966) published a compilation of heat flow measurements from the Sea of Japan across the Japanese mainland to the main basin of the Pacific (Fig. 2). As a result of this compilation, it became clear that the line of the island arc corresponds to a prominent discontinuity in the heat flow pattern; on the ocean side of the arc, the heat flow is typically oceanic (1.0–1.3 μ cal/cm^2 sec), with slightly lower values in trenches. In the sea behind the arc, the heat flow is high (1.0–3.3 μ cal/cm^2 sec, mean 1.56 μ cal/cm^2 sec) for distances of 300 to 700 km from the trench axis.

Subsequently, Sclater and Menard (1967) found a similar pattern on the Fiji plateau, an area of crust that is bounded on the west by the New Hebrides trench-arc system, on the northeast by the Vitiaz trench, and on the southeast by the Tonga trench (Fig. 3). They report values for the heat flux as high as 5.6 μ cal/cm^2 sec on the plateau contained within the arc-trench systems and a mean heat flux of 2.43 μ cal/cm^2 sec. Detailed studies on other arc-trench systems have not been published; the quantitative analysis given in this paper will, therefore, be based on the two areas discussed above, although the results should be applicable to all active arc-trench systems. Menard (1967) points out that in some of the basins between an island arc and the adjacent continent the sedimentation rates may be too high for equilibrium heat flow to be established; that is, measured values may be expected to be too low.

Volcanic activity occurs or has occurred in every known island arc system. The distribution of active volcanoes in the Western Pacific is shown in Figures 2 and 3; active volcanoes form a semicontinuous chain along the line of the island arcs parallel to the adjacent ocean trenches. The active volcanic belt is usually less than 300 km wide and may be considerably narrower; it occupies the oceanward side of the zone of high heat flow.

The heat carried to the surface by volcanic materials is, for the most part, lost directly to the atmosphere and thus represents a quantity that must be added to the measured surface heat flux when estimating the gross areal heat flux. Sugimura and others (1963) have estimated the rates of extrusion on the main Japanese islands. During the Quaternary, they estimate the volume extruded to be 5 × 10^{3}km^3 (extrusion rate of 5 × 10^{3}km^3/ million years), during the upper Miocene until the end of the Tertiary, a volume of 20 × 10^{3}km^3 (10^{3}km^3/m.y.) and during the lower and middle Miocene, a volume of 150 × 10^{3}km^3 (25 × 10^{3}km^3/m.y.). Taking values of 5 × 10^{15}cm^2 for the area of Japan, 400 cal/gm for the total heat released by crystallization and cooling of the extruded material, and 3 gm/cm^3 as its density, we find that the high rate of extrusion during the early Miocene is equivalent to a surface heat flux of about 0.2 μ cal/cm^2 sec averaged over the area of Japan. This equivalent heat flux is relatively small compared with the measured heat flux of about 2 μ cal/cm^2 sec, and we may, for the purpose of constructing a thermal model, ignore the contribution of volcanic activity to the gross regional heat flux.

SEISMIC OBSERVATIONS

All known island arcs are seismically active to some extent. Gutenberg and Richter (1941) showed that in island arc areas the greater part of the seismicity is confined to an inclined fault zone that meets the surface at the adjacent oceanic trench and dips away from the trench under the island arc at an angle near 45°. All observed deep-focus earthquakes appear to be associated with such zones. Subsequent studies (*see* Gutenberg and Richter, 1954) have confirmed this distribution of seismic activity. In 1955, Benioff proposed that the inclined fault zones defined by the seismicity were major tectonic dislocations that, in the case of the Pacific, decoupled the Pacific Basin from its continental margins; the inclined zones of seismicity have since been called Benioff zones.

In a study of the seismicity of the Tonga-Kermadec, Kuril-Kamchatka, and Caribbean trench-arc systems, Sykes (1966) recomputed the locations of earlier earthquakes and com-

97

bined these with observations of recent seismicity. In the Tonga-Fiji area, he demonstrated that the Benioff zone was 50 to 100 km thick and that seismic activity could be continuous to depths of 650 km. In some areas the Benioff zone appears to be planar; elsewhere it is either slightly convex upward, or convex downward. The contours defining the Benioff zone in Japan are shown in Figure 1, and it is shown schematically for the whole of the Southwestern Pacific in Figure 3.

A comparison of Figures 1 and 2 shows a striking correlation between the area that is at present seismically active and the area with abnormally high heat flow. The area of high heat flow on the Fiji Plateau (Fig. 3) is confined between two arc systems, and the Benioff zones dip inward beneath the plateau.

Seismic observations have also provided other information on the deep structure of island arc areas. Cleary and Hales (1967) have shown that there is an anomalous station term associated with earthquakes received in the Aleutians, and Cleary (1967) has demonstrated an azimuth-dependent source term for phenomena originating there. Cleary concludes that these features can be explained by the presence of an inclined slab of material with high seismic velocity, dipping northward under the island arc parallel to the Benioff zone. It follows from the sea-floor spreading hypothesis that this should be the form and attitude of a cold slab of crust and coupled upper mantle, as it slides down beneath the island arc. The abnormally high velocity associated with the slab can probably be attributed to its low temperature (*see* below), although the seismic anisotropy proposed by Cleary could be of local importance.

Additional evidence for the existence of an inclined zone of cool material dipping beneath the island arcs comes from the work of Oliver and Isacks (1967). These authors argue for the existence of a zone of high-Q (that is, low attenuation of seismic waves) beneath the Tonga-Kermadec arc in order to explain an anomalous pattern of body-wave amplitudes. Since Q is strongly temperature dependent, a zone of high-Q indicates the presence of cold material. The high-Q slab appears to *underlie* the Benioff zone.

Evidence that this cold slab-like zone is developed by the underriding process outlined above comes also from the study of focal mechanisms of shallow earthquakes. These studies (Aki, 1966; Stauder, 1968) indicate a relative downward motion of the oceanward block relative to the island arc block. Earlier proposals that the main movement on the Benioff zone around the Pacific is strike-slip in character have been abandoned.

MODEL FOR ISLAND ARC–OCEAN TRENCH SYSTEMS

On the basis of the evidence presented above, we propose the two-dimensional model for island arc-ocean trench systems shown in Figure 4. On the oceanward side of the trench, we have typical oceanic crust and upper mantle. The cool oceanic crust and upper mantle above the seismic low-velocity zone are referred to as the lithosphere. The thickness of the lithosphere is taken to be the thickness of the cold conduction boundary layer formed by the cooling of the hot mantle material ascending to the surface at the mid-ocean ridge (Oxburgh and Turcotte 1968a; Turcotte and Oxburgh, 1969). An estimate for the thickness of the lithosphere δ_1 can be obtained from the measured oceanic heat flux q_0, the temperature difference through the lithosphere ΔT, and the thermal conductivity of the ocean crust and upper mantle k. Assuming $q_0 = 1 \ \mu \ \text{cal/cm}^2 \ \text{sec}$, $\Delta T = 1400 \ °C$, and $k = 0.007 \ \text{cal/cm sec} \ °K$ we find

$$\delta_1 = \frac{k \Delta T}{q_0} = \frac{0.007 \times 1400}{10^{-6}} = 10^7 \ \text{cm} = 100 \ \text{km}$$

It also follows from the sea-floor spreading hypothesis that the uppermost mantle (the mantle part of the lithosphere) has undergone

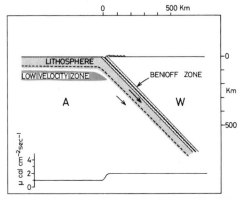

Figure 4. Sectional diagram of the relationships assumed between the various tectonic elements in an island arc area. The trench and arc are shown at the surface, and the heavy line lies along the center of the Benioff zone. Heat flow across the arc is shown at the base.

partial fusion and has been depleted of its low-melting basaltic fraction when it was in the vicinity of the oceanic ridge (Oxburgh and Turcotte, 1968a). We thus envisage that the lithosphere is composed of the usual three-layered oceanic crust (about 5 km thick) overlying nearly 100 km of peridotitic upper mantle, depleted in basalt. We regard the lithosphere as decoupled from the deeper mantle at the low-velocity zone.

At the ocean trench the cold lithosphere slides under a passive and inert wedge-shaped mass (W *in* Fig. 4). The zone of sliding is taken as the Benioff zone. The lithosphere moves downward because it is cool and, therefore, dense. A rough estimate for the shear stress that the lithosphere can exert, τ_c, is obtained by equating the shear stress to the gravitational body force on the descending lithosphere. This balance gives

$$\tau_c = \tfrac{1}{2}\,\beta \rho \delta_1 \Delta T\, g \sin \theta\,,$$

where β is the coefficient of thermal expansion, ρ the density, θ the angle of inclination of the Benioff zone, and g the acceleration of gravity. Taking $\beta = 3 \times 10^{-5}\ {}^\circ K^{-1}$, $\rho = 3.3$ gm/cm^3, $\theta = 45°$, and g $= 980$ cm/sec^2, we find

$$\tau_c = \tfrac{1}{2} \times 3 \times 10^{-5} \times 3.3 \times 10^7 \times 1400$$
$$\times 980 \times 0.7 = 5 \times 10^8 \text{ dynes/cm}^2$$
$$= 500 \text{ bars}\,,$$

which is in order of magnitude agreement with the shear stress associated with major earthquakes on the Benioff zones. The lithosphere is heated as it descends and eventually becomes physically indistinguishable from the surrounding mantle. The descending lithosphere is believed to retain its identity to depths of at least 700 km; this is the depth of the deepest earthquakes on the Benioff zone and roughly corresponds by projection to the limit of the zone of high surface heat flow.

In terms of mantle convection, the lithosphere beneath the Benioff zone corresponds to the descending flow just as the rising, spreading flow associated with ocean ridges corresponds to the ascending flow. The body forces associated with the ascending and descending flow drive a counter flow at depth. Such a counter flow in the mantle is necessary to provide the required mass balance. Clearly the rigidity and brittleness of the lithosphere have a strong influence on the over-all flow pattern.

HEATING BY CONDUCTION

We now examine possible explanations for the presence of volcanism and high heat flow in island arc areas. It is clear that as the lithosphere descends into the mantle, it will be heated by conduction. The temperature of the descending lithosphere, however, will always be lower than that of the adjacent mantle if conduction is the only heating process. Thus, the lithosphere would not melt unless its melting temperature were lower than that of the adjacent mantle.

The crustal part of the lithosphere is different in composition from the adjacent static block of mantle. Layer 1 of the oceanic crust is composed of unconsolidated sediments; these may either be carried down with the lithosphere or scraped off at the surface and added to the front of the static block; or a combination of these processes may occur. Any such material that was carried down, however, would be closest to the fault and would be both siliceous and/or calcareous in composition and have a high entrapped water content. As such material is heated, it should procede through a series of low-grade metamorphic reactions and could melt at temperatures as low as 700° C (Winkler, 1967; Boettcher and Wyllie, 1967). Basalt, greenschist, or amphibolite and any sedimentary material of layers 2 and 3 of the oceanic crust would also melt at lower temperatures than peridotitic material at the same pressure. Thus, it appears possible that conductive heating could produce magmas by the partial fusion of oceanic sediments, and basalts or their hydrated equivalents.

This conclusion does, however, presuppose a steady thermal state in the static block above the lithosphere; although this is a limiting solution to the problem, it is an unreasonable assumption. The wedge-shaped static slab is a thermal reservoir of finite size—the only heat that it receives is from radioactivity within it and from lateral heat conduction to the base of the wedge, that is, from the right-hand side of Figure 4. Clearly the heat loss to both the upper surface and the descending lithosphere would significantly cool the static block. One result of this cooling would be a reduction in the heat flux to the surface, the opposite of the observed increase in the surface heat flux behind island arcs. Once movement had started, the heat loss from the static wedge to the cold lithosphere sliding beneath it should rapidly lower temperatures along the Benioff zone so that production of magmas would be impossible after the first few million years. We conclude that the observed continuing thermal phenomena in island arc-ocean trench systems cannot be explained by simple conductive heating.

HEATING DUE TO RADIOACTIVITY

As already discussed, the mantle part of the lithosphere should have been depleted of its low-melting fraction that would be expected to contain most of the original uranium, thorium, and potassium. The lower lithosphere should be virtually free of radioactivity, therefore, and have a composition close to that of the oceanic lherzolites, analyzed by Wakita and others (1967a, 1967b), with a low-volume heat release of near 8×10^{-16} cal/gm sec.

However, there could be heating due to the radioactive elements in the crustal part of the descending lithosphere. If the oceanic crust has the composition of basalt, then a volume heat release of 10^{-13} cal/gm sec should be a good approximation. Assuming no loss of heat by conduction, an increase in temperature at the rate of $10°$ K/m.y. can be expected. At a velocity of 10 cm/year, the lithosphere moves 500 km along the fault zone in 5 m.y., and has a temperature increase of $50°$ K. Even though the downward velocity may be only a half or a third of the value taken and though any oceanic sediments that were carried down along the fault zone could have three times the heat release of basalt, it appears that the heating of the lithosphere by radioactive decay is insignificant.

HEATING DUE TO PHASE CHANGES

The possibility also exists that some heat could be evolved by exothermic phase changes occurring in the descending lithosphere. The work of Green and Ringwood (1967a) and MacGregor (1968) shows that to pressures of ~50 kb the variation in phase assemblage for ultramafic compositions is controlled largely by recombination of alumina in various ways, that is, under different circumstances in plagioclase, in spinel, in orthopyroxene, in clinopyroxene, and in garnet. The mantle rocks of the lithosphere should, however, have less than 1 per cent Al_2O_3 due to removal of the low-melting fraction and should be composed largely of olivine and orthopyroxene. These phases should be stable to depths of at least several hundred kilometers and in the near absence of alumina no significant changes of phase are to be expected.

If the oceanic crust is largely of basaltic composition, it may undergo a series of changes of phase assemblage at depths of less than 50 km. The experimental studies of Green and Ringwood (1967c) suggest that anhydrous basalt is unstable under conditions typical of the oceanic crust where inversion to the stable eclogitic (garnet-pyroxene) assemblage is inhibited presumably by the slow reaction rates at low crustal temperatures. Taking a mean value of 21 bars/° K for the slope of the equilibrium curve between the gabbro and eclogite stability fields (in reality a number of different reactions occur over different pressure-temperature ranges) and a value of 3 cm³ for the reduction in molar volume, the phase change should give off a heat of transformation of about 13 cal/gm. Assuming no conductive heat losses, this heat gives a temperature increase within the descending oceanic crust of about 40° K due to the phase change. This temperature rise probably represents the maximum to be expected from changes in phase; whether this phase change occurs or not depends upon the precise way in which the temperature varies with pressure in the descending material close to the fault and the partial pressure of water in the system. In the presence of water, hydrous phases (such as amphiboles) are stable below the liquidus at pressures to 25 kb, and the gabbro-eclogite transition would not occur. Phase changes in other materials such as the sedimentary components of layers 1 and 2 of the oceanic crust should have relatively small effects, and it is concluded that heats of transformation are unlikely to make any significant contribution to raising the temperature of the lithosphere.

FRICTIONAL HEATING—THE FAULT ZONE

It has been shown elsewhere (Oxburgh and Turcotte, 1968b; Turcotte and Oxburgh, 1968; McKenzie and Sclater, 1968) that frictional heating associated with slippage along the Benioff zone could be sufficient to account for the observed heat flow behind the island arcs of the northwestern Pacific. We consider this question in detail below; for the present we assume that this is the heating mechanism and examine some of the features of the fault zone (Benioff zone).

Volcanologists working in active island arc areas have long recognized a spatial and temporal relationship between earthquakes and surface volcanic eruptions. It has been commonly suggested that the Benioff zone is itself the source of the magmas (*see* Kuno, 1959, 1965; Coats, 1962, Dickinson, 1968). Both Kuno and Dickinson have correlated the lateral variation in the composition of surface extrusives in a direction normal to the arc, with the progressively greater depth of the Benioff zone in that direction. Beyond this, however, there is little direct evidence that the Benioff zone is the source of the magmas. Nevertheless, the proposal is, subject to certain limitations, entirely plausible, and we accept it in the subsequent discussions.

For the thermal model of island arc structure, the most important implication of this assumption is that it allows the temperature along the fault zone to be prescribed. Near the surface, the fault would have the characteristics of a brittle fracture; heating should occur by the grinding together of the opposite sides with fracturing and granulation of material. As the temperature on the fault zone rose, creep processes would become important and ultimately there would be a transition to the viscous heating process analyzed by Turcotte and Oxburgh (1968). The nature of the fault zone, however, will be somewhat modified where the temperatures are high enough for melting to begin; whether this occurs within the zone of frictional granulation or hot creep, it appears certain that melting will locally reduce friction along the zone of movement and the production of heat will diminish. We thus visualize an unsteady process operating along the fault zone; within the depth range for formation of magmas (say 20–200 km), heating should occur by dissipation until melting takes place; subsequently, that part of the fault zone is lubricated by interstitial liquid, and there will be little heating until the liquid is removed. It is also possible to imagine a steady process in which the magmas are drained continuously from the fault zone as they are generated; the sporadic seismic and volcanic activity on the Benioff zones, however, suggest that this is not the case. The unsteady process outlined above is similar to the unsteady stick-slip, shear melting process proposed by Griggs and Baker (1968) to explain seismicity on Benioff zones.

In three dimensions the Benioff zone may be regarded as a hot tabular zone containing irregularly distributed patches of melt at any instant in time. Motion in the zone may be caterpillar-like, with not all parts of one side moving simultaneously. As some patches of

melt drain to the surface or to intermediate magma chambers, others form elsewhere.

In the deeper part of the Benioff zone, it appears that no magmas are formed; at any rate there is little surface evidence to support their existence. With increasing depth, the thickness of the heated zone increases by lateral heat conduction; as a result, differential motion should be distributed over a wider zone with a consequent reduction in velocity gradients and heat production. Since temperatures of beginning of melting increase with depth, the absence of magmas is not surprising.

The hot-creep phenomena that precede melting on the fault zone, in general, operate at rates that are strongly dependent on T/T_m where T is the absolute temperature of the material and T_m its melting temperature. Thus, as the temperature rises, the width of the zone within which movement occurs should be influenced by the compositional layering of the upper lithosphere, with most of the movement occurring initially within layers 1 and possibly 2 of the old oceanic crust, as these would probably have a lower melting temperature than layer 3 as shown in Figure 5. With increasing depth and the removal of the low melting fractions from layers 1 and 2, the

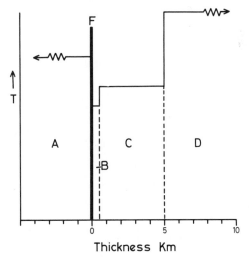

Thickness Km

Figure 5. Diagrammatic representation of temperature of beginning of melting as a function of distance normal to the fault zone (η, Fig. 6); A, peridotite of the stationary wedge; F, the center of the fault zone; B, siliceous and water-rich sedimentary material (parts of layers 1 and 2); C, basalt and amphibolite of layers 2 and 3; D, basalt-depleted peridotite. B, C, and D comprise the upper part of the descending lithosphere and the old oceanic M discontinuity occurs between C and D.

movement could begin to affect layer 3. At any depth, however, the temperature on the fault zone is either below the melting temperature or is buffered at the melting temperature of the most fusible material remaining.

In the model analyzed in the next section, we set the temperature on the fault at 20-km depth to be 700° C and at 200 km, to be 1400° C. Both temperatures are subject to large uncertainties. We assume that at 20 km entrapped and combined water in the uppermost lithosphere will have a major effect in lowering melt temperatures and that the lowest melting component (sediment, with or without amphibolite) will just be starting to melt. At 200 km the influence of water should be much less; nearly all the water in the upper lithosphere should have entered earlier melts. The selected temperature represents an extrapolation of the experimental results of Green and Ringwood (1966, 1967b) on andesite liquidus temperatures, with a subjective allowance for the presence of some water. We thus assume with O'Hara (1965), Dickinson (1968), and others that the important members of the calc-alkaline suite are formed directly by some melting process within the mantle (but not necessarily of ultramafic materials). The model would not be greatly affected by increasing the temperature at 200 km by several hundred degrees to a value appropriate for basalt generation.

In some places, Japan, for example, there is occasional volcanic activity above the deeper parts of the Benioff zone at distances of 500 or 600 km behind the island arc. Ringwood (*in* Kuno, 1965) has pointed out that the compositions of the lavas are such that they appear to have been in equilibrium with crystal residues at depths not greater than 200 km. Their temperatures of eruption confirm this observation. It is thus unlikely that these volcanics result from fusion on the deeper part of the fault zone; we return to this problem later.

FRICTIONAL HEATING—A THERMAL BALANCE

Frictional heating along the Benioff zone should certainly be a source of heat. An analysis will be given here to determine the amount of heat required to give the observed thermal phenomena. Our first assumption is that a steady-state heat balance is applicable—in the static wedge above the Benioff zone, the temperature distribution does not change with

time. The heat that is produced in the Benioff zone must be conducted away, either downward into the descending cool lithosphere or upward into the static wedge, to be lost at the surface. Some heat is also lost by upward magmatic transport, but this point will not be considered here.

In the two-dimensional thermal model, we represent the Benioff zone as a line heat source with the source strength a function of depth. The Benioff zone is inclined to the surface at an angle θ as shown in Figure 6. We assume that the descending lithosphere is moving along the Benioff zone at a constant velocity u_0. The heat conducted from the Benioff zone into the descending lithosphere forms a thin, warm thermal boundary layer in the lithosphere adjacent to the Benioff zone. If the heat flux away from the Benioff zone and the thermal properties of the lithosphere are known, the temperature distribution in this thermal boundary layer can be determined. Or, alternatively, if the temperatures on the fault zone are known, the corresponding heat flux into the lithosphere can be determined. We will consider the latter problem because, as discussed above, the volcanism permits reasonable estimates for the temperatures on the Benioff zone, whereas we have relatively little knowledge of the heat generated along it, except that it must be sufficient to heat the upper edge of the lithosphere to the specified temperatures.

For the stationary wedge above the fault, we neglect heat production due to radioactive decay within the wedge, and the temperature

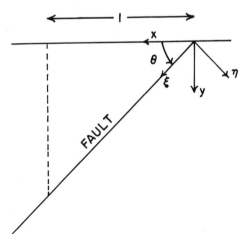

Figure 6. The coordinate system used in the analysis; *see* **text.**

distribution within it is determined from a simple conduction solution. The temperature on the Benioff zone again serves as a boundary condition that determines the heat flux into the wedge and the resultant heat flux at the surface.

TEMPERATURE DISTRIBUTION IN THE DESCENDING LITHOSPHERE

The lithosphere is probably composed of typical oceanic crust and upper mantle material depleted of its basalt. When the lithosphere reaches an ocean trench, its surface temperature is taken to be $0°$ C with a near-surface temperature gradient of $14.3°$ C/km. This gradient corresponds to a surface heat flux of 1 μ cal/cm^2 sec and a thermal conductivity of 0.7×10^{-2} cal/cm sec$°$ K. The cold descending lithosphere is assumed to be moving as a slab with a velocity u_0.

The heat produced by frictional heating on the Benioff zone will heat the upper part of the lithosphere. The governing equation for the conduction-convection, heat-transfer problem is

$$u_0 \frac{\delta T}{\delta \xi} = \kappa \left(\frac{\delta^2 T}{\delta \xi^2} + \frac{\delta^2 T}{\delta \eta^2} \right), \tag{1}$$

where κ is the thermal diffusivity, and the coordinate system is shown in Figure 6. However, it is appropriate to make the boundary layer approximation, $\delta^2/\delta \eta^2 \gg \delta^2/\delta \xi^2$ and (1) reduces to

$$u_0 \frac{\delta T}{\delta \xi} = \kappa \frac{\delta^2 T}{\delta \eta^2}. \tag{2}$$

The solution of this equation for a specified heat flux at $\eta = 0$ with $T = 0$ at $\xi = 0$, and $T \to 0$ as $\eta \to \infty$ has been given by Carslaw and Jaeger (1947) and is

$$T = \left(\frac{1}{\pi \rho u_0 c_p k} \right)^{1/2} \int_0^\xi q(\xi - \xi')$$
$$\times \exp\left(-\frac{\eta^2 u_0}{4\kappa \xi'} \right) \frac{d\xi'}{\xi'^{1/2}}, \tag{3}$$

where $q(\xi)$ is the heat flux to the convecting slab as a function of the distance along the fault zone. The corresponding value of the temperature on the fault zone as a function of ξ is given by

$$T_{(\eta=0)} = \left(\frac{1}{\pi \rho u_0 c_p k} \right)^{1/2} \int_0^\xi q(\xi - \xi') \frac{d\xi'}{\xi'^{1/2}}. \tag{4}$$

As the governing equation for the temperature is linear, solutions may be added to give new solutions. One result is that the boundary layer profiles given above may simply be added to the linear temperature profile initially present in the descending lithosphere.

It is expected that the heat lost from the fault zone should decrease with depth. As the temperature along the fault zone increases, the resistance to movement should decrease and with it, the heat produced by friction. As an analytic dependence for the heat flux into the thermal boundary layer, we choose

$$q = \frac{q_0}{\left(1 + \dfrac{\xi}{\xi_0}\right)^{1/2}}. \tag{5}$$

From (4) the corresponding temperature on the fault zone is

$$T = 2q_0 \left(\frac{\xi_0}{\pi \rho u_0 c_p k} \right)^{1/2} \sin^{-1}\left[\frac{(\xi/\xi_0)^{1/2}}{(1 + \xi/\xi_0)^{1/2}} \right]. \tag{6}$$

While the choice of this functional form is quite arbitrary, it is a dependence that gives a finite temperature on the fault zone for large ξ and is the strongest dependence on depth that does not give a temperature maximum on the fault zone.

The two constants q_0 and ξ_0 in (5) may be determined by specifying the temperature on the fault zone at two depths. As previously discussed, we require that $T = 700°$ C at a depth of 20 km, and $T = 1400°$ C at a depth of 200 km. We also assume that $\theta = 45°$; $u_0 = 9$ cm/yr (2.86×10^{-7} cm/sec); $\kappa = 8 \times 10^{-3}$ cm^2/sec; $k = 7 \times 10^{-3}$ cal/cm sec $°$K; and $c_p = 0.27$ cal/gm $°$ K. The values of the constants in (5) that satisfy these conditions are $\xi_0 = 76.5$ km and $q_0 = 17.4$ cal/cm^2 sec. This solution is added to the linear temperature distribution initially present in the lithosphere to give the temperature in the thermal boundary layer. The dependence of the heat flux into the thermal boundary layer on ξ is given in Figure 7. It should be emphasized that this is the heat loss from the fault zone downward into the convecting material. The heat loss upward must be determined independently, and this will be done in the next section. The dependence of the temperature at the fault zone ($\eta = 0$) on ξ is given in Figure 8. And the temperature distribution in the thermal boundary layer is given in Figure 9. It is seen that the minimum temperature in the descending lithosphere increases slowly with depth, reaching $400°$ C at a depth of 200 km. Even at this depth, the thickness of the thermal boundary layer is small, about 15 km. The

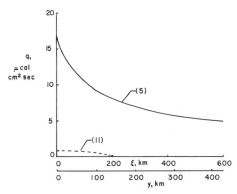

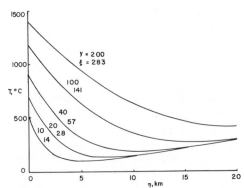

Figure 7. Calculated heat flux from the Benioff zone as a function of depth into the descending lithosphere, curve (5), and into the stationary wedge, curve (11); nearly all the heat generated along the fault goes into the heating of the lithosphere.

Figure 9. Temperature profiles in the descending lithosphere for various depths (y) in km.

analysis in this section gives the complete temperature distribution in the descending lithosphere.

TEMPERATURE DISTRIBUTION IN THE STATIC BLOCK

We assume that the wedge-shaped region above the fault zone is stationary. We also assume that the temperature distribution has reached a steady state. The applicable differential equation for the temperature is

$$\frac{\delta^2 T}{\delta x^2} + \frac{\delta^2 T}{\delta y^2} = 0 . \qquad (7)$$

In order to prescribe a solution to (7) within a closed region, it is necessary to give either the

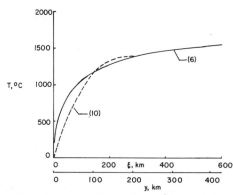

Figure 8. Assumed dependence of temperature on depth at the upper side of the descending lithosphere (6) and at the lower side of the stationary wedge (10). For an exact solution, these curves should coincide. For discussion, see text.

temperature or the heat flux on the boundaries of the region.

McKenzie and Sclater (1968) have considered this problem and have represented the fault zone by a line source of heat of constant strength with an image heat sink. Their line source extended from a depth y_1 to a depth y_2. Several objections can be made to this solution. As the authors point out, the temperatures at depth are unreasonably high. Also, a conduction solution is found on both sides of the trench. As shown above, a convection-conduction solution is required in the moving lithosphere; such a solution is parabolic, and there is no upstream influence. Therefore, the heat produced on the fault zone will not affect the heat flux on the oceanward side of the ocean trench. The solution given by McKenzie and Sclater also gives infinite temperatures at the ends of the line source, and for this reason the upper end of their line source cannot be extended to the surface in order to represent the actual physical problem. We expect that the heat produced on the Benioff zone will be a maximum at the intersection of the Benioff zone and the ocean floor. For these reasons we obtain a solution to (7) using the separation of variables method.

As in the previous section, we assume that the fault zone is inclined at an angle of 45°. We will obtain a solution within a triangular section of the wedge with a horizontal dimension l as shown in Figure 6. We require the boundary conditions $T = 0$ at $y = 0$, and $\delta T/\delta x = 0$ at $x = l$. The latter boundary condition corresponds to zero heat flux into the wedge from its base, that is, from its left-hand side. Although this boundary condition is somewhat arbitrary, it corresponds reasonably well with the physical problem considered. The general separation of variables solution of (7) that satisfies the above boundary conditions is

$$T = cy + \sum_{n=1,3,\cdots} a_n \sin\left(\frac{n\pi x}{2l}\right) \sinh\left(\frac{n\pi y}{2l}\right)$$
$$+ \sum_{m=2,4,\cdots} b_m \cos\left(\frac{m\pi x}{2l}\right) \sinh\left(\frac{m\pi y}{2l}\right). \quad (8)$$

When the temperature is fully prescribed along the fault zone side of the triangle the coefficients a_n, b_m, and c can be determined to give the complete temperature distribution.

As our knowledge of the temperature distribution on the fault zone is limited, we will keep only the first term in the infinite series expansion given in (8) with the result that

$$T = cy - a_1 \sin\left(\frac{\pi x}{2l}\right) \sinh\left(\frac{\pi y}{2l}\right). \quad (9)$$

In order to determine the corresponding temperature distribution on the fault zone, we replace the variables x and y with the variables ξ and η ($x = 2^{-1/2}(\xi - \eta)$, $y = 2^{-1/2}(\xi + \eta)$), and on the fault zone ($\eta = 0$), we obtain

$$T = \frac{c}{2^{1/2}}\xi - a_1 \sin\left(\frac{\pi\xi}{2^{3/2}l}\right) \sinh\left(\frac{\pi\xi}{2^{3/2}l}\right). \quad (10)$$

The corresponding heat flux from the fault zone into the stationary slab is given by

$$q = k\left(\frac{\delta T}{\delta \eta}\right)_{\eta=0}$$
$$= k\left[\frac{c}{2^{1/2}} - \frac{\pi a_1}{2^{3/2}l}\left\{\sin\left(\frac{\pi\xi}{2^{3/2}l}\right)\cosh\left(\frac{\pi\xi}{2^{3/2}l}\right)\right.\right.$$
$$\left.\left. - \cos\left(\frac{\pi\xi}{2^{3/2}l}\right)\sinh\left(\frac{\pi\xi}{2^{3/2}l}\right)\right\}\right]. \quad (11)$$

And the heat flux from the slab to the surface is given by

$$q = k\left(\frac{\delta T}{\delta y}\right)_{y=0} = k\left[c - \frac{\pi a_1}{2l}\sin\left(\frac{\pi x}{2l}\right)\right]. \quad (12)$$

Once the dimension of the slab l has been given, the two constants a_1 and c can be prescribed in principle by giving the temperature at two depths on the fault zone. However, if we require $T = 700°$ C at $y = 20$ km, and $T = 1400°$ C at $y = 200$ km, as in the previous section, we find that the temperature distribution has a strong maximum that is physically unrealistic. We, therefore, require $T = 1400°$ C at $y = 200$ km and take the limiting profile that does not exhibit a temperature maximum. The constants we obtain are $l = 200$ km,

$c = 16.8°$ K/km, and $a_1 = 854°$ K. The resulting temperature distribution on the fault zone is given in Figure 8. It should be emphasized that the temperature profiles on the fault zone for the two solutions, the descending lithosphere and the stationary wedge, could be made to coincide exactly if more terms were retained in the series expansion given in (8). Considering the many uncertainties, the agreement given in Figure 8 should be adequate for computational purposes.

The heat transfer from the Benioff zone into the stationary wedge, as found from (11), is given in Figure 7. It is seen that the heat transferred downward into the descending lithosphere is an order of magnitude higher than the heat transferred upward into the stationary wedge; much more heat is required to heat the descending lithosphere than to maintain the temperature of the stationary wedge.

The surface heat flux as a function of the distance from the ocean trench given by (12) is plotted in Figure 10. It is seen that the predicted heat flux is somewhat lower than the observed values behind island arcs (an average of 1.56 μ cal/cm² sec behind the Japanese arc, Vaquier and others, 1966, and an average of 2.43 μ cal/cm² sec on the Fiji plateau, Sclater and Menard, 1967). In our analysis the heat release due to radioactivity in the stationary wedge has not been considered.

Clearly it would be of interest to extend the wedge solution beyond the 200-km scale considered here. However, the variation of the thermal conductivity with depth and the

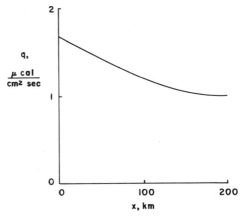

Figure 10. Predicted heat flow to the surface behind the island arc as a function of distance normal to the trench.

uncertainties regarding boundary conditions, magmatic transport, and volume heat release would make such a solution of questionable value. Because of these uncertainties, we believe the wedge solution proposed here may be in error by a considerable amount. As none of these sources of error should seriously affect the conduction-convection solution for the descending lithosphere, this solution should not be subject to such uncertainties.

STRESS ON THE FAULT ZONE

The heat dissipated on the fault zone, q, is related to the slip velocity u and the shear stress τ by

$$q = u\tau. \qquad (13)$$

This relation is valid for any type of creep process; it should also be valid when the slip is due to granulation and small-scale fracturing. The validity of (13) when a significant fraction of the motion is due to the large-scale fracturing associated with major earthquakes is open to question. The energy balance used in writing (13) does not take account of the energy radiated by an earthquake. However King and Knopoff (1968) have shown that in most earthquakes the stress drop due to the earthquake is a small fraction of the original stress. Also, Brune and Allen (1967) have concluded that only a fraction of the movement on a fault is accounted for by measured seismic activity. We assume, therefore, that the energy radiated in seismic waves represents a small fraction of the energy dissipated on the fault zone and accept (13) as the appropriate energy balance.

Taking the slip velocity to be 9 cm/year and the heat generated from Figure 7, we find from (13) that the required shear stress at the surface is 2.52×10^9 dyne/cm² (2520 bars); at a depth of 100 km it is 1.61×10^9 dyne/cm² (1610 bars), and at a depth of 200 km it is 1.17×10^9 dyne/cm² (1170 bars). These values for the shear stress are somewhat high; however, Bullen (1963) has stated that "the strength of the upper mantle cannot be much less than 10^9 dyne/cm²." While the distribution of stress with depth is dependent on the assumed variation of heat generation with depth (5), the magnitude of the stress is clearly necessary to heat the descending lithosphere to the temperatures required for the production of magmas.

The energy radiated in seismic waves has been used by many authors to determine the

stress drop and stress level associated with that particular earthquake. Brune (1968) has concluded that for a magnitude 8 earthquake the stress level is 156 bars; for a magnitude 7 earthquake the stress level is 34 bars; and for a magnitude 3 earthquake, 1 bar. It is, however, assumed that all the available energy is radiated in the form of seismic waves. Stress drops associated with a wide range of earthquakes have been tabulated by Brune and Allen (1967). The largest stress drops given are 355 bars in the Montana earthquake of 1959 (magnitude 7.1), 180 bars in the Fairview Peak earthquake of 1954 (magnitude 7.1), and 106 bars in the Niigata earthquake of 1964 (magnitude 7.5). However, King and Knopoff (1968) have concluded that, except for the largest earthquakes, the stress drop may be only a small fraction of the actual stress acting across the fault zone. Also Bullen (1963) attributes the lower values of the stress inferred from seismic observations to plastic creep prior to fracture.

Wyss and Brune (1969) have estimated apparent shear stress for earthquakes associated with oceanic trenches by using the ratio of seismic energy to seismic moment. They conclude that apparent shear stress, which should provide a lower limit to the actual average stress, rises to about 1 kb at a depth of 100 km. From this maximum, it diminishes both upward and downward to a few tens of bars near the surface and at great depth.

GRAVITATIONAL EFFECTS OF THE PROPOSED STRUCTURE

As discussed earlier, island arc systems are characterized by strong gravity anomalies (Fig. 11a). It is evident from the numerous

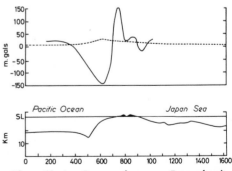

Figure 11. A. Cross section across Japan showing topography and (above) observed free-air anomaly (solid line) and computed effect of cold slab (dashed line).

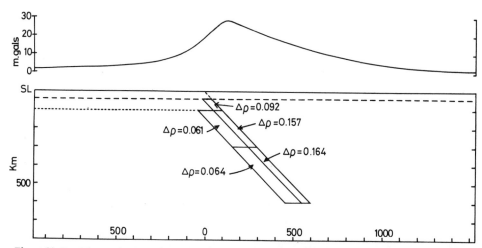

Figure 11. B. The gravity anomaly associated with the cold descending lithosphere deeper than 40 km. The model structure computed is shown below; density contrasts ($\Delta\rho$) are given in gm/cm³. Differences in density resulting from differences in thermal gradient on either side away from the slab have been ignored.

published models for island arc structure that these anomaly patterns may be satisfied by solutions that involve anomalous mass distribution no deeper than a few tens of kilometers. The present model predicts the persistence of temperature deficiencies and mass excesses to depths of hundreds of kilometers within the mantle; thus, it is interesting to examine the contribution of the cold descending lithosphere to the gravity field measured at sea level.

Density differences associated with the lithosphere below 40 km have been represented by the simple model shown in Figure 11b, and their contribution to the sea-level gravity field is shown at the top of the figure. Gravitational effects associated with the difference in geothermal gradients behind and in front of the arc have been ignored.

The anomaly had a maximum value of about +30 mgal and is of long wave length; it bears no obvious relationship to the pattern of anomalies recognized in island arc areas (Fig. 11a). These anomalies are, however, very much larger—of the order of 200 mgal, and the gravitational effects of the cold descending lithosphere could be masked by much larger near-surface density differences. One of the most important density differences could be associated with the piling up of sedimentary material from the ocean floor against the front of the arc. This hypothesis could explain the lack of correspondence between the gravity minimum and the trench axis, and displacement of the axis toward the arc where the low-

density material should be accumulating (*see* discussion in a later section).

MAGMAS IN ISLAND ARCS

We now consider some of the petrogenetic implications of the model for ocean trench–island arc systems given in this paper. It is not our aim to consider in detail the complex problems of island arc magmatic activity in this paper, but we wish to explore various magma-forming possibilities suggested by the model. In the following discussion we assume that the most abundant members of the calc-alkaline suite, the andesites and the basaltic andesites, are primary liquids, that is, they represent relatively unmodified liquids produced by partial melting except insofar as any liquid ascending through the upper mantle to the surface must equilibrate to a greater or lesser degree with olivine. As pointed out by many authors (for example, Green and Ringwood, 1967b; Dickinson, 1968), it appears that continental crust is not necessary for the development of the calc-alkaline series, and the origin by sialic contamination as proposed by Kuno (1959) is not universally applicable.

In the model outlined earlier, the zone of partial fusion gradually eats into the sinking lithosphere as its temperature rises and as the lowest melting fractions are removed. This means that unless relative melting temperatures are reversed at very high pressures of water, the first liquids must be produced from any siliceous, calcareous, or pelitic sediment that adheres to the upper surface of the lithosphere. At greater depths the lithosphere should be sufficiently heated for the partial or complete fusion of the oceanic crust. Thus, not only do the temperatures and pressures under which melting occurs vary continuously down the fault zone, but also the effective bulk composition of the melting system changes. This allows the generation of liquids whose composition varies with depth as observed by Kuno (1959) and by Dickinson and Hatherton (1967).

At shallow depths, in particular, the melting process is too complicated for its detailed behaviour to be predicted. As an illustration, we consider the possible ways in which the basaltic component of the lithosphere may melt. Figure 12 shows schematically the melting behaviour of basalt both for dry conditions and with the partial pressure of water equal to the load pressure; conditions in nature will vary within these limits and will be a function of depth. The heavy lines with arrows show the sequence of P/T conditions experienced by reference points within the descending lithosphere which lie 1, 2, and 5 km from the center of the fault zone for the model calculated. (These temperature/pressure paths would be different for a different rate of motion on the fault.) Clearly, at different distances from the fault, fusion takes place at different depths and liquids of different composition may be expected to be in equilibrium with crystal residues.

There is no direct way of estimating the de-

gree to which the descending lithosphere undergoes partial fusion. However, using observed rates of magmatic activity and movement on the fault, we can place a lower limit on the degree of fusion. In the case of the Japanese islands we can estimate, for a two-dimensional model, that crustal rocks are fed into the fault zone at a rate of 500 km²/m.y. (a 5-km crust sliding at 10 cm/yr). During the Quaternary, it is estimated that extrusives reached the surface at a rate of 2.5 km²/m.y. (5000 km³/m.y., Sugimura and others, 1963), distributed along the 2000-km-long Japanese arc. Even if only a small fraction of the liquids produced reach the surface, we must still conclude that very little of the crustal portion of the lithosphere is fused, probably less than 10 percent. This conclusion stands even if we take the much higher rate of volcanism estimated for the early Miocene; the rate of motion on the fault at that time was, however, probably different from that today (Le Pichon, 1968). We conclude, therefore, that although the model permits fusion under diverse conditions it suggests that melting is restricted to the crustal part of the lithosphere.

All gradations are found in island arc regions between dacites, andesites, basaltic andesites, and high-alumina basalts. This gradation may be interpreted either as a differentiation series, as a differential contamination series, or as a series resulting from continuously varying conditions of fusion and varying composition of

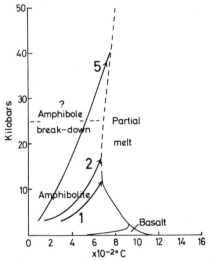 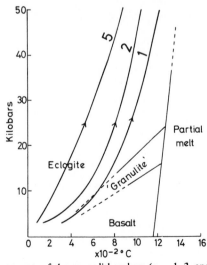

Figure 12. Calculated temperature/pressure paths for different parts of the upper lithosphere ($\eta = 1, 2,$ and 5 km) and their relation to the basaltic stability fields (left) $P_{H_2O} = P_{Total}$, (right) $P_{H_2O} = 0$. Phase relations extrapolated *from* Yoder and Tilley (1962) and Green and Ringwood (1966a).

the material fused. It seems very likely that all three mechanisms are operative to some extent and that any may assume a local importance. However, we shall be largely concerned with the last.

From a quantitative point of view, the role of sedimentary material in magma production is somewhat uncertain, insofar as it is not clear that it is carried down in significant quantities. Taylor and others (1969) emphasize the differences in trace-element abundances between the calc-alkaline suite and oceanic sediments and conclude that the latter can make no significant contribution to the genesis of the former. This may simply mean that little sediment is carried down; alternatively, it is not clear that the trace-element data for oceanic sediments, on which the conclusion is based, are truly representative. It is, in any case, to be expected that any sediment-derived siliceous liquid ascending through peridotitic upper mantle would react strongly with the country rock by dissolving olivine and, thus, undergoing a substantial modification in composition. Indeed, it is difficult to see how any liquid that reaches the surface by traversing a substantial thickness of upper mantle can fail to be in equilibrium with olivine, unless it either rises extremely rapidly or ascends by way of conduits, the walls of which have been coated by earlier eruptions.

Alternatively, Green and Ringwood (1967b) have shown that andesitic liquids may also be produced by partial fusion of amphibolite (for example, the 2-km path for $P_{H_2O} = P_{total}$ diagram *in* Fig. 12).

Although the andesitic members of the calc-alkaline suite are probably the most abundant, basaltic rocks are also important. At mid-ocean ridges, it appears that the basalts are formed by partial fusion of the peridotitic upper mantle (Oxburgh and Turcotte, 1968a). The upper mantle component of the lithosphere that descends along the Benioff zone has, however, already had some or all of its basaltic fraction removed and if, as seems plausible, the lherzolitic inclusions carried by some oceanic basalts are derived from it, it is unlikely to undergo fusion again to yield basalt.

There appear to be two alternative explanations for the production of magmas of broadly basaltic composition in this environment. Yoder and Tilley (1962) have proposed on the basis of their experimental work that, when a basaltic composition is represented by eclogitic mineralogy (that is, predominantly omphacitic

pyroxene and pyrope garnet), it is able to melt to give liquid fractions that have compositions similar to the bulk composition of the rock (basalt) that is melting. This behaviour, which is somewhat unusual in complex silicate systems, means that under suitable conditions a basalt may be derived by partial fusion of "basalt." If this is so, partial fusion of layer three of the oceanic crust at depths greater than 40 to 50 km (Fig. 12) may provide island arc basalts.

Alternatively, the peridotite of the static block above the Benioff zone may be a source of basalts. Depending upon its past history, this block may or may not have undergone the basaltic depletion characteristic of the mantle part of the lithosphere. If depletion has not occurred, basalt could be produced either by direct fusion in the vicinity of the fault zone, or, if the temperature of the fault zone is buffered at too low a temperature by the processes described above, there may be a rise, above the fault zone, of hot diapiric masses within the static block. Such diapirs could undergo partial fusion through pressure reduction as they rose. This suggestion is due to Ringwood (in the discussion *of* Sugimura, 1966). The static block would, however, ultimately be depleted of basalt; how rapidly this would occur is uncertain, but the estimates of Sugimura and others (1963) indicate that the time scale for depletion would be at least 5×10^8 yrs. Such diapiric movements within the "static wedge" would increase the surface heat flux behind the island arc, and could also be viewed in some sense as a local upward counter-flow, complimentary to the broad downward motion associated with the deeper parts of the fault zone.

We conclude our discussion of the magmatic aspects of the model with a comment on the *andesite line*. As originally envisaged, this was a line that circumscribed the Pacific and separated those areas characterized by igneous activity with continental affinities (calc-alkaline) from those typified by "oceanic" rock types, and, in this way, defined the "true" margin of the Pacific Ocean. The model, which we have adopted, suggests that in the sense outlined above the andesite line may not have any general significance. If the calc-alkaline suite is generated wherever there is a zone of descending mantle flow, the occurrence of these rocks need not necessarily provide a useful criterion for distinguishing "the true margins of the Pacific," or of any ocean, except insofar as the margins could arbitrarily be designated to coincide with the zones of descending flow. It does happen that around much of the Pacific the island arc systems are close to the margins of what, on the basis of crustal structure, may be termed continental .crust. However, in the case of the Tonga-Kermadec and New Caledonian Trenches, this is not so, and here the calc-alkaline suite is generated at large distances from any continental crust and without any simple relationship to any continental margin. It would be more useful to recognize an *andesite zone;* at any time this zone would correspond to regions of descending mantle flow and might coincidentally, or for mechanical reasons, also correspond to the margins of a major ocean basin. Calc-alkaline rocks of different ages, therefore, might be used as an indication of the locations of zones of descending flow at different times in the past.

THE GROWTH OF ISLAND ARCS

A number of authors (for example, Taylor, 1967; Taylor and White, 1965; Ringwood, 1969) have pointed out that calc-alkaline volcanism in island arc areas provided a means of generating continental crust in oceanic areas and of increasing the areas of the existing continental masses by the addition of material at their margins. It is of considerable importance to examine the rate at which material is added to the crust directly by means of volcanic activity. Using the results of Sugimura and others (1963), we find that volcanicity has increased the mean crustal thickness of Japan and its surrounding shelf area at rates ranging from 2 to 40 m per m.y. between the Miocene and the present with a total increase of about 290 m in 25 m.y. At this latter rate of volcanic addition, it would take about 3×10^9 yrs to develop the present Japanese crust, unless other factors have contributed to its growth.

The mass of island arcs could also be increased by mechanical addition of sedimentary material to that side of the arc that faces the trench. As previously discussed, it is not clear what happens to the low-density semiconsolidated and unconsolidated sediments of layer 1 of the oceanic crust when the lithosphere slides beneath the island arc system along the Benioff zone. It is difficult to see how all the sediments could accompany the descending lithosphere, although this may sometimes occur; the only alternative appears to be the accumulation of the greater part of

layer 1 at the near edge of the static block, causing a lateral growth of the island arc. Taking the sliding rate to be 10 cm/yr and a layer of sedimentary material 500 m thick, which is not carried down with the lithosphere, and assuming a subsequent compaction to 60 percent of initial volume, we find that the cross-sectional area of an arc above the M discontinuity could increase by 30 km²/m.y. The comparable rate of increase in the cross-sectional area given by the high early Miocene rate of volcanicity in Japan is 12 km²/m.y. Allowing for discontinuous motion on the fault, the variation in slip velocity along it, we take 15 km²/m.y. as an average value for sedimentary addition to the island arc and 3.5 km²/m.y. as an average value for volcanic addition. This value is the average for Japan over the last 25 m.y. and includes a probable period of reduced activity along the fault zone (Le Pichon, 1968). Taking the present cross-sectional area of Japan above the M discontinuity to be about 14×10^3 km², the present crustal mass could have accumulated in about 750 m.y. The oldest fossiliferous rocks in Japan are Silurian (Association for Geological Collaboration of Japan, 1965) and overlie an older metamorphic complex of uncertain age.

Thus, the two processes of sedimentary and magmatic addition appear to be able to account for the growth of the crust in the Japanese islands in a time that is geologically reasonable. If the sea floor spreading model is valid, the greater part of the sedimentary material of layers 1 and 2 of the oceanic crust must either directly or indirectly contribute to the building of the island arc: directly by mechanical addition, or indirectly by contributing to the magmas developed along the fault zone. A corollary of this model is that the crustal cross-sectional area of island arcs should be roughly proportional to their age, although this relationship could be strongly affected by variation in spreading rates.

It has been suggested (Taylor, 1967) that the process that occurs in island arcs today is the process by which the continental crust originally developed and has continuously grown. The rates given above are of the right order for this to be possible. However, for all the continental masses to have been generated by this means, earlier rates of movement must have been somewhat greater than during the Phanerozoic and perhaps zones of descending flow, more abundant. (The two are interdependent.) During the Phanerozoic Eon, there may have been growth of the Pacific margins of North America, but the amount is uncertain. In the northern and western Pacific, surprisingly old rocks are found in some of the island arcs; Coats (1956) reports a Palaeozoic flora from Adak Island in the central Aleutians; Permian is known from Okinawa in the Ryukyu arc. In the southwest Kurils the oldest fossiliferous beds are Senonian. Thus growth rates may have been rather variable.

REGIONAL METAMORPHISM IN ISLAND ARCS

We now consider the thermal structure of an arc that is growing by the physical application of unconsolidated sedimentary material to its leading edge. The rates discussed in the previous section indicate that in an environment of relatively rapid lithospheric destruction, oceanic sediments should pile up against the front of the arc at a rate that is too great for them to attain thermal equilibrium as they accumulate. This means that, although the sedimentary pile overlies the Benioff zone, thermal gradients within it may be extremely low and, depending upon its thickness, conditions appropriate for the glaucophane-schist facies of regional metamorphism could be attained in its lower part. It is worth noting that, although most of young glaucophane-schist belts of the Pacific margins contain a sparse recognizable fauna, in virtually none of them has it proved possible to establish a detailed regional stratigraphy because of the intensity of the internal deformation and dislocation (for example, the Franciscan Formation of California). Such disruption might be expected in sedimentary material that accumulated in the way suggested.

Around much of the Western Pacific margin, the young glaucophane-schist belts are paired on their inner side with higher temperature, lower pressure metamorphic facies belts of similar age (Miyashiro, 1961). This juxtaposition requires the existence of strong horizontal temperature gradients within the crust. Such gradients are to be expected in island arc situations; as soon as temperatures on the Benioff zone are high enough for partial melting, heat may be carried by magmas to the upper part of the crust and near-surface vertical temperature gradients could approach 100° K/km. There would be a lateral transition oceanward in a few tens of kilometers from this temperature/pressure environment to the glaucophane-schist environment described above.

A more thorough discussion of these questions will be presented elsewhere, but it is clear that although steady-state spreading allows the coexistence of two contrasting metamorphic environments close to each other, it does not provide for the preservation of parallel belts of rocks of contrasting metamorphic facies; on the contrary, with the continuous addition of material to the front of the arc, earlier glaucophane-schist zones should become part of the high temperature region. The solution to this difficulty is likely to lie in the sporadic and discontinuous nature of movements on the Benioff zone. This does not mean that spreading itself must be discontinuous, but changes in spreading direction appear to have taken place from time to time; any change in a spreading direction will result in a different component of resolved velocity down the Benioff zone. Heat generation on the zone is dependent upon this velocity and so at every major change in spreading pattern, it is to be expected that there will be a change in the thermal regime of arc systems on an ocean-wide and possibly even a worldwide basis. It is presumably to processes such as these that we must look for an explanation of the worldwide tectonic events recognized, if overemphasized, by Stille in the earlier part of this century.

RELATIONSHIPS BETWEEN ISLAND ARCS AND ARC MIGRATION

The southwest Pacific provides a number of special cases in island arc relationships. As discussed earlier the Fiji Plateau appears to be an area of oceanic crust that is bounded on opposite sides by trench and island arc systems. The relative motion of the adjacent ocean basins is toward the plateau, and the crust of the plateau should thus receive additions from below and from both sides. It is conceivable that the Fiji Plateau is an example of a continent in a very early stage of development.

The Philippine Basin is also bounded on both sides by active island arc systems, but there is an important difference between this region and the Fiji Plateau. Both the arc systems bounding the Philippine Basin, the Marianas-Bonin system on the east and the Ryukyu-Nansei Shoto-Philippine system on the west, face in the same direction—both have trenches to the east of the arcs and the Benioff planes dip westward. If both systems are active, and the evidence of volcanicity and seismicity suggests that they are, not only must the lithosphere east of the Marianas be moving

111

toward and into the Marianas Trench, but simultaneously the whole Marianas-Trench system with its associated pattern of descending flow must be moving relatively westward toward the Mindanao Trench. If this superficial interpretation of the situation is correct, it implies significant rates of mass transfer in the mantle underlying the Philippine Basin at depths of 500 or 600 km in order that mass be conserved.

Alternatively, the inner arc system is perhaps becoming inactive, and in the future the outer arc might be the locus of descending flow. The inner arc system is believed to be older than the outer (Hess, 1948) and contains rocks considerably older than any recognized in the outer. It would be necessary to assume that the crust of the Philippine Basin is no longer in motion toward the trenches to the west and that the residual volcanic and seismic activity is associated with the descending motion of cold lithosphere, which is no longer connected to the surface lithosphere, but which continues to sink due to body forces (*cf.* Isacks and others, 1968).

Another alternative explanation might invoke a complex system of transform faults breaking both inner and outer arcs into active and inactive segments. No plausible model for this alternative has been advanced.

Whether the Philippine arcs are approaching each other or not, it is a necessary consequence of the addition of sedimentary material to the front edge of an island arc system, and the outward growth of the arc, that the trench in front of the arc should migrate oceanward with time. Although it is clear that midocean ridges must be able to migrate if the ocean floor spreading hypothesis is valid, the migration of trenches with their deep structure extending to depths of 700 km or more, presents a considerable problem. However, the rate of migration required by arc growth would be rather small.

A further problem, which may have a related solution, is the variation in apparent rates of motion of different Benioff zones (Brune, 1968). If the cause of descending mass motion is the body force derived from density excesses associated with the cold lithosphere, it is to be expected that rates of descending motion should be broadly similar everywhere. Varying rates of crustal destruction along Benioff zones may be achieved, however, if a varying amount of positive overriding by the "static wedge" is able to occur. Such overriding

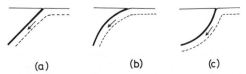

Figure 13. Possible Benioff zone cross-sectional configurations; (for discussion *see* text). The arrow indicates the direction of motion of the lithosphere beneath the Benioff zone (heavy line).

would not only cause the arc and trench to migrate relative to other features and give a varying rate of crustal destruction, but also affect the shape of the Benioff zone. Various possibilities are shown in Figure 13; (a) may be regarded as a steady state. In (b) the 'static wedge' may be overriding the trench. In (c) there may have been a recent reduction in the resolved rate of motion of material into the trench associated with a change in spreading direction.

DISCUSSION AND CONCLUSIONS

The thermal model presented in the earlier part of this paper allows the construction of a thermal profile of an ideal island arc (Fig. 14). As previously discussed, the structure on the oceanward side of the Benioff zone is subject to fewer uncertainties than the structure within the stationary wedge, and it appears certain that the descending lithosphere must retain its thermal identity as a steeply inclined cold slab to at least 500-km depth in the mantle and possibly a great deal deeper. This discussion has important seismological implications. Dissipative heating should occur along the Benioff zone, but the cold descending lithosphere is such a strong heat sink that in order for dissipation to provide the high-surface heat flow observed behind the arc in addition to heating the lithosphere, the shear stresses acting on the Benioff zone must be of the order of 1 to 2 kb. These stresses are about an order of magnitude greater than commonly accepted for such faults, but this discrepancy is not fatal to the model because the uncertainties associated with stress estimates based on seismic observations are so great.

Although dissipative heating still appears to be the most reasonable explanation for the island arc volcanicity and high heat flow, a number of problems remain. As discussed above, the inclined heat source (Benioff zone) diminishes in intensity with depth and at the same time becomes progressively further from

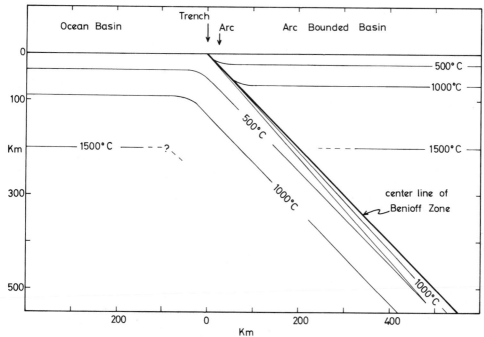

Figure 14. A possible distribution of isotherms under an island arc system based on the model described in the text. The rate of slip on the Benioff zone is taken as 9 cm/yr.; the mean heat flow to the ocean basin, as 1.0 μ cal cm^{-2} sec^{-1} and that behind the trench, as 1.6 μ cal cm^{-2} sec^{-1}.

the surface. Yet the heat flux at the surface remains relatively uniform. Thus, high values for the regional heat flow at long distances behind the arc can only be explained by the present model if processes of heat transfer other than conduction operate within the stationary wedge. Indeed, the time constant for the transfer of heat by conduction from 400-km depth to the surface is of the same order as the age of the Earth. The most likely process would be some kind of penetrative convection; it is possible that hot peridotite diapirs rise from the deeper parts of the Benioff zone to shallow depths and are the main agent of heat transport within the wedge; at present this is an *ad hoc* hypothesis.

In the present model, a steady thermal and, thus, mechanical, state has been assumed. This assumption is clearly unjustified, but it is less clear whether this may be an important source of error in the analysis. It is possible, of course, that the high heat flow behind the arc today is associated with an earlier period of much greater activity on the fault zone; how much earlier would depend on heat transfer processes within the wedge, but the heat could not be transferred to the surface by conduction alone.

The conditions along the fault zone are such that all layers of the oceanic crust undergo partial fusion to some degree; the resulting magmas should vary in composition continuously down the zone and give rise to the characteristic calc-alkaline igneous suite at the surface. Any lateral variation in rate of resolved movement down the fault, corresponding to changes in direction of surface trend of the fault or of its inclination, will affect its thermal structure and, in turn, the composition of magmas formed at any given depth. It is possible that some of the basaltic magmas observed at the surface originate through processes operating within the stationary wedge above the fault. It is very unlikely that any basalts are produced by partial fusion of the ultramafic part of the lithosphere—it should already have been depleted of its basaltic fraction and could not become hot enough to melt in the time available. Insofar as the rate of heat generation and temperatures on the fault zone are dependent upon the differential velocity down the zone, magmatic activity may occur only at times when the velocity exceeds some critical value.

The sediments of the ocean floors seem to

make the major contribution to the growth of island arcs. Any sediments that are carried down the Benioff zone are situated in the zone of highest temperature in the center of the zone and should be the first material to melt; the resultant magmas should be added to the arc at the surface. Sedimentary material not carried down in this way must be added to the side of the island arc. The variation in composition of modern deep-sea sediments with latitude and water depth suggests that the material added to island arcs should have varied in composition. If the island arc building processes outlined above are similar to those which have been important in the formation and growth of continental crust, a considerable degree of lateral inhomogeneity in crustal composition is to be expected.

ACKNOWLEDGMENTS

We thank A. E. Ringwood and T. W. Donnelly for most valuable discussions of many aspects of this paper. T. W. Donnelly also kindly made available a computer program for the gravity calculations. B. J. Isacks and G. M. Brown read the manuscript and made a number of helpful suggestions. This work was partially supported by the Air Force Office of Scientific Research. E. R. Oxburgh held a visiting Professorship at the California Institute of Technology for part of the time during which this work was done.

REFERENCES CITED

Aki, K., 1966, Earthquake generating stress in Japan for the years 1961 to 1963 obtained by smoothing first motion radiation patterns: Earthquake Research Inst. Bull., v. 44, p. 447.

Association for Geological Collaboration of Japan, 1965, The Geological Development of the Japanese Islands: Tokyo, Tsukiji Shokan, 442 p.

Benioff, H., 1955, Seismic evidence for crustal structure and tectonic activity, p. 61 *in* Poldervaart, A., *Editor*, The Crust of the Earth: Geol. Soc. America Spec. Paper 62, 762 p.

Birch, F., 1968, On the possibility of large changes in the Earth's volume: Phys. Earth Planetary Interiors, v. 1, p. 141.

Boettcher, A. L., and Wyllie, P. J., 1967, Hydrothermal melting curves in silicate-water systems at pressures greater than 10 kilobars: Nature, v. 216, p. 572.

Brune, J. N., 1968, Seismic moment, seismicity, and rate of slip along major fault zones: Jour. Geophys. Research, v. 73, p. 777.

Brune, J. N., and Allen, C. R., 1967, A low-stress-drop, low magnitude earthquake with surface faulting; the Imperial California earthquake of March 4, 1966: Seis. Soc. America Bull., v. 55, p. 501.

Bullen, K. E., 1963, An Introduction to the Theory of Seismology, 3rd Ed.: Cambridge, Cambridge Univ. Press, p. 235.

Carslaw, H. S., and Jaeger, J. C., 1947, Conduction of Heat in Solids: Oxford, Oxford Univ. Press, p. 57.

Cleary, J., 1967, Azimuthal variation of the long-shot source term: Earth Planetary Sci. Letters, v. 3, p. 29.

Cleary, J., and Hales, A. L., 1967, An analysis of the travel times of *P* waves to North American stations in the distance range 32°–100°: Seis. Soc. America Bull., v. 56, p. 467.

Coats, R. R., 1956, Geology of northern Adak Island, Alaska: U.S. Geol. Survey Bull. 1028c, p. 67.

—— 1962, Magma type and crustal structure in the Aleutian arc, p. 92 *in* The Crust of the Pacific Basin: Am. Geophys. Union Geophys. Mon. 6, 195 p.

Dickinson, W. R., 1968, Circum-Pacific andesite types: Jour. Geophys. Research, v. 73, p. 2261.

Dickinson, W. R., and Hatherton, T., 1967, Andesitic volcanism and seismicity around the Pacific: Science, v. 157, p. 801.

Green, D. H., and Ringwood, A. E., 1966, Origin of the calc-alkaline rock suite: Earth and Planetary Sci. Letters, v. 1, p. 307.

—— 1967a, The stability fields of aluminous pyroxene peridotite and garnet peridotite and their relevance in upper mantle structure: Earth and Planetary Sci. Letters, v. 3, p. 151.

—— 1967b, Crystallization of basalt and andesite under high pressure hydrous conditions: Earth and Planetary Sci. Letters, v. 3, p. 481.

—— 1967c, An experimental investigation of the gabbro-eclogite transformation and its petrological application: Geochim. et Cosmochim. Acta, v. 31, p. 767.

Griggs, D. T., and Baker, D. W., 1968, The origin of deep focus earthquakes, p. 23 *in* The Properties of Matter: New York, John Wiley & Sons, Inc., 389 p.

Gutenberg, B., and Richter, C. F., 1941, Seismicity of the earth: Geol. Soc. America Spec. Paper 34, p. 131.

—— 1954, Seismicity of the Earth and Associated Phenomena: Princeton, New Jersey, Princeton Univ. Press, 310 p.

Hales, A. L., 1936, Convection currents in the earth: Royal Astron. Soc. Monthly Notices, Geophys. Suppl., v. 3, p. 372.

Hamilton, R. M., and Gale, A. W., 1968, Seismicity and structure of North Island, New Zealand:

Jour. Geophys. Research, v. 73, p. 3859.

Heirzler, J. R., Dickson, G. O., Herron, E. M., Pitman, W. C., and Le Pichon, X., 1968, Marine magnetic anomalies, geomagnetic field reversals, and motions of the ocean floor and continents: Jour. Geophys. Research, v. 73, p. 2119.

Hess, H. H., 1937a, Geological interpretation of data collected on the cruise of U.S.S. *Barracuda* in the West Indies—preliminary report: Am Geophys. Union Trans., v. 18, p. 69.

—— 1937b, Island arcs, gravity anomalies and serpentinite intrusions; a contribution to the Ophiolite problem: 17th Inter. Geol. Cong. Proc., Moscow, v. 2, p. 263.

—— 1948, Major structural features of the western north Pacific: Geol. Soc. America Bull., v. 59, p. 417.

—— 1962, History of the ocean basins, p. 559 in Engel, A. E. J., James, H. L., and Leonard, B. F., *Editors*, Petrological Studies: A Volume in Honor of A. F. Buddington: Boulder, Colorado, Geol. Soc. America, 660 p.

Isacks, B., Oliver, J., and Sykes, L. R., 1968, Seismology and the new global tectonics: Jour. Geophys. Research, v. 73, p. 5855.

King, C. Y., and Knopoff, L., 1968, Stress drop in earthquakes: Seismol. Soc. America Bull., v. 58, p. 249.

Kuenen, P. H., 1936, The negative isostatic anomalies in the East Indies (with experiments): Leidsche geol. med., v. 8, p. 169.

Kuno, H., 1959, Origin of cenozoic petrographic provinces of Japan and surrounding areas: Bull. Volcanol., v. 20, p. 37.

—— 1965, Lateral variation of basalt magma across continental margins and island arcs: Canada Geol. Survey Paper 66–15, p. 317.

Langseth, M. G., Le Pichon, X., and Ewing, M., 1966, Crustal structure of the mid-ocean ridges. Part 5. Heat flow through the Atlantic Ocean floor and convection currents: Jour Geophys. Research, v. 71, p. 5321.

Le Pichon, X., 1968, Sea floor spreading and continental drift: Jour. Geophys. Research, v. 73, p. 3661.

Lee, W. H. K., and Uyeda, S., 1965, Review of heat flow data, p. 87 in Terrestrial Heat Flow: Am. Geophys. Union Geophys. Mon. 8, 276 p.

MacGregor, I. D., 1968, Depth of origin of basaltic magmas: Jour. Geophys. Research, v. 73, p. 3737.

McKenzie, D. P., 1967, Some remarks on heat flow and gravity anomalies: Jour. Geophys. Research, v. 72, p. 6261.

McKenzie, D. P., and Sclater, J. G., 1968, Heat flow inside island arcs of the northwestern Pacific: Jour. Geophys. Research, v. 73, p. 3173.

Menard, H. W., 1967, Transitional types of crust under small ocean basins: Jour. Geophys.

Research, v. 72, p. 3061.

Miyashiro, A., 1961, Evolution of Metamorphic Belts: Jour. Petrology, v. 2, p. 277–311.

Morgan, W. J., 1968, Rises, trenches, great faults, and crustal blocks: Jour. Geophys. Research, v. 73, p. 1959.

O'Hara, M. J., 1965, Primary magmas and the origin of basalts: Scottish Jour. Geology, v. 1, p. 19.

Oliver, J., and Isacks, B., 1967, Deep earthquake zones, anomalous structures in the upper mantle, and the lithosphere: Jour. Geophys. Research, v. 72, p. 4259.

Oxburgh, E. R., and Turcotte, D. L., 1968a, Mid-ocean ridges and geotherm distribution during mantle convection: Jour. Geophys. Research, v. 73, p. 2643.

—— 1968b, Problem of high heat flow and volcanism associated with zones of descending mantle convective flow: Nature, v. 218, p. 1041.

Ringwood, A. E., 1969, Composition and Evolution of the Upper Mantle, in the Earth's Crust and Upper Mantle: Am. Geophys. Union Geophys. Mon. 13, 735 p.

Sclater, J. G., and Menard, H. W., 1967, Topography and heat flow of the Fiji Plateau: Nature, v. 216, p. 991.

Stauder, W., 1968, Mechanism of the Rat Island earthquake sequence of 4 February 1965 with relation to island arcs and sea-floor spreading: Jour. Geophys. Research, v. 73, p. 3847.

Sugimura, A., 1966, Composition of primary magmas and seismicity of the Earth's mantle in the island arcs (a preliminary note), in Continental Margins and Island Arcs: Geol. Survey of Canada, Paper 66–15, p. 337.

Sugimura, A., Matsuda, T., Chimzei, K., and Nakamura, K., 1963, Quantitative distribution of late Cenozoic volcanic materials in Japan: Bull. Volcanol., v. 26, p. 125.

Sykes, L. R., 1966, The seismicity and deep structure of island arcs: Jour. Geophys. Research, v. 71, p. 2981.

Taylor, S. R., 1967, The origin and growth of continents: Tectonophysics, v. 4, p. 17.

Taylor, S. R., and White, A. J. R., 1965, Geochemistry of andesites and the growth of continents: Nature, v. 208, p. 271.

Taylor, S. R., Kaye, M., White, A. J. R., Duncan, A. R., and Ewart, E., 1969, Genetic significance of Co, Cr, Ni, Sc and V content of andesites: Geochim. et Cosmochim. Acta, v. 33, p. 275.

Turcotte, D. L., and Oxburgh, E. R., 1967, Finite amplitude convection cells and continental drift: Jour. Fluid Mechanics, v. 28, p. 29.

—— 1968, A fluid theory for the deep structure of dip slip fault zones: Phys. Earth and Planetary Interiors, v. 1, p. 381.

—— 1969, Convection in a mantle with variable physical properties: Jour. Geophys. Research,

v. 74, p. 1458.

Vaquier, V., Uyeda, S., Yasui, M., Sclater, J., Corrie, C., and Watanabe, T., 1966, Studies of the thermal state of the earth, 19th paper; heat flow measurements in the northern Pacific: Bull. Earthquake Research Inst., v. 44, p. 1526.

Vening Meinesz, F. A., 1932, Gravity Expeditions at Sea 1923–1930: Delft, Waltman.

—— 1937, Results of maritime gravity research, 1923–32, in International Aspects of Oceanography: Natl. Acad. Sci., Washington, p. 61.

Vening Meinesz, F. A., Umbgrove, J. H. F., and Kuenen, P. H., 1934, Gravity Expeditions at Sea 1923–1932: Delft, Waltman.

Vine, F. J., 1966, Spreading of the ocean floor; new evidence: Science, v. 154, p. 1405.

Vine, F. J., and Mathews, D. H., 1963, Magnetic

anomalies over oceanic ridges: Nature, v. 199, p. 947.

Wakita, H., Nagasawa, H., Uyeda, S., and Kuno, H., 1967a, Uranium and thorium contents in ultrabasic rocks: Earth and Planetary Sci. Letters, v. 2, p. 377.

—— 1967b, Uranium, thorium and potassium contents of possible mantle materials: Geochem. Jour., v. 1, p. 183.

Winkler, H. G. F., 1967, Die Genese der metamorphen Gesteine: Berlin, Springer, 237 p.

Wyss, N. X., and Brune, O. N., 1969, Shear stresses associated with earthquakes between 0 and 700 km: Am. Geophys. Union Trans., v. 50, p. 237.

Yoder, H. S., and Tilley, C. E., 1962, Origin of basalt magmas, an experimental study of natural and synthetic rock systems: Jour. Petrology, v. 3, p. 342.

Manuscript Originally Submitted to Editor of Caribbean Symposium in October/November 1968; Manuscript Received by The Society for Publication in the Bulletin on November 7, 1969

Present Address (Oxburgh): Department of Geology and Mineralogy, University of Oxford, Oxford, England

Erratum

On p. 116, right-hand column, the 12th line from the top should read: "Wyss, N. X., and Brune, J. N., 1969 . . ."

III
Metamorphism Related to Plate Tectonic Environments

Editor's Comments on Papers 5, 6, and 7

5 **Miyashiro, Shido, and Ewing:** *Metamorphism in the Mid-Atlantic Ridge near 24° and 30°N*

6 **Miyashiro:** *Pressure and Temperature Conditions and Tectonic Significance of Regional and Ocean-Floor Metamorphism*

7 **Coleman:** *Blueschist Metamorphism and Plate Tectonics*

Few syntheses of data exist regarding the relationship of metamorphic rocks to the new global tectonics. Nevertheless, it seems clear that, well within the interiors of stable, nonsubducted lithospheric plates, crustal recrystallization effects are slight, and annealing proceeds slowly in response to the geothermal gradient. Temperatures change imperceptibly in such environments and, inasmuch as pervasive deformation is also lacking, dynamothermal metamorphism is not produced except at profound depths; contact metamorphic aureoles are rare, too, and are confined to the vicinity of deeply generated (?) diapiric plutons.

Most crustal metamorphic processes take place in the proximity of lithospheric plate boundaries. Whereas conservative (transform-type) junctions are typified by cataclastically granulated rocks, predominantly thermally induced recrystallization at high temperatures and low pressures characterizes the subjacent portions of constructive (ridge-type) margins. Consumptive (trench-type) plate margins are the site of strong horizontal and vertical differential motions; imbricate thrusting, normal and reverse faulting, and diapiric upwelling are important tectonic processes that accompany low-temperature, high-pressure metamorphism in the subduction zone and the much higher temperature, somewhat lower pressure volcanic-plutonic arc or continental margin. These relationships, particularly the recrystallization effects, have been described in Papers 5, 6, and 7.

In Paper 5, Miyashiro, Shido, and Ewing (1971) concentrated their attention on the petrologic structure of the Mid-Atlantic Ridge, but they have documented a plausible metamorphic paragenetic sequence based chiefly on a study of submarine dredge haul samples. The more surficial metabasalts seem to have been zeolitized and spillitized, whereas the more deeply buried gabbros contain greenschist + low-pressure amphibolite (hornfels) facies mineral assemblages and are less intensely metasomatized. It is clear that coeval plate motions, probably confined to the well-known ridge–ridge transform faults, have produced mylonites, schistose rocks, and even gneisses at various localities along the Mid-Atlantic Ridge.

Miyashiro has framed regional and ocean-floor metamorphism in a general plate tectonic framework, noting a secular variation of the preserved dynamothermal environments; this summary is included as Paper 6. At least as now exposed, compared to younger rocks of equivalent depth of recrystallization, older rocks seem to have been metamorphosed at systematically higher temperatures. Miyashiro tentatively suggested that this phenomenon may reflect a change in the nature of tectonic processes with time, somehow related to lithospheric plate thickness, velocity, and/or behavior. Paired metamorphic belts also seem to be confined to past-Paleozoic, Pacific-type convergent plate junctions.

Coleman (Paper 7) and Ernst (1973) have discussed the tectonic significance of blueschist terranes. Such belts are very narrow, are generally relatively young geologically speaking, and although of controversial origin, are thought by most workers to mark convergent lithospheric plate sutures. Mineral parageneses and oxygen isotopic fractionations seem to require the attendance of very high pressures and low temperatures during the recrystallization. Lithologic associations with representatives of the ophiolite suite and structural relationships such as recumbant folds and imbricate thrust zones indicate that the generation of blueschists probably takes place within subduction zones; happily, the thermal structure of this plate tectonic locale is appropriate for the observed paragenetic sequence (see Paper 4).

Reference

Ernst, W. G. (1973) Blueschist metamorphism and P–T regimes in active subduction zones: *Tectonophysics*, **17**, 255–272.

5

Reprinted from *Phil. Trans. Royal Soc. London*, **A268**, 589–603 (1971)

Metamorphism in the Mid-Atlantic Ridge near 24° and 30° N†

By A. Miyashiro, F. Shido and M. Ewing

Lamont–Doherty Geological Observatory of Columbia University,
Palisades, New York 10964

Metabasites (metabasalts and metagabbros) occur abundantly in association with serpentinites in transverse fracture zones and on walls of the median valley. These metabasites were formed by burial metamorphism probably in deeper parts of the crust and the upper mantle beneath the Ridge crest, and were brought up to the surface of the crust probably by serpentinites rising along fracture zones and by normal faulting along the median valley.

The metabasalts are in the zeolite and greenschist facies and a transitional state from the greenschist to the amphibolite facies, whereas metagabbros tend to have been recrystallized at higher temperatures, being in the greenschist and amphibolite facies. Compositionally the metabasites are divided into two groups, I and II. Group I comprises those which retain the approximate composition of the original rocks except for water content, whereas group II comprises those which underwent marked chemical migration, as regards sodium in zeolite-facies rocks and calcium and silicon in greenschist-facies rocks. In rocks of group I, calcic igneous plagioclase remains unaltered, and albite and epidote did not form. This fact, along with the absence of epidote-amphibolite facies rocks, would be due to the low rock-pressure during metamorphism. In some rocks of group II, albite and epidote occur.

Burial metamorphism takes place probably mainly beneath the Ridge crest where the geothermal gradient is great. The resultant metamorphic rocks are probably of the low-pressure type, and move laterally by ocean-floor spreading to form the main part of the oceanic crust.

Contact metamorphic gneisses, probably derived from gabbros, have been found. Some metagabbros were subjected to cataclasis by fault movements along fracture zones and the median valley.

1. Introduction

Since 1966, metamorphosed basalts, dolerites and gabbros have been reported to occur in mid-oceanic ridges. Melson, Bowen, van Andel & Siever (1966), Melson & van Andel (1966) and Melson, Thompson & van Andel (1968) discussed a collection of more or less sheared metabasalts in the greenschist and lower facies dredged from a wall of the median valley near 22° N on the Mid-Atlantic Ridge. Aumento & Loncarevic (1969) described more or less schistose metabasalts in the greenschist facies from Bald Mountain about 60 km west of the median valley on the M.A.R. near 45° N.

Cann & Funnell (1967) have reported the occurrence of metabasalts, metadolerites and metagabbros, preserving original igneous textures but recrystallized in the lower amphibolite facies, from Palmer Ridge near 43° N and 20° W on the eastern flank of the M.A.R. Cann & Funnell (1967) believe that the pertinent metamorphic recrystallization took place beneath the crest of the M.A.R., and that the resultant rocks have been moved laterally by ocean-floor spreading.

Metabasalts in the greenschist facies, having relict igneous textures, have also been obtained from a crestal mountain adjacent to a fracture zone on the Carlsberg Ridge near 5° N by Cann & Vine (1966) and Cann (1969).

A series of dredge hauls made by one of us on the M.A.R. near 30° N during *Atlantis* cruise 150 (in 1947) was petrographically described as containing basalt, dolerite, gabbro and serpentinite (Shand 1949; Quan & Ehlers 1963; Muir & Tilley 1966). Our recent re-examination of

† Lamont–Doherty Geological Observatory Contribution No. 1579.

TABLE 1. METAMORPHIC ROCKS AND ASSOCIATED ROCK TYPES IN DREDGE HAULS FROM THE MID-ATLANTIC RIDGE NEAR 24° AND 30° N.

cruise	dredge	latitude (N)	longitude (W)	depth m	geological environment	meta-basalt	meta-gabbro	serpentinite	breccia and cataclasite	gabbro	dolerite	fresh basalt	weathered basalt
Atlantis 150	RD6	30° 06'	42° 08'	1460	c.r.	—	a	a	—	—	—	—	—
	RD7	30° 01'	42° 04'	4280	mv.-f.z.	s	—	s	—	—	—	a	s
	RD20	30° 04'	42° 16'	4144	f.z.	a	s	a	s	—	s	s	s
	RD21	30° 08'	42° 37'	4200	f.z.	—	s	a	—	—	—	—	—
Vema 25	RD5	23° 31.7'	45° 07'	3109 or 3384	m.v.	a	a	—	s	s	—	—	—
	RD6	23° 44.7'	45° 33.6'	4207	f.z.	a	a	s	a	s	s	s	a
	RD8	23° 46.7'	46° 04.2'	3841	f.z.	a	—	s	s	s	—	s	a
	RD9	23° 46.1'	46° 37'	3073	f.z.	s (?)	a	a	—	—	—	—	—

Note. c.r., on a crestal mountain adjacent to the junction of the median valley with the Atlantis Fracture Zone; m.v.–f.z. at the junctions of the median valley with fracture zones; f.z., within fracture zones across the crest province; m.v., east wall of median valley near the junction with a fracture zone. a = abundant, s = scarce.

this collection has revealed the presence of abundant metabasalts and metagabbros. A new series of dredges made on the M.A.R. near 24° N during *Vema* cruise 25 (in 1968) gave us abundant metabasalts and metagabbros as well as serpentinites. These metamorphic rocks were dredged largely from transverse fracture zones, but partly from a crestal mountain and from the median valley. Some metagabbros were found to have been subjected to later cataclasis.

This paper gives an outline of our petrologic study on these metamorphosed basic rocks. Detailed descriptive data will be published later. *Atlantis* cruise 150 and *Vema* cruise 25 will be denoted as A150 and V25 respectively. RD means rock dredge.

2. MODES OF OCCURRENCE

Our dredge stations are summarized in table 1. It is noted that most of our metamorphic rocks occur in association with serpentinites in transverse fracture zones (A150–RD20, 21; V25–RD6, 8, 9). Metagabbros from a crestal mountain adjacent to the Atlantis Fracture Zone (A150–RD6) are also associated with serpentinites. These facts suggest that the metamorphic rocks occur in and near fracture zones as inclusions in serpentinites. The metamorphic recrystallization should have taken place at some depth, and fragments of the resultant rocks would have been torn and brought up to the surface of ocean floors by intrusions of serpentinites or their parental peridotites along fracture zones (Miyashiro, Shido & Ewing 1969, 1970*a*). This interpretation is consistent with the view that the M.A.R. serpentinites were formed by hydration of upper mantle peridotite (Miyashiro *et al.* 1969). Cataclastic metagabbros occurring in a fracture zone (V25–RD6) would have been formed by shattering due to movements along the zone.

A dredge haul (V25–RD5) from a wall of the median valley near the junction with a fracture zone contains metabasalts, metagabbros and cataclastic metagabbros but no serpentinite. This mode of occurrence resembles that in the median valley near 22° N (Melson *et al.* 1968). In such cases, the rocks metamorphosed at some depths would have been exposed by normal faulting which would have caused cataclasis on some rocks.

3. BURIAL METAMORPHISM OF BASALTS

Metabasalts of A150 and V25 are usually non-schistose, preserving their original igneous textures. Recrystallization is incomplete in most cases. The structures of pillow lavas and tuff breccias are noticeable in some specimens. These suggest that basaltic eruptions produced thick volcanic piles, the lower parts of which became sufficiently hot to undergo metamorphic recrystallization, i.e. burial metamorphism.

A classification of burial metamorphic rocks from the M.A.R. near 24° and 30° N is shown in table 2. The rocks comprise metabasalts and associated metagabbros. Most of the metabasalts belong to the zeolite and greenschist facies. Some metabasalts from the Atlantis Fracture Zone (A150–RD20), however, have very small amounts of blue-green hornblende in association with greenschist-facies minerals, and are considered to be in a transitional state between the greenschist and amphibolite facies. The temperature of metamorphism for such rocks would be as high as 400 to 500 °C. Such high temperatures would have been reached somewhere in the deepest part of the crust and in the upper mantle beneath the crest of the M.A.R.

As shown in table 2, metabasalts and metagabbros that underwent burial metamorphism are compositionally classified into two groups. The members of one group, here called *group I*, preserve their approximately original igneous composition except for H_2O contents. The

50-2

members of the other group, called *group II*, suffered intense chemical migration during metamorphism. The migration is probably due to flow of a hot aqueous fluid through the crust. With emphasis on this interpretation, we may well call the latter group hydrothermally modified metabasalts and metagabbros.

TABLE 2. CLASSIFICATION OF BURIAL-METAMORPHIC BASALTS AND GABBROS
FROM THE MID-ATLANTIC RIDGE

facies	composition	
	group I compositionally virtually unchanged	group II compositionally intensely changed
zeolite facies	none	metabasalt (abundant) metagabbros (zeolitized only by retrogressive change) } Na-introduction
greenschist facies	metabasalts (abundant) metagabbros (abundant)	metabasalts (abundant) metagabbros (abundant) } Ca-decrease Si-variation
transitional between greenschist and amphibolite facies	metabasalts (abundant) metagabbros (abundant) }	none
amphibolite facies	metagabbro (abundant)	none

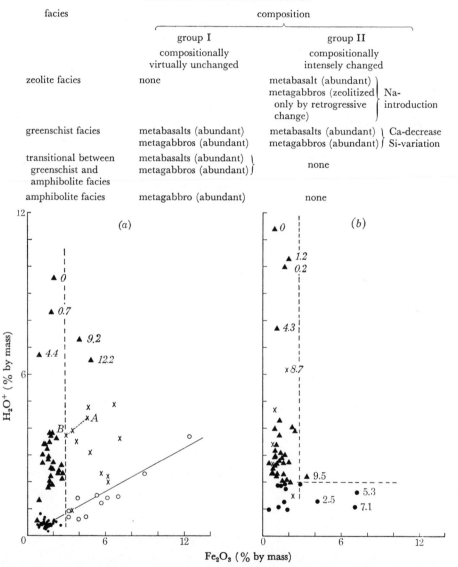

FIGURE 1. Relation between Fe_2O_3 and H_2O^+ contents in (*a*) basalts and metabasalts and (*b*) in gabbros and metagabbros. The numbers in italics represent CaO contents and those in upright type represent TiO_2 contents. In (*a*): ●, unweathered, unmetamorphosed basalt; ○, weathered basalt; ▲, metabasalt in greenschist facies; ×, metabasalt in zeolite facies. ——, trend of compositional variation in weathering. In (*b*): ●, virtually unmetamorphosed gabbro; ▲, metagabbro in greenschist and amphibolite facies; ×, metagabbro containing zeolites.

Greenschist-facies metabasalts

The greenschist-facies metabasalts of group I show chemical features of tholeiitic basalts except for H_2O content. This probably means that they were derived from abyssal tholeiites without marked migration of materials except for H_2O. It is noted that the Fe_2O_3 contents of unweathered greenschist-facies metabasalts are more or less similar to those of unweathered, unmetamorphosed abyssal tholeiites, that is, below 3.0 %, as shown in figure 1 a.

On the other hand, weathered abyssal tholeiites show higher contents of Fe_2O_3 as well as H_2O^+, and are plotted along the full line in figure 1 a. Accordingly, greenschist-facies metabasalts are clearly distinguished from weathered tholeiites by their Fe_2O_3 contents, even though they both have high H_2O^+ contents.

The greenschist-facies metabasalts of group II suffered intense chemical migration during metamorphism, commonly resulting in a decrease of CaO and an increase of H_2O^+ content. Their CaO contents approach zero in some rocks (figure 2). Such rocks are represented by metabasalts with H_2O^+ contents higher than 6 % and with Fe_2O_3 contents less than 3 % in figure 1 a. Mineralogically they are chlorite–quartz rocks. The high contents of H_2O^+ are due to high contents of chlorite. The SiO_2 contents are variable. Some metabasalts of group II, however, do not show a decrease of CaO. These are represented by metabasalts with H_2O^+ contents higher than 6 % and Fe_2O_3 contents higher than 3 % in figure 1 a, and are mineralogically epidote–chlorite rocks. They are relatively rare in the Ridge and may have been produced where oxygen pressure was high during metamorphism.

Zeolite-facies metabasalts

Zeolite-facies metabasalts are characteristically penetrated with networks of veinlets (usually less than 2 mm wide) composed mainly of one zeolite or more. Similar zeolites occur also in dispersed state in the metabasalts. In these rocks, not only the H_2O^+ but also the Fe_2O_3 contents are high, as shown in figure 1 a. All the zeolite-facies metabasalts so far dredged, however, appear to be rather intensely weathered. Their high Fe_2O_3 contents are considered to be due to weathering post-dating metamorphism for the following reason: The pair of A and B in figure 1 a represents a weathered rim and a less strongly weathered core of a zeolitized metabasalt (V 25–RD 8–T 47). It can be seen that the Fe_2O_3 content of the core is close to that of greenschist-facies metabasalts. Hence, there appears to be no great difference in oxidation condition between the greenschist-facies and zeolite-facies metamorphism.

Zeolite-facies metabasalts show Na_2O contents generally higher than those of abyssal tholeiites as shown in figure 2. The Na_2O content varies sympathetically with the H_2O^+ as shown in figure 3. Those high in Na_2O and H_2O^+ give normative nepheline, whereas those low in Na_2O and H_2O^+ approach the composition of abyssal tholeiites. All these zeolite-facies metabasalts are considered to have formed by recrystallization of abyssal tholeiites with varying amounts of Na_2O and H_2O introduced. Hence, all the zeolite-facies metabasalts belong to group II. Relict igneous pyroxene and the low TiO_2 contents (1.49 % on an average of 13 analyses) also support their tholeiitic origin. The introduction is perhaps due to permeation of an Na-containing fluid rising from greater depths.

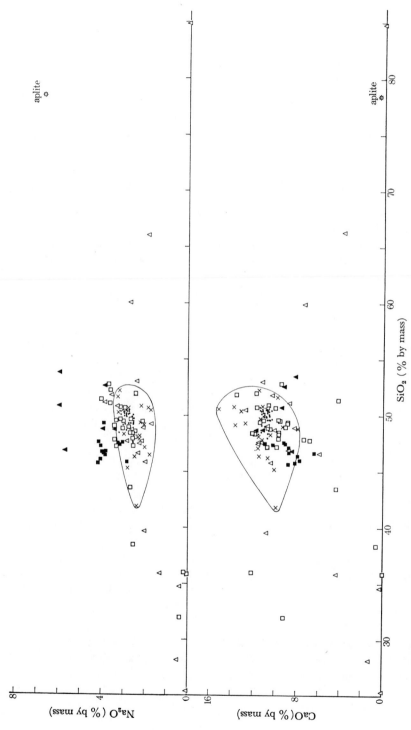

FIGURE 2. Relation between SiO$_2$, Na$_2$O and CaO contents in basalts, gabbros, metabasalts and metagabbros. The unweathered, unmetamorphosed basalts and gabbros fall in the outlined area. All the basalts now under consideration are abyssal tholeiites, and the gabbros show a much wider composition field than the basalts. ●, basalt (unweathered, unmetamorphosed); ×, gabbro (virtually unmetamorphosed); □, metabasalt in greenschist or higher facies; △, metagabbro in greenschist or higher facies; ■, metabasalt in zeolite facies; ▲, metagabbro containing zeolites.

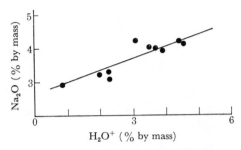

FIGURE 3. Sympathetic variation of H_2O^+ and Na_2O contents in zeolite-facies metabasalts.

4. METAGABBROS EXCEPTING THOSE OF CONTACT METAMORPHIC ORIGIN

Metagabbros are in the greenschist and amphibolite facies. However, some of them suffered retrogressive changes in the zeolite facies. They show high H_2O^+ contents but not high Fe_2O_3, as shown in figure 1b. It is noted that metabasalts are mostly in the zeolite and greenschist facies, whereas metagabbros are mostly in the greenschist and amphibolite facies. In other words, metagabbros tend to have been recrystallized at higher temperatures than metabasalts. This would mean that gabbroic intrusions are more common in deeper levels.

The typical amphibolite facies, corresponding to, say, 600 °C would be too high in temperature to be realized within the crust even beneath the crest of the M.A.R. The occurrence of such amphibolite-facies metagabbros, therefore, suggests the possibility that some metagabbros would have been consolidated and recrystallized in the upper mantle.

Most of the metagabbros approximately preserve their original texture and chemical composition, thus belonging to group I as defined above. They show a remarkable degree of magmatic differentiation in the tholeiitic trend, leading to the formation of high-iron, high-titanium gabbros in a later stage (Miyashiro, Shido & Ewing 1970b). We may consider that tholeiitic magma was intruded into volcanic piles beneath the M.A.R. to form gabbroic masses of considerable sizes, crystallization differentiation took place in them to a marked extent, and then the masses were subjected to burial metamorphism together with the surrounding volcanic piles.

Some metagabbros in the greenschist facies, however, underwent intense chemical migration, thus belonging to group II. Their CaO content decreases and their H_2O^+ content increases (figures 1b and 2). This trend is the same as that observed in greenschist-facies metabasalts. It leads to enrichment in chlorite, ultimately resulting in the generation of chlorite–quartz rocks and mono-mineralic chlorite rocks. The SiO_2 content decreases down to about 25 % in some rocks and increases up to about 85 % in others, though gabbroic textures still remain. A flowing aqueous fluid would have dissolved SiO_2 in some parts of the crust and re-deposited it in others during greenschist-facies metamorphism.

Metagabbros retrogressively modified in the zeolite facies show an increase of Na_2O content just as zeolite-facies metabasalts. Thus, we may consider that Na enrichment is characteristic of zeolite-facies metamorphism in the M.A.R.

5. MINERALOGY OF BURIAL METAMORPHIC ROCKS

Igneous plagioclase is very resistant against recrystallization in rocks of group I. In greenschist-facies metabasites of group I, igneous plagioclase (labradorite or bytownite) commonly remains intact after most of the igneous mafic minerals have been replaced by new metamorphic minerals. Therefore, albite is lacking in metabasites of group I. Probably relevant to it is the fact that epidote is absent in group I metabasites.

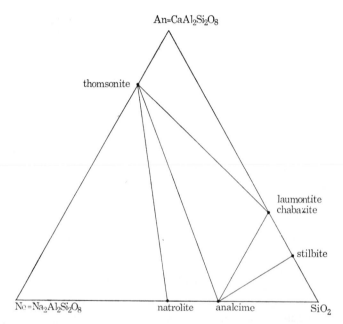

FIGURE 4. Zeolite parageneses in zeolite-facies metabasalts.

Igneous plagioclase is not so resistant in group II metabasites. Albite occurs in some greenschist-facies metabasites of group II. In zeolite-facies metabasalts of group II, plagioclase remains intact in some cases, but is partly replaced by analcime and other zeolites in many cases.

The following zeolites have been identified in zeolite-facies metabasites: natrolite, thomson-ite, analcime, chabazite, laumontite, and stilbite, among which analcime is the most wide-spread (figure 4). Zeolite-bearing rocks contain no quartz. Hence, natrolite and thomsonite, both incompatible with quartz, occur commonly. Natrolite, being poor in silica, does not coexist with stilbite rich in silica. The zeolite assemblages are considered to be close to chemical equilibrium. The zeolites occurring in deep-sea sediments are phillipsite and clinoptilolite, neither of which was found in our zeolite-facies metabasites. The zeolite-facies rocks were formed at higher temperatures than deep-sea sediments (Miyashiro & Shido 1970).

Chabazite and laumontite have the same chemical composition except for H_2O content. Rising temperature would cause dehydration of chabazite to form laumontite. Therefore, the zeolite-facies metamorphism in the Ridge would include two stages: a lower temperature one characterized by chabazite and a higher temperature one characterized by laumontite.

All the identified clay minerals of the metabasites of the Ridge belong to the three component system: chlorite–smectite–vermiculite (figure 5). Weathered basalts contain mixed-layer mineral smectite–vermiculite. Zeolite-facies metabasites contain mixed-layer mineral chlorite–smectite with or without vermiculite layers, but they never contain chlorite as discrete crystallites. The appearance of discrete chlorite marks the entrance to the greenschist facies. Thus, the progressive change of clay minerals is consistent with the recrystallization temperatures suggested by other minerals. Smectite was found in metabasites of the zeolite and greenschist facies. However, it may be an unstable relict mineral in the high-temperature part of the greenschist facies.

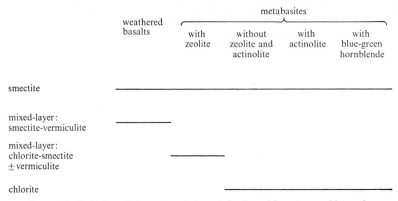

FIGURE 5. Variation of clay minerals in metabasites with metamorphic grade.

6. CHARACTERISTICS OF BURIAL METAMORPHISM IN THE MID-ATLANTIC RIDGE

The petrologic characteristics of burial metamorphism will be summarized below.

(1) The metabasalts and metagabbros belong to the zeolite, greenschist and amphibolite facies together with transitional states between them. We did not find rock groups that should be properly assigned to the prehnite-pumpellyite or to the epidote-amphibolite facies.

Prehnite was found in an unusual schistose amphibolite (A150–RD8–AM1), in which prehnite has replaced plagioclase. The prehnitization represents retrogressive change after the formation of the amphibolite. Pumpellyite was mentioned by Melson & van Andel (1966). However, the occurrence of these minerals so far known is not enough to warrant the presence of prehnite-pumpellyite facies rocks.

It is known that a zone of the epidote-amphibolite facies as defined by Eskola (1939) is lacking in regions of low-pressure type regional metamorphism in the island arc (Miyashiro 1961). The absence of this facies in the M.A.R. may also be a result of relatively low pressure.

(2) In all the metabasites of group I, calcic igneous plagioclase is very strongly resistant against recrystallization, and persists in rocks where mafic minerals were mostly or entirely recrystallized into low-temperature metamorphic minerals. These rocks, therefore, contain no albite and no epidote.

Such persistence of calcic plagioclase in group I rocks may be due simply to a very slow rate of recrystallization of plagioclase for some unknown cause. An alternative possibility is that calcic plagioclase may be stable in what have been so far called greenschist facies. If so, this facies cannot be the greenschist facies as defined by Eskola (1939), and should be a distinct metamorphic facies characterized by the mineral assemblage actinolite–chlorite–calcic plagioclase.

A metamorphic zone characterized by the assemblage actinolite–calcic plagioclase has been known to exist in the contact aureole of the Iritono area (Shido 1958), Arisu area (Seki 1961) and Sierra Nevada (Loomis 1966). It has been regarded as suggesting the existence of a new metamorphic facies under relatively low pressures (Shido 1958; Miyashiro 1961, p. 307). It is conceivable that the M.A.R. metabasites with the assemblage actinolite–chlorite–calcic plagioclase belong to this or closely related low-pressure metamorphic facies. Epidote is a mineral having a high density. At low rock-pressures, the composition field of rocks to produce epidote may become so small that calcic plagioclase may remain stable instead in ordinary meta-basites. This is consistent with the absence of epidote-amphibolite facies rocks.

Since the heat flow at the crest of the Ridge is generally very high, the metamorphism beneath the crest would be of the low-pressure type. If the triple point of Al_2SiO_5 minerals lies at about 600 °C and 6 kbar (0.6 GN m^{-2}) (Richardson, Gilbert & Bell 1969), the average geothermal gradient to result in the coexistence of the three polymorphs at a depth in the earth is about 30 K km^{-1}. The low-pressure type metamorphism, including the andalusite–sillimanite type defined by Miyashiro (1961), should correspond to geothermal gradients higher than it. The crest of the M.A.R. probably has such geothermal gradients.

The basalt–serpentinite association in the M.A.R. was compared with ophiolites by some authors (Hess 1965; Thayer 1968). Typically developed ophiolites such as those in the Alps and the Franciscan formation of California, however, are commonly accompanied by high-pressure type regional metamorphism (Miyashiro 1968, p. 827) which has never been observed in mid-oceanic ridges. It appears that there is an essential difference in the condition of forma-tion between the mid-oceanic ridge rocks and orogenic ophiolites.

(3) Calcic plagioclase is not so resistant in metabasites of group II. In some group II metabasites of the greenschist facies, epidote and albite are rather abundant, but are not accom-panied by amphibole. They may be regarded as having changed toward spilites. In zeolite-facies metabasites (group II), calcic igneous plagioclase is commonly partly replaced by analcime and other zeolites.

Melson et al. (1968) and Cann (1969) reported the occurrence of spilitic rocks, i.e. epidote–albite–chlorite metabasalts, from the Mid-Atlantic and Carlsberg Ridges respectively. Probably these rocks correspond or are related to our group II metabasites.

(4) Carbonate minerals are scarce in metabasites as well as in associated serpentinites from the M.A.R. near 24° and 30° N. This, together with development of zeolite-facies rocks, suggests that the chemical potential of CO_2 was low during metamorphic recrystallization and serpent-inization in these regions. It is not clear, however, whether such a condition holds in other parts of the M.A.R.

(5) Probably the flow and permeation of a hot aqueous fluid were widespread, resulting in marked migration of Na_2O, CaO and SiO_2. It is noted that the kinds of chemical migration are closely related to metamorphic facies. Introduction of Na_2O is characteristic of the zeolite facies, whereas migration of CaO and SiO_2 of the greenschist facies. SiO_2 was dissolved, moved and re-deposited at least partly in other parts of the crust, whereas CaO does not appear to have been re-deposited and, may have been discharged into ocean water.

(6) Almost all metabasites from the M.A.R. near 24° and 30° N are non-schistose except for the contact metamorphic rocks to be mentioned later, though some of them show a slight degree of preferred orientation. In other parts of the Ridge, schistose metabasites occur in association with non-schistoseones (Melson & van Andel 1966; Aumento & Loncarevic 1969).

Thus, metabasites of the M.A.R. are partly non-schistose and partly schistose. If we are allowed to generalize from the scarce data available at present, the proportion of non-schistose rocks is high, and the schistosity when present is not strong in most cases. We may well use the term burial metamorphism in order to emphasize the poor development of schistosity on a large scale.

7. Contact metamorphism

Dredge haul A 150–RD 6 was made on the top of a crestal mountain adjacent to the junction of the median valley with the Atlantis Fracture Zone. It is composed of serpentinites and metagabbros (Miyashiro *et al.* 1969, p. 124). The metagabbros differ from all other ones in the following respects:

(*a*) Ordinary metagabbros are devoid of parallel structures, whereas the A 150–RD 6 metagabbros are gneissic and banded. The texture suggests granulation followed by metamorphic recrystallization. Big crystals of ortho- and clinopyroxenes and plagioclase in the original gabbros remain as porphyroclasts.

(*b*) Brown hornblende occurs as minute grains produced by recrystallization in interstices between granulated pyroxenes. Hence, these metagabbros are considered to be in a transitional state between the amphibolite and a higher facies (i.e. the granulite or pyroxene-hornfels facies). In other words, they are higher in recrystallization temperature than other metagabbros.

The serpentinite–metagabbro association in this case resembles the association of serpentinized peridotite and contact metamorphic rocks of the Lizard area, England (Green 1964). In the Lizard, the innermost contact aureole is made up of banded ortho- and clinopyroxene gneisses. Therefore, we consider that the metagabbros of A 150–RD 6 were formed by contact metamorphism due to high-temperature peridotite intrusion, which were later hydrated to become serpentinites.

8. Cataclastic metamorphism

Cataclastic metagabbros have been found to occur in a fracture zone (V 25–RD 6) and on a wall in the median valley (V 25–RD 5). The degree of cataclasis in these rocks is variable, but not extreme. The cataclasis postdated metamorphic recrystallization, and is probably due to fault movements.

9. A model for the Mid-Atlantic Ridge

Recent investigations of magnetic lineations in oceanic regions have revealed that the magnetic layer responsible for the anomalies is as thin as about a half to a few kilometres, and that the underlying layer is virtually demagnetized (Vine 1966; Talwani, Windisch, Langseth & Heirtzler 1968; Heirtzler 1968). The depth of demagnetization is too shallow to be ascribed to the Curie point. It is probable that the relatively thin magnetic layer is composed of basalts, dolerites and gabbros, whereas the underlying demagnetized layer is composed of metamorphic derivatives of such rocks (van Andel 1968; Miyashiro *et al.* 1970*a*).

Basalts, dolerites and gabbros commonly have strong thermoremanent magnetism. In metamorphic recrystallization the strong magnetism disappears and instead weak chemical remanent magnetism appears, which is negligible as compared with thermo-remanent magnetism (table 3). Hence, most of the seismological layers 2 and 3 appears to be composed of metamorphosed basalts, dolerites and gabbros. The densities of metabasites are shown in figure 6.

TABLE 3. MAGNETIC PROPERTIES OF ROCKS IN DREDGE HAULS
OF *ATLANTIS* CRUISE 150

rock type	intensity of n.r.m., $10^3 M_n/\text{e.m.u.cm}^{-3}$	susceptibility, $10^3\chi/\text{e.m.u.cm}^{-3}$	$Q'_n = M_n/\chi$
unmetamorphosed basalts	12–36	0.37–1.11	10–81
metabasalts	0.0003–0.01	0.02–0.07	0.004–0.2
metagabbro	0.31	0.15	2.1
amphibolite	0.0003	0.08	0.004
serpentinite	2.3	3.85	0.6

Note: compiled from Opdyke & Hekinian's (1967) data with modification of rock names so as to agree with our classification (Miyashiro *et al.* 1970*a*).

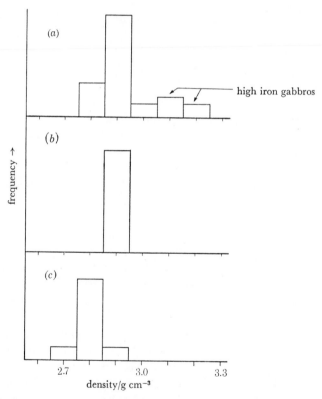

FIGURE 6. Densities of metabasites: (*a*) metagabbros; (*b*) metabasalts (greenschist facies); (*c*) metabasalts (zeolite facies).

The crust beneath normal oceanic basins is about 6 km thick. Assuming a geothermal gradient of 20 or 30 K km^{-1}, the temperature at its base is about 120 to 180 °C. Fissure filling and slight recrystallization would occur there, but the temperature appears to be too low for intense metamorphism.

131

On the other hand, in the crest of the M.A.R., the crust is thinner, but the heat flow and geothermal gradient are much greater so that extensive metamorphic recrystallization would take place at depth as was first suggested by Cann & Funnell (1967).

However, the heat flow values measured on the crest are not uniform. Values as high as 300 mW m^{-2} (8 μcal cm^{-2} s^{-1}) occur mixed with those as low as 40 to 80 mW m^{-2}(1–2 μcal cm^{-2} s^{-1}) (von Herzen & Uyeda 1963; Langseth 1967). If all these values are accepted, the high values in the crest would depend not only on the general heat conduction but also on some more localized events. Igneous intrusion is a possible event. Flow of a hot aqueous fluid is another. It was mentioned before that some of the greenschist-facies rocks were probably selectively subjected to hydrothermal modification. The aqueous fluid responsible for it would reach the surface of the ocean floor, causing localized high heat flow. If this is the case, the isothermal surfaces beneath the crest must be highly uneven and irregular. As is suggested by the widespread introduction of Na into zeolite-facies metabasites, permeation of a hot aqueous fluid would have been widespread. Even the apparently general increase of heat flow on the crest would be due not only to heat conduction but also to permeation of a fluid.

Thus, the temperature at the base of the crust beneath the crest would be conceivably variable to some extent within a range, say, between 150 and 450 °C. The upper limit would correspond to a transitional state between the greenschist and amphibolite facies. This agrees with the highest metamorphic facies observed in metabasalts. It supports the view that the whole thickness of oceanic crust is mainly composed of basaltic volcanic piles, the bulk of which are metamorphosed to varying degrees to result in demagnetization.

If the temperature in the uppermost mantle may be assumed to be 150 to 500 °C, basaltic magma intruded in small masses there may be chilled to form an ordinary basaltic texture. It is possible that some metabasalts were crystallized and then recrystallized in the uppermost mantle.

Gabbro masses intrusive into the crust should be subjected to metamorphic recrystallization under the same conditions as those prevailing in the surrounding metabasalt complexes. However, a large part of the metagabbros is in the amphibolite facies, that is, higher in metamorphic temperature than metabasalts. This suggests that a considerable part of gabbros was crystallized and recrystallized in the upper mantle as stated before.

The lithosphere would move laterally by ocean-floor spreading. The rocks recrystallized beneath the crest should be gradually cooled with the movement, and would be subjected to retrograde metamorphism. Some metagabbros evidently show such effects.

In our previous paper (Miyashiro et al. 1970a), we have assumed that the very rugged topography of the crest of the M.A.R. is a surface expression of fracturing within the lithosphere, and that the fracturing causes decrease in seismic velocity. Volcanic piles and their metamorphic derivatives probably have seismic velocities corresponding to layers 2 and 3 respectively. Owing to the fracturing, however, the seismic velocity of the metamorphosed volcanics beneath the crest would have been decreased to a value corresponding to layer 2, with resultant disappearance of layer 3 beneath the Ridge crest.

As the oceanic crust moves laterally away from the axis of the Ridge by ocean-floor spreading, fracturing would gradually wane and stop, while mineral veins would fill the fracture systems in the crust from beneath, to lead to the formation of a coherent plate that represents layer 3. For a more detailed discussion of this problem, refer to Miyashiro et al. (1970a).

10. NOMENCLATURE

The burial metamorphism beneath the M.A.R. appears to be on a grand scale. However, to call it by the name of regional metamorphism is undesirable, because this name has been used since the nineteenth century to represent metamorphism taking place in orogenic belts. In many cases, the regional scale of isothermal curves (isograds) and the widespread presence of schistosity also have been considered intrinsic to it (e.g. Harker 1932).

If the mid-oceanic ridge represents the zone where lithospheric plates are created, the orogenic belt would represent a zone related to the disappearance of a plate. Therefore, metamorphism on the ridge differs entirely in geologic setting from regional metamorphism in orogenic belts. Moreover, schistosity is lacking or weak in most of the metamorphic rocks of the ridge. Therefore, the use of the name regional maetamorphism for mid-oceanic ridges will be confusing.

The term burial metamorphism (Coombs 1961) may well be used for ridges as a non-commital name to emphasize the poorness of the development of schistosity. Though this name has been used for zeolite-facies rocks in orogenic belts, there is no rigid historical tradition in its usage.

When we wish to emphasize the geologic setting of the metamorphism in mid-oceanic ridges, it is probably proper to call it by the name of mid-oceanic ridge metamorphism or of ocean-floor metamorphism.

In this paper, the classical names of metamorphic facies adopted by Eskola (1939) have been used with two additional ones by Coombs (1961). It has not been well established to what extent ocean-floor metamorphism resembles regional metamorphism mineralogically. It is regrettable that some authors used such names as the almandine–amphibolite facies or quartz–albite–epidote–almandine subfacies on the assumption that they were dealing with regional metamorphism. These names were originally proposed on the basis of the mineralogical features in regions of Barrovian zones. The burial metamorphism of the Mid-Atlantic Ridge would be of a lower pressure type than that of the Barrovian zones.

We are grateful to Drs E. J. W. Jones and D. Ninkovich who read an early draft of this paper and offered valuable advice. The present study was made with financial support from the G. Unger Vetlesen Foundation.

REFERENCES (Miyashiro et al.)

Aumento, F. & Loncarevic, B. D. 1969 The Mid-Atlantic Ridge near 45° N. III. Bald Mountain. *Can. J. Earth Sci.* **6**, 11–23.

Cann, J. R. 1969 Spilites from the Carlsberg Ridge, Indian Ocean. *J. Petrology* **10**, 1–19.

Cann, J. R. & Funnell, B. M. 1967 Palmer Ridge: a section through the upper part of the ocean crust? *Nature, Lond.* **213**, 661–664.

Cann, J. R. & Vine, F. J. 1966 An area on the crest of the Carlsberg Ridge: petrography and magnetic survey. *Phil. Trans. Roy. Soc. Lond.* A **259**, 198–217.

Coombs, D. S. 1961 Some recent work on the lower grade metamorphism. *Aust. J. Sci.* **24**, 203–215.

Eskola, P. 1939 Die metamorphen Gesteine. In *Die Entstehung der Gesteine* (ed. C. W. Correns), pp. 263–407. Berlin: Julius Springer.

Green, D. H. 1964 The metamorphic aureole of the peridotite at the Lizard, Cornwall. *J. Geol.* **72**, 543–563.

Harker, A. 1932 *Metamorphism*. London: Methuen.

Heirtzler, J. R. 1968 Sea-floor spreading. *Scient. Am.* **219**, No. 6, 60–70.

Hess, H. H. 1965 Mid-oceanic ridges and tectonics of the sea-floor. In *Submarine geology and geophysics* (eds. W. F. Whittard and R. Bradshaw), pp. 317–333. London: Butterworths.

Langseth, M. G. 1967 Review of heat flow measurements along the mid-oceanic ridge system. In *The World Rift System* (Report of Symposium, Ottawa, Canada, 1965), pp. 349–362. *Geol. Surv. Can. Pap.* no. 66–14.

Loomis, A. A. 1966 Contact metamorphic reactions and processes in the Mt. Tallac roof remnant, Sierra Nevada, California. *J. Petrology* **7**, 221–245.

Melson, W. G., Bowen, V. T., van Andel, Tj. H. & Siever, R. 1966 Greenstones from the central valley of the Mid-Atlantic Ridge. *Nature, Lond.* **209**, 604–605.

Melson, W. G., Thompson, G. & van Andel, Tj. H. 1968 Volcanism and metamorphism in the Mid-Atlantic Ridge, 22° N latitude. *J. geophys. Res.* **73**, 5925–5941.

Melson, W. G. & van Andel, Tj. H. 1966 Metamorphism in the Mid-Atlantic Ridge, 22° N latitude. *Marine Geol.* **4**, 165–186.

Miyashiro, A. 1961 Evolution of metamorphic belts. *J. Petrology* **2**, 277–311.

Miyashiro, A. 1968 Metamorphism of mafic rocks. In *Basalts* (eds. H. H. Hess and A. Poldervaart), Vol. 2, pp. 799–834. New York: Interscience.

Miyashiro, A. & Shido, F. 1970 Progressive metamorphism in zeolite assemblages. *Lithos* **3**, 251–260.

Miyashiro, A., Shido, F. & Ewing, M. 1969 Composition and origin of serpentinites from the Mid-Atlantic Ridge near 24° and 30° north latitude. *Contr. miner. Petrol.* **23**, 38–52.

Miyashiro, A., Shido, F. & Ewing, M. 1970a Petrologic models for the Mid-Atlantic Ridge. *Deep Sea Res.* **17**, 109–123.

Miyashiro, A., Shido, F. & Ewing, M. 1970b Crystallization and differentiation in abyssal tholeiites and gabbro sfrom mid-oceanic ridges. *Earth Planet. Sci. Lett.* **7**, 361–365.

Muir, I. D. & Tilley, C. E. 1966 Basalts from the northern part of the Mid-Atlantic Ridge. II. The Atlantis collections near 30° N. *J. Petrology* **7**, 193–201.

Opdyke, N. D. & Hekinian, R. 1967 Magnetic properties of some igneous rocks from the Mid-Atlantic Ridge. *J. geophys. Res.* **72**, 2257–2260.

Quan, S. H. & Ehlers, E. G. 1963 Rocks of the northern part of the Mid-Atlantic Ridge. *Bull. geol. Soc. Am.* **74**, 1–7.

Richardson, S. W., Gilbert, M. C. & Bell, P. M. 1969 Experimental determination of kyanite–andalusite and andalusite–sillimanite equilibria; the aluminum silicate triple point. *Am. J. Sci.* **267**, 259–272.

Seki, Y. 1961 Calcareous hornfelses in the Arisu district of the Kitakami Mountains, northeastern Japan. *Jap. J. Geol. Geogr.* **32**, 55–78.

Shand, S. J. 1949 Rocks of the Mid-Atlantic Ridge. *J. Geol.* **57**, 89–92.

Shido, F. 1958 Plutonic and metamorphic rocks of the Nakoso and Iritōno districts in the central Abukuma Plateau. *J. Fac. Sci. Univ. Tokyo*, Sec. II, **11**, 131–217.

Talwani, M., Windisch, C., Langseth, M. & Heirtzler, J. R. 1968 Recent geophysical studies on the Reykjanes Ridges (abstract). *Trans. Am. geophys. Un.* **49**, 201.

Thayer, T. P. 1968 Continental alpine-type mafic (ophiolitic) complexes as possible keys to the geology of mid-oceanic ridges (abstract). *Trans. Am. geophys. Un.* **49**, 365.

van Andel, Tj. H. 1968 The structure and development of rifted midoceanic rises. *J. mar. Res.* **26**, 144–161.

Vine, F. J. 1966 Spreading of the ocean floor: new evidence. *Science, N.Y.* **154**, 1405–1415.

von Herzen, R. P. & Uyeda, S. 1963 Heat flow through the eastern Pacific Ocean floor. *J. geophys. Res.* **68**, 4219–4250.

Reprinted from *Tectonophysics*, **13**, 141–159 (1972)

PRESSURE AND TEMPERATURE CONDITIONS AND TECTONIC SIGNIFICANCE OF REGIONAL AND OCEAN-FLOOR METAMORPHISM

AKIHO MIYASHIRO

Department of Geological Sciences, State University of New York at Albany, Albany, N.Y. (U.S.A.)

(Received August 10, 1971)

ABSTRACT

Miyashiro, A., 1972. Pressure and temperature conditions and tectonic significance of regional and ocean-floor metamorphism. In: A.R. Ritsema (Editor), *The Upper Mantle. Tectonophysics*, 13(1–4): 141–159.

A $p-t$ scale of metamorphism based on experimental investigations has been established in the past ten years. The scale suggests that the temperature of regional metamorphism is higher than was formerly considered, that partial melting takes place commonly within a continental-type crust with possible generation of granitic masses and large-scale layering of the crust, and that eclogite is stable in a wide range of conditions in the crust and mantle. High-pressure metamorphism (glaucophanitic metamorphism) represents an unusually low geothermal gradient which can be realized presumably in relation to tectonic descent of relevant geologic masses.

Regional metamorphism takes place in orogenic belts on the convergent edges of lithospheric plates, whereas ocean-floor metamorphism takes place beneath mid-oceanic ridges on the divergent edges. In many orogenic belts, petrographically mapped areas have been enlarged with resultant clarification of the regional distribution of $p-t$ conditions in metamorphic belts. It makes a connecting link between mineralogic and petrographic investigations and tectonic interpretations of the belts. Low-pressure metamorphic terranes (with andalusite) are more widespread than medium- and high-pressure ones in the world. A large part of the Precambrian terranes, belonging to the low-pressure type, should represent relatively shallow original depths. Detailed field work and dating indicated that the metamorphic history of many areas was more complex than had formerly been considered.

A high-pressure and a low-pressure regional metamorphic belt occur in a pair in many parts of the circum-Pacific regions. The high-pressure belt is claimed to have been formed by the influence of the descent of a lithospheric slab into depths of the upper mantle, whereas the low-pressure belt, being accompanied by granitic and andesitic rocks, is claimed to have formed by the thermal and other effects of materials rising from a Benioff zone. The apparent increase of glaucophane-schist facies rocks with decreasing geologic age may be a result of secular variation in the characteristics of plate movements.

It has recently been found that a newly created oceanic crust, except for its thin surface layer, is subjected to ocean-floor metamorphism beneath a mid-oceanic ridge. The metamorphosed crust moves laterally with ocean-floor spreading. The metamorphism demagnetizes the main part of the crust. Related aqueous fluids should transfer heat with resultant increase of heat-flow values on the mid-oceanic ridge.

135

INTRODUCTION

Researches in metamorphic geology have been greatly diversified for the past decade. A number of new branches have grown up to prove their importance. In the present review, however, I wish to choose only such branches that should exert strong influence on our understanding of large-scale tectonic processes of the earth. This is because a greater number of geologists and geophysicists will be more interested in the tectonic significance of metamorphism than in the problems of purely mineralogical or petrological importance.

Regional and ocean-floor metamorphism are the only important classes of metamorphism from the tectonic angle. *Regional metamorphism* takes place in orogenic belts, which are considered to form on convergent edges of lithospheric plates from the viewpoint of plate tectonics. *Ocean-floor metamorphism* takes place beneath mid-oceanic ridges which represent divergent edges of lithospheric plates (Miyashiro et al., 1971, p. 602). Investigation of such metamorphism gives us important information on the tectonic processes taking place in the relevant places. Ordinary textbooks of petrology give many other categories of metamorphism, which, however, are of little tectonic interest. The present review is concerned almost exclusively with regional and ocean-floor metamorphism.

In particular, the following three fields of progress will be treated in some detail: (1) recent establishment of the $p-t$ scale of metamorphism and its petrogenetic and tectonic significance; (2) nature and tectonic significance of regional metamorphism; and (3) finding of ocean-floor metamorphism and its significance.

PRESSURE AND TEMPERATURE OF METAMORPHISM

Recent progress of experimental investigations

From the beginning of the 20th century, many pioneers in metamorphic petrology dreamt of the days when the $p-t$ conditions of metamorphism would be determined on an experimental basis. Recent progress has resulted in a considerable degree of realization of those dreams. Practically all the metamorphic minerals have been synthetically investigated. Metamorphic conditions can be discussed largely on the basis of experimental work. Winkler's (1967) book of metamorphic petrology, giving particular emphasis on the experimental side of this science, is a timely triumphant song of experimental petrology.

The equilibrium curves for the reactions: 2 jadeite = albite + nepheline and jadeite + quartz = albite, were determined by Robertson et al. (1957) and Birch and LeComte (1960). These were among the early successful examples of high-pressure experiments. Fig.1 shows the curve for the latter reaction, which is widely believed to be near the high-pressure low-temperature limit of metamorphic conditions.

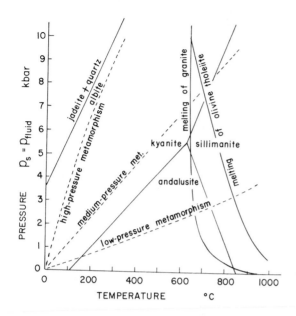

Fig.1. Stability fields of jadeite associated with quartz (Birch and LeComte, 1960) and the three polymorphs of Al_2SiO_5 (Richardson et al., 1969) as well as the curves for the beginning of melting of granite and olivine tholeiite (Yoder and Tilley, 1962). Geothermal gradients for low-, medium- and high-pressure metamorphism are shown by broken lines.

Other examples of experimentally investigated solid–solid reactions are the phase transformations between the polymorphs of Al_2SiO_5 (andalusite, kyanite and sillimanite) and the formation of eclogite, both of which will be commented upon later (Fig.1, 2). The equilibrium curves for such solid–solid reactions are indispensable indicators of p_s and t, since they are independent of p_{H_2O} and p_{CO_2}.

Most of the progressive metamorphic reactions, however, involve the liberation of H_2O and CO_2. A large number of reactions of this category were experimentally investigated, and are well reviewed by Winkler (1967), Ernst (1967), Sobolev (1970), and Hewitt and Wones (1971).

Since metamorphism means changes of rocks or rock complexes in an essentially solid state, the melting temperature of rocks indicates the high-temperature limit of metamorphism. As shown in Fig.1, granitic rocks begin to melt at about 620–700°C in the pressure range of 2–10 kbar, if p_s = p_{fluid} (Goranson, 1931; Tuttle and Bowen, 1958; Yoder and Tilley, 1962; Boettcher and Wyllie, 1968). Pelitic rocks begin to melt at a slightly higher temperature (Wyllie and Tuttle, 1961). With decreasing p_{H_2O}, the melting temperature increases. Under dry conditions, the temperature of the beginning of melting of granite is about 950°C at zero pressure, and should increase with increasing pressure. Since the value of p_{H_2O} during metamorphism is not known for certain, the high-temperature limit of metamorphism is ambiguous to a considerable extent.

Relevant theoretical analysis

Theoretical analysis of the factors controlling the conditions of metamorphism is a necessary basis for the application of experimental studies to the interpretation of petrographic data. A framework of such an analysis was laid down in the 1950's by Thompson (1955) and Korzhinskii (1959). The nature of dehydration and decarbonation reactions in metamorphic processes was clarified in the same period (Danielsson, 1950; Thompson, 1955; Harker, 1958; Greenwood, 1961). The effect of p_{O_2} on the stability relations of ferromagnesian silicates was realized for the first time around 1960 (Eugster, 1959; Mueller, 1960; Miyashiro, 1964).

The occurrence of metamorphic minerals, such as cordierite and staurolite, depends to some extent on the $p-t$ conditions but also to no less extent on the bulk chemical composition of the pertinent rocks. The stability range of a mineral varies with the associated minerals. The application of experimental results to natural rocks with diversified compositions and mineral assemblages could be facilitated through theoretical petrologic grids based on the Schreinemakers rule. Attempts to derive such grids were made by Albee (1965), Hess (1969), and Hoschek (1969).

Pressure–temperature assignment for metamorphic facies

The theoretical framework of metamorphic petrology is largely based on the principle of metamorphic facies (Eskola, 1920, 1939) and the zonal mapping of progressive metamorphic terranes (Harker, 1932). Thus the relative temperatures of metamorphic facies were known from petrographic studies made before World War II (Eskola, 1939). However, there were little reliable data on the numerical values of temperature and pressure till about 1960.

Cautious authors in the 1950's refrained from speaking of numerical values of $p-t$ conditions (e.g., Thompson, 1955; Francis, 1956). Other authors, however, gave an estimate that the greenschist facies corresponds to 0–250°C, and the amphibolite facies to 350–600°C (Ramberg, 1952, p.137; Hietanen, 1956; Barth, 1962, p.332). Recent experimental works indicate that the temperature corresponding to these two facies are probably 100–350°C higher than the estimate, as shown in Fig.2. Since the total range of temperature in ordinary metamorphism is only several hundred degrees, the change of estimation by, say, 300°C has an important petrogenetic significance.

This change resulted mainly from two causes. One is the finding of the zeolite and prehnite–pumpellyite facies (Coombs et al., 1959; Coombs, 1961), followed by the experimental determination of the upper stability limit of pertinent zeolites (Campbell and Fyfe, 1965; Liou, 1970; Thompson, 1970, 1971). Before the finding of the zeolite facies, it had been commonly believed that the greenschist facies represented the lowest temperature range of metamorphism, and that its low-temperature limit was close to the temperature on the surface of the earth (cf., Harker, 1932, p.209). After the establishment of the zeolite and prehnite–pumpellyite facies, it was recognized that the

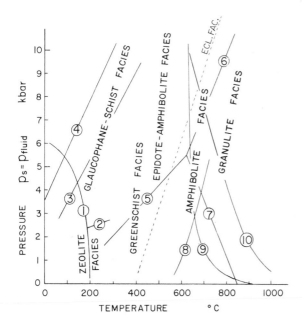

Fig.2. Temperature and pressure corresponding to metamorphic facies. Equilibrium curves for important metamorphic reactions and melting curves are shown as follows:
(1) analcime + quartz = albite + H_2O (Thompson, 1971).
(2) lawsonite + 2 quartz + 2 H_2O = laumontite (Thompson, 1970).
(3) aragonite = calcite (MacDonald, 1956; Crawford and Fyfe, 1964; Johannes and Puhan, 1971).
(4) jadeite + quartz = albite (Birch and LeComte, 1960; Newton and Kennedy, 1968).
(5) kyanite = andalusite (Althaus, 1967; Richardson et al., 1969).
(6) kyanite = sillimanite (Althaus, 1967; Richardson et al., 1969).
(7) andalusite = sillimanite (Althaus, 1967; Richardson et al., 1969).
(8) muscovite + quartz = K feldspar + Al_2SiO_5 + H_2O (Evans, 1965; Day, 1970).
(9) beginning of melting of granite (Yoder and Tilley, 1962).
(10) beginning of melting of olivine tholeiite (Yoder and Tilley, 1962).
ECL. FAC.: low-pressure limit of the eclogite facies (Green and Ringwood, 1967).

greenschist facies should represent a temperature range on the higher side of these two facies. Some mineral assemblages of the zeolite facies are stable around 250–350°C. Even the analcime + quartz assemblage, which is stable up to the middle of the zeolite facies, was found to persist to about 200°C in the 0–2 kbar range (Curve *1* in Fig.2). Thus, the establishment of the zeolite and prehnite–pumpellyite facies drastically increased our temperature estimates of the low-grade metamorphism.

The other cause is that the temperature of the triple point of Al_2SiO_5 minerals was experimentally shown to be much higher than had been expected. The researches made in the first half of the 1960's gave values of $t = 300–390°C$ and $p = 8–9$ kbar for the point. In the second half, the determined temperature shifted to higher values and the determined pressure to lower ones. Thus, Althaus (1967) and Richardson et al. (1969) gave temperatures around 600°C and pressures around 6 kbar (Fig.1, 2). Since this triple point lies probably at a lower part of the amphibolite facies, the above values mean that the amphibolite facies represent temperatures around and higher than 600°C and pressures covering a wide range on both sides of 6 kbar.

It is to be noted that oxygen isotope geothermometry has made progress and can now give significant figures for the temperature of metamorphism. The results are generally consistent with those derived from synthetic works (Epstein and Taylor, 1967; Garlick and Epstein, 1967; Taylor and Coleman, 1968).

Petrogenetic and tectonic significance of the recent p–t assignment for metamorphic facies

The p–t scale we have gained is naturally incomplete and will be modified to some extent in the future. However, now we are certainly on a much stronger basis than a decade ago. Assuming the approximate validity of the scale, we will consider its significance. I wish to outline four relevant problems below.

Granite and granulite problems

In the late 1940's and early 1950's, a considerable number of authors maintained a metamorphic origin for granitic rocks. Among them Ramberg (1952) gave a most thorough and ingenious analysis of this hypothesis. His discussions, however, were based partly on the old, too low estimates for the temperature of metamorphic facies (e.g., Ramberg, 1952, p.244), which led him to preclude the possibility of magma generation by partial melting in the crust; hence his belief in the metamorphic origin of granite.

Our recent estimate gives temperatures of about 600–700°C for the amphibolite facies. Partial melting begins in this facies if p_{H_2O} is equal or close to p_s (Fig.2). Therefore, the possibility of partial melting in the crust itself is no longer the main point of dispute. Whether partial melting can produce ordinary orogenic granites with observed characteristics such as Sr isotope ratios, remains an unsolved problem, as will be discussed later.

Since the granulite facies represents a higher temperature than the amphibolite facies, partial melting should be more common in the former. Melts can contain a relatively high percentage of H_2O, and hence the formation and increase of a melt within metamorphic rocks could greatly decrease the prevailing p_{H_2O}. It might contribute to the dehydration of mineral assemblages. Granulite facies metamorphic rocks show systematic deviations in chemical composition from lower facies rocks (e.g., Ramberg, 1951, 1952; Eade et al., 1966; Lambert and Heier, 1968; Sighinolfi, 1969). It may mean that such rocks are solid residues of partial melting. Partial melting and the rise of a liquid part may have been a main cause for the formation of crustal layering, as granulite facies rocks appear to be widespread in deeper crust.

Cause of high-pressure regional metamorphism

Miyashiro (1961a) emphasized the diversity of the geothermal gradient in regional metamorphism and proposed a three-fold classification: (1) low-pressure metamorphism (originally called andalusite–sillimanite type) corresponding to a geothermal curve running on the low-pressure side of the triple point of Al_2SiO_5 minerals as shown in Fig.1; (2) medium-pressure metamorphism (originally called kyanite–sillimanite type) corresponding to a geothermal curve only slightly on the high-pressure side of the triple point; and (3) high-pressure metamorphism (originally called jadeite-glaucophane type) corresponding to a geothermal curve approaching or crossing the low-pressure stability limit of jadeite + quartz. Each category has a number of characteristic metamorphic facies series. The term "baric types" will be used to designate the low-, medium- and high-pressure types in metamorphism and metamorphic facies series.

The average geothermal gradient for high-pressure regional metamorphism is of the order of $10°C/km$. Such an abnormally low gradient cannot represent a stationary state. It is produced probably by tectonic descent and is maintained only for a short period after the halt of descent. This is consistent with the finding that individual phases of recrystallization are relatively short (Suppe, 1969), and with the view that high-pressure metamorphism takes place in relation to underthrusting of oceanic crust along the Benioff zone (e.g., Miyashiro, 1967; Hamilton, 1969).

Eclogite problem

Eclogite was traditionally regarded as being formed only under extremely high pressures at great depths, probably within the mantle. This view was denied by Green and Ringwood (1967), who showed that the eclogite-facies conditions could be realized not only within the mantle but also in the crust. The $p-t$ range of the eclogite facies overlaps that of the epidote-amphibolite, glaucophane-schist and other facies, though the eclogite facies represent a lower p_{H_2O} than the latter (Fig.2). The majority of the lower crust may be in the granulite and eclogite facies. The melting relations of eclogitic rocks may play an important role in the generation of crustal materials. A considerable number of papers were published in recent years on experimental and theoretical investigations of eclogite rocks (e.g., Yoder and Tilley, 1962; Green and Lambert, 1965; Ringwood and Green, 1966a, b; Banno, 1967; Banno and Green, 1968; Green and Ringwood, 1968; Ringwood, 1969).

NATURE AND TECTONIC SIGNIFICANCE OF REGIONAL METAMORPHISM

Progress in the petrographic mapping of regional metamorphic belts

Rapid enlargement of the extent of petrographically mapped areas in recent years clarified the distribution of metamorphic facies and facies series in pertinent orogenic belts. Such works are a connecting link between the investigation of metamorphic minerals and mineral assemblages on the one hand and that of tectonic development of the orogenic belts on the other. Early attempts of such studies were made particularly in Scotland (Kennedy, 1948) and in Japan (Miyashiro, 1961a).

In the past decade, large-scale petrographic mapping has been made, for example, in the Scottish Highlands (e.g., Johnson, 1963; Chinner, 1966), the northern Appalachians (e.g., Albee, 1968; Thompson and Norton, 1968), the Swiss Alps (e.g., Niggli, 1970), and New Zealand (e.g., Landis and Coombs, 1967). In the Scottish Highlands and the northern Appalachians, a low-pressure metamorphic terrane grades into a medium-pressure one. Correlation of thermal events, tectonic phases and isotopic ages has been attempted in the Scottish Highlands (e.g., Dewey and Pankhurst, 1970).

Dobretsov and co-workers in the Siberian Branch of the U.S.S.R. Academy of Sciences, published a *Map of Metamorphic Facies of the U.S.S.R.* at a scale of 1 : 7,500,000 in 1966 (Dobretsov et al., 1966; Sobolev et al., 1967).

The International Union of Geological Sciences has the "Working Group for the Cartography of the Metamorphic Belts of the World" with H.J. Zwart as chairman (Zwart et al., 1967). This group has been making an effort to prepare petrographic maps of the metamorphic belts on a unified scheme. A fine new *Metamorphic Facies Map of Japan* at a scale of 1 : 2,000,000 (Hashimoto et al., 1970) is a result of this project. The metamorphic maps of western Europe, the Soviet Union and Australia in this project are now in preparation under the leadership of Zwart, V.S. Sobolev and T.G. Vallance, respectively.

Important facts clarified by petrographic mapping

World-wide attempts of petrographic mapping have clarified some facts of great petrologic and tectonic importance.

Predominant occurrence of low-pressure regional metamorphic terranes

Judging from the results of world-wide mapping, low-pressure regional metamorphic terranes are much more widespread than the metamorphic regions of the other two baric types (H.J. Zwart, personal communication, 1971). They occur in various parts of Japan (e.g., Miyashiro, 1961a; Hashimoto et al., 1970), northeast China, many parts of Australia (e.g., Vallance, 1967; Offler and Fleming, 1968), New Zealand (Landis and Coombs, 1967), the Svecofennian and Karelian regions of the Baltic Shield, the Hercynian belt of Europe (Zwart, 1967a,b, 1969), the northern Appalachians (e.g., Thompson and Norton, 1968), Colombia (T. Feininger and B. Doolan, personal communication, 1970), and Chile (Gonzalez-Bonorino, 1971).

Apparent secular change in the baric type of regional metamorphism and the nature of Precambrian terranes

Low-pressure regional metamorphism took place probably in all geologic times, at least from the Middle Precambrian (Svecofennides) to Tertiary (Hidaka belt). On the other hand, high-pressure metamorphic terranes appear to be very rare in the Precambrian (e.g., de Roever, 1956, 1965). Most of the exposed high-pressure terranes are Late Paleozoic or younger (down to Tertiary) as summarized in Table I. The meaning of this relation will be discussed on a later page.

TABLE I

Baric types of regional metamorphism and geologic ages

	Precambrian	Paleozoic	Mesozoic–Cenozoic
Low-pressure metamorphism	Svecofennides Karelides Canada (partly) Australia N.E. China	Hercynides Appalachians (partly) eastern and South Australia Hida belt (Japan) Pichilemu series (Chile)	Ryoke-Abukuma belt (Japan) Hidaka belt (Japan)
Medium-pressure metamorphism	Canada (partly)	Caledonides Appalachians (partly)	North American Cordillera (partly)
High-pressure metamorphism		Sangun belt (Japan) Curepto series (Chile)	Alps Franciscan group (California) Sanbagawa belt (Japan) Kamuikotan belt (Japan) New Caledonia

Whatever the meaning of the age relation may be, it is important that the greater part of the Precambrian shields is made up of metamorphic rocks of the low-pressure type (Zwart, 1967a,b; Lambert and Heier, 1968). The greater part of such rocks is in the amphibolite or lower facies. It was long imagined that the Svecofennian and many other Precambrian terranes with abundant granitic bodies represented the deepest erosion level on the earth (e.g., Eskola, 1932). This view is at variance with the above petrographic evidence. Fig.1 indicates that some of the medium- and high-pressure metamorphic terranes should represent higher pressures and greater depths of formation than such Precambrian rocks.

Widespread occurrence of low-temperature metamorphic rocks

Before the middle of the 1950's, it was customary to regard only the areas of well-recrystallized rocks as metamorphic terranes. However, many of the so-called unmetamorphosed areas that were situated on the low-temperature side of such well-recrystallized terranes were subsequently found to have been recrystallized to variable extents at low temperatures (i.e., in the zeolite, prehnite–pumpellyite, glaucophane-schist and greenschist facies). Thus, the recognized areal extent of metamorphic terranes has been greatly enlarged.

This finding not only enlarged the width of metamorphic belts but also had an influence on genetical views of metamorphism. It was a predominant view around 1950 that glaucophane schist was formed by local contact effect of ultrabasic (and basic) bodies on the surrounding rocks, but recently the regional-scale distribution of glaucophane-schist facies rocks has been well established (e.g., Ernst, 1965; Ernst et al., 1970). The widespread occurrence, accompanied by a great diversity, of zeolite facies rocks was established (e.g., Miyashiro and Shido, 1970). A zone of the prehnite–

pumpellyite facies occurs in terranes ranging from the low- to the high-pressure type (e.g., Coombs, 1961; Hashimoto, 1968; Seki et al., 1969; Hashimoto et al., 1970). Recently prehnite—pumpellyite facies rocks were found to occur at the lowest temperature part of the well-known metamorphic terranes of the Swiss Alps and northern Appalachians (Coombs et al., 1970; Niggli, 1970; Zen, 1971).

Complexity of metamorphic history

In the past, a coherent metamorphic terrane was usually regarded as being produced (i.e., recrystallized) at a single metamorphic event. Recent progress of detailed mapping combined with intensive use of isotopic dating has shown that this is not the case in many regions. The metamorphic terranes of the northern Appalachians, for example, are a result of a long complicated development involving a number of metamorphic events. Mineral assemblages in different parts were formed in different ages, but our knowledge on the age distribution is still very uncertain (Rodgers, 1967; Albee, 1968; Lyons and Faul, 1968). The existence of a relatively simple pattern of temperature distribution does not guarantee simple progressive metamorphism in a limited period.

Recrystallization in a part of a metamorphic terrane may be followed by deposition in another part, which in turn may be subjected to a later phase of recrystallization. In cursory observation, all parts of such a metamorphic terrane may appear to have been recrystallized in a single phase. Such a relation of repeated metamorphism and intervenient sedimentation in an apparently single metamorphic terrane was demonstrated in the Haast Schist of New Zealand (Landis and Coombs, 1967) and in Tertiary metamorphic terranes of Japan (Matsuda and Kuriyagawa, 1965; Seki et al., 1969). Suppe (1969) showed by K/Ar dating that the high-pressure metamorphic terrane of the Franciscan group was a complex including a number of areas deposited and recrystallized at different ages.

Plate tectonics and regional metamorphism

Paired metamorphic belts in circum-Pacific regions

Paired metamorphic belts occur in many parts of the circum-Pacific regions such as Honshu and Hokkaido in Japan, Celebes, New Zealand and Chile. The Franciscan group and the Sierra Nevada zone of California may also be regarded as representing a similar relation. A pair is composed of a high-pressure and a low-pressure regional metamorphic belt which run in parallel. The former is usually on the ocean side (Miyashiro, 1961a).

Prior to the beginning of plate tectonics and even of the hypothesis of ocean-floor spreading, some Japanese geologists had accounted for the origin of paired metamorphic belts by the underthrusting of the Pacific Ocean floor along a Benioff zone beneath island arcs and continental margins (Miyashiro, 1961a,b, 1965, 1967; Matsuda, 1964; Takeuchi and Uyeda, 1965). The high-pressure belt was regarded as corresponding to the zone of a present-day trench, whereas the low-pressure belt corresponded to a zone of island arc volcanism (Fig.3). These ideas were widely accepted in recent plate tectonic interpretations.

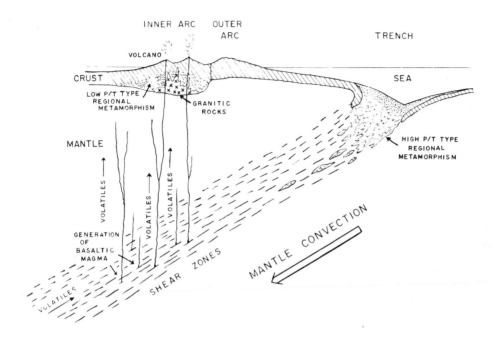

Fig.3. An early model for the origin of paired metamorphic belts. Here, low- and high-pressure metamorphism are called low and high *p/t* type regional metamorphism respectively (Miyashiro, 1967).

Present-day island arcs usually have a double arc structure which is composed of a volcanic inner and a non-volcanic outer arc. In other words, a non-volcanic arc is situated between a volcanic arc and a trench zone. The non-volcanic arc was treated as a part of the trench zone in Miyashiro's (1967) model for the formation of paired belts (Fig.3), whereas Matsuda (1964) and Matsuda and Uyeda (1971) maintained the independence of this arc.

High-pressure metamorphic belts
The Franciscan high-pressure metamorphic terrane in California in particular was discussed by a large number of recent authors as having been metamorphosed in relation to Mesozoic underthrusting of the Pacific plate beneath the North American one. The meta-sediments of the terrane were considered as having been deposited on the Pacific Ocean floor and a trench along the then west coast of North America. Ophiolitic rocks were regarded as fragments of oceanic crust and upper mantle caught in the disturbed sedimentary pile (Hamilton, 1969; Ernst, 1970; Bailey et al., 1970; Coleman, 1971). Hamilton (1969) further proposed that the volcanics in the Paleozoic and Mesozoic eugeosynclinal piles of western North America include not only fragments of oceanic crust and mantle but also volcanic island arcs that formed at some distance offshore and afterwards moved to the continental margin to be welded cn it.

145

Some high-pressure metamorphic belts in the world may not appear to be associated with low-pressure ones. A well-known example of such belts is in the Alps of western Europe. However, Ernst (1971) pointed out the structural and petrologic similarity between the high-pressure metamorphic belt of the Alps and those of Japan and California, suggesting a similarity in the process of their formation. He discussed that the metamorphic grade increases generally in the direction of plate decent in all cases, and that a metamorphic belt with a greater width tends to show a higher maximum pressure of metamorphism and hence a greater maximum subduction depth.

Belts of low-pressure metamorphism, granitic plutonism and andesitic volcanism

Comparison of various parts of the circum-Pacific regions suggests that a zone of intense andesitic volcanism would be underlain by a low-pressure regional metamorphic belt, and that granitic plutons would rise through the metamorphic complex to reach the deeper part of overlying andesitic piles. When erosion is not so advanced as to expose the metamorphic terrane, the area is characterized by granitic plutons and andesitic volcanics.

The high temperature at shallow depths which causes low-pressure metamorphism is created probably by the thermal effects of the rise of aqueous fluids and magmas (Miyashiro, 1967; Hamilton, 1969) in cooperation with the convective and diapiric rise of peridotitic materials through an upper mantle above the Benioff zone. The aqueous fluid and magmas originating from the Benioff zone would transport heat to induce the convective and diapiric movements. The magmas would form by partial melting of the oceanic crust within the downgoing lithosphere. The rising convective and diapiric masses would undergo partial melting at some depths, resulting in the production of another group of magmas. It is conceivable that the magmas originating in the Benioff zone lead to the formation of calc-alkali volcanic rocks mainly composed of ordinary andesite and dacite together with granitic rocks, whereas the magmas formed in convective and diapiric masses of the upper mantle lead to the formation of tholeiites, high-alumina basalts, and alkali basalts together with their derivatives (Ringwood, 1969).

There are some apparently unpaired low-pressure metamorphic belts in the world; e.g., the Hercynides and Svecofennides. Such a belt may have been initially paired with a high-pressure metamorphic belt, which, however, disappeared by later recrystallization except for small remnants such as the glaucophane schists of Ile de Groix off Bretagne in the Hercynides. Alternatively it is possible that such low-pressure metamorphic terranes are intrinsically unpaired belts that were formed by different tectonic processes.

A grand-scale tectonic process controls the formation of granitic magmas, leading to the regional distribution of granitic masses and metamorphic terranes. The temperature distribution in individual terranes, however, appears to be related to some extent to associated granitic masses (e.g., Hamilton and Myers, 1967).

Important recent progress in the granite problem is the finding that the $^{87}Sr/^{86}Sr$ ratios of granitic rocks at the time of emplacement are usually very close to the same ratios of the mantle materials (Hurley et al., 1962). There are a number of geologic models that are consistent with this relation, as exemplified below. First, a new eugeosynclinal pile containing much mantle-derived material may be remelted to produce granitic magmas with such $^{87}Sr/^{86}Sr$ ratios (e.g., Peterman et al., 1967). Secondly, an oceanic crust within a lithospheric slab descending along a Benioff zone may undergo phase transformation and partial melting to produce such granitic magmas (Green and Ringwood, 1968; Ringwood, 1969). Thirdly, granitic magmas may be directly generated in a deep mantle under high-pressure phase relations (Matsumoto, 1965, 1968).

Secular variation of tectonic processes

Empirically we know that the rocks of the glaucophane—schist and the zeolite facies are rare in old metamorphic terranes and especially in the Precambrian, whereas those of the granulite facies are rare in younger terranes. A number of discussions were published on this problem (e.g., De Roever, 1956, 1965; Miyashiro, 1961a; Sobolev et al., 1967; Zwart, 1967a; Dobretsov, 1968), but its meaning is not clear yet. A difficulty in the interpretation comes from the fact that we do not know whether the rocks of the whole high-pressure series are rare, or those only of the glaucophane-schist facies (or the low-temperature part of the high-pressure series) are rare, in the Precambrian. Even if we may assume the validity of the former possibility, still there will remain a question whether rocks of the high-pressure series were hardly formed in the Precambrian, or once-formed rocks of the series were recrystallized afterwards to be converted into lower-pressure types.

However, a possibility exists that such empirical rules may reflect the secular change in the nature of tectonic processes in the history of the earth. The thickness, the velocity and the behavior of lithospheric plates may have changed with geologic age. Many Precambrian orogenic belts are more or less similar to the Hercynides in their great areal extent, abundant granitic rocks, and the apparently unpaired low-pressure type of regional metamorphism. Such orogenies appear to take place in any geologic age. On the other hand, the type of orogeny that produces paired metamorphic belts may be confined to younger geologic ages (Paleozoic to the present).

The significance of metamorphism and associated igneous activity in plate tectonics will be discussed more boldly in another paper (Miyashiro, 1972).

FINDING OF OCEAN-FLOOR METAMORPHISM AND ITS SIGNIFICANCE

Finding of metamorphic rocks in ocean floors

The occurrence of metamorphic rocks derived from basalt, dolerite and gabbro in ocean floors has been described by several authors since 1966. It is confined to the Mid-Atlantic and Mid-Indian Ocean Ridges, that is, the ridges with a well-developed median valley and a low rate of spreading (Melson, et al., 1966, 1968; Cann and Funnell, 1967; Aumento and Loncarevic, 1969; Miyashiro et al., 1970, 1971). Intense fault movement and serpentinite intrusion in such ridges would increase the chance of exposure of metamorphic rocks that had originally formed at some depths below the ocean floor. Metamorphic rocks were dredged so far from the walls of the median valley and fault block mountains as well as from transverse fracture zones across ridges. In the fracture zones, they are always accompanied by serpentine and other ultramafics. Probably some ocean-floor serpentinites are also metamorphosed rocks.

Mid-oceanic ridges that have a smooth surface and no median valley, such as the East Pacific rise, do not have metamorphic rocks exposed on the surface. Conceivably metamorphic rocks are present beneath such ridges also but have little chance of exposure for the weaker fault movement and the absence of serpentinite intrusion.

Most of the basic metamorphic rocks are nearly or entirely lacking in schistosity, though there exist some other rocks showing considerable preferred orientation. The metamorphic rocks usually belong to the zeolite, greenschist or amphibolite facies. Zeolite facies rocks were found only from a part of the Mid-Atlantic Ridge (Miyashiro et al., 1971).

Since the oceanic crust is much thinner than the continental one, such a temperature rise that causes greenschist facies or higher metamorphism can occur only beneath mid-oceanic ridges where the existence of high thermal gradients is suggested by high heat flow values. Even zeolite-facies metamorphism should take place mainly beneath mid-oceanic ridges. Some of the amphibolite-facies metamorphic rocks were probably recrystallized in the upper mantle and subsequently brought up to the surface by serpentinite intrusion.

Metamorphosed crust should move laterally from a mid-oceanic ridge to a normal ocean basin by ocean-floor spreading. Hence, it is likely that the major part of the oceanic crust is composed of metamorphic rocks.

Significance of ocean-floor metamorphism

The stripes of magnetic anomalies observed on the ocean surface appear to be due to a thin layer, about 0.5—2.0 km thick, of magnetized basaltic and gabbroic rocks (Vine, 1966; Heirtzler, 1968). The major part of the oceanic crust underlying the above layer is virtually demagnetized. It is natural to presume that the demagnetization is due to metamorphic recrystallization (Miyashiro et al., 1970).

Metasomatic introduction of Na and removal of Ca is intense in rocks subjected to the low-temperature recrystallization in such metamorphism. The chemical migration is caused probably by moving aqueous fluids. The rise of such fluids should have resulted in a great increase in the apparent heat-flow value on mid-oceanic ridges. The fluid containing much Ca would be ultimately emitted into sea water to promote carbonate deposition.

High heat flows observed on mid-oceanic ridges were usually interpreted as suggesting the presence of convective upwelling of mantle materials beneath the ridges. Alternatively, however, they may simply be a result of igneous intrusion and rise of aqueous fluids in ridges (Miyashiro et al., 1971). It is yet to be decided whether we have to presume an active role of mantle convection in the framework of plate tectonics or not.

If ophiolite complexes in orogenic belts represent fragments of oceanic crust and upper mantle, they should show similarity in chemical compositions. Common basic metamorphic rocks of the ophiolite complexes in high-pressure metamorphic terranes show higher alkali contents than abyssal tholeiites and associated gabbros which are the most common rocks of the surface of oceanic crust. However, it is not inconceivable that the majority of the basic rocks of oceanic crust underwent intense metasomatism so as to gain compositions very different from abyssal tholeiites. Regional metamorphism in orogenic belts also would have affected the composition of ophiolites.

REFERENCES

Albee, A.L., 1965. A petrogenetic grid for the Fe-Mg silicates of pelitic schists. *Am. J. Sci.*, 263: 512–536.

Albee, A.L., 1968. Metamorphic zones in northern Vermont. In: E-an Zen, W.S. White, J.B. Hadley and J.B. Thompson (Editors), *Studies of Appalachian Geology: Northern and Maritime*. Interscience, New York, N.Y., pp. 329–341.

Althaus, E., 1967. The triple point andalusite-sillimanite-kyanite. *Contrib. Mineral. Petrol.*, 16: 29–44.

Aumento, F. and Loncarevic, B.D., 1969. The Mid-Atlantic Ridge near 45°N. III. Bald Mountain. *Can. J. Earth Sci.*, 6: 11–23.

Banno, S., 1967. Effect of jadeite component on the paragenesis of eclogitic rocks. *Earth Planet. Sci. Lett.*, 2: 249–254.

Banno, S. and Green, D.H., 1968. Experimental studies on eclogites: the role of magnetite and acmite in eclogitic assemblages. *Chem. Geol.*, 3: 21–32.

Bailey, E.H., Blake Jr., M.C. and Jones, D.L., 1970. On-land Mesozoic oceanic crust in California Coast Ranges. *U.S. Geol. Surv. Prof. Pap.*, 700C: C70–C81.

Barth, T.F.W., 1962. *Theoretical Petrology*. Wiley, New York, N.Y., 2nd edition, 416 pp.

Birch, F. and LeComte, P., 1960. Temperature-pressure plane for albite composition. *Am. J. Sci.*, 258: 209–217.

Boettcher, A.L. and Wyllie, P.J., 1968. Melting of granite with excess water to 30 kilobars pressure. *J. Geol.*, 76: 235–244.

Campbell, A.S. and Fyfe, W.S., 1965. Analcime-albite equilibria. *Am. J. Sci.*, 263: 807–816.

Cann, J.R. and Funnell, B.M., 1967. Palmer ridge: a section through the upper part of the ocean crust? *Nature*, 213: 661–664.

Chinner, G.A., 1966. The distribution of pressure and temperature during Dalradian metamorphism. *Q. J. Geol. Soc. Lond.*, 122: 159–186.

Coleman, R.G., 1971. Plate tectonic emplacement of upper mantle peridotites along continental edges. *J. Geophys. Res.*, 76: 1212–1222.

Coombs, D.S., 1961. Some recent work on the lower grades of metamorphism. *Aust. J. Sci.*, 24: 203–215.

Coombs, D.S., Ellis, A.J., Fyfe, W.S. and Taylor, A.M., 1959. The zeolite facies, with comments on the interpretation of hydrothermal synthesis. *Geochim. Cosmochim. Acta*, 17: 53–107.

Coombs, D.S., Horodyski, R.J. and Naylor, R.S., 1970. Occurrence of prehnite-pumpellyite facies metamorphism in northern Maine. *Am. J. Sci.*, 268: 142–156.

Crawford, W.A. and Fyfe, W.S., 1964. Calcite-aragonite equilibrium at 100°C. *Science*, 144: 1569–1570.

Danielsson, A., 1950. Das Calcit-Wollastonitgleichgewicht. *Geochim. Cosmochim. Acta*, 1: 55–69.

Day, H.W., 1970. Redetermination of the stability of muscovite + quartz. *Geol. Soc. Am., 1970 Ann. Meet., Abstr. with Programs*, p. 535.

De Roever, W.P., 1956. Some differences between post-Paleozoic and older regional metamorphism. *Geol. Mijnbouw*, 18: 123–127.

De Roever, 1965. On the cause of the preferential distribution of certain metamorphic minerals in orogenic belts of different age. *Geol. Rundsch.*, 54: 933–943.

Dewey, J.F. and Pankhurst, R.J., 1970. The evolution of the Scottish Caledonides in relation to their isotopic age pattern. *Trans. R. Soc. Edinb.*, 68: 361–389.

Dobretsov, N.L., 1968. Lawsonite-glaucophane and glaucophane-schists of the U.S.S.R. and some problems of metamorphism in orogenic belts. *Rep. Sov. Geol., Int. Geol. Congr. 23rd Sess., Probl. 3, Acad. Sci. U.S.S.R.*, pp. 31–39 (in Russian with English abstract).

Dobretsov, N.L., Reverdatto, V.V., Sobolev, V.S., Sobolev, N.V., Ushakova, Ye.N. and Khlestov, V.V., 1966. Distribution of regional metamorphism facies in U.S.S.R. *Int. Geol. Rev.*, 8: 1335–1346.

Eade, K.E., Fahrig, W.F. and Maxwell, J.A., 1966. Composition of crystalline shield rocks and fractionating effects of regional metamorphism. *Nature*, 211: 1245–1249.

Epstein, S. and Taylor Jr., H.P., 1967. Variation of $^{18}O/^{16}O$ in minerals and rocks. In: P.H. Abelson (Editor), *Studies in Geochemistry*, 2. Wiley, New York, N.Y., pp. 29–63.

Ernst, W.G., 1965. Mineral parageneses in Franciscan metamorphic rocks, Panoche Pass, California. *Geol. Soc. Am., Bull.*, 76: 879–914.

Ernst, W.G., 1967. Experimental metamorphic petrology. *Trans. Am. Geophys. Union*, 48: 661–666.

Ernst, W.G., 1970. Tectonic contact between the Franciscan mélange and the Great Valley Sequence– crustal expression of a Late Mesozoic Benioff zone. *J. Geophys. Res.*, 75: 886–901.

Ernst, W.G., 1971. Metamorphic zonations on presumably subducted lithospheric plates from Japan, California and the Alps. *Contrib. Mineral. Petrol.*, 34: 43–59.

Ernst, W.G., Seki, Y., Onuki, H. and Gilbert, M.C., 1970. Comparative study of low-grade metamorphism in the California Coast Ranges and the outer metamorphic belt of Japan. *Geol. Soc. Am. Mem.*, 124, 276 pp.

Eskola, P., 1920. The mineral facies of rocks. *Nor. Geol. Tidsskr.*, 6: 143–194.

Eskola, P., 1932. On the origin of granitic magmas. *Mineral. Petrogr. Mitt.*, 42: 455–481.

Eskola, P., 1939. Die metamorphen Gesteine. In: T.F.W. Barth, C.W. Correns and P. Eskola, *Die Entstehung der Gesteine*. Springer, Berlin, pp. 263–407.

Eugster, H.P., 1959. Reduction and oxidation in metamorphism. In: P.H. Abelson (Editor), *Research in Geochemistry*. Wiley, New York, N.Y., pp. 397–426.

Evans, B.W., 1965. Application of a reaction-rate method to the breakdown equilibria of muscovite and muscovite plus quartz. *Am. J. Sci.*, 263: 647–667.

Francis, G.H., 1956. Facies boundaries in pelites at the middle grades of regional metamorphism. *Geol. Mag.*, 93: 353–368.

Garlick, G.D. and Epstein, S., 1967. Oxygen isotope ratios in coexisting minerals of regionally metamorphosed rocks. *Geochim. Cosmochim. Acta*, 31: 181–214.

Gonzalez-Bonorino, F., 1971. Metamorphism of the crystalline basement of central Chile. *J. Petrol.*, 12: 149–175.

Goranson, R.W., 1931. The solubility of water in granite magmas. *Am. J. Sci.*, 22: 481–502.

Green, D.H. and Lambert, I.B., 1965. Experimental crystallization of anhydrous granite at high pressures and temperatures. *J. Geophys. Res.*, 70: 5259–5268.

Green, D.H. and Ringwood, A.E., 1967. An experimental investigation of the gabbro to eclogite transformation and its petrological applications. *Geochim. Cosmochim. Acta*, 31: 767–833.

Green, T.H. and Ringwood, A.E., 1968. Genesis of the calc-alkaline igneous rock suite. *Contrib. Mineral. Petrol.*, 18: 105–162.

Greenwood, H.J., 1961. The system $NaAlSi_2O_6-H_2O$–argon: total pressure and water pressure in metamorphism. *J. Geophys. Res.*, 66: 3923–3946.

Hamilton, W., 1969. Mesozoic California and the underflow of Pacific mantle. *Geol. Soc. Am. Bull.*, 80: 2409–2430.

Hamilton, W. and Myers, W.B., 1967. The nature of batholiths. *U.S. Geol. Surv. Prof. Pap.*, 554C, 29 pp.

Harker, A., 1932. *Metamorphism*. Methuen, London, 360 pp.

Harker, R.I., 1958. The system $MgO-CO_2$–A and the effect of inert pressure on certain types of hydrothermal reactions. *Am. J. Sci.*, 256: 128–138.

Hashimoto, M., 1968. Glaucophanitic metamorphism of the Katsuyama district, Okayama Pref., Japan. *J. Fac. Sci. Univ. Tokyo, Sect.* II, 17: 99–162.

Hashimoto, M., Igi, S., Seki, Y., Banno, S. and Kojima, G., 1970. *Metamorphic Facies Map of Japan (Scale 1/2,000,000)*. Geol. Surv. Japan, Kawasaki.

Heirtzler, J.R., 1968. Sea-floor spreading. *Sci. Am.*, 219 (6): 60–70.

Hess, P.C., 1969. The metamorphic paragenesis of cordierite in pelitic rocks. *Contrib. Mineral. Petrol.*, 24: 191–207.

Hewitt, D.A. and Wones, D.R., 1971. Experimental metamorphic petrology. *Trans. Am. Geophys. Union*, 52: I.U.G.G. 73–82.

Hietanen, A., 1956. Kyanite, andalusite, and sillimanite in the schist in Boehls Butte quadrangle, Idaho. *Am. Mineral.*, 41: 1–27.

Hoschek, G., 1969. The stability of staurolite and chloritoid and their significance in metamorphism of pelitic rocks. *Contrib. Mineral. Petrol.*, 22: 208–232.

Hurley, P.M., Hughes, H., Faure, G., Fairbairn, H.W. and Pinson, W.H., 1962. Radiogenic strontium-87 model of continent formation. *J. Geophys. Res.*, 67: 5315–5334.

Johannes, W. and Puhan, D., 1971. The calcite-aragonite transition, reinvestigated. *Contrib. Mineral. Petrol.*, 31: 28–38.

Johnson, M.R.W., 1963. Some time relations of movement and metamorphism in the Scottish Highlands. *Geol. Mijnbouw*, 42: 121–142.

Kennedy, W.Q., 1948. On the significance of thermal structure in the Scottish Highlands. *Geol. Mag.*, 85: 229–234.

Korzhinskii, D.S., 1959. *Physicochemical Basis of the Analysis of the Paragenesis of Minerals*. Consultants Bureau, New York, N.Y., 142 pp.

Lambert, I.B. and Heier, K.S., 1968. Geochemical investigations of deep-seated rocks in the Australian shield. *Lithos*, 1: 30–53.

Landis, C.A. and Coombs, D.S., 1967. Metamorphic belts and orogenesis in southern New Zealand. *Tectonophysics*, 4: 501–518.

Liou, J.G., 1970. Synthesis and stability relations of wairakite, $CaAl_2Si_4O_{12} \cdot 2H_2O$. *Contrib. Mineral. Petrol.*, 27: 259–282.

Lyons, J.B. and Faul, H., 1968. Isotope geochronology of the northern Appalachians. In: E-an Zen, W.S. White, J.B. Hadley and J.B. Thompson (Editors), *Studies of Appalachian Geology: Northern and Maritime*. Interscience, New York, N.Y., pp. 305–318.

MacDonald, G.J.F., 1956. Experimental determination of calcite-aragonite equilibrium relations at elevated temperatures and pressures. *Am. Mineral.*, 41: 744–756.

Matsuda, T., 1964. Island arc features and the Japanese Islands. *Chigaku-zasshi*, 73: 271–280 (in Japanese with English abstract).

Matsuda, T. and Kuriyagawa, S., 1965. Lower grade metamorphism in the eastern Akaishi Mountains, Central Japan. *Bull. Earthquake Res. Inst., Tokyo Univ.*, 43: 209–235 (in Japanese with English abstract).

Matsuda, T. and Uyeda, S., 1971. On the Pacific-type orogeny and its model–extension of the paired belts concept and possible origin of marginal seas. *Tectonophysics*, 11: 5–27.

Matsumoto, T., 1965. Some aspects of the formation of primary granitic magmas in the upper mantle. *Upper Mantle Symp. New Dehli, 1964*; 112–126.

Matsumoto, T., 1968. A hypothesis on the origin of the Late Mesozoic volcano-plutonic association in East Asia. *Pac. Geol.*, 1: 77–83.

Melson, W.G., Bowen, V.T., Van Andel, Tj.H. and Siever, R., 1966. Greenstones from the central valley of the Mid-Atlantic Ridge. *Nature*, 209: 604–605.

Melson, W.G., Thompson, G. and Van Andel, Tj.H., 1968. Volcanism and metamorphism in the Mid-Atlantic Ridge, 22°N latitude. *J. Geophys. Res.*, 73: 5925–5941.

Miyashiro, A., 1961a. Evolution of metamorphic belts. *J. Petrol.*, 2: 277–311.

Miyashiro, A., 1961b. Metamorphism of rocks. In: C. Tsuboi (Editor), *Constitution of the Earth*. Iwanami-shoten, Tokyo, pp. 243–268 (in Japanese).

Miyashiro, A., 1964. Oxidation and reduction in the earth's crust with special reference to the role of graphite. *Geochim. Cosmochim. Acta*, 28: 717–729.

Miyashiro, A., 1965. *Metamorphic Rocks and Metamorphic Belts*. Iwanami-shoten, Tokyo, 458 pp. (in Japanese).

Miyashiro, A., 1967. Orogeny, regional metamorphism, and magmatism in the Japanese Islands. *Medd. Dan. Geol. Foren.*, 17: 390–446.

Miyashiro, A., 1972. Metamorphism and related magmatism in plate tectonics. *A. J. Sci.*

Miyashiro, A. and Shido, F., 1970. Progressive metamorphism in zeolite assemblages. *Lithos*, 3: 251–260.

Miyashiro, A., Shido, F. and Ewing, M., 1970. Petrologic models for the Mid-Atlantic Ridge. *Deep-Sea Res.*, 17: 109–123.

Miyashiro, A., Shido, F. and Ewing, M., 1971. Metamorphism in the Mid-Atlantic Ridge near $24°$ and $30°$N. *Philos. Trans. R. Soc. Lond. Ser. A*, 268: 589–603.

Mueller, R.F., 1960. Compositional characteristics and equilibrium relations in mineral assemblages of a metamorphosed iron formation. *Am. J. Sci.*, 258: 449–497.

Newton, M.S. and Kennedy, G.C., 1968. Jadeite, analcite, nepheline, and albite at high temperatures and pressures. *Am. J. Sci.*, 266: 728–735.

Niggli, E., 1970. Alpine Metamorphose und alpine Gebirgsbildung. *Fortschr. Mineral.*, 47: 16–26.

Offler, R. and Fleming, P.D., 1968. A synthesis of folding and metamorphism in the Mt. Lofty Ranges, South Australia. *J. Geol. Soc. Aust.*, 15: 245–266.

Peterman, Z.E., Hedge, C.E., Coleman, R.G. and Snavely, P.D., 1967. ^{87}Sr/^{86}Sr ratios in some eugeosynclinal sedimentary rocks and their bearing on the origin of granitic magma in orogenic belts. *Earth Planet. Sci. Lett.*, 2: 433–439.

Ramberg, H., 1951. Remarks on the average chemical composition of granulite and amphibolite-to-epidote amphibolite facies gneisses in west Greenland. *Medd. Dan. Geol. Foren.*, 12: 27–34.

Ramberg, H., 1952. *The Origin of Metamorphic and Metasomatic Rocks*. Univ. Chicago Press, Chicago, Ill., 317 pp.

Richardson, S.W., Gilbert, M.C. and Bell, P.M., 1969. Experimental determination of kyanite-andalusite and andalusite-sillimanite equilibria; the aluminum silicate triple point. *Am. J. Sci.*, 267: 259–272.

Ringwood, A.E., 1969. Composition and evolution of the upper mantle. In: P.J. Hart (Editor), *The Earth's Crust and Upper Mantle*. *Geophys. Monogr.*, 13. Am. Geophys. Union, Washington, D.C., pp. 1–17.

Ringwood, A.E. and Green, D.H., 1966a. An experimental investigation of the gabbro-eclogite transformation and some geophysical implications. *Tectonophysics*, 3: 383–427.

Ringwood, A.E. and Green, D.H., 1966b. Petrological nature of the stable continental crust. In: J.S. Steinhart and T.J. Smith (Editors), *The Earth Beneath the Continents*. *Geophys. Monogr.* 10. Am. Geophys. Union, Washington, D.C., pp. 611–619.

Robertson, E.C., Birch, F. and MacDonald, G.J.F., 1957. Experimental determination of jadeite stability relations to 25,000 bars. *Am. J. Sci.*, 255: 115–135.

Rodgers, J., 1967. Chronology of tectonic movements in the Appalachian region of eastern North America. *Am. J. Sci.*, 265: 408–427.

Seki, Y., Oki, Y., Matsuda, T., Mikami, K. and Okumura, K., 1969. Metamorphism in the Tanzawa Mountains, central Japan. *J. Jap. Assoc. Mineral., Petrol. Econ. Geol.*, 61: 1–29, 50–75.

Sighinolfi, G.P., 1969. K–Rb ratio in high grade metamorphism: a confirmation of the hypothesis of a continual crust evolution. *Contrib. Mineral. Petrol.*, 21: 346–356.

Sobolev, V.S. (Editor), 1970. *The Facies of Metamorphism*. Nedra, Moscow, 432 pp. (in Russian).

Sobolev, V.S., Dobretsov, N.L., Reverdatto, V.V., Sobolev, N.V., Ushrakova, E.N. and Khlestov, V.V., 1967. Metamorphic facies and series of facies in the U.S.S.R. *Medd. Dan. Geol. Foren.*, 17: 458–472.

Suppe, J., 1969. Time of metamorphism in the Franciscan terrain of the northern Coast Ranges, California. *Geol. Soc. Am. Bull.*, 80: 135–142.

Takeuchi, H. and Uyeda, S., 1965. A possibility of present day regional metamorphism. *Tectonophysics*, 2: 59–68.

Taylor, H.P. and Coleman, R.G., 1968. O^{18}/O^{16} ratios of coexisting minerals in glaucophane-bearing metamorphic rocks. *Geol. Soc. Am. Bull.*, 79: 1727–1756.

Thompson, A.B., 1970. Laumontite equilibria and the zeolite facies. *Am. J. Sci.*, 269: 267–275.

Thompson, A.B., 1971. Analcite-albite equilibria at low temperatures. *Am. J. Sci.*, 271: 79–92.

Thompson Jr., J.B., 1955. The thermodynamic basis of the mineral facies concept. *Am. J. Sci.*, 253: 65–103.

Thompson Jr., J.B. and Norton, S.A., 1968. Paleozoic regional metamorphism in New England and adjacent areas. In: E-an Zen, W.S. White, J.B. Hadley and J.B. Thompson (Editors), *Studies of Appalachian Geology: Northern and Maritime.* Interscience, New York, N.Y., pp. 319–327.

Tuttle, O.F. and Bowen, N.L., 1958. Origin of granite in the light of experimental studies in the system $NaAlSi_3O_8 - KAlSi_3O_8 - SiO_2 - H_2O$. *Geol. Soc. Am., Mem.*, 74: 153 pp.

Vallance, T.G., 1967. Paleozoic low-presure regional metamorphism in southeastern Australia. *Med. Dan. Geol. Foren.*, 17: 494–503.

Vine, F.J., 1966. Spreading of the ocean floor: new evidence. *Science*, 154: 1405–1415.

Winkler, H.G.F., 1967. *Petrogenesis of Metamorphic Rocks.* Springer-Verlag, New York, N.Y., 2nd edition, 237 pp.

Wyllie, P.J. and Tuttle, O.F., 1961. Hydrothermal melting of shales. *Geol. Mag.*, 98: 56–66.

Yoder, H.S. and Tilley, C.E., 1962. Origin of basalt magmas: An experimental study of natural and synthetic rock systems. *J. Petrol.*, 3: 342–532.

Zen, E-an, 1971. Pumpellyite-bearing metamorphic rocks from the west side of the northern Appalachian region. *Geol. Soc. Am. Northeastern Section, 6th Ann. Meet., Abstr. with Programs,* pp. 64–65.

Zwart, H.J., 1967a. The duality of orogenic belts. *Geol. Mijnbouw*, 46: 283–309.

Zwart, H.J., 1967b. Orogenesis and metamorphic facies series in Europe. *Medd. Dan. Geol. Foren.*, 17: 504–516.

Zwart, H.J., 1969. Metamorphic facies series in the European orogenic belts and their bearing on the causes of orogeny. *Geol. Assoc. Can., Spec. Pap.*, 5: 7–16.

Zwart, H.J., Corvalan, J., James, H.L., Miyashiro, A., Saggerson, E.P., Sobolev, V.S., Subramaniam, A.P. and Vallance, T.G., 1967. A scheme of metamorphic facies for the cartographic representation of regional metamorphic belts. *Int. Union Geol. Sci., Geol. Newsletter*, 1967, (2): 57–72.

7

Reprinted from *24th Internat. Geol. Congr.*, Rept. Sec. 2, 19–26 (1972)

Blueschist Metamorphism and Plate Tectonics*

R. G. COLEMAN,
U.S.A.

ABSTRACT

The tectonic conditions that allow the formation of blueschists can be related in space and time with geosutures developed on the edges of interacting lithospheric plates. Distribution of blueschists within the circum-Pacific orogenic belts and the Alpine mountain system within the mobile Tethys zone of Europe and Asia clearly show a close relationship between deformation and metamorphism. Blueschists form narrow belts parallel to the structural trend. In many places these belts are closely associated with, and paralleled by, ophiolite zones. Blueschists are not preserved within old cratons but invariably occur along ancient continental edges. Experimental petrology and complementary oxygen isotope studies are consistent and show that blueschists form under conditions of high pressure and low temperature. These P-T conditions clearly separate blueschists from other metamorphic facies series in requiring a low thermal gradient combined with unusually high pressures. Compressional zones associated with lithospheric plate boundaries, where subduction and obduction are active, provide a geologic environment capable of forming blueschists. Radiometric age determinations on blueschists demonstrate a recrystallization penecontemporaneous with the accumulation of their protolith sediments and volcanic material. Associated ophiolites, invariably older than the blueschists and representing oceanic crust, are commonly found on top of blueschist metamorphic terranes. Where continental accretion has been continuous, blueschists of contrasting ages are found in mélange zones. High pressure may be produced by "tectonic overpressures" directly under obducted oceanic crust or within a cold descending prism of trench sediments carried to great depths.

INTRODUCTION

BLUESCHIST METAMORPHISM has long held an arbitrary position in the scheme of metamorphic petrology. Eskola (1920), in his original presentation of metamorphic facies, considered glaucophane schists to have formed under moderate temperatures and high pressures. Since his original designation of glaucophane schist as a separate metamorphic facies, there have been numerous conflicting papers regarding the validity of the facies and conditions of its formation (Schurmann, 1951; 1953; Turner and Verhoogen, 1951, p. 472-473; Fyfe and others, 1958, p. 224-225; Miyashiro, 1961; Ernst, 1963; Ernst and others, 1970; Miyashiro and Banno, 1958; Taliaferro, 1943; Vance, 1968; Brothers, 1970; Gresen, 1969; de Roever, 1970; Kaaden, 1966). Much of the disagreement concerns geologic conditions necessary to produce blueschists. Experimental petrologic studies on the stabilities of critical minerals of the blueschists have clearly shown that total pressures accompanied by equally high water pressures are required. In contrast, field studies on blueschist terranes have produced equivocal evidence that has led to the development of divergent views on the origin of these rocks (Ernst, 1970;

---

*Publication authorized by the Director, U.S. Geological Survey.

Authors' addresses are given at the back of this book.

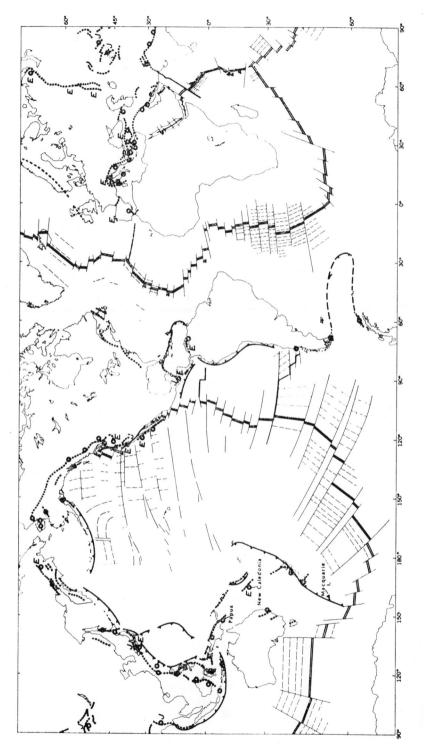

FIGURE 1 — World map showing distribution of ophiolite belts (solid squares) and associated blueschists (open circles) in orogenic zones and occurrences of crustal eclogites type C (E) and pure jadeite pods (J) in serpentinite. Heavy double lines represent spreading ridge, heavy line with barbs represents active subduction zone, heavy and light solid lines represent transform faults and rises, and dashed lines indicate approximate position of magnetic anomalies with respect to the ridges. Dredge hauls of ultramafic rocks from oceanic ridges are shown as solid triangles.

Ernst and others, 1970; Suppe, 1969; Blake and others, 1969; Brothers, 1970; Cogulu, 1967; Dobretsov, 1964). The purpose of this paper is to relate blueschist metamorphism to the concept of plate tectonics whereby reasonable geologic conditions for the formation of blueschists can be developed.

DISTRIBUTION OF BLUESCHISTS

Blueschists have a restricted world distribution and invariably are located within tectonic belts that are inferred to be former continental edges (Figure 1). The most commonly encountered protoliths for the blueschists are graywackes, mafic volcanics, shale and cherts characteristic of trench accumulations. Up until now, blueschist metamorphic terranes have not been discovered within Precambrian shields or within stable Phanerozoic continental areas where low-pressure high-temperature metamorphic rocks are present. Blueschist metamorphism is closely associated with ancient linear ophiolite belts (oceanic crust) (Coleman, 1971) (Fig. 1). Convergence of oceanic and continental plates is thought to have produced major suture zones that contain blueschist zones parallel to the elongate tectonic trends (Bailey and Blake, 1969; Ernst and others, 1970; Brothers, 1970; Davies, 1968, 1971; Cogulu, 1967). In California, a crude inverted mineral zoning within the blueschist terranes grades upward in places to the Coast Range thrust, now considered part of an ancient exhumed subduction zone (Ernst, 1970; Bailey and others, 1970). The Coast Range fault or subduction zone had placed oceanic crust (ophiolites) and overlying unmetamorphosed sediments of the Great Valley on top of the metamorphosed Franciscan trench sediments. Continued post-metamorphic near-surface thrusting along the Coast Range fault zone and post-Valanginian tear faults in the upper plate have obscured the structural relations developed during the blueschist metamorphism (Jones and Irwin, 1971). Brothers (1970) describes a situation in New Caledonia where ultramafic slabs (ophiolites) without a thick sedimentary cover have been emplaced on top of Eocene and Cretaceous sediments recrystallized to blueschist assemblages. In California and New Caledonia, the tectonic emplacement of oceanic crust is apparently penecontemporanous with formation of blueschist (Coleman, 1967). A recent discovery of blueschists in the Owen-Stanley Range of New Guinea has been related to the obduction of a large slab of oceanic crust (ophiolites). Here the blueschist minerals have developed directly under the thrust contact between the ophiolites and the sediments making up the Owen-Stanley Range (Davies and Smith, 1971). Cogulu (1967) and Kaaden (1966) describe similar relations in Turkey, and blueschists in the Pennine nappes of the Alps can be related to emplacement of ophiolite nappes (Nicolas, 1969). Further field work on blueschist terranes is necessary to document the apparent near contemporaneity between collision of oceanic and continental edges or two continental edges, and the development of blueschists in suture zones; however, the present evidence strongly supports this contemporaneity.

AGE OF BLUESCHIST METAMORPHISM

Where reliable radiometric ages are available, it appears that blueschist metamorphism often is only slightly younger than its protolith (Suppe, 1969; Coleman, 1967; Coleman and Lanphere, 1971). For instance, in-situ blueschists from California and Oregon give two distinct ages of approximately 110 and 130 m.y. Both of these ages of metamorphism are very close to the ages of fossils contained within their protolith sediments. From these age relations one must conclude that sedimentation was penecontemporaneous with metamorphism along the western edge of the American continental plate during Jurassic-Cretaceous time. Presence of older (150 m.y.) high-grade blueschists and eclogites as tectonic blocks, with mélange zones separating younger blueschist terranes, illustrates the tectonic transport after metamorphism. Metamorphosed Eocene and Cretaceous sediments from New Caledonia give radiometric ages of 21 to 38 m.y. The period of deformation and blueschist metamorphism in New Caledonia coincides with the period during which the overlying ophiolites were emplaced (Routhier, 1953). Ages of blueschists from other circum-Pacific orogenic belts show that the onset of metamorphism was close to the end of sedimentation, except for the outer metamorphic belt of Japan (Ernst and others, 1970). The close temporal spacing of blueschist metamorphic events in western North America indicates that the metamorphic events are much shorter (< 20 m.y.) than previously thought.

Some problems exist in the interpretation of radiometric ages on blueschists, as many of these rocks show discordant age patterns which indicate that post-metamorphic deformation has reduced the K/Ar ages of the glaucophanes but has had little effect on the micas (Coleman and Lanphere, 1971). In other areas where blueschist metamorphism has been followed by low-pressure and higher temperature metamorphism, radiometric ages may represent these later metamorphic events, as the high-pressure phases of the blueschists are very sensitive to thermal events (Ellenberger, 1960; de Roever, 1970). Where radiometric data are available, the associated ophiolites are always found to be older than the associated blueschists or their protolith sediments (Cogulu, 1967; Coleman and Lanphere, 1971; Lanphere, 1971). If we accept the idea that the ophiolites represent oceanic crust that has been transported to the continental edge from a spreading ridge, these rocks must be older than the trench sediments deposited on any part of the oceanic crust being consumed or accreted to the leading edge of the continental plate. Miyashiro (1961) proposed the concept of paired metamorphic belts where blueschist facies metamorphic rocks formed at the continental edges and high-temperature low-pressure facies metamorphic rocks developed synchronously, associated with granitic plutons toward the interior of the continents. This concept has been employed by numerous proponents of plate tectonics and has been considered as part of the dynamics within subduction zones (Oxburgh and Turcotte, 1971; Bird, 1970; Hamilton, 1966).

Radiometric dating from California does not support the synchroneity of blueschist metamorphism with metamorphism and plutonism to the east (Evernden and Kistler, 1970; Kistler and others, 1971). In fact, development of paired metamorphic belts contemporaneously within a subduction zone produces some first-order tectonic probems in allowing final juxtaposition of blueschists against granite and high-temperature metamorphic rocks of the same age as in Japan (Ernst and others, 1970) and New Zealand (Landis and Coombs, 1967) Zwart (1967) clearly shows a separation in time and space between the Hercynian (low-pressure metamorphism) and Alpine (high-pressure metamorphism) orogens of Europe. Further detailed work is necessary to relate paired belts by radiometric data in order to solve this first-order tectonic problem of post-metamorphic juxtaposition.

CONDITIONS FOR THE FORMATION OF BLUESCHISTS

As stated earlier, experimental work carried out on the critical blueschist minerals, such as jadeite, lawsonite, aragonite, phengite and glaucophane, clearly shows that the formation of assemblages such as glaucophane + lawsonite + chlorite + sphene (metabasalts) or quartz + jadeite pyroxene + lawsonite + white mica (metasediments) requires high pressures (5-7 Kb) and moderate temperatures (150°-300°C). The arguments supporting this experimental work, together with a review of the existing data, have been recently presented by Ernst and others

(1970) and need not be repeated here. The problem of producing high pressures while maintaining low temperatures has not yet been satisfactorily resolved. However, the discovery of aragonite marbles (Coleman and Lee, 1962) suggests to me that if rocks undergoing metamorphism are strong enough, "tectonic overpressure" developed during metamorphism might provide the required pressures, without extreme depth of burial. Recent experimental work on rock strengths (Brace and others, 1970; Robertson, in press) has shown that the graywackes from the Franciscan Formation in California do indeed have the necessary strength to develop sufficient "tectonic overpressures" when pore pressure is low (pore pressure, λ, expressed as $[P_{H2O}/\sigma_3]$; $\lambda = 1.0$ high pore pressure, $\lambda = < .04$ low pore pressure). When λ increases, rock strength falls off rapidly, and rocks having $\lambda > 0.7$ are unlikely to develop significant overpressures (Brace and others, 1970). Because hydrated metamorphic phases such as lawsonite - chlorite - white mica are present to some degree in all blueschists, the metamorphic conditions require that P_{H2O} nearly equal P total. Thus, in order to produce the hydrated phases characteristic of the blueschist facies, it appears the λ must be high, and this will result in apparent concomitant weakening of the rocks.

This argument has led Brace and others (1970) and Ernst (1971) to propose that it is unlikely that "tectonic overpressures" can be obtained and that the high pressures required to produce blueschist assemblages are developed by submerging trench sediments deep within subduction zones (Ernst, 1970). The rapid descent down subduction zones would allow cold sediments to reach great depths (i.e., high pressures with moderate temperatures).

This ingenious scheme has many merits and provides for P-T conditions necessary to form blueschists consistent with plate tectonics. However, this model has certain flaws that should be discussed. (1) No plausible tectonic mechanism can be given to elevate blueschists after they were formed deep within a subduction zone and before they become consumed in the upper mantle. (2) The presence of blueschists directly under oceanic crust showing inverted metamorphic zonation (New Caledonia, New Guinea) is not compatible with the model of metamorphism deep in a subduction zone.

If all pressure is to be developed by depth of burial or tectonic thickening, then the mechanics of exhuming the blueschists within geosutures becomes a first-order tectonic problem. The presence of mélanges and the juxtaposing of blueschists of different ages in California suggest that large-scale horizontal movements have indeed occurred, but no convincing tectonic field evidence has been given to show that blueschists have been exhumed from depths of 20-30 km. The elevation of blueschists formed deep in a subduction zone would require the cessation of subduction followed by tectonic movement capable of rotating the original metamorphic zonation upside down or at least rotating the section 90 degrees. Radiometric evidence suggests blueschist metamorphism at ~150, ~130 and ~100 m.y. in California and Oregon and requires interrupted subduction and vertical uplift or transport at 20-m.y. intervals in order to preserve the metamorphic rocks from being consumed deep within the mantle. Such interruptions may relate to the 30-m.y. plutonic intervals within the Sierra Nevada recorded at ~140, ~180 and ~85 m.y. by Kistler and others (1971). Dickinson (1970) documents variations in Great Valley sediments that are related to these magmatic episodes; however, there is nothing in the sedimentary record to account for large vertical uplifts following blueschist metamorphism at 20-m.y. intervals.

Another way to produce blueschists is by deep-seated thrusting of oceanic crust over continental edges to produce "tectonic overpressures." This overcomes the first-order tectonic problem of elevating and rotating blueschists produced by deep burial. However, producing "tectonic overpressures" in rocks while still maintaining high partial pressures of water remains a difficult problem in light of the experimental data on rock strengths (Brace and others, 1970).

In California, metagraywackes and metabasalts (blueschist) actually contain more water ($+110°C$) than many of their presumed equivalents — unmetamorphosed Franciscan graywackes and basalts. Thus there is evidence that water is consumed by the hydrated metamorphic mineral structures during metamorphism, rather than being expelled — as it would be if recrystallization resulted from increased temperatures and produced progressive dehydration. If water weakening is the main argument against "tectonic overpressures," consumption of fluids by recrystallization could effectively reduce pore pressures and bring about increased strengths as recrystallization proceeds. Porosity decreases as Franciscan graywackes are metamorphosed under blueschist conditions. Confining pressures reduce the initial porosity as water is consumed by crystallization of hydrated metamorphic minerals. Improved grain bonding may allow incremental fluid overpressures to develop in step hardening during synkinematic deformation. If such crystallization hydration can be documented, then one could call upon this mechanism to effectively lower λ with concomitant increase of total pressure within a stress field. Experimental work on rock strengths has not measured the strengthening effect caused by crystallizing hydrated metamorphic minerals. Brace and others (1970) have drawn attention to the fact that shale layers within the graywackes would further reduce rock strengths during metamorphism; however, the amount of shale within the Franciscan sediments is quite variable, and shale is absent in some sections. Until further quantitative mapping is carried out, the weakening effect of shale in the thick massive metagraywacke units may be considered to be of only local importance and could explain the presence of mélange units.

Deep burial and modest tectonic overpressures are probably both necessary for the development of blueschist under reasonable geologic conditions. The P-T conditions of blueschist metamorphism certainly are distinct and separate from other types of metamorphism. Broad diffuse metamorphic zonation is usually subparallel to the regional tectonic trends and confined to narrow elongate belts. Post-metamorphic thrusting and recumbent folding commonly obscure evidence of thermal- or pressure-gradient trends, and the geometry during blueschist metamorphism remains obscure.

The development of blueschists as part of the interaction between lithospheric plates seems to me to be the most acceptable model yet proposed for this particular metamorphic facies series. The common change of blueschists to higher temperature and lower pressure metamorphic facies within some geosutures shows that the preservation of blueschists within Precambrian cratons is quite unlikely, as later high-temperature and low-pressure metamorphism or plutonism would tend to destroy blueschist minerals (de Roever, 1970). Integration of studies on blueschist metamorphism with other facets of orogenic history in ancient geosutures should provide guides as to the early periods of lithospheric plate collision and subsequent deformation.

REFERENCES

Bailey, E. H., and Blake, M. C., Jr., 1969. Late Mesozoic sedimentation and deformation in western California. Geotektonika, 3 & 4.
Bailey, E. H., Blake, M. C., Jr., and Jones, D. L., 1970. On-land Mesozoic oceanic crust in California Coast Ranges. *In* Geological Survey research, 1970. U.S. Geol. Surv. Prof. Pap. 700-C, p. C70-C81.

Bailey, E. H., Irwin, W. P., and Jones, D. L., 1964. Franciscan and related rocks, and their significance in the geology of western California. California Div. Min. and Geol. Bull., no. 183, 171 p.

Bird, J. M., 1970. General concepts of orogenesis in terms of lithosphere plate tectonics. Geol. Soc. Am. Abs. with Programs, 2, no. 7, p. 733-734.

Blake, M. C., Jr., Irwin, W. P., and Coleman, R. G., 1969. Blueschist-facies metamorphism related to regional thrust faulting. Tectonophysics, 8, p. 237-246.

Brace, W. F., Ernst, W. G., and Kallberg, R. W., 1970. An experimental study of tectonic overpressure in Franciscan rocks. Geol. Soc. Am. Bull., 81, p. 1325-1338.

Brothers, R. N., 1970. Lawsonite-albite schists from northernmost New Caledonia. Contr. Mineralogy and Petrology, 25, p. 185-202.

Cogulu, E., 1967. Etude pétrographique de la région de Mihaliccik (Turquie). Bull. Suisse Mineral. Petrogr., 47, p. 683-824.

Coleman, R. G., 1967. Glaucophane schists from California and New Caledonia. Tectonophysics, 4, p. 479-498.

Coleman, R. G., 1971. Plate tectonic emplacement of upper mantle peridotites along continental edges. J. Geophys. Res., 76, p. 1212-1222.

Coleman, R. G., and Lanphere, M. A., 1971. Distribution and age of high-grade blueschists, associated eclogites, and amphibolites from Oregon and California. Geol. Soc. Am. Bull., 82, p. 2397-2412.

Coleman, R. G., and Lee, D. E., 1962. Metamorphic aragonite in the glaucophane schists of Cazadero, California. Am. J. Sci., 260, p. 577-595.

Davies, H. L., 1968. Papuan ultramafic belt. Int. Geol. Congr., Prague, 1968. Rep., Proc. Sec. 1, p. 209-220.

Davies, H. L., 1971 (in press). Peridotite-gabbro-basalt complex in eastern Papua: An overthrust plate of oceanic mantle and crust. Bur. Miner. Resour. Aust. Bull. 128.

Davies, H. L., and Smith, I. E., 1971. Geology of Eastern Papua: A synthesis. Geol. Soc. Am. Bull., 82, p. 3299-3312.

de Roever, W. P., 1970. Some problems concerning the origin of glaucophane and lawsonite. *In* Problems of petrology and genetic mineralogy (Sobolev Volume), "Nauka" Moscow, v. 11, p. 24-40 (in Russian).

Dickinson, W. R., 1970. Relations of andesites, granites, and derivative sandstones to arc-trench tectonics. Rev. Geophys. and Space Phys., 8, no. 2, p. 813-860.

Dobretsov, N. L., 1964. The jadeite rocks as indicators of high pressure in the Earth's crust. XXII Int. Geol. Congr., India, Vol. of Abstracts, p. 231-232.

Ellenberger, F., 1960. Sur une paragénèse éphémère à lawsonite et glaucophane dans le métamorphisme alpin en Haute-Maureienne (Savoie). Soc. Geol. France Bull., 7, p. 190-194.

Ernst, W. G., 1963. Petrogenesis of glaucophane schists. J. Petrol., 4, p. 1-30.

Ernst, W. G., 1970. Tectonic contact between the Franciscan mélange and the Great Valley sequence — crustal expression of a late Mesozoic Benioff zone. Geophys. Res., 75, p. 886-901.

Ernst, W. G., 1971. Do mineral parageneses reflect unusually high-pressure conditions of Franciscan metamorphism. Am. J. Sci., 270, p. 81-108.

Ernst, W. G., Seki, Y., Onuki, H., and Gilbert, M. C., 1970. Comparative study of low-grade metamorphism in the California Coast Ranges and the Outer Metamorphic Belt of Japan. Geol. Soc. Am. Mem. 124, 280 p.

Eskola, P., 1920. The mineral facies of rocks. Norsk Geol. Tidsskr., 6, p. 143-194.

Essene, E. J., Fyfe, W. S., and Turner, F. J., 1965. Petrogenesis of Franciscan glaucophane schists and associated metamorphic rocks, California. Contr. Mineral. and Petrol., 11, p. 695-704.

Evernden, J. F., and Kistler, R. W., 1970. Chronology of emplacement of Mesozoic batholithic complexes in California and western Nevada. U.S. Geol. Surv. Prof. Pap. 623, 42 p.

Fyfe, W. S., Turner, F. J., and Verhoogen, J., 1958. Metamorphic reactions and metamorphic facies. Geol. Soc. Am. Mem. 73, 260 p.

Gresens, R. L., 1969. Blueschist alteration during serpentinization. Contr. Mineral. and Petrol., 24, p. 93-113.

Hamilton, W., 1969. Mesozoic California and the underflow of Pacific mantle. Geol. Soc. Am. Bull., 80, p. 2409-2430.

Jones, D. L., and Irwin, W. P., 1971. Structural implications of an offset Early Cretaceous shoreline in northern California. Geol. Soc. Am. Bull., 82, p. 815-822.

Kaaden, G. V. d., 1966. The significance and distribution of glaucophane rocks in Turkey. Turkey Miner. Res. and Explor. Inst. Bull., no. 67, p. 36-67.

Kaaden, G. V. d., 1969. Zur Entstehung der Glaukophan-Lawsonite und glaukophanitischen Grünschierterfazies, Geländebeobachtungen und Mineral-Synthesen. Fortschr. Mineral., 46.

Kistler, R. W., Evernden, J. F., and Shaw, H. R., 1971. Sierra Nevada plutonic cycle: Part I. Origin of composite granitic batholiths. Geol. Soc. Am. Bull., 82, p. 853-868.

Landis, C. A., and Coombs, D. S., 1967. Metamorphic belts and orogenesis in southern New Zealand. Tectonophysics, 4, p. 501-518.

Lanphere, M. A., 1971. Age of the Mesozoic oceanic crust in California Coast Ranges. Geol. Soc. Am. Bull., 82, p. 3209-3212.

Miyashiro, A., 1961. Evolution of metamorphic belts. J. Petrol., 2, p. 277-311.

Miyashiro, A., and Banno, S., 1958. Nature of glaucophanitic metamorphism. Am. J. Sci., 256, p. 97-110.

Nicolas, A., 1969. Tectonique et métamorphisme dans les stura di lanzo (Alpes Piémontaises). Bull. Suisse Mineral. Petrogr., 49, p. 359-377.

Oxburgh, E. R., and Turcotte, D. L., 1971. Origin of paired metamorphic belts and crustal dilation in island arc regions. J. Geophys. Res., 76, p. 1315-1327.

Robertson, E. C., 1971 (in press). Strength of metamorphosed graywacke and other rocks. Birch Volume.

Routhier, P., 1953. Etude géologique du versant occidental de la Nouvelle-Calédonie entre le Cola de Boghen et la Pointe d'Arama. Soc. Géol. France Mém., 67, 271 p.

Schürmann, H. M. E., 1953. Beiträge zur Glaucophanfrage. Neues Jahrb. Mineral. Abh., 85, p. 303-394.

Schürmann, H. M. E., 1951. Beiträge zur Glaucophanfrage. Neues Jahrb. Mineral. Monatsh., p. 49-68.

Seki, Y., 1958. Glaucophanitic regional metamorphism in the Kanto Mountains, central Japan. Jap. J. Geol. and Geogr., 29, p. 233-258.

Suppe, John, 1969. Times of metamorphism in the Franciscan terrain of the Northern Coast Ranges, California. Geol. Soc. Am. Bull., 80, p. 135-142.

Taliaferro, N. L., 1943. Franciscan-Knoxville problem. Am. Assoc. Pet. Geol. Bull., 27, p. 109-219.

Turner, F. J., and Verhoogen, J., 1951. Igneous and metamorphic petrology. McGraw-Hill, New York, 602 p.

Vance, J. A., 1968. Metamorphic aragonite in the prehnite-pumpellyite facies, Northwest Washington. Am. J. Sci., 266, p. 299-315.

Zwart, H. J., 1967. The duality of orogenic belts. Geol. en Mijnbow, 46, p. 283-309.

IV
Petrologic Problems

Metamorphic terranes seem to be most broadly and intensively developed in the region of convergent plate margins. Here the thermal structure of the earth is most dramatically displaced from the quasi-steady-state condition characteristic of the interior of the plates; in addition, crustal segments of compressive portions are the locale of a broad, complex orogenic zone, the deformational history of which is largely a function of the dynamics and geometry of lithospheric plate motions.

Eclogites, ophiolite suites, and blueschists are typical of many, if not most, recognizable phanerozoic convergent plate sutures. Inasmuch as the origins of such rock types are still being debated among petrologists, they constitute important problems on which to focus: because of their common plate tectonic association, an appreciation of the petrogenesis of these lithologies to some extent provides answers to questions regarding the relationship between metamorphism and lithospheric plate interactions.

Editor's Comments on Papers 8 Through 11

8 **Coleman, Lee, Beatty, and Brannock:** *Eclogites and Eclogites: Their Differences and Similarities*

9 **Ringwood and Green:** *An Experimental Investigation of the Grabbo–Eclogite Transformation and Some Geophysical Implications*

10 **Banno:** *Classification of Eclogites in Terms of Physical Conditions of Their Origin*

11 **Fry and Fyfe:** *On the Significance of the Eclogite Facies in Alpine Metamorphism*

Eclogites

Eclogites, essentially bimineralic garnet + clinopyroxene rocks of roughly basaltic composition, have been a continuous source of mineralogic and petrologic interest for the past 150 years (Haüy, 1822). Their great density (i.e., approximately 3.5) compared to other metabasaltic rocks led Eskola (1939) to recognize them as representatives of a separate high-pressure, high-temperature metamorphic facies, in spite of the fact that field occurrences are enigmatic. Such rocks, although quite rare, are widely distributed around the world, where they occur invariably as xenoliths, pods, or tectonic blocks in largely noneclogitic terranes.

Geologic and petrochemical relationships of eclogites have been investigated by numerous researchers. Paper 8, by Coleman, Lee, Beatty, and Brannock, was selected for this volume because of the rather complete and general treatment. These authors have distinguished three main groups of eclogites, as shown in Table 1. Because of the

Table 1

Eclogite Type	A	B	C
Geologic occurrence	Inclusions in kimberlites, basalts, peridotites	Lenses and pods in gneisses	Tectonic blocks in blueschist terranes
Bulk rock composition	Olivine basalt	Intermediate basalt	Tholeiitic basalt
Garnet composition	>55 pyrope	30–55 pyrope	<30 pyrope
Clinopyroxene composition	Diopsidic	Diopsidic omphacite	Jadeitic
Inferred physical conditions	Very high T, High P	Moderately high T, high P	Low T, high P

broad range of physical conditions (inferred from the geologic association) over which eclogitic rocks seem to be produced, Coleman, Lee, Beatty, and Brannock favored abandonment of the concept of an eclogitic metamorphic facies. However, they have clearly shown that the mineralogy of such rocks is a function of the presumed physical conditions; hence these mineralogic parameters may be utilized to characterize the metamorphic environment. To this extent, eclogites are as valuable as any other rock type to metamorphic petrologists.

Experimental-phase equilibrium investigations regarding transformation of various metabasaltic assemblages and rock bulk compositions to predominantly garnet + clinopyroxene have been carried out in several different laboratories. Among the published studies may be mentioned those by Yoder and Tilley (1962); Ringwood and Green (Paper 9); Green and Ringwood (1967); Cohen, Ito, and Kennedy (1967); Ito and Kennedy (1970, 1971); and Lambert and Wyllie (1972). All workers have shown that low-pressure basaltic phase associations are separated in P–T space from high-pressure eclogitic mineral assemblages by a transition zone of considerable width, within which garnet, plagioclase, and clinopyroxene coexist (= garnet granulite). The 1966 contribution by Ringwood and Green has been selected as Paper 9 because of its comprehensiveness and early publication. Moreover, these authors invoked a rather prescient petrotectonic model which employed aspects of the sea-floor-spreading hypothesis and the known gravitative instability of eclogite in the uppermost mantle to account for the origin of andesitic volcanism in island arcs.

Controversy has surrounded the experimental studies (see, e.g., Green and Ringwood, 1972; Kennedy and Ito, 1972), with two principal problems in sharp focus: (1) what are the P–T coordinates of the transition zone in the far-subsolidus region, and (2) what is the nature of the systematic mineralogic variations across the transition zone? Yoder and Tilley, and Cohen, Ito, and Kennedy, extrapolated the basalt-eclogite transition to pressures on the order of 10 ± 3 kb at 400° C, whereas at this temperature Ringwood and Green placed this transition at 2 ± 2 kb. Although it is clear that, within the transition zone, the plagioclase decreases in amount and becomes more sodic with increasing pressure, whereas the amount of garnet concomitantly increases, Ito and Kennedy believed that these changes take place chiefly at discrete P–T values; in contrast, Green and Ringwood claimed that there is a gradual, continuous change in phase proportions. In spite of these interpretive differences, it is clear that eclogitic assemblages are generated over a broad range of relatively high pressure physical conditions, in harmony with the conclusions of Coleman, Lee, Beatty, and Brannock.

Because of the extensive P–T stability field for garnet-clinopyroxene rocks, it is desirable to obtain a measurable parameter that will allow delineation of the physical conditions attending the formation of specific eclogitic assemblages. In Paper 10, Banno demonstrated the utility of measuring the iron-magnesium fractionation between garnet and clinopyroxene in this regard. Similar results were presented by Ernst, Seki, Onuki, and Gilbert (1970, Chaps. V and IX). The distribution constant varies regularly with grade: the most extreme fractionation occurs in eclogites from blueschist terranes, whereas partitioning tends toward unit values in garnet + clinopyroxene pairs from the successively higher-temperature granulite and kimberlite environments.

The problem of the occurrence of eclogites versus amphibolites was addressed in Paper 11 by Fry and Fyfe. They noted that where a separate aqueous fluid phase (largely H_2O in composition) is present in metabasaltic systems at pressures approaching lithostatic values, experimental phase-equilibrium studies (e.g., Yoder and Tilley, 1962; see also Lambert and Wyllie, 1972, and Allen, Modreski, Haygood, and Boettcher, 1972) indicate that amphibolitic assemblages are stable relative to garnet +

clinopyroxene. Fry and Fyfe suggested that the activity of H_2O is low in regions where eclogite is formed either due to dilution of the fluid phase by other components such as alkali chlorides, or because of the absence of a separate H_2O-rich phase.

References

Allen, J. C., Modreski, P. J., Haygood, C., and Boettcher, A. L. (1972) The role of water in the mantle of the earth: the stability of amphiboles and micas: *24th Intl. Geol. Congress, Montreal,* Sec. 2, p. 231–240.

Cohen, L. H., Ito, K., and Kennedy, G. C. (1967) Melting and phase relations in an anhydrous basalt to 40 kilobars: *Am. Jour. Sci.,* **265,** 475–518.

Ernst, W. G., Seki, Y., Onuki, H., and Gilbert, M. C. (1970) Comparative study of low-grade metamorphism in the California Coast Ranges and the Outer Metamorphic Belt of Japan: *Geol. Soc. America Mem.,* **124,** 276 p.

Eskola, P. (1939) Die metamorphen Gesteine, in *Die Entstehung der Gesteine* by T. F. W. Barth, C. W. Correns, and P. Eskola, Springer, Berlin, 422 p.

Green D. H., and Ringwood, A. E. (1967) An experimental investigation of the gabbo to eclogite transformation and its petrological applications: *Geochim. Cosmochim. Acta,* **31,** 767–833.

Green, D. H., and Rignwood, A. E. (1972) A comparison of recent experimental data on the gabbo-garnet granulite-ecologite transition: *Jour. Geol.,* **80,** 277–288.

Haüy, R. J. (1822) *Traité de minéralogie,* 2nd ed., v. 2: Bachelier, Paris, 594 p.

Ito, K., and Kennedy, G. C. (1970) The fine structure of the basalt-eclogite transition: *Mineral. Soc. Amer. Spec. Paper 3,* 77–83.

Ito, K., and Kennedy, G. C. (1971) An experimental study of the basalt-garnet granulite-eclogite transition: *Geophys. Monogr. Ser.* **14,** 303–314.

Kennedy, G. C., and Ito, K. (1972) Comments on: "A comparison of recent experimental data on the gabbro-garnet granulite-eclogite transition": *Jour. Geol.,* **80,** 289–292.

Lambert, I. B., and Wyllie, P. J. (1972) Melting of gabbo (quartz eclogite) with excess water to 35 kilobars, with geological applications: *Jour. Geol.,* **80,** 693–708.

Yoder, H. S., Jr., and Tilley, C. E. (1962) Origin of basalt magmas: an experimental study of natural and synthetic rock systems: *Jour. Petrol.* **3,** 342–532.

8

Reprinted from *Geol. Soc. America Bull.*, **76**, 483–508 (May 1965)

R. G. COLEMAN
D. E. LEE
L. B. BEATTY
W. W. BRANNOCK

U. S. Geological Survey, Menlo Park, Calif.

Eclogites and Eclogites:
Their Differences and Similarities

Abstract: Eclogites are divisible into three groups based on mode of occurrence: Group A, inclusions in kimberlites, basalts, or layers in ultramafic rocks; Group B, bands or lenses within migmatite gneissic terrains; Group C, bands or lenses within alpine-type metamorphic rocks. The compositions range from olivine basalt for Group A to tholeiitic basalts for Group C. New analytical data on six eclogites from glaucophane schist terrains in California and New Caledonia now permit comparisons among the three eclogite types. The pyrope content of the garnets is distinctive for each group as follows: Group A, greater than 55 per cent py; Group B, 30–55 per cent py; Group C, less than 30 percent py. Pyroxenes coexisting with these garnets also reflect a compositional change related to their occurrence. The jadeite content progressively increases from Group A through Group B, whereas the diopside content decreases. A comparison of eclogites from different geologic occurrences but with similar bulk compositions demonstrates variation in Ca-Mg partition between coexisting garnet and pyroxene. The Ca/Mg ratio increases in garnet

and decreases in pyroxene from Group A through Group B eclogites. This obvious difference in the Ca-Mg partition between coexisting garnet-pyroxene in eclogites of the same bulk composition indicates a broad range of pressure-temperature conditions obtained during crystallization. Experimental synthesis of eclogite-like material at high pressures and temperatures demonstrates that some eclogites may form in the earth's mantle, but naturally occurring Group C eclogites have coexisting garnet-pyroxene with distinct Ca/Mg ratios when compared to Group A or B eclogites of similar bulk composition. This difference in the Ca/Mg ratio must reflect the pressure-temperature conditions characterizing the glaucophane schist facies.

The formation of eclogites within different metamorphic facies is strong evidence of the divergent pressure-temperature conditions that allow basalts to recrystallize into garnet-pryoxene rocks. In view of the rather compelling field evidence, it would seem advisable to discontinue the concept of an eclogite metamorphic facies.

CONTENTS

INTRODUCTION

In the past decade an emphasis on learning more about the earth's crust and mantle has brought forth numerous new facts concerning its history. As yet, however, we cannot say with any certainty what could be the chemical composition of the mantle or subcrustal material. Geophysical data have revealed that the material making up the earth's mantle at or directly below the M-discontinuity (under continental areas) has a density of approximately 3.23 g/cc at temperatures of about 500°C (Kennedy, 1959). This geophysical information dictates that the material forming the upper mantle must consist of very dense minerals. The suggestion by Fermor (1913) that eclogite could satisfy the requirements for upper mantle material has received continuing support from geologists and geophysicists. Eclogite, as originally defined by Haüy (1822), is a rock composed of grass-green pyroxene (omphacite) and red garnets. Since this early description of eclogite, occurrences of the rock have been recorded, but its limited outcrops, combined with its peculiar geological occurrence, have strengthened the argument that these eclogites represent isolated fragments of mantle material. Recent laboratory studies have shown that rocks of basaltic composition can be converted to an eclogite-like rock (garnet and pyroxene) at pressures and temperatures expected in the upper mantle of the earth (Kennedy, 1959; Yoder and Tilley, 1962). These very intriguing relationships have therefore enhanced the importance of eclogites as they relate to the earth's history.

The purpose of this paper, however, is to demonstrate that eclogites, as originally defined by Haüy, may form under diverse geologic conditions. Eclogites have been described from metamorphic terrains belonging to the glaucophane schist, granulite, and amphibolite facies as well as from kimberlites, and as ultramafic segregations. This diversity indicates that eclogites may not represent any particular P–T range such as the "eclogite facies" but may form over a range of conditions characteristic of either the Earth's crust or upper mantle. The present use of the rock-term "eclogite" in the geological literature needs further qualification, and it is hoped that this study will provide guidelines for the use of the term. Our interest in this subject stems from our studies of the glaucophane schists and associated eclogites in California (Coleman and Lee, 1962; 1963; Lee and others, 1963a).

ACKNOWLEDGMENTS

We are indebted to Richard Erd for his help and advice on the study of the pyroxenes. Single crystal investigations now under way by Joan Clark have been of great value to us. The spectrographic work was very carefully carried out by Robert Mays. The manuscript has been improved by the reviews of E-an Zen and David Wones. All these persons are members of the U.S. Geological Survey. We have also benefited greatly from discussions with many geologists who have participated in field conferences on the California eclogites and glaucophane schists.

ECLOGITE NOMENCLATURE

The nomenclature of eclogite-like rocks has become complex over the years for numerous rock types have been described as eclogites from many geologic environments. It is not clear from Haüy's (1822) original description that eclogites comprise only those bimineralic rocks consisting of pyroxene and garnet. Since he listed rutile, sphene, quartz, epidote, amphibole, and sulfides as accessories, it would seem justifiable to distinguish eclogite varieties according to the presence or absence of certain minerals. There are many excellent studies on the mineralogy and petrology of eclogites that provide abundant data concerning the presently known types (Briere, 1920; Eskola, 1921; Williams, 1932; Alderman, 1936; Tilley, 1936; Switzer, 1945; Subramaniam, 1956; Angel, 1957; O'Hara, 1960; Yoder and Tilley, 1962; Nixon and others, 1963). These studies show that eclogites have bulk compositions ranging from olivine basalts to tholeiitic basalts. This range in composition is reflected by the occurrence of other primary minerals in addition to garnet and pyroxene. Furthermore, there are important variations in the composition of the coexisting garnet and pyroxene. As with any common rock type, such as granite, varieties are distinguished by the presence or absence of certain primary minerals. These varieties may reflect changes in bulk composition or different pressure-temperature conditions of crystallization. Without considering genesis, the authors recommended that the eclogite varieties (Table 1) be included in the eclogite family as being petrographically similar. These mineralogic varieties should not be confused with the genetic groups discussed later.

These varieties may not include all those rocks that previously have been called eclogite, but they do cover a broad basaltic compositional range. Such an inclusive use of the term "eclogite" is preferable since discussion of this problem would be hampered if only bimineralic garnet-pyroxene rocks were considered as "true eclogites."

The origin of eclogites has provoked considerable controversy, and as yet the subject has not been fully elucidated. The best summation of the debate is given by Yoder (1950) and need not be repeated here. The underlying cause for the eclogite controversy is the anomalous geologic setting of these rocks combined with their very restricted outcrops. In many areas eclogites cannot be directly related

amphibolite and show retrograde metamorphism.

Group C. Bands or lenses within the metamorphic rocks of the alpine-type orogenic zones as defined by Hess (1955) and Kundig (1956), locally forming isolated blocks when associated with glaucophane schists.

Eskola (1921) recognized four groups of eclogite occurrences, but we have consolidated his first two groups under our Group A, that is, eclogites from diamond pipes and layered ultramafics are considered under one heading. Group A seems to be truly of deep-seated igneous or metamorphic origin and may indeed

TABLE 1. ECLOGITE VARIETIES AND MINERALOGY

Eclogite variety	Essential minerals*	Accessory minerals
Eclogite (*sensu stricto*)	Garnet, pyroxene	Rutile, sphene, clinozoisite
Kyanite eclogite	Garnet, pyroxene, kyanite	Rutile, quartz, sphene, mica
Amphibole eclogite	Garnet, pyroxene, amphibole	Rutile, sphene, clinozoisite, mica, quartz
Orthopyroxene eclogite	Garnet, pyroxene, orthopyroxene	Rutile, olivine, diamond
Plagioclase eclogite	Garnet, pyroxene, plagioclase	Amphibole, spinel, quartz

* In this compilation only the primary minerals are listed; it must be pointed out that retrograde alteration of these primary minerals is more the rule than the exception. The garnets of eclogites consist of almandine-grossularite-pyrope with the pyrope content ranging from 7 to 70 mol per cent; at the expense of the almandine, grossularite ranges between 20 and 35 mol per cent. The pyroxenes are essentially mixtures of jadeite (37–6 mol per cent) and diopside (88–10 mol per cent) with minor variations in acmite and hedenbergite molecules.

to their enclosing rocks—for example, those in diamond pipes (Williams, 1932) or those that often appear as retrograded lenses in high-grade metamorphic terrains. However, it is possible to divide the eclogite occurrences into three main groups for the purposes of this discussion.

The following grouping of eclogites is not meant to be genetic but is rather an attempt to subdivide the various types into geologically similar occurrences. Future workers may want to subdivide the various groups; for instance, the eclogites described by Briere (1920) could perhaps be better placed as a subgroup in Group C rather than Group B. Therefore these groups should be considered only as a first attempt to categorize these rocks:

Group A. Inclusions in kimberlites, basalts, or layers in ultramafic rocks. These often contain diamond, orthopyroxene, or olivine and appear to have a very deep-seated igneous or metamorphic origin. O'Hara and Yoder (1963) divide this group into three separate units according to occurrence.

Group B. Bands or lenses within migmatite gneissic terrains. These are surrounded by

represent mantle material; whereas Groups B and C are difficult to assess properly because they seem to be geologically out of place in their associated rocks, assuming of course that *all* eclogites represent the highest grade of metamorphism. Recent articles by Yoder and Tilley (1962), Subramaniam (1956), and O'Hara (1960) do not clearly distinguish between the three groups nor do they entertain the possibility that eclogites may form as a result of diverse geological conditions. In a later article, O'Hara and Yoder (1963) also suggested that the eclogites of our Groups B and C may represent recrystallized basaltic material within deep orogenic zones rather than material formed in the upper mantle. In the following sections of this paper a comparison of eclogites from the glaucophane schist terrains (Group C eclogites) will be made with those eclogites from Groups A and B. The basis of this comparison will be the mineral compositions of the coexisting garnet and pyroxene as related to the bulk composition of these eclogites. Again, the purpose of such a comparison is to present evidence to show that certain eclogites may represent crustal metamorphism, whereas other eclogites may have been formed as mantle material.

ECLOGITES FROM GLAUCOPHANE SCHISTS

OCCURRENCE AND MINERAL ASSEMBLAGES: Nine samples of pyroxene-bearing rocks from

that these eclogites are not in place and may have been transported to their present position as inclusions in diapiric serpentines (Coleman, 1961) or in shear zones of major faults. Small-scale interlayering of eclogite with glaucophane schists within individual tectonic blocks, however, demonstrates the contemporaneity of the two rock types. It is not possible therefore to

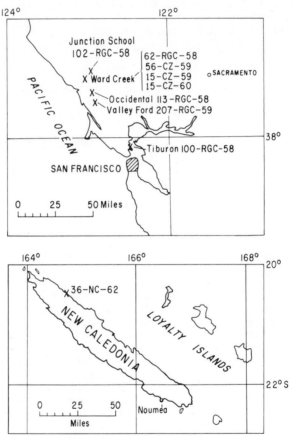

Figure 1.   Index maps showing locations of analyzed eclogites
and pyroxene-bearing metamorphic rocks

the glaucophane schist facies of California and New Caledonia were selected for this study (Fig. 1). Six of these are referred to as eclogites, and others represent various metamorphic grades within the California glaucophane schists. The eclogite specimens were taken from isolated tectonic blocks within glaucophane schist terrains. The unusual nature of these isolated tectonic blocks has been discussed earlier by Coleman and Lee (1963). It seems

relate the eclogites directly to those lower-grade glaucophane schists found in place; however the metamorphic events that produced these various metamorphic grades within the glaucophane schist terrains of the Cazadero area are closely related in time, as shown by K/Ar ages (Lee and others, 1963b). The following descriptions are of the eclogites and pyroxene-bearing metabasalts used in this study; two of the low and intermediate metamorphic-grade,

TABLE 2. MODAL ANALYSES OF ECLOGITES IN VOLUMETRIC PER CENT

See Figure 1 for location of samples.

	Occidental 113-RGC-58	Tiburon 100-RGC-58	Valley Ford 207-RGC-59	Ward Creek 62-RGC-58	Junction School 102-RGC-58	New Caledonia 36-NC-62	Loch Duich* Group B	KaO Pipe† Group A
Omphacite	22.7	42.6	89.4	40.5	67.2	34.4	36.8	70
Garnet	16.6	28.9	6.2	9.4	16.0	28.9	45.9	27
Glaucophane	53.6	..	..	21.6	0.8	19.9	..	..
Epidote	5.1	1.4	0.1	7.4	..	3.9	present	..
Muscovite	1.2	2.2	..	..	0.9	0.7	..	..
Chlorite	0.8	14.1	4.3	9.1	12.2	0.5	..	..
Sphene	..	3.2	..	10.2	2.5	1.5	present	..
Rutile	..	1.0	..	0.2	0.4	2.0	present	0.5
Lawsonite	..	..	..	1.6	..	..	present	..
Quartz	..	6.6	..	..	..	8.2	present	..
Sum	100.0	100.0	100.0	100.0	100.0	100.0	..	..

*Eclogite from Loch Duich, Scotland, 7.8 volume per cent altered pyroxene and 9.4 volume per cent consisting of rutile sphene iron ores, and small amounts of biotite, quartz, epidote, and apatite. From Yoder and Tilley (1962, p. 476)

†Eclogite from Basutoland, South Africa, 0.5 volume per cent consists of natrolite. From Nixon and others (1963, p. 1101)

pyroxene-bearing metabasalts near Ward Creek, California, have been described earlier by Coleman and Lee (1963). These pyroxene-bearing glaucophane schists are within a continuous schist sequence, whereas the eclogite-bearing rocks are isolated tectonic blocks. Initially the rock was crushed to −80 mesh, and splits for rock analysis and mineral separation were taken from this original crushed sample. Analyses of these rocks are given in Table 2 (modal analyses) and Tables 3 and 4 (chemical and spectrographic analyses).

ECLOGITE, OCCIDENTAL QUARRY (113-RGC-58): Individual tectonic block partially destroyed by quarrying, situated alongside Occidental Road, approximately 1.7 miles east of Occidental, SW¼SW¼sec. 36, T. 7 W., R. 10 W., Mount Diablo Meridian (part of Canada De Jonive Spanish Land Grant), lat. 38°28'30" N., long. 122°55' W., Camp Meeker, 7½-minute quadrangle, Sonoma County, California. This locality shown as garnet quarry on geological map of this area by Travis (1952).

The rock has a dense, green groundmass containing reddish-brown garnets up to 5 mm in diameter. Microscopically the texture is porphyroblastic; the groundmass is granoblastic and consists of equigranular omphacite with numerous fine-grained sphene trains. The garnet porphyroblasts contain numerous random inclusions in their central portions, but the outer rims are generally free of inclusions. Subhedral clinozoisite is a common groundmass mineral along with minor amounts of intersertal muscovite, as shown in Figure 2. Alteration has produced irregular areas of epidote overgrowths on clinozoisite and replacement of omphacite and garnet throughout the rock. Chlorite has crystallized during the period of epidote replacement. An earlier, more complete description of the various rock types characterizing this block has been given by Switzer (1951, p. 69).

ECLOGITE, TIBURON PENINSULA (100-RGC-58): Blocks (5' × 5' × 3' or smaller) resting on a soil surface within small eucalyptus grove approximately 400 feet north of railroad tunnel near Reed Station, Tiburon Peninsula, lat. 37°54' N., long. 122°29'30"W, San Quentin 7½-minute quadrangle, Marin County, California.

The rock is green with numerous reddish-brown euhedral garnet porphyroblasts. The groundmass consists of anhedral omphacite with minor amounts of muscovite, epidote, sphene, and rutile (Fig. 3). The cores of the garnets contain numerous inclusions of groundmass minerals, whereas their rims are nearly free of such inclusions. Chlorite and lawsonite are retrogressive replacements of some garnets. Two chlorite types can be recognized; the chlorite replacing the garnet has a strong dispersion in contrast to the higher birefringent chlorite of the groundmass. Glaucophane is sparsely developed, along with the chlorite and lawsonite, as a retrograde mineral. Rutile is present as a common accessory mineral and is often partly replaced by sphene.

ECLOGITE, VALLEY FORD (207-RGC-59): Large block (12' × 25' × 8') resting on soil surface approximately 300 feet west of second bridge along Ebabias Creek[1] ½ mile from Valley Ford-Freestone highway, ½ mile south of Bodega Bay road junction, lat. 38°21'30" N., long. 122°53'15" W., Valley Ford 7½-minute quadrangle, Sonoma County, California. Area shown as glaucophane schist terrain on geologic map of Travis (1952).

The rock is dark green and dense, containing numerous porphyroblastic reddish-brown garnets (up to 6 mm in diameter). The groundmass consists of anhedral omphacite grains with numerous subhedral sphene trains. The garnet cores contain numerous inclusions consisting of the groundmass minerals, whereas the rim is nearly free of such inclusions. Muscovite forms only rarely as an intersertal mineral. Evidence of retrogressive metamorphism is seen where chlorite and glaucophane replace the omphacite-garnet assemblage. Strain shadows and shear zones indicate late tectonic movements. Petrographic and chemical data have been given earlier by Bloxam (1959).

ECLOGITE, WARD CREEK (62-RGC-58): Boulders along Ward Creek Stream Canyon, NE¼NW¼sec. 17, T. 8 N., R. 11 W., Mount Diablo meridian, lat. 38°32' N., long. 123°06' W., Cazadero 7½-minute quadrangle, Sonoma County, California.

The rock has a greenish-blue, dense groundmass with reddish-brown porphyroblastic garnets. The groundmass consists of omphacite with large porphyroblastic glaucophane (Fig. 4). Sphene forms subhedral grains which often contain rutile cores. Anhedral epidote is ubiquitous throughout the groundmass. The

---

[1] Although the eclogites selected for this study may appear to be erratics and isolated boulders by description, they belong to the general metamorphic assemblage of the Franciscan Formation. Mass wasting and later tectonic movements have, however, obscured the relationships with the immediate country rock. A detailed discussion of these relationships has been given by Bailey and others (1964, p. 89).

TABLE 3. CHEMICAL AND SPECTROGRAPHIC ANALYSES OF ECLOGITES (*See* Figure 1 for location of samples.)

	Occidental 113-RGC-58	Tiburon 100-RGC-58	Valley Ford 207-RGC-59	Ward Creek 62-RGC-58	Junction School 102-RGC-58	New Caledonia 36-NC-62	Loch Duich Group II*	Roberts-Victor Mine Group I†
	CHEMICAL ANALYSES**							
SiO_2	47.1	44.6	49.8	45.8	47.4	47.2	42.60	43.56
TiO_2	1.4	1.8	1.8	3.2	1.5	2.1	3.84	.70
Al_2O_3	16.2	16.3	11.0	13.2	13.4	15.9	12.41	12.80
Fe_2O_3	3.8	5.9	5.2	6.4	5.6	5.2	3.50	8.72
FeO	6.5	8.8	5.4	9.1	7.6	4.5	16.94	9.43
MnO	0.19	0.26	0.26	0.23	0.20	0.15	0.36	.21
MgO	4.7	5.9	7.5	5.4	7.1	6.1	7.23	14.40
CaO	16.1	11.7	12.3	11.0	10.6	7.3	11.18	8.68
Na_2O	3.4	3.1	5.1	4.3	4.4	3.5	1.09	nil
K_2O	0.18	0.22	0.40	0.04	0.20	1.6	0.15	nil
H_2O^+	0.60	1.5	1.4	1.7	1.9	5.4	0.48	1.3
H_2O^-	0.06	0.08	0.14	0.09	0.06	0.62	nil	nil
P_2O_5	0.14	0.12	0.14	0.17	0.05	0.43	0.18	.09
CO_2	0.09	0.07	<0.05	<0.05	0.08	0.36	..	.15
FeS_2	..	..	..	<0.05	..	..	..	..
Sum	100.46	100.35	100.44	100.63	100.09	100.36	99.96	100.04
Specific gravity	3.42	3.39	3.18	3.32	3.36	3.11	..	..
	SEMIQUANTITATIVE SPECTROGRAPHIC ANALYSES (ppm)††							
Ga	10	15	15	7	7	20	..	..
Cr	700	150	300	15	70	300	..	..
V	500	700	700	300	700	500	..	..
Ni	300	50	150	70	70	150	..	..
Co	70	30	50	30	50	30	..	..
Sc	50	50	200	30	50	20	..	..
Zr	100	150	150	150	150	200	..	..
Y	50	70	100	70	70	50	..	..
Yb	5	10	10	7	10	5	..	..
Sr	200	300	200	150	300	200	..	..
Ba	100	100	300	15	150	200	..	..
Cu	100	50	150	30	15	70	..	..
B	10	10	10	<10	0	15	..	..
Mo	..	..	..	3	..	..	..	..
Nb	..	..	..	..	..	50	..	..

* Eclogite from Scotland. Analysis shows 0.47 weight per cent S not included in the summation. Yoder and Tilley (1962, p. 480)

† Eclogite from South Africa kimberlite pipe. Analysis shows 0.01 weight per cent S not included in the summation. Williams (1932)

** Chemical analyses performed by methods similar to these described in U. S. G. S. Bull. 1036-C. Analysts: Paul Elmore, Sam Betts, Gillison Chloe, Lowell Artis, H. Smith

†† Analysts: Chris Heropoulos and Katherine V. Hazel

173

TABLE 4. CHEMICAL AND SPECTROGRAPHIC ANALYSES
OF PYROXENE-BEARING GLAUCOPHANE SCHISTS

Chemical analyses performed by methods similar to those described in U. S. Geol. Survey Bull. 1036-C. Analysts: Paul Elmore, Sam Botts, Gillison Chloe, Lowell Artis, and H. Smith. *See* Figure 1 for location of samples.

	Type II* 15-CZ-60	Type III* 15-CZ-59	Type IV* 56-CZ-59
CHEMICAL ANALYSES			
SiO_2	48.9	48.5	59.0
TiO_2	1.7	1.5	0.82
Al_2O_3	14.0	16.0	10.90
Fe_2O_3	5.5	1.6	5.2
FeO	8.6	6.8	11.0
MnO	0.26	0.16	1.1
MgO	6.2	7.0	4.4
CaO	10.0	8.2	3.6
Na_2O	3.2	3.3	3.2
K_2O	0.13	2.0	0.25
H_2O^+	0.78	3.4	0.74
H_2O^-	0.09	0.12	0.04
P_2O_5	0.21	0.18	0.06
CO_2	<0.05	0.15	0.07
Sum	99.62	98.91	100.38
Specific gravity	3.26	3.20	3.30
SEMIQUANTITATIVE SPECTROGRAPHIC ANALYSES (ppm)[†]			
Ga	15	30	20
Cr	70	500	150
V	700	500	200
Ni	100	150	150
Co	50	30	50
Sc	70	70	30
Zr	100	100	200
Y	70	50	70
Yb	7	5	10
Sr	200	50	30
Ba	20	300	150
Cu	70	50	100
B	..	30	10
Nb	..	..	10
Ag	5	..	..

[*] Type refers to metamorphic grade of Coleman and Lee (1963).

[†] Semiquantitative spectrographic analyses by Chris Heropoulos

garnets exhibit extensive retrogression with the rims usually completely replaced by chlorite, glaucophane, and sphene. All the garnets contain numerous inclusions consisting of groundmass minerals. Lawsonite with minor pumpellyite forms in late fractures and also partially replaces some garnets (Coleman and Lee, (1963).

ECLOGITE, JUNCTION SCHOOL (102-RGC-58): Isolated tectonic block, partially destroyed by quarrying, near caretaker's house on Mill Creek

Road, approximately 0.4 mile east of Junction School. Sotoyme Spanish Land Grant covers area, and standard land survey sections are not plotted, lat. 38°36′30″ N., long. 122°53′45″ W., Guerneville 7½-minute quadrangle, Sonoma County, California. Area featured as glaucophane schist zone on geologic map of Gealey (1951).

The rock is dark, green, dense, and contains numerous porphyroblasts of reddish-brown garnets (up to 5 mm in diameter). The groundmass consists of anhedral grains of omphacite with minor amounts of glaucophane and muscovite. Sphene forms irregular blebs interstitial to the pyroxene and often has a core of rutile. The garnets contain numerous inclusions of groundmass minerals. Evidence for retrograde effects is seen where chlorite replaces garnet or, in rarer instances, glaucophane mantles some omphacite grains (Fig. 5). Muscovite is present as a minor intersertal mineral. Pyrite forms small euhedra throughout the rock. Earlier descriptions accompanied by

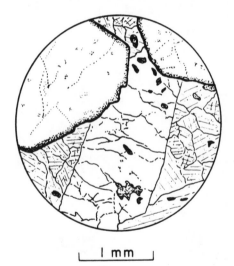

|__ 1 mm __|

Figure 2. Eclogite, Occidental quarry—113-RGC-58, showing porphyroblastic garnets associated with equigranular omphacite and a large tabular clinozoisite. Irregular dark grains are sphene.

mineralogical and chemical data have been given by Switzer (1945) and Borg (1956).

ECLOGITE, NEW CALEDONIA (36-NC-62): Boulder (2′ × 3′ × 4′) resting on coral beach, approximately 1 km northeast of Amos Stream where a side road descends to beach from Amos-Balade Road, northeast coast of New Caledonia, lat. 20°18′ S., long. 164°28′ E. New Caledonia is a French island between the New Hebrides and Australia. Boulder is in terrain mapped as glaucophane schists by Routhier (1953).

The rock is dark green and dense and contains numerous grains of reddish-brown garnet (up to 6 mm in diameter) producing a distinct porphyroblastic texture (Fig. 6). The groundmass consists of omphacite and a blue-green amphibole producing a lepidoblastic texture. The paragenetic relationships between the omphacite and amphibole are conflicting, but we assume that they formed simultaneously.

¼sec. 17, T. 8 N., R. 11 W., Mount Diablo Meridian, lat. 38°32′ N., long. 123°06′ W., Cazadero 7½-minute quadrangle, Sonoma County, California. Geologic map of Ward Creek glaucophane schist sequence shows relationships of this tectonic block to surrounding rocks (Coleman and Lee, 1963, Fig. 2).

The rock is coarse-grained, banded, and contains layers of varying mineral assemblages. The following assemblages are present and represent individual bands: (1) glaucophane-epidote-muscovite-garnet, (2) omphacite-

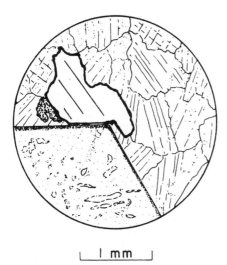

Figure 3. Eclogite, Tiburon Peninsula—100-RGC-58, showing a large single garnet euhedra porphyroblast set in an equigranular omphacite groundmass. Juxtaposed with the garnet is a rutile anhedra partly replaced by sphene.

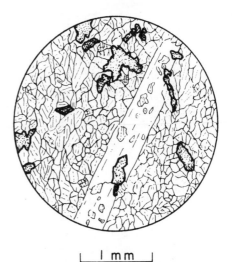

Figure 4. Eclogite, Ward Creek—62-RGC-58, showing a large prismatic glaucophane coexisting with equigranular omphacite. Dark irregular grains are sphene.

Epidote occurs as anhedral grains irregularly present throughout the rock. Quartz commonly formed in the pressure shadows near the garnet porphyroblasts. The garnets are generally crowded with numerous inclusions consisting predominantly of sphene, epidote, and rutile. Chlorite with minor amounts of epidote has formed around the rims of the garnets. Rutile, where present, is always mantled by sphene. Muscovite forms in intersertal areas often associated with quartz (Fig. 6). Briere (1920) has described similar eclogites from New Caledonia.

GLAUCOPHANE SCHIST, WARD CREEK (56-CZ-59): Large block (30′ × 20′ × 10′) of coarsely crystalline gneissic rock in Ward Creek canyon, NE¼NW

garnet (Fig. 7), (3) epidote-garnet, and (4) garnet-glaucophane-omphacite. Minor accessories include sphene, rutile, and apatite. In each of the individual compositional bands, the minerals show no mutual replacement, and the prevailing lepidoblastic texture suggests only one period of crystallization.

METABASALT, WARD CREEK (15-CZ-59): Schist exposed in place along Ward Creek stream bed, SW¼NE ¼sec. 18, T. 8 N., R. 11 W., Mount Diablo Meridian, lat. 38°32′ N., long. 123°06′ W., Cazadero 7½-minute quadrangle, Sonoma County, California. Featured in geologic map of Ward Creek schist sequence (Coleman and Lee, 1963, Fig. 2).

The rock is a dense, fine-grained, bluish-green schist with poorly developed foliation and lineation. Glaucophane, pumpellyite, pyroxene, muscovite, and chlorite make up most of the rock, producing a lepidoblastic microscopic texture. Sphene, lawsonite, pyrite, and quartz are present in minor amounts. This schist belongs to the Type III grade of metamorphism, as described by Coleman and Lee (1963).

METABASALT, WARD CREEK (15-CZ-60): Massive outcrop, in place, of metabasalt forming small falls across Ward Creek, NW¼NW¼sec. 17, T. 8 N.,

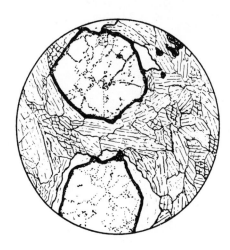

⌊ 2 mm ⌋

Figure 5. Eclogite, Junction School—102-RGC-58, showing porphyroblastic garnets contained within a granoblastic groundmass of omphacite. Rims of the garnets are partially replaced by retrograde chlorite. Sphene forms small irregular dark grains.

R. 11 W., Mount Diablo Meridian, lat. 38°32′ N. long. 123°06′ W., approximately at point where section line dividing secs. 17 and 18 bisects Ward Creek in north-south direction, Cazadero 7½-minute quadrangle, Sonoma County, California. This area featured in area map of Ward Creek schist sequence as given by Coleman and Lee (1963, Fig. 1).

The rock is fine-grained, greenish-gray, and seems to be a vesicular basalt in the hand specimen; however microscopic investigation shows that even though the igneous textures are partly preserved, the rock has completely recrystallized. Lawsonite, pyroxene, glaucophane, and chlorite are the main metamorphic minerals. Minor amounts of aragonite, mica, quartz, sphene, and montmorillonite make up the remainder. Metamorphic rocks of this

grade in the Ward Creek glaucophane schist sequence belong to the Type II as defined by Coleman and Lee (1963).

SUMMARY: The rocks described heretofore as eclogites characteristically consist of 40–90 volume per cent garnet-omphacite and have strong variations in the relative amounts of the other minerals. These variations in mineralogical composition are noteworthy and seem to be controlled by retrograde metamorphic effects or by differences in bulk composition. Primary minerals assumed to have formed contemporaneously with garnet and omphacite are epidote-clinozoisite, muscovite, amphibole, rutile, sphene, and quartz. Retrograde minerals are chlorite, glaucophane, lawsonite, and sphene. Sphene may be both primary and retrograde in these rocks; the possible reactions have been considered by Coleman and Lee (1963, p. 286).

⌊ l mm ⌋

Figure 6. Eclogite, New Caledonia—36-NC-62, consisting of large irregular garnet grains, omphacite anhedra associated with quartz (white), and a single muscovite grain.

The role of amphibole is not clear in these particular rocks since glaucophane and other amphiboles are present in apparent stable equilibrium with the garnet and omphacite (36-NC-62, Table 3; 56-CZ-59, Table 4), whereas in other cases amphibole seems to be clearly retrograde, replacing the primary minerals. Information concerning the amphibole compositions is needed to further elucidate this question. In summary then, the following assemblages appear typical of eclogites associated with glaucophane schists:

(1) Omphacite-garnet-rutile (sphene)
(2) Omphacite-garnet-epidote-rutile (sphene)
(3) Omphacite-garnet-muscovite-rutile (sphene)
(4) Omphacite-garnet-amphibole-rutile (sphene)

In comparing the California and New Caledonia eclogites with those eclogites from Groups A and B several points should be emphasized.

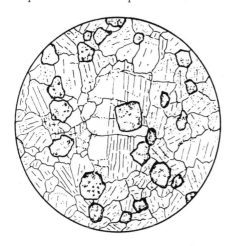

| 2 mm |

Figure 7. Eclogite band, Ward Creek—56-CZ-59, within a glaucophane-epidote-muscovite-schist (Type IV *of* Coleman and Lee, 1963) showing garnet porphyroblasts contained within granoblastic groundmass of omphacite.

phasized. The eclogites from glaucophane schists seem to be characterized by the common occurrence of amphibole (glaucophane) and epidote. The complete absence of kelyphite or symplectite coronas around garnets is noteworthy because such coronas are common in eclogites from Group A and less so in Group B. The appearance of omphacitic pyroxene in the lower-grade glaucophane schists is unusual in that pyroxenes are more commonly developed in the highest grade of metamorphism. The sections to follow will make further comparisons between the minerals and bulk composition of the eclogites from the three groups.

MINERALOGY

General Statement

A detailed mineralogical study has been made of the eclogites and the pyroxene-bearing metabasalts just described. The remainder of this paper will concentrate on the coexisting pyroxenes and garnets in eclogites, on pyroxenes in lower-grade rocks, and on amphibole from the New Caledonia eclogite. Each analyzed mineral was carefully purified by centrifuging in heavy liquids and magnetic fractionation in the Frantz isodynamic separator. These mineral purifications are splits of the material used for rock analyses, and further splits of the purified mineral were used for optical, X-ray, and chemical determinations. The chemical analyses were carried out by Leoniece B. Beatty and W. Wallace Brannock.

Pyroxenes

Six pyroxenes from eclogites and three pyroxenes representing the various grades of metamorphism in the glaucophane schist facies form the basis for this discussion. The chemical and optical data for these pyroxenes are given in Table 5. The eclogite pyroxenes are characterized by a very regular Na_2O content, averaging 6.5 weight per cent. This is somewhat higher than the average of 4 weight per cent Na_2O given by Yoder (1950) for 18 eclogitic pyroxenes. In contrast, the Na_2O content of the pyroxene from the lowest-grade glaucophane schist is 9.6 weight per cent Na_2O (15-CZ-60, Table 5). Even though the Group C eclogites and the associated pyroxene-bearing glaucophane schists exhibit a range in Na_2O content, the calculated $jd(NaAlSi_2O_6)$ component is remarkably uniform for this group of rocks. Carrying this comparison further, pyroxenes from Group C eclogites consistently contain more jd than do the pyroxenes from Group A and B eclogites (Fig. 8). Perhaps one of the most important variations in the omphacite composition for the Group C eclogites is $ac(NaFe^{3+}Si_2O_6$, 5–20 mol per cent); however $di(CaMgSi_2O_6)$ is the predominant component with lesser amounts of $he(CaFe^{+2}Si_2O_6)$ and Tschermak's molecule $tsch(Ca,MgAlSi_2O_6)$. Very little silica is replaced by alumina in fourfold coordination.

No attempt is made here to correlate the optical data with the chemical composition. Wide variations in these properties, particularly 2V and the extinction angle made it impossible to place much confidence in the results. No exsolution or intergrowths were observed, but mottled extinction and strong variations within a single grain were found. Noteworthy are the large $Z \wedge C$ angles recorded for these pyroxenes as compared to the compilation of Deer

TABLE 5. CHEMICAL AND PHYSICAL PROPERTIES OF PYROXENES FROM ECLOGITES AND GLAUCOPHANE SCHISTS
See Figure 1 for location of samples.

	113-RGC-58	100-RGC-58	207-RGC-59	62-RGC-58	102-RGC-58	36-NC-62	15-CZ-60	15-CZ-59	56-CZ-59
				CHEMICAL ANALYSES*					
SiO_2	54.6	54.3	53.4	53.2	54.9	54.7	51.8	54.8	..
TiO_2	0.52	0.20	0.32	0.21	0.15	0.18	2.8	0.59	0.20
Al_2O_3	10.7	10.0	10.0	9.8	9.7	10.3	11.2	10.2	..
Fe_2O_3	2.0	3.5	5.5	7.4	4.3	3.8	11.3	2.4	10.0
FeO	3.8	3.2	3.0	4.1	3.1	3.60	2.4	4.0	13.8
MnO	0.04	0.02	0.19	0.14	0.05	0.03	0.11	0.14	..
MgO	7.7	8.2	7.4	5.6	7.7	7.5	3.6	7.6	..
CaO	15.0	13.3	13.3	13.2	13.1	13.4	5.1	13.5	..
Na_2O	5.9	6.8	6.6	6.5	6.8	6.5	9.6	6.3	7.2
K_2O	0.02	0.02	0.04	<0.02	<0.02	<0.02	0.38	0.05	0.05
H_2O^+	0.33	0.12	0.30	N.D.	0.11	0.28	1.9	0.56	..
H_2O^-	0.02	0.06	0.04	..	0.05	0.07	..	0.10	..
Sum	100.6	99.7	100.1	100.2	100.0	100.4	100.2	100.2	..
Specific Gravity	3.32	3.34	3.34	3.35	3.33	3.32	3.30	3.30	3.36
				OPTICAL CONSTANTS†					
α	1.679	1.682	1.684	1.694	1.684	1.681	1.677	1.677	1.697
β	1.684	1.689	1.690	1.700	1.690	1.688	1.679	1.682	1.706
γ	1.697	1.705	1.706	1.714	1.701	1.700	1.685	1.694	1.725
$2V$	73°	69°	67°	71°	70°	71°	74°	69°	70°
$Z \wedge C$	47°	48°	48°	55°	48°	48°	49°	45°	56°

Ion	1	2	3	4	5	6	7	8	9
Si	1.95	1.96	1.94	1.93	1.98	1.96	1.91	1.97	..
Al	0.05	0.04	0.06	0.07	0.02	0.04	0.09	0.03	..
(Σ)	2.00	2.00	2.00	2.00	2.00	2.00	2.00	2.00	..
Al	0.40	0.39	0.37	0.35	0.39	0.40	0.40	0.41	..
Ti	0.01	0.01	0.01	0.01	..	0.01	0.08	0.02	..
Fe^{+3}	0.06	0.10	0.16	0.20	0.12	0.10	0.31	0.07	..
Mg	0.41	0.44	0.40	0.30	0.43	0.40	0.20	0.41	..
Fe^{+2}	0.11	0.10	0.09	0.12	0.09	0.11	0.07	0.12	..
Mn	..	..	0.01	..	..	..	..	..	..
(Σ)	0.99	1.04	1.04	0.98	1.03	1.02	1.06	1.03	..
Na	0.41	0.48	0.46	0.46	0.48	0.45	0.69	0.44	..
Ca	.57	.51	.51	.51	.51	.52	.20	.52	..
K	..	..	..	..	..	..	0.01	..	..
(Σ)	0.98	0.99	0.97	0.97	0.99	0.97	0.90	0.96	..

MOLECULAR PROPORTIONS OF PYROXENE END-MEMBERS

	1	2	3	4	5	6	7	8	9
Jd	37.3	37.7	29.9	27.7	35.8	35.1	38.9	37.1	..
Ac	5.3	9.9	15.9	20.5	11.9	10.0	31.3	7.0	..
Hd	10.8	9.9	14.5	12.3	7.3	10.0	8.1	11.0	..
Di	41.3	38.7	32.9	30.8	42.8	40.1	13.1	41.1	..
Tsch	5.3	3.7	6.7	8.7	2.2	4.8	8.6	3.8	..

* Chemical analyses by L. B. Beatty, A. C. Bettiga, and W. W. Brannock
† Optical determinations by sodium light, refractive indexes good to ±0.003, $2V$ and $Z \wedge C$ very erratic—taken as an average ± 5°.

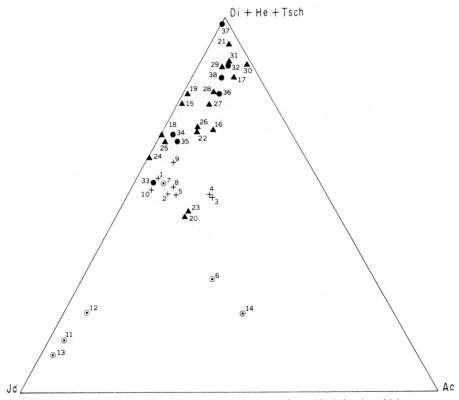

Figure 8.　Triangular diagram featuring the end members, jd, ac, (di + he + tsch) in pyroxenes from eclogites and glaucophane schist facies. Circles, Group A eclogites; triangles, Group B eclogites; crosses, Group C eclogites; dots, glaucophane schist facies.

Key for pyroxene analyses:

Group C eclogites and glaucophane schists

1. Occidental Quarry, Calif. (113-RGC-58), Table 4
2. Tiburon Peninsula, Calif. (100-RGC-58), Table 4
3. Valley Ford, Calif. (207-RGC-59), Table 4
4. Ward Creek, Calif. (62-RGC-58) Table 4
5. Junction School, Calif. (102-RGC-58), Table 4
6. Ward Creek, Calif. (15-CZ-60), Table 4. Metabasalt
7. Ward Creek, Calif. (15-CZ-59), Table 4. Metabasalt
8. Amos Stream, New Caledonia (36-NC-62), Table 4
9. Valley Ford, Calif.; Bloxam (1959)
10. Shubina Village deposit, Southern Urals, USSR; Chesnokov (1959)
11. Angel Island, Calif. (unpub. data). Metagraywacke
12. Valley Ford, Calif.; Bloxam (1956)
13. Celebes Island, De Roever (1955). Metagraywacke
14. Tokusima Prefecture, Japan; Iwasaki (1960)

Group B eclogites

15. Silberbach, Germany; Yoder and Tilley (1962)
16. Glenelg, Scotland; Yoder and Tilley (1962)
17. Loch Duich, Scotland; Yoder and Tilley (1962)
18. Glenelg, Scotland; Alderman (1936)
19. Weissenstein, Bavaria; Tilley (1936)
20. Duen Type, Sondmore, Norway; Eskola (1921)

and others (1963). Further precise analytical data are needed before correlations between optics and chemical composition can be attained.

A detailed single-crystal study is presently being made by the X-ray crystallography group of the U.S. Geological Survey (Dr. Joan R. Clark, written communication, 1964) to determine the unit cell and space group as related to possible inhomogeneity of these pyroxenes. Preliminary data on one crystal from the Tiburon eclogite 100-RGC-58 show the true space group to be $P2/m$ or $P2$, but only a few weak reflections violate $C2/c$ symmetry. The unrefined cell dimensions for this same pyroxene are a = 9.572Å, b = 8.773Å, c = 5.253Å, β = 106°50'. O'Hara (1960, p. 146) is apparently the only previous investigator who attempted to determine the homogeneity of eclogitic pyroxenes. He reported that an unpublished study by Bown and Gay on single crystals of a pyroxene from a garnet-hornblende-two pyroxene rock (eclogite) from Glenelg shows it to be a single monoclinic phase. The optical variations noted earlier, combined with certain indexing difficulties encountered with the preliminary X-ray study of these pyroxenes, suggested the possibility of unmixing. Continued study on this problem could produce important information bearing on the history of these rocks.

A comparison based on the proportions of jd, ac, (di + he + tsch) components of the California and New Caledonia pyroxenes from Group C eclogites with those pyroxenes from Groups A and B eclogites is revealing. As pointed out previously, the eclogite pyroxenes have a variable jd content, with those pyroxenes from Group C eclogites showing a more consistent composition and a higher jd content. The pyroxene compositions from Groups A and B eclogites are broadly overlapping and much less consistent than Group C eclogites (Fig. 8). The omphacites from eclogites range continuously from approximately 1 to 37 molecular per cent jd, forming a distinct compositional field. The apparent compositional gap between omphacites and jadeitic pyroxenes (Coleman, 1961) is significant, particularly in relationship to the P–T conditions of eclogite formation. Presumably the amount of the jd content present in the pyroxene could be a measure of the relative total pressure acting during crystallization (Kennedy, 1959; Robertson and others, 1957). On the basis of the comparison between Group C and Group A pyroxenes, it would appear that those eclogites forming within the glaucophane schist facies require higher pressures than those pyroxenes in eclogites that may be mantle material. Unfortunately this comparison cannot be defended, as there is no other sodium-bearing essential mineral present in these eclogites, and in general the total Na₂O content increases in the eclogites from Groups A to C. Nonetheless this observation is consistent with the current contention that the glaucophane schist facies is characterized by high pressures and low

---

Figure 8 Key *continued*

21. Rodhaugen, Sondmore, Norway; Eskola (1921)
22. Silden, Nordfjord, Norway; Eskola (1921)
23. Fay, France; Briere (1920)
24. Bielice No. 1, East Sudetes, Poland; Smulikowski (1960)
25. Nowawies No. 4, East Sudetes, Poland; Smulikowski (1960)
26. Gertrusk, Karntar, Austria; Angel and Schaider (1950)
27. Silberbach, Fichtelgibirge, Germany; von Wolff (1942)
28. Silberbach, Fichtelgibirge, Germany; von Wolff (1942)
29. Hurry Inlet, East Greenland; Sahlstein (1935)
30. Mampong, Ghana, Africa; von Knorring and Kennedy (1958)
31. Glen More, Glenelg, Scotland; O'Hara (1960)

Group A eclogites

32. Salt Lake Crater, Oahu, Hawaii; Yoder and Tilley (1962)
33. Jagersfontein Pipe, South Africa; Williams (1932)
34. Kao Pipe, Basutoland, South Africa; Nixon and others (1963)
35. Kaalvallei Pipe, South Africa; Nixon and others (1963)
36. Gongen-yama, Bessi district, Japan; Shido (1959)
37. Gongen-yama, Bessi district, Japan; Miyashiro and Seki (1958)
38. Eclogite in kimberlite 37079 Cambridge Collection; O'Hara (1963)

temperatures (Ernst, 1963; Coleman and Lee, 1963). The increase in ac content in certain omphacites along with the extreme concentration of ac in the low-grade pyroxene-bearing glaucophane schist 15-CZ-60 (31 mol per cent) appears significant. An increase in the partial pressure of oxygen could easily change the Fe^{-3}/Fe^{-2} ratio and allow an ac-rich pyroxene to form at considerably lower pressures. Iwasaki (1960) described such ac-rich jadeitic py-

tain even less pyrope than do garnets from these high-grade metamorphic rocks. The apparently low py content of these California garnets was not generally known until Tröger's (1959) compilation and the more recent analyses of Lee and others (1963a).

The garnets from Group C eclogites are predominantly almandine-pyrope-grossular, with the py content much less than 15 per cent. The unusual feature of these garnets is the

TABLE 6. CHEMICAL ANALYSES AND PHYSICAL CONSTANTS OF GARNETS FROM ECLOGITES

Analyses by L. B. Beatty, A. C. Bettiga, W. W. Brannock, Paul Elmore, Ivan Barlow, Samuel Botts, and Gillison Chloe. FeO determinations by Robert Meyrowitz. Methods similar to these described *in* U. S. Geol. Survey Bull. 1036C. See Figure 1 for location of samples.

	113-RGC-58	100-RGC-58	207-RGC-59	62-RGC-58	102-RGC-58	36-NC-62	56-CZ-59
SiO_2	38.4	37.6	37.4	37.5	37.9	37.8	37.5
Al_2O_3	20.7	20.5	20.4	20.1	20.4	20.5	20.1
Fe_2O_3	1.3	1.6	1.8	2.3	0.7	0.7	1.8
TiO_2	0.26	0.37	0.29	0.0	0.38	0.54	0.06
FeO	22.7	25.7	24.2	27.6	27.1	26.7	27.6
CaO	13.7	10.2	11.1	9.5	9.9	10.2	6.0
MnO	0.57	0.58	2.0	1.3	0.58	0.64	4.4
MgO	2.5	3.1	2.4	1.7	2.4	2.5	2.4
Sum	100.1	99.7	99.6	100.0	99.4	99.6	99.9
Almandine	50.3	57.3	54.0	62.3	60.7	59.6	62.6
Spessartite	1.3	1.3	4.6	3.0	1.2	1.2	10.0
Pyrope	9.7	12.4	9.7	6.7	9.7	10.0	9.7
Andradite	4.5	5.5	6.0	6.5	3.0	3.4	4.5
Grossularite	34.2	23.5	25.7	21.5	25.4	25.8	·13.2
S. G.	3.94	3.99	4.02	4.04	4.00	4.03	4.08–4.12
Refractive Index	1.786	1.792	1.793	1.802	1.793	1.795	1.800–1.806
Unit Cell in Å	11.651	11.617	11.628	11.62	11.63	11.61	11.582–11.606

roxenes from the glaucophane schists of Japan. Further speculation on the effect of ac, di, or the other pyroxene components on the relative stability of jd-bearing pyroxenes is fruitless until additional experimental evidence is presented.

Garnets

Seven analyses of garnets were made in connection with the present study; of these, five have previously been described (Lee and others, 1963a) (Table 6). In this earlier publication Lee and others have shown that the py($Mg_3Al_2Si_3O_{12}$) component of the garnets from the California eclogites is low in comparison with the garnets from Groups A and B eclogites. In fact, when one considers the average py contents of garnets from charnockites, granulites, and amphibolites, one finds that the California eclogite garnets con-

large amount of the ugrandite moles present in the pyralspites. This admixture may be indicative of the pressure-temperature environment that produced the eclogites, as suggested by Lee and others (1963a). This strong admixture of ugrandite within the pyralspites appears to be characteristic for the garnets from Group A and B eclogites, amphibolites, charnockites, and granulites, as brought out by Tröger's (1959) compilation (Fig. 9). Miyashiro (1953) had previously shown that calcium-poor pyralspites from pelitic rocks decrease in MnO content and increase in FeO during progressive metamorphism. Increasing grades of metamorphism in the glaucophane schist facies produce a similar relationship in the FeO and MnO content of the garnets, but in addition the ugrandite molecule is usually in excess of 27 molecular per cent. This increase in ugrandite must somehow be related to the

182

pressure-temperature conditions of formation and not to bulk composition, since even the garnets from silica-rich granulites and charnockites contain unusually large amounts of ugrandite. It would seem that extreme pressures may be most effective in producing such

A deep-seated eclogites contain garnets whose py content usually exceeds 55 molecular per cent; garnets of Group B eclogites, from migmatitic gneisses, contain 30–55 molecular per cent py; and garnets of Group C, from

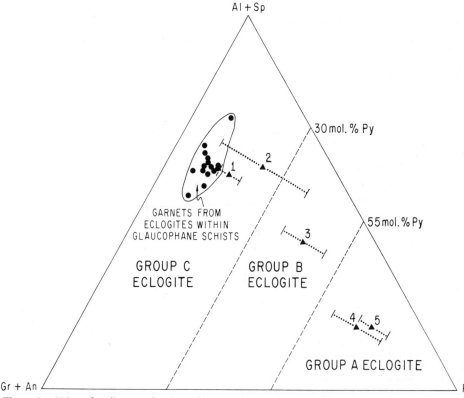

Figure 9.   Triangular diagram showing relative proportions of (al + sp), (gr + an), py garnet end members for garnets in eclogites and related rock types. Dots represent all known analyses of garnets from Group C eclogites; triangles, average composition of garnets given by Tröger (1959) for garnets from certain rock types; dotted lines, the range in composition for the following averages: 1—garnets from amphibolites, 2—garnets from charnockites and granulites, 3—garnets from eclogites occurring in gneissic or migmatite metamorphic terrains, 4—garnets from eclogites associated in kimberlite pipes, 5—garnets from eclogites within ultramafic rocks such as dunite and peridotite. Diagonal dashed lines set off compositional ranges based on the pyrope content of the garnet as related to the geological groupings of the eclogites.

sures may be most effective in producing such extensive mixing of ugrandite with pyralspites, as even those garnets from the low-grade glaucophane schists (Type III of Coleman and Lee, 1963) contain up to 33 molecular per cent ugrandite.

The comparison (Fig. 9) of garnet compositions from the three main groups of eclogites is related more directly to the mode of occurrence than to the bulk rock composition. The Group

glaucophane schists, contain less than 30 molecular per cent pyrope. The effect of bulk composition on such a comparison will be discussed in the section on petrogenesis.

According to the experimental work of Yoder and Chinner (1960), substitution of pyrope in almandine is more a function of pressure than temperature. Assuming that these experimental results approximate the conditions of formation of eclogitic garnets without consideration of the effect of other phases present, the increase in pyrope content from Group C to Group A eclogites implies that these garnets may have formed under wide variations of pressure. On the other hand, when we consider the trend of the jd component in the coexisting pyroxenes, as an indicator of pressure, there seems to be a complete reversal. That is, the py content of the garnet suggests that Group A eclogites formed at the highest pressures, whereas the jd content of the pyroxenes indicates that Group C eclogites represent the highest pressures. Such disparities between the experimental work and the present observations may be the result of applying simple experimental systems to the more complicated natural system represented by these rocks. Clearly more experimental work is needed to clarify the possible pressure-temperature stability fields represented by the garnets from the various eclogites.

Other Minerals

As suggested initially, the restriction of eclogites to only those bimineralic rocks consisting of pyroxene and garnet is not realistic. We will discuss here the minerals commonly present as primary accessory minerals in the California and New Caledonia eclogites. Mica is present as a common primary mineral in those eclogites containing K_2O above 0.2 weight per cent. The pyroxenes were found to contain less than 0.05 weight per cent K_2O, and it would seem that nearly all the potassium in these eclogites is contained within the mica. Preliminary mineralogical work indicates that this mica is similar to the phengites described by Ernst (1963) from the California glaucophane schists. These micas are the $2M$ structural type containing considerable MgO and FeO; their composition is similar to the silica-rich celadonites.

The calcium-rich eclogites often contain abundant clinozoisite-epidote as a primary mineral. The iron content of these epidotes appears to be variable but does not exceed approximately 10 weight per cent Fe_2O_3. The variability of the iron content can be discerned optically; often a single grain will show a strong zoning resulting from variation in iron content.

Amphiboles in these eclogites show conflicting evidence as to their paragenesis. In some instances the amphibole is clearly primary, as in the case of the New Caledonia eclogite, for which chemical analysis and optical data are given in Table 7. The composition is similar to

glaucophanes except for its unusually high calcium content. It appears to occupy a composition between glaucophane and actinolite components but consists predominantly of the glaucophane end member. Glaucophane in the

TABLE 7. CHEMICAL AND PHYSICAL PROPERTIES OF AMPHIBOLE FROM NEW CALEDONIA ECLOGITE (36-NC-62)

Analyses by L. B. Beatty, A. C. Bettiga, and W. W. Brannock

CHEMICAL ANALYSIS		
SiO_2	52.1	
Al_2O_3	10.5	
TiO_2	0.31	
Fe_2O_3	2.4	
FeO	9.5	
MgO	11.8	
CaO	6.7	
Na_2O	4.5	
K_2O	0.18	
H_2O^+	1.4	
H_2O^-	0.10	
MnO	0.02	
Sum	99.5	
OPTICAL PROPERTIES		
α	$1.622 \pm .002$	
β	1.649	
γ	1.653	
$Z \wedge C$	$11°–22°$	
S.G.	$3.14 \pm .02$	
N	Colorless	
Y	Pale smoky blue	
Z	Pale greenish blue	
NUMBER OF IONS TO THE BASE 24 (O,OH)		
Si	7.45	8.00
Al	0.55	
Al	1.22	
Ti	0.03	
Fe	0.26	5.17
Mg	2.52	
Fe^{+2}	1.14	
Na	1.25	
Ca	1.03	2.31
K	0.03	
OH	1.34	1.34

eclogite from Ward Creek (62–RGC–58) is only partly retrograde, whereas in the Valley Ford eclogites, all the glaucophane is definitely retrograde. Our previous work has shown that glaucophane and actinolite may coexist in the high-grade glaucophane schists. It is not clear how extensive the solid solution of the glaucophane component with actinolite may be in these rocks. The reason for the presence or absence of primary amphiboles in eclogites is not yet completely understood, and further work is needed to determine the compositional ranges.

Rutile and sphene are present in California eclogites as primary minerals. Secondary sphene is often found mantling or replacing rutile; this replacement may be regarded as part of the retrograde effects that are so often present.

There is much evidence of retrogression in the California eclogites, often manifested by replacement of garnet by chlorite, lawsonite, stilpnomelane, and pumpellyite. In some instances, as reported heretofore, glaucophane may be present as a retrograde mineral replacing pyroxene. Noteworthy is the absence of kelyphite rims around the garnets; such reaction rims are often reported from Group A and B eclogites.

Minor Elements

Spectrographic determinations of the minor elements contained in the eclogites and the coexisting garnet-pyroxene were made to determine the partition of those elements (Table 8). The results demonstrate further the general similarity among the eclogites from Group C. No strong variations in the minor-element content could be established from one eclogite sample to the other. Several relationships are, however, worthy of comment and could characterize Group C eclogites. Both Ni and Sr are strongly concentrated within the pyroxene structure, whereas Y is preferentially concentrated within the garnets. Vanadium apparently becomes more concentrated in the pyroxene than in the garnet; however, in those garnets from silica-rich metasediments of the glaucophane schists where pyroxene is absent, the V content may be as high as 0.51 weight per cent (Lee and others, 1963a). Minor-element data on eclogites from different environments are generally lacking. However, Nixon and others (1963) presented data on eclogite materials within the Basutoland kimberlites; although coexisting pairs are not compared with whole-rock determinations, the results are interesting to compare with our Group C determinations. The pyroxenes and garnets from Basutoland contain much more chromium than do the Group C garnets and pyroxenes. A considerably higher Ni content ($\sim$100 ppm) in the pyrope garnets of Basutoland is noteworthy and may be related to their pressure-temperature environment: Group C garnets usually contain less than 5 ppm Ni. On the other hand, the Group C garnets have a higher concentration of Y ($\sim$130 ppm) than do those garnets from Basutoland (<70 ppm).

Future determinations of the partition of minor elements in coexisting garnet-pyroxene from eclogites of various origins could provide another chemical means of distinguishing these groups.

PETROGENESIS

The previous discussion has shown that the pyroxenes and garnets from the Group C eclogites are distinct from these same coexisting minerals in Group A and B eclogites. Modal analyses of the Group C eclogites have shown that considerable amounts of epidote, amphibole, and muscovite may also be present (Table 2). Chemical analyses of the Group C eclogites reveal that they are basaltic and their compositions overlap those of other eclogites, but in general the Group C eclogites are less mafic and are somewhat higher in alkalies (Table 3; Fig. 10). Furthermore, the California eclogites are similar in composition to unmetamorphosed basalts of the Franciscan Formation (Coleman and Lee, 1963). It is interesting to note that Group A eclogites are generally more mafic than are the other two eclogite groups; however there is a widespread overlap in composition between Group A and B eclogites (Fig. 10). The average SiO_2 and MgO content of the three eclogite groups is as follows:

Group A—46 weight per cent SiO_2, 13 weight per cent MgO (29 samples); Group B—48 weight per cent SiO_2, 9 weight per cent MgO (38 samples); Group C—48 weight per cent SiO_2, 6 weight per cent MgO (13 samples).

The Alk,F,M plot suggests that the three groups of eclogites follow a compositional trend from olivine basalt through tholeiitic basalt. This rather pronounced variation in chemical composition between the three eclogite groups may account for the variations in the garnet and pyroxene noted earlier. As can be seen, Group C eclogites have a more restricted chemical composition than do the Group A and B eclogites.

Normative ne and ol are present in the California eclogites, whereas the nearly chemically equivalent unmetamorphosed Franciscan basalts are not as undersaturated with respect to silica and often contain hy and qtz in their norm. This difference in normative minerals suggests that a possible loss of silica occurred during the formation of these eclogites, as-

TABLE 8. SPECTROGRAPHIC ANALYSES OF ECLOGITES AND THEIR COEXISTING GARNETS AND PYROXENES (PPM)

Analyses by Robert E. Mays. *See* Figure 1 for location of samples.

	113-RGC-58 Rock	Gar	Pyr	100-RGC-58 Rock	Gar	Pyr	207-RGC-59 Rock	Gar	Pyr	62-RGC-58 Rock	Gar	Pyr	102-RGC-58 Rock	Gar	Pyr	36-NC-62 Rock	Gar	Pyr	Amp	56-CZ-59 Rock	Gar	Pyr
Ga	10	4	14	15	5	13	15	6	10	7	9	30	7	5	11	20	9	24	22	20	7	17
Cr	700	360	500	150	43	70	300	110	200	15	36	40	70	22	60	300	14	20	20	150	120	20
V	500	140	400	700	160	800	700	190	500	300	270	900	700	140	440	500	220	450	370	200	120	60
Ni	300	<4	240	50	<4	19	150	5	35	70	<4	13	70	<4	26	150	14	32	110	150	<4	90
Co	70	46	55	30	25	30	50	18	14	30	20	23	50	19	16	30	28	24	60	50	21	22
Pb	:	<20	:	:	<20	:	:	80	:	:	<20	:	:	:	:	:	34	:	:	:	<20	:
Cu	100	7	240	50	19	42	150	28	38	30	29	37	15	24	19	70	34	32	36	100	9	80
Sc	50	50	95	50	49	60	200	75	65	30	55	90	50	70	50	30	85	70	30	30	44	37
Zr	100	80	100	150	70	110	150	70	110	150	160	230	150	60	60	200	50	40	60	200	160	60
Y	50	10	<20	70	120	<20	100	140	50	70	190	60	70	150	20	50	120	<20	<20	70	160	30
Yb	5	<4	3	10	15	3	10	25	4	7	28	6	10	16	<2	5	12	<2	<2	10	14	4
Sr	200	<4	60	300	4	60	200	<4	200	150	46	340	300	<4	220	200	<4	190	44	30	<4	90
Ba	100	<4	3	100	<4	2	300	4	10	15	4	8	150	<4	<4	200	<4	<4	2	150	<4	6
B	10	:	:	10	:	:	10	:	:	<10	:	:	:	:	:	15	:	:	13	10	:	:
Nb	:	:	:	:	:	:	:	:	:	:	:	:	:	:	<2	50	<2	<2	<2	:	:	:
Be	:	:	:	:	:	:	:	:	:	3	<4	:	:	:	:	:	:	:	:	:	:	3
Mo	:	<4	:	:	<4	:	:	<4	:	:	:	:	:	:	:	:	:	:	:	:	<4	:

Note: Gar—Garnet; Pyr—Pyroxene; Amp—Amphibole

suming that they are indeed metamorphosed Franciscan basalts. The high calcium content of the California eclogites could also be a possible contributing factor in producing different garnet-pyroxene compositions. The high-calcium eclogite from Occidental quarry (113-RGC-58) does contain a garnet with excep-

relationships can be demonstrated for the three groups of eclogites. Initially, comparisons are made by using a triangular CFM plot as employed by Yoder and Tilley (1962), O'Hara (1960), and Subramaniam (1956) (Fig. 11). The molecular proportions of CaO, (FeO + MnO), MgO are recalculated to 100 per cent

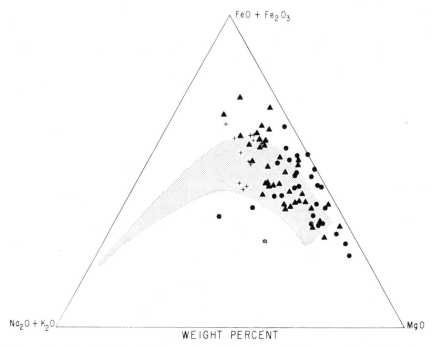

Figure 10. Alk, F,M triangular diagram in weight per cent of all known analyses of eclogites. The weight per cents of FeO + Fe₂O₃ and Na₂O + K₂O are simply added together. Circles, Group A eclogites; triangles, Group B eclogites; crosses, Group C eclogites. Shaded area represents alkali, tholeiitic, and nepheline basalt trend from the Hawaiian Islands (Kuno and others, 1957).

tional grossular component (34.2 molar per cent), thus changing the Fe/Mg ratio in the garnet, but the Fe/Mg ratio in the pyroxene is unaffected. Yoder and Tilley (1962) suggest that an eclogite exists for each major compositional type of basalt. This was originally shown by Angel (1957), but it does not necessarily demonstrate that any particular eclogite may represent basaltic material crystallized within the mantle. There is little doubt, however, that all eclogites have a restricted chemical composition—that is, either basaltic or gabbroic.

If we now consider the problem of mineral compositions versus rock compositions, several

for the garnet and pyroxene to represent variations in these components on a triangular diagram. Plotting the molecular proportions of garnet and pyroxene on the diagram and connecting the points by tie lines allow a geometric comparison to be made whereby changes in mineral composition can be related to pressure-temperature variations and geologic occurrence. As pointed out by Yoder and Tilley (1962, p. 485), the tie lines should not cross one another but should assume a continuous sweep across the diagram if eclogites of varying bulk composition formed under the same pressure-temperature conditions—assuming, of course, that these are bimineralic rocks or that other minerals present do not contain components that could easily partition between the pyroxene or garnet. Yoder and Tilley (1962) demonstrated that eclogites selected for their study did show such a continuous sweep of tie lines, and therefore they concluded that "two-phase" eclogites crystallize in a limited pressure-temperature range.

Unfortunately the "progressive" compositional trend from Group A to Group C eclogites does not allow a realistic comparison between the deep-seated group and the alpine group.

The nearly identical composition of the Loch Duich eclogite (Yoder and Tillery, 1962) and the Tiburon eclogite (Table 3) does, however, allow a good comparison between Group B and C eclogites. The CFM plot (Fig. 12) of the tie lines for the Loch Duich and Tiburon shows a striking variation in the Fe/Mg ratios in the garnets and less in the pyroxenes. If one uses the Yoder and Tilley argument, the crossed tie lines must indicate that the pressure-temperature conditions were different for the formation of these two eclogites. Other projections (not

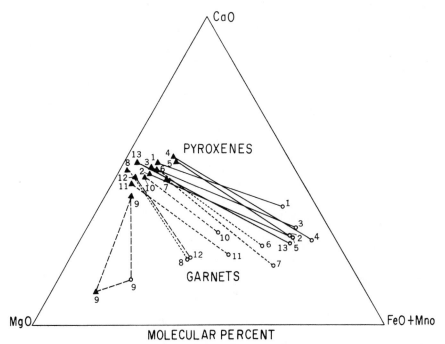

Figure 11.   CFM plot of coexisting garnet-pyroxene pairs from individual eclogites. Tie line connects these pairs. Dashed line, Group A eclogites; Dotted line, Group B eclogites; Solid line, Group C eclogites. Rock compositions are not plotted, to avoid diagram confusion.

Key for coexisting garnet-pyroxene pairs:

 1. Occidental quarry, Calif. (113-RGC-58), Tables 2,4,5
 2. Tiburon Peninsula, Calif. (100-RGC-58), Tables 2,4,5
 3. Valley Ford, Calif. (207-RGC-59), Tables 2,4,5
 4. Ward Creek, Calif. (62-RGC-58), Tables 2,4,5
 5. Junction School, Calif.; Borg (1956), Switzer (1951)
 6. Glenelg, Scotland; Yoder and Tilley (1962)
 7. Loch Duich, Scotland; Yoder and Tilley (1962)
 8. Silberbach, Germany; Yoder and Tilley (1962)
 9. Salt Lake Crater, Oahu, Hawaii; Yoder and Tilley (1962)
10. Bielice No. 1, East Sudetes, Poland; Smulikowski (1960)
11. Nowa Wies No. 4, East Sudetes, Poland; Smulikowski (1960)
12. Kao Pipe, Basutoland, South Africa; Nixon and others (1963)
13. Shubina Village deposit, southern Urals, U.S.S.R.; Chesnokov (1959)

shown) using various combinations of components have verified the tie-line crossing for other geometric perspectives. The presence of muscovite (primary) and chlorite-lawsonite (retrograde) will not materially affect the position of the Tiburon tie line. Further comparison of the California eclogites (Group C)

and pyroxene is different for the various eclogite groups, and the geologic nature of their occurrence strongly suggest that eclogites must form under a wide range of pressure-temperature conditions.

Contrary to O'Hara (1960) and Yoder and Tilley (1962) the partition of Ca, Fe, and Mg

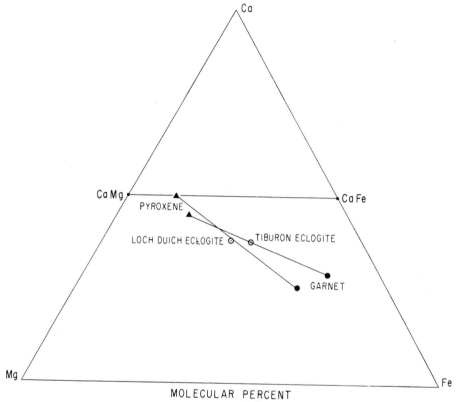

Figure 12. CFM plot of two eclogites (open circles) and their coexisting garnet and pyroxene. Tie lines connect the garnet-pyroxene-eclogite compositions. Loch Duich analytical data is *from* Yoder and Tilley (1962); the Tiburon analyses are given *in* Tables 2, 4, and 5.

in Figure 11 shows that these tie lines cross and are subparallel to the tie lines shown for the Group B eclogites. Furthermore, a crossing of the tie lines within the Group B eclogites is apparent. Admittedly a selection of certain tie lines within each group of eclogites could produce a continuous sweep of noncrossing tie lines, but the subparallelism within the separate eclogite groups and nonparallelism from group to group is significant. The combination of chemical evidence, which shows that the partition of Fe-Mg-Ca between coexisting garnet

between the garnet and pyroxenes is not similar within the eclogite facies. This observation in chemistry is concordant with the geologic grouping of the eclogites and raises certain questions concerning the origin of eclogites. Coleman and Lee (1963) have documented the progressive metamorphism of Franciscan basalts from low-grade glaucophane-lawsonite-pumpellyite-jadeitic pyroxene schists up to high-grade glaucophane-epidote-garnet-omphacite schists containing conformable layers of eclogite (Group C). Thus, in California and New

Caledonia, eclogites have formed within the highest grade of metamorphism of the glaucophane schist facies. In a recent paper by Smulikowski (1960) on eclogites from the Stary Gieraltow complex, Snieznik Mountains of East Sudetes, Poland, convincing evidence is presented that shows conformable bands of eclogite (Fig. 11, Group B, No. 11) formed contemporaneously with the gneisses of the amphibolite facies. Group B eclogites have also been described within the charnockitic rocks of Madras, India, by Subramaniam (1956). These geological examples demonstrate supracrustal metamorphism of different facies, yielding eclogites derived from rocks of basaltic composition. More recently O'Hara and Mercy (1963, p. 301) have stated:

"Conversion of gabbro, basalt, or amphibolite to eclogite within the crust would presumably be possible given sufficient excess pressure due to high stress, and low enough temperatures. Several authors (Backlund 1936; Chevenoy 1957; Udovchina 1959) have claimed to observe such a conversion to eclogite. To establish such an interpretation it is essential to show unity of the eclogitic and basic igneous material, both by field observation in well exposed ground and by chemical analyses. It is also necessary to show from textural evidence that the change is progressing towards eclogite, and not retrograding from it due to regional or thermal metamorphism. Conversion of gabbro or basalt to eclogite is a likely process in the upper mantle. To establish that it has happened within the crust requires proof that the entire basic mass does not have tectonic contacts to the rocks among which it occurs. Desirable confirmatory evidence of the change would be a parallel conversion of sedimentary or acid igneous rock types to a new high pressure assemblage. Judged by these criteria, no case of conversion of basic rock to eclogite within the crust can be regarded as established. Even if it does occur, it is remarkable that there are so few instances of partial conversion among the many known eclogite occurrences. It must be concluded that most eclogites which occur as lenses in metamorphic terrains have been introduced there as crystalline masses by tectonic action."

These rigorous criteria and observations are a fair statement of the problem involved in proving that eclogites form as a result of crustal metamorphism of basaltic rocks. The chemical evidence presented in this paper, combined with the published field evidence of Coleman and Lee (1962; 1963), fulfills as nearly as possible the rigorous criteria set down by O'Hara and Mercy. Certainly the pronounced change in Ca-Mg-Fe partition discovered for the alpine-type eclogites, combined

with the compelling evidence of high pressures and low temperatures characterizing the glaucophane schist facies, overcomes the most difficult problems. The erratic and limited distribution of eclogites within the glaucophane schist facies indicates that the pressure-temperature conditions favorable for the formation of these rocks is localized and not regional, having resulted perhaps from "tectonic overpressures" during metamorphism (Coleman and Lee, 1962). Furthermore, the alpine-type eclogites from the glaucophane schists of California have a basaltic composition notably differentiated, making it unlikely that they could have developed within the mantle; O'Hara and Yoder (1963) recognized this point.

The question of whether or not there is an eclogite facies of metamorphism arises from these observations. Obviously, if rocks of similar bulk composition and mineral assemblage are present as part of a conformable sequence in glaucophane schist, amphibolite, and granulite facies rocks, they could not have formed under similar pressure-temperature conditions. Furthermore, the demonstrated variation in the Ca-Mg-Fe partition between the pyroxene and garnet is strong chemical evidence supporting the concept that eclogites form over a pressure-temperature range that is much too broad to be characteristic of a metamorphic facies. It would seem more in keeping with the geological distribution to discontinue the eclogite facies since it cannot be mapped and since it transgresses into other metamorphic facies. The discontinuation does not preclude in any way the useful concept that certain eclogites may indeed form within the mantle and may represent partial fusion products (O'Hara and Yoder, 1963). Recognition that eclogites may have multiple origins could lead to a more critical evaluation of the observed field relationships.

Continued detailed field and laboratory studies may permit certain estimates to be made concerning the pressure-temperature conditions that characterize the various eclogite groups. The experiments of Kennedy (1959), Yoder and Tilley (1962), and O'Hara (1963) have shown that basalt can be converted to a garnet-pyroxene rock at very high pressures and temperatures. We can only suggest that at somewhat lower pressures and temperatures other eclogites also form and their garnet-pyroxene compositions are distinct from those formed at higher pressure-temperature conditions.

REFERENCES CITED

Alderman, A. R., 1936, Eclogites in the neighbourhood of Glenelg, Invernesshire: Geol. Soc. London Quart. Jour., v. 92, p. 488–530

Angel, Franz, 1957, Einige ausgewählte Probleme eklogitischer Gesteinsgruppen der osterreichischen Ostalpen: Neues Jahrbuch Mineralogie Abh., v. 91, p. 151–192

Angel, Franz, and Schaider, Ferdinand, 1950, Granat und omphazit aus dem Eklogit des Gertrusk (Saualpe, Kärnten): Carinthia II, Naturw. Beitr. Kärntens, Jahrg. 58–60, p. 33

Backlund, H. G., 1936, Zur genetischen Deutung des Eklogits: Geol. Rundschau, v. 27, p. 47–61

Bailey, E. H., Irwin, W. P., Jones, D. L., 1964, Franciscan and related rocks, and their significance in the geology of Western California: Calif. Div. Mines Bull. 183, p. 1–176

Bloxam, T. W., 1956, Jadeite-bearing metagraywackes in California: Am. Mineralogist, v. 41, p. 488–496

—— 1959, Glaucophane-schists and associated rocks near Valley Ford, California: Am. Jour. Sci., v. 257, p. 95–112

Borg, I. Y., 1956, Glaucophane schists and eclogites near Healdsburg, California: Geol. Soc. America Bull., v. 67, p. 1563–1584

Briere, P. Y., 1920, Les eclogites françaises—leur composition minéralogique et chemique; leur origine: Soc. France Mineralogie Bull., v. 43, p. 72–222

Chesnokov, B. V., 1959, Rutile-bearing eclogites from the Shubino village deposit in the Southern Urals: Izvestiya Vysshikh Uchebykh Zavendeniy, Geologiya i Razvedka, no. 4, p. 124–136 (Translation in Internat. Geology Rev., 1960, v. 2, p. 936–945)

Chevenoy, M., 1957, Sur l'origine des éclogites de Sauviat, (Creuse): Compt. Rend., Acad. Sci. Paris, v. 241, p. 426–428

Coleman, R. G., 1961, Jadeite deposits of the Clear Creek area, New Idria District, San Benito County, California: Jour. Petrology, v. 2, no. 2, p. 209–247

Coleman, R. G. and Lee, D. E., 1962, Metamorphic aragonite in the glaucophane schists of Cazadero, California: Am. Jour. Sci., v. 60, p. 577–595

—— 1963, Glaucophane-bearing metamorphic rock types of the Cazadero area, California: Jour. Petrology, v. 4, no. 2, p. 260–301

Deer, W. A., Howie, R. A., and Zussman, J., 1963, Chain silicates, Volume 2 of Rock-forming minerals: London, Longmans, Green and Co., Ltd., 379 p.

De Roever, W. P., 1955, Genesis of jadeite by low-grade metamorphism: Am. Jour. Sci., v. 253, p. 283–298

Ernst, W. G., 1963, Significance of phengitic micas from low-grade schists: Am. Mineralogist, v. 48, p. 1357–1373

Eskola, Pentti, 1921, On the eclogites of Norway: Vidensk.-selskapets Skr., Kristiania, I. Matematisk-Naturvidenskapelig Kl., no. 8, p. 1–118

Fermor, L. L., 1913, Preliminary note on garnet as a geological barometer and on an infra-plutonic zone in the earth's crust: Records Geol. Survey India, v. 43, pt. 1, p. 41–47

Gealey, W. K., 1951, Geology of the Healdsburg quadrangle, California: Calif. Div. Mines Bull., v. 161, p. 1–50

Haüy, R. J., 1822, Traité de minéralogie 2d ed., v. 2,: Paris, Bachelier, 594 p.

Hess, H. H., 1955, Serpentines, orogeny, and epeirogeny, p. 391–408 in Poldervaart, Arie, Editor, Crust of the Earth: Geol. Soc. America Speical Paper 62, 762 p.

Iwasaki, M., 1960, Clinopyroxene intermediate between jadeite and aegirin from Suberi-dani, Tokusima Prefecture, Japan: Geol. Soc. Japan Jour., v. 66, no. 776, p. 334–340

Kennedy, G. C., 1959, The origin of continents, mountain ranges, and ocean basins: Am. Scientist, v. 47, p. 491–504

Knorring, O. von, and Kennedy, W. Q., 1958, The mineral paragenesis and metamorphic status of garnet-hornblende-pyroxene-scapolite gneiss from Ghana (Gold Coast): Mineralog. Mag., v. 31, p. 846

Kundig, Ernst, 1956, The position in time and space of the ophiolites with relation to orogenic metamorphism: Geol. Mijnb. (NW ser.) 18, no. 4, p. 106–114

Kuno, Hisashi, Yamasaki, Kazuo, Iida, Chuzô, and Nagashima, Kozo, 1957, Differentiation of Hawaiian magmas: Japanese Jour. Geology and Geography, v. 28, no. 4, p. 179–218

Lee, D. E., Coleman, R. G., and Erd, R. C., 1963a, Garnet types from the Cazadero area, California: Jour. Petrology, v. 4, no. 3, p. 460–492

Lee, D. E., Thomas, H. H., Marvin, R. F., and Coleman, R. G., 1963b, Isotope ages of glaucophane schists from Cazadero, California: Art. 142 in U. S. Geol. Survey Prof. Paper 475-D, p. 105–107

Miyashiro, Akiho, 1953, Calcium-poor garnet in relation to metamorphism: Geochim et Cosmochim. Acta, v. 4, p. 179–208

Miyashiro, Akiho, and Seki, Yotaro, 1958, Mineral assemblages and subfacies of the glaucophane schist facies: Japanese Jour. Geology and Geography, v. 29, p. 199

Nixon, P. H., Knorring, O. von, and Rooke, J. M., 1963, Kimberlites and associated inclusions of Basutoland: A mineralogical and geochemical study: Am. Mineralogist, v. 48, p. 1090–1132

O'Hara, M. J., 1960, A garnet-hornblende-pyroxene rock from Glenelg, Invernesshire: Geol. Mag., v. 97, p. 145–146

—— 1963, Melting of bimineralic eclogite at 30 kilobars: Carnegie Inst. Wash. Yearbook, v. 62, p. 76–77

O'Hara, M. J., and Mercy, E. L. P., 1963, Petrology and petrogenesis of some garnetiferous peridotites: Royal Soc. Edinburgh Trans., v. 65, no. 12, p. 251–314

O'Hara, M. J., and Yoder, H. S., Jr., 1963, Partial melting of the mantle: Carnegie Inst. Wash. Yearbook, v. 62, p. 66–71

Robertson, E. C., Birch, Francis, and MacDonald, G. J. F., 1957, Experimental determination of jadeite stability relations to 25,000 bars: Am. Jour. Sci., v. 255, p. 115–137

Routhier, Pierre, 1953, Étude géologique du versant occidental de la Nouvelle Caledonie entre le Col de Boghen et la Pointe D'Arana: Geol. Soc. France Memoir, v. 67, 266 p.

Sahlstein, T. G., 1935, Petrographie der Eklogiteinschlüsse in dem Gneisen des südwestlichen Liverpool-Landes in Öst-Gronland: Medd. om Gronland, v. 95, no. 5, p. 1–43

Shido, Funiko, 1959, Notes on rock forming minerals (9). Hornblende-bearing eclogite from Gongenyama of Higasi-Akaisi in the Bessi District, Sikoku: Geol. Soc. Japan Jour., v. 65, p. 701

Smulikowski, Kazimierz, 1960, Petrographic notes on some eclogites of the East Sudetes: Bull. de l'Acad. polonaise des sciences, Série des sciences géologique et géographique, v. 8, no. 1, p. 11–19

Subramaniam, A. P., 1956, Mineralogy and petrology of the Sittampundi complex, Salem district, Madras State, India: Geol. Soc. America Bull., v. 67, no. 3, p. 317–390

Switzer, George, 1945, Eclogite from the California glaucophane schists: Am. Jour. Sci., v. 243, p. 1–8

—— 1951, Mineralogy of the California glaucophane schists: Calif. Div. Mines Bull. 161, p. 51–70

Tilley, C. E., 1936, The paragenesis of kyanite eclogites: Mineralog. Mag., v. 24, p. 422–432

Travis, R. B., 1952, Geology of the Sebastopol quadrangle, California: California Dept. Nat. Res., Div. Mines Bull. 162, 33 p.

Tröger, W. E., 1959, Die Granatgruppe: Beziehungen zwischen Mineralchemismus und Gesteinsart: Neues Jahrbuch Mineralogie, v. 93, p. 1–44

Udovchina, N. G., 1959, K voprosy ob eklogitisatsii uletrosnovneich porod v. yuzhnoj chasti khreeta marunkeu: Voprosy Magmatizma Urala, v. 32, p. 5–18

Williams, A. F., 1932, The genesis of the diamond: London, Ernest Benn, Ltd., 636 p.

Wolff, T., von, 1942, Methodischus zur quantitativen Gestems-und Mineral-Untersuchung mit Hilfe der Phase analyse (Am. Beispiel der magischen Komponenten des Eklogits von Silberbach): Tschermaks mineralogische und petrographische Mitt., v. 54, (1–3) p. 1–122

Yoder, H. S., Jr., 1950, The jadeite problem: Am. Jour. Sci., v. 248, p. 225–248, 312–334

Yoder, H. S., Jr., and Chinner, G. A., 1960, Almandite-pyrope-water system at 10,000 bars, p. 81–84 in Annual Report of the Director, Geophysical Laboratory for 1959-60: Carnegie Inst. Wash. Yearbook, v. 59, 515 p.

Yoder, H. S., Jr., and Tilley, C. E., 1962, Origin of basalt magmas: An experimental study of natural and synthetic rock systems: Jour. Petrology, v. 3, no. 3, p. 342–532

MANUSCRIPT RECEIVED BY THE SOCIETY MAY 18, 1964

$\mathcal{9}$

Reprinted from *Tectonophysics*, **3**(5), 383–427 (1966)

AN EXPERIMENTAL INVESTIGATION OF THE GABBRO–ECLOGITE TRANSFORMATION AND SOME GEOPHYSICAL IMPLICATIONS

A.E. RINGWOOD and D.H. GREEN

Department of Geophysics and Geochemistry, Australian National University, Canberra, A.C.T. (Australia)

(Received May 9, 1966)

SUMMARY

A detailed experimental investigation of the gabbro–eclogite transformation in several basalts has been carried out. More than 200 runs at temperatures between 900 and 1,250°C and pressures up to 30 kbar were made, and the resultant phase assemblages studied by optical and X-ray techniques. In all the basalts studied, the transformation from the low-pressure gabbroic assemblage (pyroxene + plagioclase) into the high-pressure eclogite assemblage (garnet + pyroxene) was found to proceed via a transitional mineral assemblage characterised by co-existing garnet, pyroxene and plagioclase. This mineral assemblage closely resembles that displayed in nature by basic rocks in the garnet–clinopyroxene–granulite sub-facies. The width of the transitional assemblage (garnet granulite) varies from 3.4 to 12 kbar at 1,100°C in different basalts and the pressures required for incoming of garnet and elimination of plagioclase likewise vary widely according to the particular basaltic composition studied. These variations are caused by small differences in chemical composition, and are interpreted in terms of the principal mineral equilibria which are involved in the transformation. In passing from gabbro to eclogite via the garnet granulite assemblage, the proportion of garnet is observed to increase regularly with pressure, whilst the proportion of plagioclase decreases regularly. Thus the increase in density and seismic velocity from gabbro ($\rho \sim 3.0$ g/cm^3, $V_P \sim 6.9$ km/sec) to eclogite ($\rho \sim 3.5$ g/cm^3, $V_P \sim 8.3$ km/sec) is uniformly smeared out over the entire garnet granulite transition interval.

The effect of temperature upon the pressures required for the gabbro–eclogite transformation is investigated. Our experimental results, together with the results of other workers on simple systems closely related to the gabbro–eclogite transformation, strongly indicate that eclogite would be stable relative to gabbro and garnet granulite throughout large regions of the normal continental crust. This conclusion has important tectonic implications.

The bearing of the experimental results upon the hypothesis that the continental Mohorovičić discontinuity is caused by an isochemical transformation from gabbro to eclogite is examined. It is concluded that this hypothesis must be rejected on the following grounds:

(1) Eclogite, not gabbro, appears to be stable throughout the continental crust.

(2) The experimental temperature gradient of the transformation is incompatible with inferred geophysical relationships between the temperature at the base of the crust and thickness of the crust.

(3) The transformation cannot explain the seismic velocity distribution in the crust.

(4) Minor changes in chemical composition strongly affect the pressure required for eclogite stability. This makes it difficult to understand the uniformity of crustal thickness in stable continental regions.

(5) The density of eclogite (3.5 g/cm^3) is higher than the density of the upper mantle inferred from gravity observations.

(6) An upper mantle of eclogitic composition would be gravitationally unstable.

The chemical composition of the continental crust is discussed in the light of the widely accepted geophysical model that it consists of a layer of granitic material overlying or passing gradually downwards into a gabbroic layer. Our experimental results are in conflict with the view that the lower crust is generally of gabbroic composition. It is more likely that the lower crust consists on the average of intermediate rocks in the eclogite facies.

Although the gabbro–eclogite transformation is not believed to play a significant role in the structure of stable continental regions, it may be of major importance in tectonically active areas where the Mohorovičić discontinuity cannot be clearly recognized, e.g., regions of recent orogenesis, continental margins, island arcs and mid-oceanic ridges. It is shown that the basalt–eclogite transformation may provide a tectonic engine of great orogenic significance. Large volumes of basalt, when extruded and intruded at and near the earth's surface may become transformed to eclogite under suitable circumstances on cooling. Because of the high density of eclogite, such large scale transformations would generate gravitational instability. Large blocks of eclogite would sink through the crust, dragging it down initially into a geosyncline, and later causing extensive deformation (folding). Because the density of eclogite (3.5 g/cm^3).is greater than that of the ultramafic rocks which make up most of the mantle (3.3 g/cm^3) blocks of eclogite will sink deep into the mantle, and may undergo partial fusion, leading to generation of andesitic and granodioritic magmas which rise upwards and intrude the folded geosyncline.

INTRODUCTION

It is well-known that rocks of basaltic chemical composition may exhibit several distinct·mineral assemblages. Two such assemblages are those of gabbro (plagioclase + pyroxene) and eclogite (garnet + pyroxene). One of the first workers to appreciate the relationships between these rock types was Fermor (1913, p.41, 1914) who pointed out that eclogite ($\rho \sim 3.5$ g/cm^3) must be a high pressure form of gabbro or basalt ($\rho \sim 3.0$ g/cm^3). This led him to propose that the earth's outer mantle was composed of eclogite and that the transformation between basaltic crustal rocks and the eclogitic mantle (infra-crust) would have some important tectonic implications. Similar views were strongly advocated in the 1920's by Holmes (1926a, b, 1927) and Goldschmidt (1922). Holmes showed that the seismic velocity of eclogite was similar to that observed in the upper mantle and argued that

the crust–mantle boundary, now defined as the Mohorovičić (M) discontinuity was caused by a transformation from gabbro to eclogite. Holmes also
emphasized the tectonic consequences of this model - changes in temperature
at the crust–mantle boundary would cause transformation of gabbro to
eclogite or vice versa, resulting in crustal uplift or depression.

In the period 1930–1950, the concept of an eclogitic mantle lost ground
to the long-standing rival hypothesis that the upper mantle was generally of
peridotitic composition, and that accordingly the crust–mantle boundary
represented a change in chemical composition. However, an important paper
by Birch (1952) re-opened the whole question. Basing his arguments on geochemical grounds and upon the elastic properties of minerals, Birch was
led to favour an eclogitic rather than a peridotitic mantle. Although the
principal argument against peridotite subsequently proved to be unfounded,
owing to inaccuracies in some of the early experimental data on elastic
properties of minerals, this paper resulted in a revival of interest in the
subject. This was stimulated by the first syntheses by Coes (reported by
Roy and Tuttle, 1956) of several eclogitic mineral components in a high
pressure–temperature apparatus. Apparatus of this type was also developed
by Birch and his students, and by Kennedy, with the objective of determining
the important equilibria involved in the basalt–eclogite transformation by
direct experiment. Pioneering results in this field were published by
Robertson et al. (1957) and Kennedy (1956, 1959). These authors concluded
that their preliminary experimental results were generally favourable to the
hypothesis that the M-discontinuity is a phase change.

A novel argument was introduced by Lovering (1958) on the basis of
an achondritic earth model. According to this model the earth's mantle is
dominantly ($\sim$ 60%) composed of material similar to basaltic achondrites,
which under the p, t conditions obtaining, would crystallize as eclogite.
Since this material occurred mainly in the outer part of the mantle, it was
argued that the crust–mantle boundary was probably caused by a phase
change from basalt to eclogite. However this model was based upon an assumed relationship between chondritic and achondritic meteorites which
its author has since abandoned (Lovering, 1962) and accordingly its status
is questionable.

The geochemical, geothermal and tectonic consequences of the gabbro–
eclogite transformation were briefly explored during this period by Sumner
(1954) and more thoroughly by Lovering (1958) and Kennedy (1959). Interest
in this subject has cascaded in recent years and a large number of papers
dealing either directly or indirectly with the properties and consequences
of the transformation have appeared, e.g., Harris and Rowell (1960),
MacDonald and Ness (1960), Bullard and Griggs (1961), Wetherill (1961),
Yoder and Tilley (1962), Broecker (1962), Wyllie (1963), Stishov (1963).
Some of these authors favour the phase change hypothesis; others oppose it.
We will review specific arguments in later sections.

The prime cause of the wide divergences of opinion on this important
subject is lack of detailed experimental data on the nature of the transformation and on the variations in physical properties accompanying the transformation. The objective of the present investigation, which has been in
progress for 3 years, has been to fill this gap. We have chosen several
representative basalts, subjected them to a wide range of closely spaced
p, t conditions, and then investigated the chemical and physical nature of the

Tectonophysics, 3 (5) (1966) 383–427

mineral phases produced and their stability fields. Over 200 runs at high
pressures and temperatures have been carried out. These have led to a
detailed understanding of many important aspects of the gabbro–eclogite
transformation, particularly relating to the transitional mineral assemblages
which intervene between the gabbro and eclogite stability fields. The present
paper deals with the geophysical implications of these results. An accompanying paper, dealing with the geochemistry, petrology and mineralogy of
the transformation is being published elsewhere (Green and Ringwood, 1966).
A preliminary account of our early results has already been published
(Ringwood and Green, 1964).

Before proceeding, a brief discussion of our usage of the term "eclogite"
is desirable. In this paper we regard "eclogite" as a rock of basaltic chemical composition characterised by the mineral assemblage, garnet +
pyroxene $\pm$ quartz. Plagioclase is *not* present. The density of such a rock is
> 3.3 g/cm^3 and its P-wave velocity is > 8 km/sec. This definition is
somewhat loose from the petrologic point of view but is convenient and
adequate for the discussion of many geophysical problems covered herein.
In the accompanying paper dealing with the petrology of the transformation,
a more rigid definition based upon detailed mineralogic criteria is used.
We emphasize that our present usage of "eclogite", "garnet granulite" and
"pyroxene granulite" is in the form of descriptive rock terms, without
implications as to the metamorphic classification of these rocks.

EXPERIMENTAL

Previous experimental investigations

The transformation of gabbro to eclogite is a complex equilibrium between
mineral solid solutions involving many components. One approach is to study
related equilibria in simple systems between individual components which
participate in the gabbro–eclogite transformation. This approach is exemplified in the pioneering work by Robertson et al. (1957) who investigated
the equilibrium:

$$\text{Na Al Si}_3\text{ O}_8 \quad + \quad \text{Na Al Si O}_4 \quad = 2 \text{ Na Al Si}_2\text{ O}_6$$

albite   nepheline   jadeite

Such investigations have yielded much important information which will be
referred to in later sections. However, they are not capable in themselves
of solving many of the major geophysical problems which are associated
with the gabbro–eclogite transformation. These require direct experiments
upon complex chemical systems of basaltic compositions.

Such experiments were first undertaken by Kennedy (1956) who crystallized basaltic glasses at high temperatures (800–1,000°C) and pressures.
At pressures below 10–15 kbar, depending upon temperature, he found that
basaltic glass crystallized to a felspar-bearing assemblage. Above these
pressures, felspars were not found and Kennedy suggested that the mineral
assemblages might be eclogitic. However no positive identification of the
mineralogy of the high-pressure assemblage was reported. Subsequently,
Kennedy (1959) described the crystallization of basaltic glass at 500°C.

Tectonophysics, 3 (5) (1966) 383–427

Below 10 kbar, felspar was observed. At pressures above 10 kbar, the amount of felspar decreased and finally a "rock made up dominantly of jadeitic pyroxene" was observed. The suggestion that this assemblage was "eclogitic" cannot be accepted in the absence of garnet as a major phase. Furthermore, from our experiments on the kinetics of crystallization of basaltic glasses at high pressure (Green and Ringwood, 1966) we strongly suspect that Kennedy's pyroxene was metastable and not relevant to the gabbro–eclogite transformation.

The first definite conversion of basalt to eclogite was obtained by Boyd and England (1959) at 1,200°C in the pressure range 33–40 kbar. This was followed by the well-known work of Yoder and Tilley (1962) who crystallized a number of basalts and eclogites at pressures of 10 kbar, 20 kbar and 30 kbar in the temperature range 1,200–1,400°C. This work showed clearly that basalt was stable relative to eclogite at 10 kbar, whereas eclogite appeared to be stable above about 20 kbar. At 30 kbar, eclogite was definitely

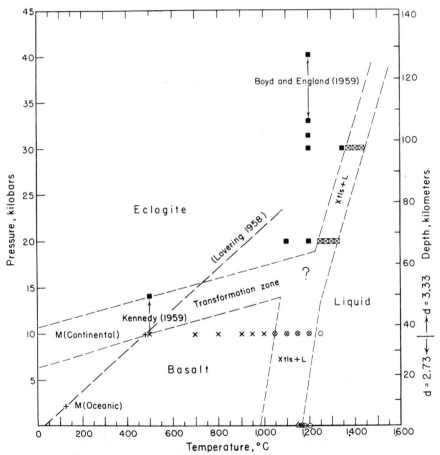

Fig.1. Stability fields for basalt and eclogite as proposed by Yoder and Tilley (1962, p.498). (Symbols used are explained in their paper.)

Tectonophysics, 3 (5) (1966) 383–427

stable. The work at 20 kbar was not completely definite since only one run
was made in which a basalt was converted into an eclogite assemblage at
that pressure. The remaining runs were conducted with eclogite as a starting
material, and in view of our subsequent experience with this starting material, it is difficult to be confident that equilibrium was attained.

Yoder and Tilley (1962) constructed a diagram for the gabbro–eclogite
transformation based upon their results around 1,200°C and Kennedy's
results at 500°C (Fig.1). On the basis of this diagram they discussed the
continental M-discontinuity and concluded that it might be caused by a
gabbro–eclogite transformation. This diagram has frequently been reproduced by other workers and cited as the experimental back-ground for
their theoretical discussions. Unfortunately this diagram is misleading and
does not permit valid conclusions to be drawn concerning the nature of the
M-discontinuity. Any discussion of the latter depends critically upon a
knowledge of the p, t gradient of the transition and the transition interval.
The gradient and transition zone in Fig.1 are seen to be dependent upon
Kennedy's results at 500°C which are unacceptable for reasons discussed
earlier.

A detailed study of the transformation at 1,100°C on two basalts was
reported by Ringwood and Green (1964) who concluded that the continental
M-discontinuity was not caused by an isochemical gabbro–eclogite transformation. Their arguments are developed in greater detail in the present
paper.

The present investigation

Although basalts constitute an easily recognisable and distinct petrologic
class of rocks, their chemical compositions vary over a substantial range,
necessitating further classification into various sub-classes. One of the
principal objectives of the present investigation was to study the effects of
varying chemical compositions (if any) upon the gabbro–eclogite transformation. Typical representatives of the principal types of basalt were accordingly chosen for investigation (Table I). These were as follows: typical quartz
tholeiite (2), a quartz tholeiite rather poorer than usual in alkalis (3), a
typical alkali olivine basalt (4), and an alkali-poor olivine tholeiite (6). In
addition, a typical oceanic high-alumina olivine basalt (1) has been studied
by Mr. Trevor Green who has kindly permitted us to refer to his results.
An investigation was also made of the effects of certain other chemical
parameters on the transformation. The alkali olivine basalt (4) was oxidised
to an Fe_2O_3/FeO ratio of 2 : 1, to elucidate the effects of a high oxidation
state (5). The effects of changes in FeO/MgO ratios on the transformation
were also studied in a separate series of experiments. The chemical and
normative compositions of the principal basaltic compositions which were
investigated are given in Table I.

Our experimental procedure is described in detail in the accompanying
paper (Green and Ringwood, 1966), and accordingly a brief summary will
suffice in the present case. The basalt compositions in Table I were obtained
either from natural analyzed basalts and/or minerals or from synthetic
starting materials. After fine grinding and intimate mixing, the components
were melted in a controlled atmosphere and quenched to a homogeneous

TABLE I

Chemical compositions of basaltic glasses used in experimental work[1]

	(1) High-alumina basalt	(2) Quartz tholeiite	(3) Alkali-poor quartz tholeiite	(4) Alkali olivine basalt	(5) Oxidised alkali olivine basalt	(6) Alkali-poor olivine tholeiite
SiO_2	49.90	52.16	49.88	45.39	45.39	46.23
TiO_2	1.29	1.86	2.14	2.52	2.52	0.07
Al_2O_3	16.97	14.60	13.89	14.69	14.69	14.52
Fe_2O_3	1.52	2.46	2.84	1.87	9.82	0.54
FeO	7.60	8.39	9.65	12.42	4.18	11.80
MnO	0.16	0.14	0.16	0.18	0.18	0.30
MgO	8.21	7.36	8.48	10.37	10.37	12.45
CaO	11.41	9.44	10.82	9.14	9.14	13.00
Na_2O	2.78	2.68	1.84	2.62	2.62	0.81
K_2O	0.16	0.73	0.08	0.78	0.78	0.01
P_2O_5	–	0.18	0.22	0.02	0.02	0.03
Cr_2O_3	–	–	–	–	–	0.24
	100.00	100.00	100.00	100.00	99.71	100.00
$\dfrac{100 \times Mg \text{ (mol)}}{Mg + Fe^{2+}}$	68	61.0	61.0	60	82	66
Norms						
Qz	–	2.5	2.8	–	–	–
Or	1.0	4.8	0.5	4.5	4.5	–
Ab	23.5	22.1	15.4	18.0	22.0	6.8
Ne	–	–	–	2.2	–	–
An	33.4	25.5	29.3	26.2	26.2	35.9
Di	18.9	17.1	19.6	15.7	15.0	23.5
Hy	9.4	20.6	23.7	–	6.8	9.6
Ol	9.3	–	–	25.8	8.4	23.0
Ilm	2.5	3.6	4.2	4.8	4.8	0.2
Mt	2.2	3.6	4.2	2.9	6.7	0.7
Ap	–	0.4	0.5	–	–	–
Haem	–	–	–	–	5.2	–
Chrom	–	–	–	–	–	0.3
Normative Olivine	Fo_{62}	–	–	Fo_{66}	Fo_{100}	Fo_{65}
Normative Plagioclase	An_{59}	An_{52}	An_{64}	An_{58}	An_{54}	An_{83}

[1]Analyst: A.J. Easton, Department of Geophysics and Geochemistry, Australian National University, Canberra.

Tectonophysics, 3 (5) (1966) 383–427

glass. The contents of FeO and Fe_2O_3 in the glass were then determined by chemical analysis and are given in Table I.

Glass samples so prepared were sealed into platinum capsules and subjected to various desired temperatures and pressures between 900°C and 1,250°C and pressures between 5 kbar and 40 kbar in an internally heated, high-pressure apparatus. The apparatus used was a replica of that described by Boyd and England (1960). The reader is referred to their paper for details of experimental techniques and operation. Temperatures were measured by a Pt–Pt 10% Rh thermocouple in contact with the charge, and are accurate to ± 15°C. No correction for the effect of pressure on thermocouple EMF was applied. The thermocouple was also used in conjunction with a suitable controller to regulate power input into the graphite furnace so that recorded temperature was kept steady within ± 5°C of the desired setting. Total pressure delivered by the high-pressure piston is given within 0.5% by a Heise gauge in the hydraulic circuit. A correction of - 10% was applied to the total pressure so obtained in order to determine the pressure actually operating on the sample. This correction is required because of friction and imperfect pressure transmission in the apparatus and talc pressure medium. The magnitude of the pressure correction was evaluated in a separate series of experiments (Green et al., 1966). Absolute pressures so obtained are probably correct to ± 3%, whilst relative pressures over limited pressure intervals are undoubtedly much more precise than this.

Runs lasted between 1–6 h, the majority being of 4 h duration. After completion of a run, the charge was quenched in a few seconds by terminating the power to the furnace. After removal from the apparatus it was investigated by optical and X-ray methods. Relative proportions of phases present in runs were determined by comparison of X-ray diffractometer records and powder films of samples with those of specially prepared standards.

One of the problems in experimental work of this nature is whether or not chemical equilibrium was reached in the runs. This problem was investigated in considerable detail. The sequence of crystallization of basalt glass at 1,100°C was followed in a series of runs ranging in length from 1/4 to 12 h. Runs of short period were dominated by the appearance of metastable complex pyroxenes, which broke down on longer runs to form stable pyroxene plus garnet. At 1,100°C, runs of 2 h sufficed to achieve equilibrium in most cases. Most of our runs were nevertheless carried out for 4 h. The question of reversibility was studied in detail in one of the quartz tholeiites (2). The maximum pressure at which plagioclase could be identified in runs initially of glass was the same as that at which plagioclase was grown from starting material consisting of garnet + pyroxene but of identical chemical composition. Similar reversibility was demonstrated for the incoming of garnet.

Changes in oxidation state of the charge and loss of iron to the platinum container during runs were also investigated by chemically analyzing for FeO and Fe_2O_3 after completion of runs. It was demonstrated that under the conditions employed, loss of iron and changes in oxidation state were of minor proportions.

RESULTS

A detailed chemical and mineralogical discussion of the experimental runs

is given in the accompanying paper (Green and Ringwood, 1966). Our aim
in this section is to outline the major generalizations which can be drawn
from the experiments. In Fig.2 we have compared in a simplified form, the
phase assemblages which were found in the typical basalts studied at 1,100°C,
as a function of pressure. Although the pressures required for appearance
and disappearance of phases vary between individual basalts, there is an
important qualitative resemblance between the sequence of phase assemblages displayed by all rocks with increasing pressure. For each basalt,
there are clearly three principal mineral stability fields, corresponding
closely with naturally observed mineral assemblages. The low-pressure
assemblage is that of gabbro or pyroxene granulite. It is characterised by
the presence of pyroxene and plagioclase ± olivine ± quartz ± spinel, according to the particular bulk chemistry. Garnet is not present. Within the stability field of this mineral assemblage it is observed that the ratio of plagioclase to pyroxene decreases with increasing pressure. This is caused in
part by solid solution of plagioclase components (principally anorthite) in
pyroxene as Tschermak's molecule, which increases with pressure. It is
also caused by reaction of plagioclase with olivine to form aluminous
pyroxenes ± spinel.

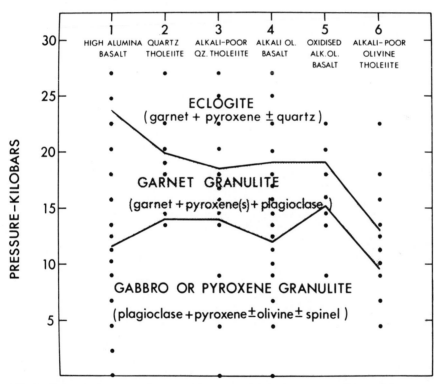

Fig.2. Principal mineral assemblages displayed by several basaltic compositions (Table I) as a function of pressure at 1,100°C. Solid circles denote
experimental runs.

Tectonophysics, 3 (5) (1966) 383–427

In each basaltic composition, as pressure increases, a point is reached at which garnet appears in the mineral assemblage. With further increase in pressure, the proportion of garnet steadily increases, whilst the proportion of plagioclase correspondingly decreases. Thus, we have here a field of co-existing garnet, pyroxene and plagioclase. The pyroxenes in the garnet granulite field are rich in Al_2O_3 in the form of Mg and Ca Tschermak's molecules. With further increase of pressure, these components are rejected and crystallize as garnet and accordingly the garnet/pyroxene ratio rises. Plagioclase becomes more sodic as it decreases in abundance. At the highest pressures, sodic plagioclase breaks down to form jadeite which enters into solid solution in the pyroxene. Free quartz may or may not be formed, depending upon the silica saturation of the rock. A further increase in garnet/pyroxene ratio occurs as the Tschermak's molecule component of the pyroxene is reduced to low levels. These transformations mark the beginning of the eclogite mineral assemblage, characterized by the co-existence of pyrope-rich garnet, omphacite ± quartz.

Thus we see that in all the basalt compositions studied, the transformation from gabbro or pyroxene granulite to eclogite proceeds through an intermediate mineral assemblage characterised by co-existing garnet, pyroxene(s) and plagioclase. This possesses an extensive stability field varying from 3.5 to 12 kbar in width, and is identical with the natural garnet–clinopyroxene granulite subfacies recognizable in some metamorphic terranes (De Waard, 1965; Green and Ringwood, 1966).

The results plotted in Fig.2 show that rather modest changes in chemical composition cause large changes in the pressures and width of the gabbro–eclogite transformation. Thus, the pressures at which garnet first appears vary between 9.6 and 15.2 kbar, whilst the pressure required to cause the final disappearance of plagioclase varies between 13.0 and 23.3 kbar. These variations are of considerable geophysical significance and will be further discussed in the section on the nature of the M-discontinuity. The reasons for the variation are readily explicable in terms of the chemical and mineralogical equilibria involved in the transformation, and are discussed in detail in the accompanying paper (Green and Ringwood, 1966). The pressures required for the incoming of garnet are smaller in undersaturated rocks (Fig.2, Table I, 1, 4, 6) than in oversaturated rocks (Fig.2, Table I, 2, 3, 5) whilst the pressure required for the final disappearance of plagioclase is decreased in basaltic compositions which are poorer than usual in soda (Fig.2, Table I, 3, 6). Changes in FeO/MgO ratio have an important influence over the pressure required for the incoming of garnet, these pressures being smaller the higher the FeO/MgO ratio of the rock. Such changes do not appear to markedly affect the pressure required for the disappearance of plagioclase, although our data relating to this point are rather scanty. Neither do changes in oxidation state affect the pressure at which plagioclase disappears (Table I, 5). However, the entrance of garnet requires higher pressures in highly oxidised rocks (Table I, 5).

Finally, in Fig.3, the results of experiments on a quartz tholeiite (2) over a range of temperatures are given. These results establish a temperature gradient for the transformation which permits extrapolation to lower temperatures. The slopes of the boundaries for the incoming of garnet and disappearance of plagioclase tend to converge at lower temperatures. This appears to be a real phenomenon, although it may not be as marked as it

Tectonophysics, 3 (5) (1966) 383–427

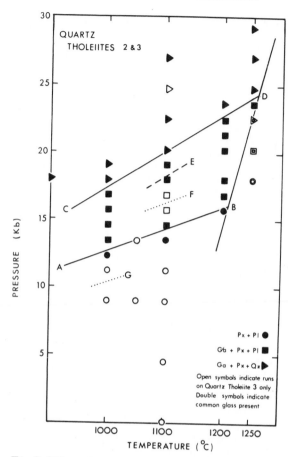

Fig.3. Mineral assemblages observed in quartz tholeiites (Table I, 2 and 3) over a range of temperature and pressure conditions. The eclogite field lies above CD and the pyroxene granulite field lies below AB. Area between AB and CD represents the garnet granulite stability field. Boundary BD is the solidus. Px = pyroxene, Pl = plagioclase, Ga = garnet, Qz = quartz. Plagioclase is absent on the high-pressure side of CD in quartz tholeiite 2, but is absent on the high-pressure side of the line E in quartz tholeiite 3 composition. The line G marks the appearance of garnet in highly iron-enriched quartz tholeiite; F marks the first appearance of garnet in magnesian quartz tholeiite.

appears in Fig.3. The question of extrapolation of mineral assemblage boundaries to lower temperatures is considered in the section on the stability of eclogite within the crust.

CHANGES IN PHYSICAL PROPERTIES ASSOCIATED WITH THE TRANSFORMATION

A considerable amount of information is available on the densities and seismic velocities of basalts and gabbroic rocks (Birch, 1960, 1961). The densities of most fresh rocks of this type are close to 3.00 g/cm³, whilst the corresponding P-wave velocities (corrected for the effect of initial porosity) mostly range from 6.8 to 7.1 km/sec. On the other hand, less is known about the seismic velocities and densities of eclogites. Birch (1960) attempted to determine these quantities but his samples had clearly been subjected to considerable secondary alteration and his results (ρ = 3.33 - 3.44 g/cm³; V_P, 7.3 - 7.8 km/sec) cannot be held to be characteristic of fresh eclogites. Indeed most whole-rock samples of eclogite which have been measured for density by other workers appear to have suffered secondary alteration, resulting in systematic underestimates of the densities of eclogite. Thus, Kennedy (1959) states that the mean density of eclogite is 3.3 g/cm³, whilst according to Lovering (1958) the mean density is 3.4 g/cm³. We shall see below that these densities are too low and that the geophysical consequences of a substantially higher mean density ($\sim$ 3.5 g/cm³) for eclogite are far-reaching.

Some of the best determinations of the densities of eclogite were obtained by Eskola (1921). The rock descriptions indicate a small amount of secondary alteration so that the densities which he obtained are probably slightly low (Table II). The average density of this group is 3.54 g/cm³. Alternatively the densities of eclogites can be calculated directly from their chemical compositions, which are cast into "eclogite norms", the compositions of garnets and pyroxenes being made similar to those in natural eclogites. The densities of individual minerals can either be measured directly on unaltered, separated samples, or else they can be calculated with adequate precision from knowledge of densities of their end–member components and mineral compositions. The density of the whole rock is then obtained by combining the densities of individual minerals. Results of such calculations for some of the basaltic compositions which we have investigated and also for some natural eclogites are given in Table II and III. Finally, the density of a sample of alkali olivine basalt (4) which had been subjected to a pressure of 30 kbar at 1,100°C was measured directly and found to be 3.54 g/cm³. From these results we see that the densities of undersaturated rocks of basaltic composition are usually between 3.5 and 3.6 g/cm³, whereas the densities of quartz tholeiites in eclogitic facies are somewhat lower. Quartz tholeiite (2) with 13% free quartz (ρ = 2.65 g/cm³) has a density of 3.43 g/cm³. It is possible that some quartz-rich tholeiites may have somewhat lower densities than this. However, enrichment in quartz is usually accompanied by increase in the FeO/MgO ratio of the rock. This leads to an increase in density which compensates for the higher amount of quartz. Indeed, the quartz tholeiite (2) has an abnormally low FeO/MgO ratio for its degree of silica saturation and its density is therefore probably minimal for this type of rock.

Considering all the available data, the average density of eclogites of quartz tholeiite composition is probably close to 3.45 g/cm³, whilst that of eclogites of olivine tholeiite composition is probably around 3.55 g/cm³. The average for eclogites generally may be taken as 3.50 g/cm³.

Seismic velocities of eclogites may be obtained from the velocity–density relationship of Birch (1961). These are also given in Table II and III. An alternative method is to calculate the seismic velocities from the elastic properties of end-member mineral components (Ringwood, 1966a).

Tectonophysics, 3 (5) (1966) 383–427

TABLE II

Compositions, densities and seismic velocities of some eclogites[1]

	(1) Rodhaugen	(2) Duen	(3) Lyngenes	(4) Silden	(5) Glenelg	(6) Loch Duich	(7) Hawaiian magma
SiO_2	48.7	41.5	45.33	52.4	50.05	42.60	46.9
Al_2O_3	11.7	15.9	13.06	11.9	13.37	12.41	12.9
Cr_2O_3	0.2	0.0	tr	–	–	–	–
Fe_2O_3	1.4	3.4	1.58	2.4	3.71	3.50	2.1
FeO	6.8	16.2	12.73	8.0	10.39	16.94	9.6
MnO	0.2	0.4	0.22	0.2	0.25	0.36	–
MgO	16.7	4.2	17.28	10.6	6.49	7.23	14.6
CaO	13.9	12.4	7.57	11.5	11.00	11.18	9.9
Na_2O	0.4	2.6	0.79	2.3	2.38	1.09	1.5
K_2O	0.2	0.1	0.16	0.5	0.36	0.15	0.4
TiO_2		1.2	1.35	0.4	1.55	3.84	2.3
P_2O_5		2.4	0.19		0.12	0.18	
H_2O		0.2	0.22	0.1	0.45	0.48	
					$(S, CO_2$ $= 0.22)$	$(S =$ $0.47)$	
	100.3	100.7	100.5	100.3	100.34	100.43	100.2
Garnet	55	55.1	55	34.1	30.2	45.9	48.4
Clinopyroxene	45	38.1	40	57.7	53.6	44.6	48.3
Quartz	–	–	–	7.9	8.1	–	–
Rutile	–	0.3	1.4	0.о	1.5	3.8	·2.3
Accessories and secondary minerals	–	6.5	3.6	–	6.6	5.6	–
Density (g/cm^3)	3.51	3.61	3.57	3.42	3.44	3.66	3.55
V_P(km/sec)	8.4	8.3	8.5	8.1	8.2	8.4	8.5

[1] No.1, 2, 3, 4 were measured directly on relatively fresh rocks (Eskola, 1921). Nevertheless these rocks contained a few percent of secondary minerals. No.5 and 6: Densities were calculated from known compositions and proportions of minerals. Analyses from Yoder and Tilley (1962). No.7: Composition of parental Hawaiian magma type (MacDonald and Katsura, 1961) Composition was calculated into an eclogite norm and the rock density obtained from densities of individual minerals. P velocities for no.1, 3, 4, 5, 7 were calculated from densities according to method given by Birch (1961), solution 7. Velocities of no.2 and 6 are lower than are given by this solution, because of their high iron content.

For the Kilauea Iki "eclogite", V_P obtained by this method is 8.41 km/sec (Table II). This compares with 8.48 km/sec obtained from the Birch relationship. As would be anticipated, the seismic velocities of eclogites decrease markedly with an increasing amount of free quartz.

Tectonophysics, 3 (5) (1966) 383–427

TABLE III

Mineralogical compositions, densities and seismic velocities of eclogitic facies of some of the basaltic compositions given in Table I*

	(1) High Al. olivine basalt	(2) Quartz tholeiite	(4) Alkali olivine basalt	(6) Alkali-poor olivine basalt
Garnet	51.8	43	46	55
Clinopyroxene	38.0	43	48	45
Quartz	8.5	13	–	–
Rutile	1.7	1	2	–
Olivine	–	–	4	–
Density (g/cm^3)	3.49	3.43	3.54	3.61
V$_P$(km/sec)	8.3	8.2	8.4	8.6
(Birch, 1961)				

*No.1, 2, 6: Compositions were cast into eclogite norms and densities were calculated from these. No.4: Density was measured directly upon a synthetic eclogite made by holding glass of this composition at 30 kbar and 1,100°C for 2 h.

Having discussed the densities and seismic velocities of gabbros and eclogites, we can now consider the manner in which these vary as gabbro is transformed to eclogite with increasing pressure. To investigate these questions, a series of closely spaced runs was made on the basalt compositions (Fig.2). The proportions of phases present were estimated by comparison of powder patterns and diffractometer records of runs with specially prepared standards. Although the precision of this method is not high, the principal conclusions which follow are beyond reasonable doubt.

In all cases, as the pressure increases from zero up to the point at which garnet first appears, there is an increase in the pyroxene/plagioclase ratio caused by solid solution of plagioclase components in pyroxenes which become notably aluminous. The rocks in the higher pressure range are perhaps better regarded as pyroxene granulites rather than gabbro. The change in the pyroxene/plagioclase ratio would necessarily cause a slight increase in densities with pressure. The density of the pyroxene granulites just before entry of garnet are probably about 3.05–3.10 g/cm^3. The incoming

TABLE IV

Results of runs at 1,100°C on alkali-poor olivine tholeiite

Pressure (kbar)	Phases present	ρ (g/cm^3) (estimated)	V$_P$ (km/sec) (estimated)
9.0	pyroxenes, plagioclase, olivine spinel		
10.1	pyroxenes, plagioclase, rare garnet	3.1	7.3
11.3	pyroxenes, plagioclase, 20% garnet		
12.4	pyroxene 60, plagioclase 10, garnet 30	3.38	8.0
13.5	pyroxene 60, garnet 40, trace plagioclase?		
15.8	garnet 45, pyroxene 55		
18	garnet 50, pyroxene 50		
22.5	garnet 55, pyroxene 45	3.61	8.6

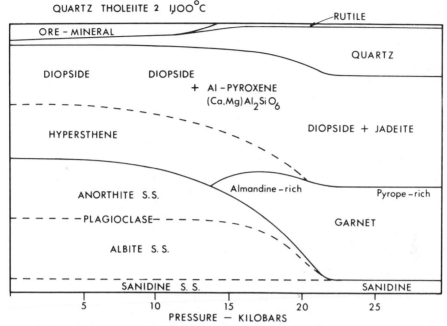

Fig.4. Approximate proportions of mineral phases present in quartz tholeiite (*2*) in the pressure range 0–30 kbar at 1,100°C.

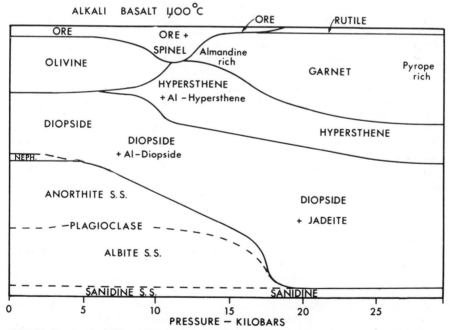

Fig.5. Approximate proportions of mineral phases present in the alkali basalt (*4*) in the pressure range 0–30 kbar at 1,100°C.

Tectonophysics, 3 (5) (1966) 383–427

of garnet marks the beginning of the garnet granulite field which is transitional between pyroxene granulite and eclogite. As pressure increases throughout this field, the proportion of garnet steadily increases, whilst the proportion of plagioclase steadily decreases. To a first approximation the changes in mineralogy across the garnet granulite zone occur at a uniform rate (Fig.4,5) and accordingly the corresponding changes in seismic velocity and density will also change regularly throughout the garnet granulite zone. With the exception of the alkali-poor olivine tholeiite which is discussed below, there is no suggestion of the presence of a narrow pressure interval within the garnet granulite field, characterised by an abnormal increase of garnet and/or decrease of plagioclase.

Yoder and Tilley (1962) previously discussed the question of whether a wide transitional field (which we have now established) would be an objection to the hypothesis that the M-discontinuity is a phase change. They argue that although the total width of the transition zone may be large, the "effective change in seismic velocity will be realized over a much smaller pressure range" because of the large contribution to the seismic velocity change caused by the incoming of garnet. Accordingly, they believe that the "Mohorovičić discontinuity under the continents may be the result of the transformation of basalt to eclogite."

With one partial exception, the results which we have described on a variety of basalts effectively contradict this interpretation.

The results on the alkali-poor olivine tholeiite (6) are in contrast to the others. Here, the incoming of garnet and the disappearance of plagioclase occur within a pressure interval of only 3.4 kbar. Much of this change occurs within 2.3 kbar. Results are given in Table IV. The chemical and mineralogical reasons for this behaviour are discussed in the accompanying paper (Green and Ringwood, 1966). Although the transformation of this basalt into a garnet–pyroxene rock free of felspar is nearly complete at 12.4 kbar, the rock contains only 30% of garnet, and the pyroxene is highly aluminous. With increasing pressure between 12.4 and 22.5 kbar, there is a large increase in garnet/pyroxene ratio and a corresponding large increase in seismic velocity and density (Table IV).

The transitional interval in the oxidised alkali-olivine basalt (5) is smaller than in the unoxidised basalt, owing to a higher pressure required for the appearance of garnet. However, there is an extensive field of spinel present in this basalt between 5 kbar and 25 kbar, and in fact the change in density and seismic velocity between the gabbroic and eclogite facies of this basalt are more "smeared out" than in any of the basalts which we have studied.

STABILITY OF ECLOGITE WITHIN THE CRUST

The accompanying paper (Green and Ringwood, 1966) contains a detailed comparison of our experimental results on the gabbro–eclogite transformation, with petrological observations of basaltic rocks which have been metamorphosed in the granulite facies in the earth's crust. Under such conditions, basaltic rocks are transformed successively with increasing pressure into aluminous pyroxene–plagioclase assemblages (pyroxene granulites), then into garnet–aluminous pyroxene–plagioclase assemblages and finally into eclogites. The mineralogy of the transformations observed in our experiments

Tectonophysics, 3 (5) (1966) 383–427

in the range 1,000–1,250°C corresponds extremely closely to the sequence of mineralogical changes observed in the granulite facies, which were probably established at temperatures in the vicinity of 500–800°C. The correspondence between experimental and natural mineral equilibria justifies extrapolation of experimental results to lower temperatures in order to interpret the conditions of formation of natural granulites and eclogites.

The pyroxene granulite, garnet granulite and eclogite stability fields have been delineated between 1,000 and 1,250°C for the quartz tholeiite (*2*) (Fig.3). These experiments, including reversal of the boundaries shown, have provided a reliable value for the width of the transition zone in this temperature interval. Because of the small experimental temperature interval, the gradients of the two boundaries of the garnet granulite field have an appreciable uncertainty. Accordingly, we have used the average of these two gradients, which are based upon twice the number of experiments as were available for either single boundary, as the gradient for downward extrapolation of the transition zone (Fig.6). The mean gradient thus obtained is 21 bar/°C.

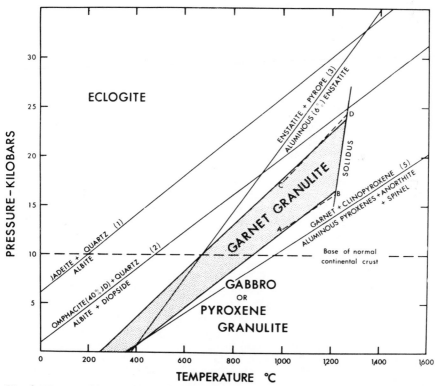

Fig.6. Extrapolated stability fields of eclogite, garnet granulite and pyroxene granulite–gabbro for the quartz tholeiite (*2*) composition. The broken lines *AB* and *CD* are the experimental boundaries from Fig.3. Extrapolation is based upon the average of the gradient *AB* and *CD* together with the assumption that the width of the garnet granulite zone is proportional to absolute temperature. (See text.)

Tectonophysics, 3 (5) (1966) 383–427

The simplest procedure for extrapolation to lower temperatures would be to use this mean gradient, together with the mean transition interval as obtained from Fig.3. The use of this procedure would not alter any of the conclusions reached in subsequent sections. In Fig.3 and 6 it is seen that although the experimental boundaries are approximately parallel, they nevertheless converge toward lower temperatures. This is probably a real feature but the small temperature range covered by experiments makes the degree of convergence difficult to estimate with any confidence. It would perhaps be more realistic to use an extrapolation which allows for a modest decrease in transition width with falling temperature. Accordingly, in Fig.6, we have set the width of the transition zone to be proportional to absolute temperature, using a mean gradient of 21 bar/°C as discussed above. This procedure is somewhat arbitrary, but probably represents a slight improvement upon the assumption of parallel boundaries. When experimental uncertainties are taken into account, it is consistent with the experimental boundaries established between 1,000 and 1,250°C (Fig.6).

With these assumptions, the granulite field boundaries have been extrapolated to low pressures and temperatures (Fig.6). It is seen that the temperature on the garnet granulite/eclogite boundary at a pressure corresponding to the base of the normal continental crust (10 kbar) is 670°C. If the temperature at the base of the crust is lower than 670°C, then eclogite would be the stable form of a basalt of this composition throughout the crust.

The temperature distribution in the crust is determined principally by the surface heat flux and the depth distribution of radioactivity. Numerous models have been investigated by Birch (1955), Clark (1961, 1962), Clark and Ringwood (1964) and others. Their results show that in stable continental regions of normal crustal thickness, and characterised by heat flows between 0.8 and 1.5 μcal/cm^{2}sec the temperature at the base of the crust is usually less than 670°C for most reasonable assumptions concerning radioactivity distribution. This conclusion is practically certain for Precambrian shields characterised by mean heat flows of 1.0 μcal/cm^{2}sec. The temperature at the base of the crust of these regions is probably less than 450°C. Such temperature distributions, taken in conjunction with the experimental results discussed above, indeed imply that eclogite is the stable modification of quartz tholeiite throughout very large regions of normal continental crust. This conclusion was unexpected by us when this project was commenced, and if correct, would have profound tectonic consequences. For this reason it deserves careful scrutiny. One is tempted to ask whether uncertainties in experimental data and in the extrapolation to lower temperatures are sufficiently large to invalidate this conclusion. We have examined our data with this in mind. If the maximum reasonable allowance for possible experimental errors is made the temperature of the eclogite/garnet granulite boundary for quartz tholeiite at 10 kbar could be brought down to 500°C. However, this is almost certainly higher than the temperature at the M-discontinuity beneath Precambrian shields, and consequently eclogite would be stable in the crust in such regions. The only escape from this conclusion would be to introduce a much larger curvature into the extrapolated phase boundaries than appears plausible.

Quartz tholeiite is the most common type of continental basaltic rock. However, there are also other important classes of continental basalts which have been investigated in our experiments. Of the six basalts which we

Tectonophysics, 3 (5) (1966) 383–427

investigated, four required smaller pressures to reach the eclogite stability field than the quartz tholeiite (2) discussed above. If the gradient of the transformation in these basalts is similar to the one which we investigated, then our conclusion regarding eclogite stability in the crust is generalized and reinforced. In the case of the alkali-poor olivine tholeiite (Table IV) this conclusion may be considered practically certain. The only basalt which required a higher pressure to reach the eclogite stability field than quartz tholeiite (2) is the high-alumina olivine basalt, a typical oceanic tholeiite. If the transformation in this basalt has the same slope, the eclogite/garnet granulite boundary (Fig.6) would intersect the continental M-discontinuity at a temperature of 540°C, which is probably higher than the temperature at this depth in many continental regions. In such regions eclogite would be the stable form of this basalt also.

An alternative method of establishing the gradient of the transition zone is by comparison with the known gradients in simple systems which are closely related to the gabbro–eclogite transformation. It is reasonable to assume that the gradients in these simple systems will be generally similar to that of the gabbro–eclogite transformation. As was discussed briefly in the section "Results" and more fully by Green and Ringwood (1966), the principal equilibria occurring near the garnet granulite/eclogite boundary are the breakdown of sodic plagioclase to form jadeite + quartz (jadeite enters into solution in pyroxene) and the breakdown of aluminous pyroxenes form garnet and low-alumina pyroxenes. The gradients of several of these equilibria (Table V, no.1,2,3, and 5) have been established experimentally and are plotted on Fig.6. It is seen that they are generally parallel to the

TABLE V

Gradients of simple equilibria related to the garnet granulite–eclogite transition[1]

No.	Equilibrium	dp/dt (bar/°C)
1	albite = jadeite + quartz	20
2	albite + diopside = omphacite (40% Jd) + quartz	19
3	aluminous enstatite (6% Al_2O_3) = enstatite ($< 6\%$ Al_2O_3) + pyrope	34
4	albite + nepheline = 2 jadeite	18
5	clinopyroxene + orthopyroxene + anorthite + spinel = garnet + clinopyroxene (Overall composition: 1 forsterite + 1 anorthite)	16
6	clinopyroxene + orthopyroxene + forsterite + spinel = garnet + forsterite + clinopyroxene (Overall composition: 1 forsterite + 2 anorthite)	25
7	clinopyroxene + orthopyroxene + quartz = garnet + quartz + pyroxene (Overall composition: 2 enstatite + 1 anorthite)	10?*
8	4 enstatite + spinel = forsterite + pyrope	17

[1]References: (1) Birch and Le Compte (1960); (2) Kushiro (1965); (3) Boyd and England (1964); (4) Robertson, Birch and MacDonald (1957); (5), (6) Kushiro and Yoder (1965); (7) Kushiro and Yoder (1964); (8) MacGregor (1964).
*This gradient is based upon very limited experimental data and has a large uncertainty.

Tectonophysics, 3 (5) (1966) 383–427

experimentally determined slope of the basalt–eclogite transformation. The slopes of several other simple equilibria which are related to the basalt–eclogite transformation are also given in TableV, no.4,6,7 and 8. The average slope of the eight equilibria in Table V is 20 bar/°C compared to our mean slope for the basalt–eclogite transformation of 21 bar/°C. (Fig.3,6). Thus we see that if the slopes of the simple equilibria are used as the basis for extrapolation, the conclusion regarding stability of eclogite in the normal continental crust is reinforced.

The experimental observations discussed above, when considered as a whole, strongly suggest that eclogite is the stable modification of most rocks of basaltic composition under dry conditions within large regions of continental crust. In the case of undersaturated, alkali-poor basaltic rocks occurring in continental regions with a heat flow $\leqslant 1.0 \ \mu \, cal/cm^2 sec$, this conclusion is practically certain. Such compositions are in fact, characteristic of many natural eclogites (Table II).

It is important to consider the bearing of geological field evidence upon the above conclusion. Eclogites occurring in the crust clearly have more than one mode of origin. Those occurring in diamond pipes and perhaps those occurring as inclusions or segregations in ultramafic bodies appear to have been originally derived from the mantle. However, there is another important class of eclogite occurrences as small lenses, inclusions and sometimes as conformable bodies in regional metamorphic rocks of amphibolite, greenschist and glaucophane schist facies. The origin of these eclogites is not settled, but many competent geologists consider that they have been formed in situ in the crust, by metamorphism of basaltic rocks under conditions of relatively low water vapour pressure, and in some cases, of localized shear stress (Backlund, 1936; Korzhinsky, 1937; Kozlowski, 1958; Smulikowski, 1960; Bearth, 1959; Coleman et al., 1965). This conclusion is in complete harmony with the experimental evidence. A detailed review of geological and petrological evidence relating to the occurrence of granulitic and eclogitic rocks in the crust is given by Green and Ringwood (1966).

Although there is strong observational evidence that basalt may be transformed into eclogite in the crust under appropriate conditions, it is well known that in the exposed crust, eclogites are extremely rare rocks compared to gabbros, dolerites and basalts. It might be held that this fact argues against the conclusion advanced above. There are, however, additional factors which may be responsible for the rarity of eclogite:

(a) The vast majority of rocks of basaltic composition have been emplaced in the crust originally as magmas, which complete their crystallization above 1,000°C. At these temperatures, most basaltic rocks would crystallise dominantly to plagioclase and pyroxene at all levels within the crust. On cooling in a normal crustal environment, the basalts would pass through the garnet granulite stability field into the eclogite stability field, as previously discussed. Whether a given basaltic rock will actually undergo these transformations depends upon kinetic factors. The transformations are apparently extremely sluggish at low temperatures and under dry conditions, so that if a gabbroic assemblage can be brought to a temperature below about 200–300°C under conditions of low water vapour pressure and shear stress, it may be preserved indefinitely. The common occurrence of unaltered basalts, dolerites and gabbros in the crust is thus ascribed to their kinetic stability and not to their thermodynamic stability.

Tectonophysics, 3 (5) (1966) 383–427

Similar arguments apply to the survival of basic granulites in the crust. Such rocks are often produced by the intense regional metamorphism of basic rocks under fairly dry conditions. Regional metamorphism is usually a consequence of an abnormal rise in temperature in large areas of the crust due to processes which are poorly understood. Basaltic rocks are then converted to pyroxene and/or garnet granulites under equilibrium conditions in the abnormal geothermal field. Whether they will then revert to eclogite when crustal temperatures cool to the normal state depends again upon kinetic factors.

(b) When subjected to regional metamorphism at low to moderate temperatures and high water vapour pressures, basaltic rocks are usually converted into various hydrated mineral assemblages, e.g., amphibolites, greenschists, glaucophane schists. These assemblages are stable for basaltic rocks in such environments. Neither eclogite nor gabbro is stable in the crust under conditions of high water vapour pressure. (The previous conclusion that eclogite is the stable form of basalt specifically referred only to relatively dry conditions.)

(c) A third factor contributing to the rarity of crustal eclogite is the high density of this type of rock (section "Changes in physical properties associated with the transformation"). If a large volume of basaltic rock should be transformed into eclogite within the crust, an acute gravitational instability would be caused. The contrast in density between eclogite (~ 3.5 g/cm^3) and crustal rocks (~ 2.85 g/cm^3) would probably result in the body of eclogite sinking physically through the crust into the mantle. The situation would be strictly analogous but opposite in sense to the diapiric intrusion of salt domes and gneiss domes in the crust. This topic is reopened in the section "Possible tectonic consequences of the gabbro–eclogite transformation".

NATURE OF THE MOHOROVIČIĆ DISCONTINUITY

The hypothesis that the Mohorovičić discontinuity is caused by a phase change from basalt to eclogite is currently receiving a great deal of attention. In its most general form, this hypothesis has been applied as an explanation both of the continental and oceanic M-discontinuities. There are, however, some serious objections to this general statement of the hypothesis.

Harris and Rowell (1960), Bullard and Griggs (1961) and Wetherill (1961) have shown that temperature and pressure distributions beneath oceans and continents are fundamentally inconsistent with the postulate that oceanic and continental M-discontinuities are caused by the *same* phase change. The inconsistency is a consequence of the crossing of oceanic and continental geotherms at shallow depth owing to the different distributions of radioactivity beneath these contrasting areas. A second objection raised by Bullard and Griggs and Wetherill (and noted also by MacDonald and Ness, 1960) concerned the wide variation of heat flux observed at the earth's surface. This in turn implied the presence of substantial temperature variations at similar depths in the crust and upper mantle, which, according to the phase change hypothesis should be closely correlated with depth to the M-discontinuity. The required correlation is not observed.

To survive these objections, a more restricted form of the phase change hypothesis must be introduced. Thus, it might be assumed (1) that a gabbro–eclogite transformation is responsible for the M-discontinuity beneath continents, but not beneath oceans, or vice versa, and (2) that the pressure at which the gabbro–eclogite transformation occurs is almost independent of temperature.

Stishov (1963) has reviewed the relevant evidence and has accepted the above limitations. He prefers to adopt a restricted form of hypothesis which maintains that the continental M-discontinuity is caused by the gabbro–eclogite transformation, whereas the oceanic M-discontinuity is caused by a chemical change to ultramafic material. This view has also been taken by Yoder and Tilley (1962), Wyllie (1963) and Van Bemmelen (1964). It is probable that most of those (e.g., Kennedy, 1959) who wish to retain the phase-change hypothesis would accept this choice since the objections to a gabbro–eclogite transformation at the oceanic M-discontinuity are generally regarded as being stronger than for the same transition at the base of the continental crust. Our experimental results are directly applicable to this restricted version of the phase-change hypothesis. Indeed, they appear to confront the hypothesis with several insuperable difficulties which are discussed below.

Stability of eclogite within the crust

In the previous section strong evidence was presented supporting the conclusion that eclogite is thermodynamically stable under the p, t conditions existing in very large regions of normal continental crust. Conversely, gabbro and basalt are thermodynamically unstable under these conditions. This conclusion, if accepted, absolutely invalidates the hypothesis that the M-discontinuity in such regions is caused by a reversible phase change from gabbro to eclogite.

Relationship between temperature at M-discontinuity and thickness of crust

Numerous measurements of surface heat flux have disclosed that this quantity varies by a factor of two or more in stable continental regions (Lee and Uyeda, 1965). The geothermal models of Birch (1955), Clark (1961, 1962) and Clark and Ringwood (1964) have shown that these variations imply the existence of temperature differences of 200°C or more at a depth of 37 km. If the M-discontinuity is caused by a phase change, such temperature differences might be expected to result in large variations in crustal thickness. (Bullard and Griggs, 1961). Geophysical data on the Australian continent are particularly well suited for evaluation of this effect. There is a large systematic difference in the heat flux between the oldest areas of the west Australian Precambrian shield (mean value 1.0 μcal/cm^{2}sec) and the off-shield areas (mean value 2.0 μcal/cm^{2}sec). Geothermal calculations (Howard and Sass, 1964) imply that the *minimum* temperature difference at 37 km beneath the two regions is 200°C. Seismic measurements of crustal thickness have been made in both provinces, in regions adequately covered by heat-flow data. The latest results (Everingham, 1965) give a crustal thickness beneath the west Australian Shield of 42 km which is the same as the thickness of the

Tectonophysics, 3 (5) (1966) 383–427

crust in the Snowy Mountains of eastern Australia (Doyle et al., 1966). There is a substantial margin of error in these determinations and the close agreement is partly coincidental. Nevertheless the evidence for similarity of crustal structures in both regions is very strong. Most of the possible errors in these determinations are systematic and it is highly probable that the average crustal thicknesses within these areas are similar within 7 km (J.R. Cleary, H.A. Doyle and R. Underwood, personal communication, 1966). Accordingly if the M-discontinuity is caused by a phase change, the pressure at which this occurs must be relatively insensitive to temperature ($\mathrm{d}p/\mathrm{d}t <$ 9 bar/degree).

In the diagram (Fig.1) of the basalt–eclogite transformation constructed by Yoder and Tilley (1962) the gradient was placed at 7 bar/degree. If this value is correct, then observations on crustal thickness and surface heat flow would not be in conflict with the phase-change hypothesis. However, we have already pointed out that this diagram is almost certainly incorrect. In the previous section we have detailed the direct and indirect experimental evidence for the gradient of the transition and obtained a value of 21 bar/degree (also: section "Results"; Fig.3 and 6). This value, even allowing for a generous uncertainty on the basis of possible experimental error, cannot be reconciled with the requirements of the phase-change hypothesis. It would result in large variations in crustal thickness with surface heat flow. For example the crustal thickness in eastern Australia would be at least 16 km greater than on the west Australian Shield.

Seismic velocity distribution in the crust

A critical requirement of the phase change hypothesis is its ability to explain the seismic velocity distribution in the crust. Determination of the velocity profile in the crust is a difficult and incompletely solved problem. Generalisations are dangerous owing to wide variations in crustal structure in different areas. Nevertheless, if we restrict our attention to stable continental regions, where indeed the M-discontinuity appears to be defined most clearly, some valid general statements which are highly relevant to the present investigation can be made.

It is widely agreed in such areas that seismic velocity increases significantly with depth. In some areas there may be a rapid increase in velocity around 20 km (Conrad discontinuity) caused by crustal layering, whereas in other areas this may be absent. The lower crust is commonly characterised by velocities between 6.5 and 7.0 km/sec (Tuve et al., 1954; Tatel and Tuve, 1955; Gutenberg, 1955, 1959; Steinhart et al., 1961). The nature of the transition from the lower crust to the mantle ($V_P \sim 8.2$ km/sec) is not well understood, and may well vary widely from place to place. In some regions there is evidence for reflections from the M-discontinuity (Dix, 1965). In such regions the transition would be sharp ($<$ 0.25 km wide). There is still some doubt about the reality of these reflections (Steinhart et al., 1961). However, it is widely believed that a large part of the increase in velocity from about 7.0 km/sec to 8.2 km/sec occurs in a limited depth range which is probably smaller than 5 km thick (Tuve et al., 1954; Tatel and Tuve, 1955; Nakamura and Howell, 1964; Phinney, 1964). The former authors have published a range of permissible velocity distributions for the

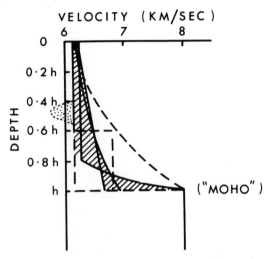

Fig.7. Permissible range (shaded) of velocity–depth relationships for the Maryland region. (After Tuve et al., 1954.)

crust in Maryland (Fig.7). It appears probable that similar limits would apply to crustal velocity profiles in many other stable continental areas.

Accordingly it is important to see whether the experimental observations on the transformation are consistent with the probable form of the crustal velocity profile. The experimental results at 1,100°C (Fig.2) indicated that in most basaltic compositions the transformation between gabbro and eclogite proceeded via a transition zone of garnet granulite varying in width between 4 and 12 kbar. Furthermore it was shown (section "Changes in physical properties associated with the transformation") that the change in seismic velocity associated with the transformation was approximately linear across the entire width of the transition zone. The width of the transition zones would be smaller at lower temperatures. In the previous section it was suggested that the width might be approximately proportional to absolute temperature. To place the phase-change hypothesis in its most favourable light, we will ignore all of the experimental data relating to the gradient of the transition, and will *assume* that the M-discontinuity is in fact caused by the basalt–eclogite transformation, and that therefore the eclogite field is entered at a depth of approximately 37 km. This assumption is unrealistic; however, our principal aim at this point is to test the internal consistency of the phase-change hypothesis. We will assume that the temperature at the M-discontinuity is 550°C. This is a median value in relation to many geothermal calculations (Birch, 1955; Clark, 1961, 1962; Clark and Ringwood, 1964). The particular value is based upon a continental geothermal model of the latter authors corresponding to a surface heat flow of 1.2 μcal/cm^{2}sec. Substantial changes in the assumed temperature at the M-discontinuity would not alter the arguments which follow.

The combination of our experimental results at 1,100°C with the above assumptions suffice to define the stability fields of gabbro, garnet granulite and eclogite for each of the basalt compositions experimentally investigated (Fig.8). We have constructed diagrams for four of the six basalts studied.

Tectonophysics, 3 (5) (1966) 383–427

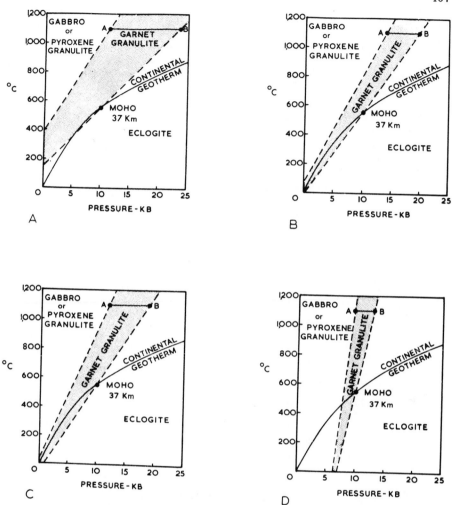

Fig.8. Hypothetical eclogite, garnet granulite and pyroxene granulite–gabbro stability fields for several basalts based upon experimental observations at 1,100°C and upon the assumption that the continental Mohorovičić discontinuity is caused by a "phase change" and that eclogite is therefore stable below Moho. A typical temperature–depth distribution, corresponding to a surface heat flow of 1.2 $\mu cal/cm^2 sec$, as given by Clark and Ringwood (1964) is also represented. The corresponding diagrams for quartz tholeiite (3) and oxidised alkali–olivine basalt (5) would closely resemble the diagram for quartz tholeiite (2).

A. High alumina basalt (1). B. Quartz tholeiite (2). C. Alkali olivine basalt (4). D. Alkali-poor olivine tholeiite (6).

The diagram for the alkali-poor quartz tholeiite would be almost identical to that of the normal quartz tholeiite, as would the oxidised alkali-olivine basalt. (The garnet granulite field of the latter is slightly smaller, but the physical transition zone is effectively widened by the presence of abundant spinel, both on the high- and low-pressure sides of the garnet granulite field. This has the effect of "smearing" out the transformation.)

Turning to Fig.8A which applies to the high-alumina olivine basalt, we see that the average geotherm lies in the eclogite field throughout the crust, except for a grazing entry into the garnet-granulite field around 5 kbar. Since the geotherm does not even enter the gabbro field, there is no possible way in which the rapid velocity increase around 37 km which is responsible for the M-discontinuity, can be caused along this geotherm. With the alkali-olivine basalt (Fig.8B) the geotherm commences near the boundary of the gabbro–garnet granulite fields ($V_P \sim 7.0$ km/sec) and gradually penetrates deeper into the garnet granulite field with increasing depth, until the eclogite field is reached at the M-discontinuity. The velocity change is thus smeared out through the entire depth of the crust for rocks of this composition. The velocity distribution in the case of the quartz tholeiite is intermediate between those of the high-alumina basalt and alkali basalt (Fig.8C)

Thus we see that for five out of the six rocks investigated, velocity distributions resembling those in Fig.7 cannot be explained in terms of a gabbro–eclogite transformation. The difficulty is of a general nature, and is caused jointly by the occurrence of a substantial transition interval at constant temperature and by the fact that geotherms tend to cross the stability field boundaries at low angles, resulting in a great enlargement of the effective transition intervals.

The only basaltic composition which is not subject to the above objection is the alkali-poor olivine tholeiite (6). The transformation interval of this rock is small and the geotherm intersects it at a high angle, resulting in the velocity increase at the M-discontinuity being confined to an interval of 9 km. This is somewhat high, but perhaps not impossible.

Although an explanation of the M-discontinuity in terms of a phase change in this particular rock is not subject to the difficulties noted in this section for other basalts, it is even more open to other objections. The assumptions employed in the construction of Fig.8D lead to a gradient for the transformation of 4.5 bar/°C. This cannot be reconciled with the experimental data quoted in the previous section. Of all the rocks which were investigated, the eclogitic assemblage of this rock is the most certain to be stable throughout the continental crust. Further difficulties in explaining the M-discontinuity in terms of the transformation of this particular rock will be discussed later in this section.

Effect of changes of chemical composition upon transition parameters

One of the principal objectives of the experimental investigations was to investigate the effect of changes in chemical composition upon the pressures and temperatures required for the basalt–eclogite transformation. The relevant information is given in Fig.2. We see that the pressures required to stabilise eclogite vary by about 11 kbar at 1,100°C for the basaltic compositions studied. Although the corresponding pressure range at the average temperature at the M-discontinuity (550°C) might be somewhat smaller than this, a major reduction is not likely since the gradients of the transformations in individual basalts are probably rather similar.

Tectonophysics, 3 (5) (1966) 383–427

If the lower crust and upper mantle is generally of basic composition, it is reasonable to expect that all of the common classes of basaltic compositions would be represented. The large effect on transition pressures caused by rather minor changes in chemical compositions would lead to wide fluctuations in depth to the M-discontinuity, or to further "smearing-out" of the velocity distribution in the transformation zone, depending upon the scale of occurrence of particular basaltic compositions. This would make it difficult to interpret the comparative uniformity of crustal thicknesses in stable continental regions and the nature of the seismic velocity distribution in the crust (Fig.7).

Density of the upper mantle

Limitations on the density of the upper mantle arise from the interpretation of gravity observations and from the theory of isostasy. Most interpretations of gravity data are made on the assumption of a mantle density close to 3.3 g/cm^3, although Talwani et al. (1959) found that a density of 3.4 g/cm^3 yielded a satisfactory solution for the gravity profile over the Puerto Rico Trench. (However, their interpretation is not unique – Morgan, 1965). Worzel and Shurbet(1955) reviewed available data on continental and oceanic crustal structures and densities and concluded that isostatic balancing of standard oceanic and continental sections required a mantle density of 3.27 g/cm^3 if the crustal density was 2.84 g/cm^3. Actually, it is the density contrast between crust and mantle which is determined by gravity observations. Drake et al. (1959) concluded that the average density difference between crust and mantle was close to 0.43 g/cm^3. Limitations on the density of the crust are obtained from direct observation of the occurrence and densities of crustal rocks combined with geologic inferences concerning their abundances. Another method of obtaining the near crustal density is from the seismic velocity distribution in the crust, combined with knowledge of the relationship between seismic velocity and density for common rock types (Birch, 1961). Arguments based upon the above methods have led to the widely accepted view that the mean density of the normal continental crust is between 2.8 and 2.9 g/cm^3. From these values, together with the density contrast between crust and mantle as given by gravity data, we may conclude that the density of the upper mantle is usually between 3.3 and 3.4 g/cm^3. This is further supported by an independent method of density determination based upon inversion of surface wave data. (Dorman and Ewing, 1962).

The densities of eclogites were reviewed in the section "Changes in physical properties associated with the transformation". They ranged between 3.42 and 3.66 g/cm^3. The average density is about 3.5 g/cm^3. Densities in the lower part of the range are usually caused by the presence of quartz in the eclogite. Eclogites which are believed on strong evidence to be derived from the mantle (e.g., inclusions in diamond pipes and peridotites) do not contain quartz, and their mean density is substantially higher than 3.5. Accordingly, if the upper mantle beneath the M-discontinuity is of eclogitic composition, it is unlikely that the density of this region would be smaller than 3.5 g/cm^3. This is substantially higher than the probable density of the upper mantle which was concluded above to lie between 3.3 and 3.4 g/cm^3. On these grounds the hypothesis of an eclogitic upper mantle

Tectonophysics, 3 (5) (1966) 383–427

does not appear probable. Woollard (1962) has reached a similar conclusion on analogous grounds.

Gravitational instability of an eclogitic upper mantle

Most of those who have advocated the presence of an eclogitic layer beneath the crust have also maintained that this passed downwards into a zone of ultramafic or perioditic composition (Goldschmidt, 1922; Holmes, 1926a,b, 1927; Birch, 1952; Lovering, 1958; MacDonald, 1959; Wyllie, 1963; Stishov, 1963). Opinions regarding the thickness of the eclogite layer have varied widely, however, recent advocates, e.g. Wyllie (1963), have inclined to the view that the eclogite zone is only of the order of a few tens of kilometers thick. This seems required if the much greater abundance of peridotite inclusions compared to eclogite inclusions in diamond pipes is to be explained.

Thus, according to these authors, the upper mantle consists of a layer of eclogite, with a density of 3.5 g/cm^3 or greater, *overlying* a zone of peridotite (mean density 3.32 g/cm^3). The extreme gravitational instability of this model should be remarked. It appears highly improbable that such an unstable configuration could have been established and maintained in the earth's gravitational field for 4.5 billion years. We will return to this aspect in the section on possible tectonic consequences of the gabbro–eclogite transformation.

Conclusion

In this section we have discussed a number of objections to the hypothesis that the Mohorovičić discontinuity in normal continental regions is caused by an isochemical phase transformation from gabbro (cr basalt) to eclogite. These were as follows:

(1) Our experimental evidence strongly suggests that eclogite is thermodynamically stable under dry conditions throughout the normal continental crust and that basalt is thermodynamically unstable. The M-discontinuity cannot then be caused by an equilibrium basalt–eclogite transformation.

(2) Experimental evidence on the effect of temperature on the pressure required for the gabbro–eclogite transformation cannot be reconciled with the rather small differences in crustal thickness in normal continental areas characterised by widely differing surface heat flows, and by inference, widely different temperatures at the base of the crust.

(3) In most rocks of basaltic composition, the transformation from gabbro to eclogite occurs over a broad pressure interval, and the rate of change of seismic velocity is approximately uniform across this interval. Furthermore the effective breadth of the transformation in the earth would be greatly expanded owing to the tendency of geotherms to cross phase boundaries at low angles. The large effective width of the transformation in the earth makes it impossible to explain the M-discontinuity which requires a substantial velocity increase within a depth of approximately 5 km.

(4) Small changes in basaltic chemical composition have a large effect upon the pressure required for a gabbro–eclogite transformation. On a small scale, this would cause further smearing of the transformation zone

Tectonophysics, 3 (5) (1966) 383–427

if the M-discontinuity was caused by a phase change. On a large scale it would lead to improbably large fluctuations in the thickness of the crust.

(5) The average density of eclogites is 3.5 g/cm^3, whereas the density of the upper mantle is generally believed to lie between 3.3 and 3.4 g/cm^3.

(6) Most current advocates of the phase change model argue that the eclogite layer (density 3.5 g/cm^3) immediately below the continental crust passes downwards into peridotite (density 3.3 g/cm^3). Such a configuration would possess a high degree of gravitational instability and is inherently improbable.

The above arguments, when considered together, can lead to only one conclusion: *The Mohorovičić discontinuity beneath normal continental areas is not caused by an isochemical phase change between gabbro (basalt) and eclogite.* Conversely, this implies that the only alternative explanation of the M-discontinuity, i.e., that it is caused by a change in chemical composition, must be accepted. The most probable explanation of the M-discontinuity is the widely accepted view that it is caused by a change from intermediate or basic rocks which characterise the lower crust, into ultrabasic peridotite, which is the predominant rock type of the upper mantle.

It is possible to postulate other types of chemical discontinuity. For example it might be suggested that the M-discontinuity is caused by a change from lower crustal rocks of intermediate composition into an upper mantle of eclogite. Such a model would forfeit most of the advantages which were previously believed to be inherent in the phase change hypothesis, e.g., changes in temperature at the M-discontinuity causing large vertical movement of the earth's crust. This model is also in conflict with points 5 and 6 above.

Finally, some comments on the nature of sub-oceanic Mohorovičić discontinuity may be appropriate. Because of the smaller temperature extrapolation and the knowledge that the basic chemical equilibria were identical, our experimental results are more applicable to a direct discussion of the nature of the sub-continental M-discontinuity than to its oceanic counterpart. Nevertheless, many of the arguments previously used are of a general nature and may be applied in the oceanic case.

Most advocates of the hypothesis that the M-discontinuity is caused by a gabbro–eclogite transformation have believed that this hypothesis is more applicable to the sub-continental than to the sub-oceanic M-discontinuity. When it was shown by Harris and Rowell (1960) and Bullard and Griggs (1961) that the M-discontinuity in both regions could not be caused by the *same* phase transformation, it was argued that the phase change hypothesis applied to the continental crust, whereas the sub-oceanic M-discontinuity was probably caused by a change in chemical composition (Yoder and Tilley, 1962; Wyllie, 1963; Stishov, 1963; Van Bemmelen, 1964). We have previously concluded that the continental M-discontinuity *cannot* be explained in terms of a phase change. If these arguments are accepted, the proposition that the oceanic discontinuity is nevertheless caused by a phase change does not appear enticing.

The specific objections to the phase-change hypothesis for the oceanic M-discontinuity are similar to those made in the sub-continental case. The problem of a transition interval is even more severe since the increase in seismic velocity from 6.7 to 8.2 km/sec must occur within about 2 km, equivalent to a pressure interval of 0.5 kbar. Although the width of the transitional field between basalt and eclogite is probably much smaller at 150°C than at 550°C, it is almost certainly substantial. Because of the approximate parallelism of geotherms and mineral stability field boundaries, the effective width of the transition is greatly expanded, and thus it becomes extremely difficult to explain the observed seismic profile. There is also the fact that large variations of surface heat flow observed in oceanic regions (Von Herzen and Uyeda, 1963) imply the existence of temperature differences at the base of the oceanic crust of the order of 100°C. Such temperature differences, in conjunction with the experimentally inferred gradients for the gabbro–eclogite transformation, would lead to large variations in crustal thickness in oceanic regions, which should be closely correlated with heat flow. The uniform crustal thicknesses in deep oceanic basins are not consistent with these inferences. Finally there are the difficulties connected with the large density of eclogite, previously noted.

An alternative suggestion regarding the nature of the sub-oceanic upper mantle has recently been made by Kennedy (discussion by Engel et al., 1965). According to this suggestion, the oceanic M-discontinuity is caused by a non-equilibrium transition from an oceanic tholeiite crust into a low-grade, hydrous metamorphic equivalent in the glaucophane schist or greenschist facies. The suggestion possesses some merit, in that it emphasizes the probability that at the low temperatures occurring at the base of the oceanic crust, rocks of basaltic composition are likely to display hydrated mineral assemblages (see also, Ringwood, 1962b). This has always been an objection to proposals that anhydrous eclogite is the predominant rock type immediately beneath the oceanic M-discontinuity. However, it does not appear probable that the proposed low-grade metamorphic equivalents of basalt would possess the required seismic P-wave velocities of around 8.2 km/sec. The densities of rocks of basaltic composition in glaucophane schist and greenschist facies rarely if ever, exceed 3.3 g/cm^3. According to Birch's (1961) velocity–density relationship, such rocks would possess seismic velocities in the vicinity of 7.8 km/sec, which are substantially smaller than the observed mantle velocities. The densest low-grade metamorphic rock of basaltic composition known to the authors is an epidote amphibolite (Christensen, 1965). This has a density of 3.26 g/cm^3 and a seismic P-velocity (corrected for initial porosity) of 7.45 km/sec. It is just possible that rocks of basaltic composition consisting dominantly of minerals such as lawsonite, jadeite, epidote and amphibole may have higher velocities and densities than the above rock. Further data on low grade metamorphic rocks are clearly required. In the light of present evidence, Kennedy's suggestion does not appear promising.

COMPOSITION OF THE CONTINENTAL CRUST

Seismic velocity distributions in stable continental crusts vary substantially from place to place. Nevertheless some generalizations can be made. Strong evidence exists that seismic velocity shows a net increase with depth. Velocities in the upper crust are commonly in the range 6.0–6.3 km/sec whilst in the deeper crust, velocities between 6.6 and 7.0 km/sec are often inferred. The smaller velocities correspond to those of acidic igneous and metamorphic rocks and suggest that the upper crust is dominantly of "granitic"

Tectonophysics, 3 (5) (1966) 383–427

composition. The interpretation is reinforced by direct geological observation upon exposed basement complexes. The seismic velocities of the lower crust are similar to those of basalts, dolerites, diabases and gabbros (Birch, 1961) and accordingly it has become widely accepted among geophysicists that the lower crust is composed of these rocks (e.g., Birch, 1958; Gutenberg, 1955, 1959). Thus there has arisen the concept of a two-layer crust, the upper layer being granitic, and the lower layer of gabbroic composition. The transition between these layers may be gradual or relatively sharp. In the latter case, a seismic discontinuity, the Conrad, may result.

Sufficient information now exists to show that this model is seriously wrong. In the section on stability of eclogite within the crust, strong evidence was presented that gabbro was not thermodynamically stable under p, t conditions existing in the normal continental crust. It is also most improbable that gabbro could be present as a *metastable* assemblage in the lower crust. Temperatures of 400–600°C are probably characteristic of this region and with long periods of time available, thermodynamic equilibrium would almost certainly be reached. Under dry conditions, as was shown in the section above, basic rocks in the crust would then probably occur in the form of eclogite. Taking the most conservative possible view, and making the maximum allowance for possible experimental errors, it might be concluded that some basaltic rocks would occur in the higher pressure grades of the garnet granulite mineral assemblage rather than in the eclogite assemblage.

If the lower crust is composed of dry basic rocks occurring as eclogites or high-grade garnet granulites, the density of this region would lie between 3.3 and 3.6 g/cm³ and it would possess a seismic P velocity of between 7.5 and 8.5 km/sec. These physical properties cannot be reconciled with the observed and inferred properties of the lower crust in most regions and accordingly we must reject the crustal model. This appears to leave two alternatives:

(a) The lower crust is composed of rocks of intermediate chemical composition occurring in the eclogite facies. Such rocks would contain quartz and alkali felspars together with garnet and jadeitic pyroxene and would yield seismic velocities and densities which are acceptable for the lower crust (Green and Lambert, 1965).

(b) The lower crust cannot be considered dry and a substantial water vapour pressure exists. Under such conditions neither eclogite nor gabbro would be stable. Rocks of basaltic composition might then crystallise dominantly as amphibolites which might possess acceptable seismic velocities and densities (Christensen, 1965).

In our opinion, alternative (a) is much more probable as a general explanation than (b) although the latter may be applicable under certain restrictive conditions. A lower crust of amphibolite might be formed where vast thicknesses of basaltic lavas had accumulated upon an oceanic crust during the early stages of subsidence of a geosyncline. In a water-rich environment of this nature, if the basalts were sufficiently permeable, they would probably be converted to hydrated mineral assemblages (zeolite, glaucophane schist and greenschist facies). As temperatures at the base of the geosyncline rose to about 400°C, reconstitution to amphibolites would occur.

The response of amphibolites (once formed) to further metamorphism

Tectonophysics, 3 (5) (1966) 383–427

depends upon the particular combination of pressure, temperature and water vapour pressure to which they are subjected. Field observations in metamorphic terrains provide direct evidence upon this point. With increasing grade, amphibolites are observed to be transformed into pyroxene granulites. It appears that this transformation may occur at temperatures between 500°C and 700°C and at rather moderate pressures. An excellent example of the gradual transformation of amphibolite into pyroxene granulite is described by Engel and Engel (1962). They estimate that the temperature required for this transformation was about 625°C and the depth was of 10 km or greater.

In most regions in the stable continental crust, the temperatures now existing at the M-discontinuity may be in the vicinity of 450–650°C, as previously discussed. Under these conditions, basic rocks in the amphibolite facies at the base of the crust may be marginally stable. However, we must consider the earlier history of the normal continental crust. In most regions, at shallow depth crystalline basement rocks are found. Formation of the assemblage of granitic and metamorphic rocks referred to as "basement" implies that at an earlier stage of crustal evolution, the temperatures within the present upper 15 km of the crust were much higher than normal, perhaps 400–600°C, and locally even higher. At the time when these temperatures were generated, the temperatures in the lower crust must have been much higher, probably in the vicinity of 600–1,000°C. Under such conditions, amphibolites would have been converted into granulites and it is probable that the lower crust would have become rather thoroughly dehydrated. Subsequently, after its period of active evolution had proceeded to completion (next section) the crust slowly cooled to its present state. However, amphibolites have not been reformed in the lower crust simply because of lack of water. The possibility that large volumes of the lower crust consisting of impermeable granulites may become rehydrated on cooling by downward access of water from the upper crust appears remote, although it may occur on a local and restricted scale. Accordingly, we conclude that the lower regions of the normal continental crust are probably essentially "dry". Similar views have been expressed by Heier and Adams (1965) on geochemical grounds and by Den Tex (1965) on petrological grounds.

If this conclusion is accepted then, for reasons previously given, the dominant rock type cannot be of overall basic composition. It is more likely that the lower crust consists of rocks of intermediate composition occurring in the eclogite facies. The stable mineral assemblage would be quartz, K–Na felspar(s), garnet and clinopyroxene. Kozlowski (1958) has described natural examples of such rocks, whilst Green and Lambert (1965) have experimentally demonstrated the stability of a similar mineral assemblage in a more acidic composition. A stimulating discussion of the petrology of the lower crust has been provided by Den Tex (1965), who reaches a conclusion parallel to that expressed above. He regards the lower crust as consisting of a heterogeneous mixture of anhydrous acid granulites, charnockites and eclogites, with a mean composition approaching that of intermediate rocks.

POSSIBLE TECTONIC CONSEQUENCES OF THE GABBRO–ECLOGITE TRANSFORMATION

Our previous discussion of the nature of the Mohorovičić discontinuity has been restricted to regions where the latter is well-defined, for example in stable continental regions and in deep oceanic basins. There are, however, many regions on the earth where the M-discontinuity is poorly defined, and where the crust is separated from the "mantle" by layers of rock possessing P-wave velocities between 7 and 8 km/sec. Whether these intermediate velocity layers are to be regarded as crust or mantle is to some extent subjective. In fact, the very existence of an M-discontinuity in such regions is in doubt. Frequently, the seismic data are consistent with a continuous transition of velocities between crust and mantle with no evidence of a discontinuity. The existence and position of the latter is often inferred from gravity data which do not permit unique solutions.

Regions characterised by intermediate crust–mantle seismic velocities or by abnormally low upper-mantle velocities are of great tectonic importance. They include mid-oceanic ridges, some oceanic rises, island arc regions, some continental margins, continental rifts and continental regions which have undergone comparatively recent mountain building. A review of these regions is given by Cook (1962). Several explanations of the nature of the intermediate-velocity material have been proposed. Cook (1962) suggests that it is a physical mixture of mantle and crustal material. Ringwood (1962a,b) and Thompson and Talwani (1964a,b) suggest that it is undifferentiated ultramafic mantle material consisting of the mineral assemblage olivine, pyroxene and plagioclase (plagioclase pyrolite) whilst R.P. von Herzen (personal communication, 1962) attributes the lower velocities to abnormally high mantle temperatures as are indicated by the high heat flows often observed over mid–oceanic ridges. Finally Pakiser (1965) suggests that the intermediate velocity regions are caused by the gabbro–eclogite transformation.

On critical examination, none of the above hypotheses appears capable of supplying the sole explanation of the nature of the intermediate velocity material in tectonically active regions. Rather, it appears that all of the explanations may be applicable to different regions to different extents. In this section, we will discuss the possible role of the gabbro–eclogite transformation in such regions.

Previously, we have argued that the gabbro–eclogite transformation is not responsible for the M-discontinuity in stable continental regions or in deep oceanic basins. However, most of these arguments are not directly applicable in the cases of many tectonically active regions possessing intermediate crust–mantle velocities. Since these are often characterised by high heat flows and consequent high sub-surface temperatures, it is possible for gabbroic and garnet granulite assemblages to be stable in the crust, and to transform downwards to eclogite. This transformation would give rise to a wide zone of intermediate seismic velocities (section "Nature of the Mohorovičić discontinuity"). Furthermore, as we shall see, the high density of eclogite may not be an objection to its presence in active orogenic zones, as is the case in stable continental regions.

In the following discussion, we will adopt the hypothesis (Ringwood 1962a,b; 1966a,b) that the primitive composition of the upper mantle lies between basalt and peridotite, and closer to peridotite. This hypothetical parental upper-mantle rock has been called pyrolite. Fractional melting of pyrolite yields basalt magma and leaves behind a refractory residue of dunite or peridotite. There are strong reasons for believing that

Tectonophysics, 3 (5) (1966) 383–427

the continental crust has evolved by fractional melting and differentiation processes from the upper mantle over geologic time (Rubey, 1951, 1955; Bullard, 1952; Wilson, 1954; Engel, 1963). This implies that the mantle immediately beneath continents is depleted in low-melting components and easily fractionated elements, and is probably similar in composition to dunites and Alpine peridotites. Beneath oceanic areas it appears that the upper mantle has not been subjected to substantial fractionation, and accordingly the more primitive pyrolite may extend upwards to the M-discontinuity.

The mean density of the upper mantle according to the pyrolite model is 3.30 g/cm^3 (Clark and Ringwood, 1964). According to this model, the low-melting point fraction is basalt. Accordingly, in regions undergoing differentiation, large volumes of basaltic magma may be extruded at the surface of the solid earth. On cooling, this magma will crystallise initially as plagioclase and pyroxene, with a mean density of 3.0 g/cm^3. However, we have seen that in normal continental regions, the thermodynamically stable assemblage is eclogite, not gabbro. Accordingly, on further cooling under dry conditions, piles of basalt extruded at the earth's surface would pass through the garnet granulite field into the eclogite stability field. If the kinetic conditions should be favourable to transformation, basalt will be converted to eclogite with a mean density of 3.5 g/cm^3. If this should occur throughout a large volume (i.e., of the order of tens of km^3) of basalt resting on the earth's surface, an acute gravitational instability would be caused, and it is improbable that the crust would be able to withstand the accompanying stresses indefinitely. The block of eclogite would eventually sink diapirically through the crust into the mantle. Since its density is greater than that of the mantle, it would continue to sink until it reached a zone of similar density. This may well be very deep in the mantle.

The sinking of a large block of eclogite in such a manner would cause severe crustal deformation (Ramberg, 1963; Morgan, 1965). Initially the crust might be dragged down into a deep depression (trench or geosyncline) and finally, when the block became detached and sank into the mantle, the crust would be dragged together in compression (folding). Furthermore, depending upon the p, t conditions in the mantle, and the rate of sinking, the eclogite might suffer partial fusion on its downward journey. Magmas thus formed (andesites, granodiorites) would rise and penetrate the deformed crust.

According to the above speculations, the basalt–eclogite transformation may generate the driving force for a tectonic engine, which might in turn be responsible for the major crustal orogenic cycle. The conditions under which this engine might operate are varied and flexible. There are, however, certain essential prerequisites. Two of the most fundamental of these are: (1) that eclogite is thermodynamically stable relative to basalt in the normal geothermal gradient in the crust; (2) that kinetic factors will permit the transformation of large volumes of basalt into eclogite within the crust.

The validity of (1) has been extensively discussed in the section on stability of eclogite within the crust. It was concluded that this prerequisite was highly probable. The second is more doubtful since the direct evidence bearing upon the subject is limited. However, there is strong geological observational evidence that under appropriate conditions, basalt may become converted to eclogite (section above ; Green and Ringwood, 1966). The conditions required are apparently (a) temperatures of 300°C and greater, (b) low, but not zero, partial pressure of water, (c) sufficient time, (d) (perhaps) shear stress. The only condition which appears somewhat difficult to fulfil is (b). However, it might be realized in basaltic lavas which were extruded rapidly on the sea floor, forming a thick, impermeable pile. Ramberg (1963) has shown by a series of model experiments that the structural conditions which result in the *extrusion* of basic lavas are likely to be accompanied by the *intrusion* of much greater total volumes as sills and dykes within the crust. Large volumes of intruded basaltic magma are likely to crystallise to dense impermeable gabbros and dolerites which would be particularly resistant to hydration, and if subjected to the correct time–temperature conditions, susceptible to transformation to eclogite.

We will not argue this question of transformation kinetics any further. Clearly it needs detailed consideration and intensive studies of cases where eclogites are believed by geologists to have formed in situ by transformation of basalt. In the subsequent discussion we will assume that the transformation is possible under appropriate conditions, and use this assumption as the keystone of an hypothesis of orogenesis.

Proposed model of an orogenic cycle

The following speculative and simplified model (Fig.9) illustrates the development of a geosyncline near a continental margin, in the absence of an offshore island arc. A possible example is the eugeosyncline now believed to be forming off the east coast of the U.S.A. (Drake et al., 1959). This model is chosen to demonstrate the possibilities of the basalt–eclogite transformation engine. It is not unique, and numerous variants could be devised. With some relatively minor changes the model could be applied to the evolution of island arcs and trenches.

Stage 1 (Fig.9A)

Hypothetical initial state with normal continental section directly adjacent to deep oceanic section. The temperature distribution beneath deep oceanic basins probably results in a *decrease* of density with depth in the first 100 km (Clark and Ringwood, 1964). The oceanic upper mantle is therefore potentially gravitationally unstable. For the instability to become manifest, a triggering effect is necessary, e.g., a horizontal temperature gradient, such as is found near the continental–oceanic borders. The combination of these effects may localise orogenic activity near some continental margins (Ringwood, 1962a; MacDonald, 1963).

Stage 2 (Fig.9B)

Erosion on the continent with consequent deposition of sediments on the neighbouring oceanic crust (assumed to be of basaltic composition) leads to the accumulation of a thick wedge of sediments. Gravitational instability in the sub-oceanic mantle leads to the upward rise of masses of primary mantle material (pyrolite) from the low velocity–low density zone. (This process is better termed advection - Elsasser, 1963 - rather than convection.) Rising masses of pyrolite undergo fractional melting, yielding basaltic

Tectonophysics, 3 (5) (1966) 383–427

magma. Magma becomes separated from residual dunite—peridotite and rises into crust, becoming interlayered with sediments. A much larger volume of basaltic magma does not reach the surface, and intrudes the oceanic crust (Ramberg, 1963) causing a large thickening of the crust. The generation of crustal thickening by magmatic activity was discussed in some detail by Drake et al. (1959) in connection with the geosyncline off the east coast of the U.S.A. In this example, there was little evidence for the present existence of volcanic activity. However, the occurrence of earlier activity was inferred from the presence of seamounts and from the pattern of magnetic anomalies. The lack of obvious volcanism near continental margins cannot be used to infer the absence of deep-seated magmatic activity. Because of high density, basaltic magmas may not readily penetrate a thick

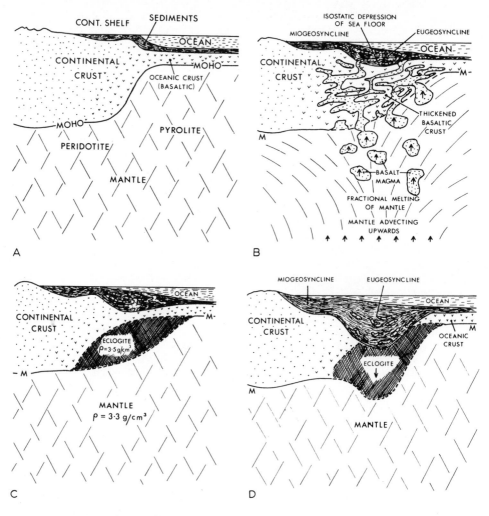

Fig.9A–D (Legend see p.419).

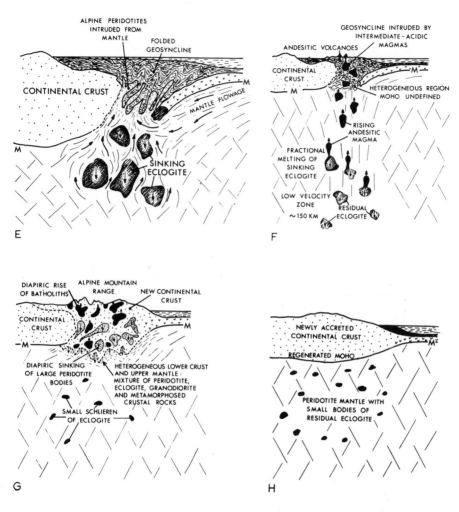

Fig.9. Model for an orogenic cycle near a continental margin (e.g., east coast of U.S.A., Drake et al., 1959) in the absence of an off-shore island arc, based upon the gabbro–eclogite transformation as a tectonic engine.
A. Hypothetical initial state. B. Mantle advection and crustal thickening. C. Transformation of thick basaltic crust to eclogite. D. Gradual sinking of eclogite lower crust leading to formation of geosyncline or trench. E. Catastrophic sinking of eclogite into mantle, folding of geosyncline, intrusion of Alpine ultramafics into folded geosyncline. F. Fractional melting of sinking eclogite to yield intermediate-acid magmas which rise and intrude geosyncline. G. Isostatic rise of folded geosyncline to form Alpine mountain chain, widespread regional metamorphism. H. Erosion of mountains, major diapiric differentiation under gravity leading to re-establishment of M-discontinuity, further cooling at depth resulting finally in a stable shield. (Not to scale.)

Tectonophysics, 3 (5) (1966) 383–427

wedge of unconsolidated sediments. Such magmas would be more likely to spread out and accumulate at the base of the sediments (Ramberg, 1963).

Stage 3 (Fig.9C)

Intruded and extruded basalts cool to the temperature of their surroundings. They are subjected to increased load pressure from the sediments which overlie them. Basaltic rocks are accordingly in the eclogite stability field. Under suitable kinetic conditions, transformation of basalt to eclogite occurs throughout much of the thickened basaltic crust.

Stage 4 (Fig.9D)

The wedge of dense eclogite begins to drag the crust downwards forming a downbuckle. Vast thicknesses of sediments may accumulate in this down-buckle forming a geosyncline. If access of sediments is prevented by other geologic factors an oceanic trench may form instead. Morgan (1965) and Elsasser (1966) have also developed the concept of oceanic trenches being caused by the falling of a dense sinker in the mantle and have shown that this model leads to a satisfactory solution of gravity and seismic data. Slumping of sediments within the geosyncline occurs as it deepens. The weight of the sediment causes increased pressure, thus pushing basaltic rocks further into the eclogite field. During geosynclinal sedimentation the temperatures throughout the crust are abnormally low. There is a long time lag before thermal blanketing by sediments causes temperatures in the crust to rise beyond the normal geothermal distribution.

Stage 5 (Fig.9E)

Downward sinking of the eclogitic lower crust becomes catastrophic. Large blocks of eclogite become detached and sink deep into the mantle. Downward sinking of eclogite causes major crustal deformation dragging the crust into compression (Morgan, 1965) and folding the sediments in the geosyncline. The folding is accompanied by net crustal shortening and thickening in the deformed zone. Foundering of lower crust causes a major disturbance in the underlying mantle, which leads to large scale intrusion of Alpine peri-dotites into the folded geosyncline.

Stage 6 (Fig.9F)

Eclogite bodies sink into the low-velocity zone of the mantle and undergo partial fusion. Andesitic and dacitic magmas are generated, and rise up-wards, intruding the folded geosyncline as bodies of quartz diorites and granodiorite, and causing extensive andesitic and dacitic volcanic activity. (The generation of the calc-alkaline suite of igneous rocks by partial fusion of basaltic material under eclogitic conditions is at present being studied experimentally by Mr. Trevor Green and one of the authors - A.E. Ringwood. Preliminary results indicate that this mechanism is extremely promising).

Stage 7 (Fig.9G)

Crustal compression relaxes as eclogitic bodies sink through low-velocity

Tectonophysics, 3 (5) (1966) 383–427

zone. The folded geosynclinal wedge rises isostatically to form an Alpine mountain chain. Temperatures in the thick crust become abnormally high due to the combined effects of radioactive heat generation in situ and conduction from large intrusions of primary calc-alkaline magmas from the mantle as discussed in stage 6. This leads to regional metamorphism in the base of the fold-mountain belt and generation of secondary granitic magmas by partial fusion of crustal rocks. The lower crust at this stage, is a heterogeneous mixture of metamorphic rocks, andesitic and dacitic material, primary unfractionated plagioclase pyrolite from the upper mantle, residual Alpine peridotites and dunites from the mantle, and rocks of basaltic composition. However, because of the abnormally high crustal temperatures, basaltic rocks would now occur in the stability fields of pyroxene granulite and garnet granulite rather than eclogite. The physical heterogeneity of the rocks of the lower crust and upper mantle cause the seismic velocity distribution between crust and mantle to be approximately continuous. The Mohorovičić discontinuity cannot be recognized at this stage.

Stage 8 (Fig.9H)

During this, the final and longest stage of crustal evolution, the mountains gradually become eroded, and the crust develops a relatively stable gravitational and rheological configuration. Large bodies of ultramafic rocks previously intruded into the crust, sink diapirically back into the mantle and large volumes of lighter acidic rocks in the mantle rise diapirically into the crust. The end result of this separation of large volumes of rocks in the gravitational field according to their densities, is the re-establishment of a distinct and recognizable boundary between generally intermediate but heterogeneous lower crustal rocks in the granulite and eclogite facies and the ultramafic rocks of the upper mantle. Thus, the Mohorovičić discontinuity which was present beneath the oceanic crust at the beginning of the cycle, but which was obliterated during the active phase of orogenesis, becomes re-established in the stable continental crust.

   The major cycle lead ultimately to strong upward concentration of radioactive elements in the upper part of the crust. Further erosion of the upper crust causes the heat producing elements to be dispersed into new geosynclines which are being formed at the continental margins. The strong upward concentration of radioactivity, combined with its net depletion by erosion in highly evolved continental crusts, ultimately results in lower subcrustal temperatures. These cause the crust and mantle beneath it to assume the state of relative gravitational stability and tectonic rigidity which is characteristic of Precambrian shields (Ringwood, 1962a,b; Clark and Ringwood, 1964). Thus the major geologic cycle completes its course.

Further applications of model

The model proposed in the previous paragraph is an illustration of the manner in which the basalt–eclogite transformation might provide an engine for driving the orogenic cycle. Numerous variations of the model are conceivable. For example the hypothesis of ocean floor renewal proposed by Hess (1962) can be restated using the basalt–eclogite transformation as the

Tectonophysics, 3 (5) (1966) 383–427

principal driving force, as in Fig.10. According to this model, gravitational
instability develops in the mantle beneath mid-oceanic ridges. Primary
mantle material (pyrolite) from the low-velocity zone advects inwards
towards the ridge and then upwards. Fractional melting of pyrolite occurs
during upward motion leading to generation of basaltic magma, together with
residual unmelted peridotite. The axes of the ridges are characterised by
high heat flow and the sub-surface temperatures are high enough to maintain stability of the basaltic mineral assemblage. The mid-oceanic ridges
thus develop as tensional features composed of heterogeneous mixtures of
gabbro, peridotite and pyrolite. The oceanic crust moves outwards from
these axial structures as in the Hess model. Conversion of basalt to eclogite
occurs near continental margins and/or island arcs, as discussed in the
previous model, providing a major source of gravitational instability. It is
not necessary that the eclogite should sink vertically. The paths of sinking
blocks of eclogite will be determined partly by the "viscosity" distribution
within the surrounding mantle, which in turn is determined by the temperature distribution. If the active orogenic belts near continental margins are
characterised by abnormally high sub-crustal temperatures, it is possible
that the sinking blocks of eclogite might move downwards and inwards
towards the continent, thus providing an explanation for the occurrence of
deep-focus earthquakes. The overall flow patterns in the upper mantle
shown in Fig.10 are not to be compared with the regular cells contemplated
in most of the conventional theories of convection. They are probably highly
complex, as in the "advection" configurations of Elsasser (1963). Furthermore they are irreversible since they involve chemical differentiation. In

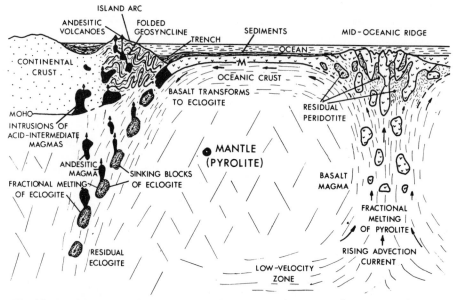

Fig.10. Modification of ocean-floor spreading hypothesis (Hess, 1962) using
basalt—eclogite transformation as tectonic engine. (Not to scale.)

Tectonophysics, 3 (5) (1966) 383—427

this sense they have affinities with the "mass–circuits" advocated by Van Bemmelen (1964).

Finally, we should consider the possible future of present geologic structures which are characterised by great thicknesses of basaltic lavas resting upon the crust, e.g., Hawaii, Iceland and mid-oceanic ridges, rises and seamounts in general. These vast structures will be stable as long as their internal temperature distributions are high enough to maintain them in the gabbro stability field or in the medium grades of the garnet granulite stability field. In fact, these structures are all young, and characterised by high heat flows. We can look forward to the time when volcanic activity subsides, and the mantle is no longer a source of advective heat. When this stage is reached, there will be a general cooling of the entire structures, and transformation into eclogite will probably occur in the deeper levels. The islands and ridges will then subside into the mantle, causing complex crustal deformation. Since the minimum melting point of common basalts in the eclogite facies is much smaller than the minimum melting point of pyrolite (unpublished data) the subsiding eclogites may undergo fractional melting during sinking, resulting in production of large volumes of inter-mediate and acidic magmas. The magmas will rise to the surface and it is possible that they may form the nucleus of new continental blocks. In such a manner a continental nucleus could be formed directly in the midst of an ocean, without requiring a supply of sedimentary material from a nearby continent. This process may indeed be presently operating in some island arc areas far removed from continents, e.g., the Aleutian Islands. It is interesting to speculate that "continental" islands such as Fiji and New Zealand may have originated from basalt piles.

ACKNOWLEDGEMENTS

The authors are grateful to Dr. B. Raleigh and Mr. Trevor Green of the Department of Geophysics and Geochemistry, Australian National Univer-sity, for helpful comments on the manuscript.

REFERENCES

Backlund, H.G., 1936. Zur genetischen Deuten der Eklogite. Geol. Rundschau, 27: 47–61.
Bearth, P., 1959. Über Eklogite, Glaukophanschiefer und metamorphe Pillolaven. Schweiz. Mineral. Petrog. Mitt., 39: 267–286.
Birch, F., 1952. Elasticity and constitution of the earth's interior. J. Geophys. Res., 57: 227–286.
Birch, F., 1955. Physics of the crust. Geol. Soc. Am., Spec. Papers, 62: 101–118.
Birch, F., 1958. Interpretation of the seismic structure of the crust in the light of experimental studies of wave velocities in rocks. In: H. Benioff, M. Ewing, B.F. Howell and F. Press (Editors) Contributions in Geophysics in Honour of B. Gutenberg. Pergamon, London, pp.158–170.
Birch, F., 1960. The velocity of compressional waves in rocks to 10 kbar, 1. J. Geo-phys. Res., 65: 1083–1102.
Birch, F., 1961. The velocity of compressional waves in rocks to 10 kbar, 2. J. Geo-phys. Res., 66: 2199–2224.
Birch, F. and Le Compte, P., 1960. Temperature–pressure plane for albite composi-tion. Am. J. Sci., 258: 209–217.

Boyd, F.R. and England, J.L., 1959. Experimentation at high pressures and temperatures. Carnegie Inst. Wash., Yearbook, 58: 82–89.

Boyd, F.R. and England, J.L., 1960. Apparatus for phase-equilibrium measurements at pressures up to 50 kbar and temperatures up to 1,750°C. J. Geophys. Res., 65: 741–748.

Boyd, F.R. and England, J.L., 1964. The system enstatite-pyrope. Carnegie Inst. Wash., Yearbook, 63: 157–160.

Broecker, W.S., 1962. The contribution of pressure-induced phase changes to glacial rebound. J. Geophys. Res., 67: 4837–4842.

Bullard, E.C., 1952. Discussion of paper by Revelle and Maxwell. Nature, 170: 200.

Bullard, E.C. and Griggs, D.T., 1961. The nature of the Mohorovičić discontinuity. Geophys. J., 6: 118–123.

Christensen, N., 1965. Compressional wave velocities in metamorphic rocks at pressures to 10 kbar. J. Geophys. Res., 70: 6147–6164.

Clark, S.P., 1961. Geothermal studies. Carnegie Inst. Washington, Yearbook, 60: 185–190.

Clark, S.P., 1962. Temperatures in the continental crust. In: C.M. Herzfeld (Editor), Temperature, its Measurement and Control in Science and Industry. Reinhold, New York, N.Y., 3: 779–790.

Clark, S.P. and Ringwood, A.E., 1964. Density distribution and constitution of the mantle. Rev. Geophys., 2: 35–88.

Coleman, R.G., Lee, D.E., Beatty, I.B. and Brannock, W.W., 1965. Eclogites and eclogites, their differences and similarities. Bull. Geol. Soc. Am., 76: 483–508.

Cook, K.L., 1962. The problem of the crust-mantle mix: lateral inhomogeneity in the uppermost part of the earth's mantle. Advan. Geophys., 9: 296–360.

Den Tex, E., 1965. Metamorphic lineages of orogenic plutonism. Geol. Mijnbouw, 44: 105–132.

De Waard, D., 1965. A proposed subdivision of the granulite facies. Am. J. Sci., 263: 455–461.

Dix, C.H., 1965. Reflection seismic crustal studies. Geophysics, 30: 1068–1084.

Dorman, J. and Ewing, M., 1962. Numerical inversion of seismic surface wave dispersion data and crust–mantle structure in the New York–Pennsylvania area. J. Geophys. Res., 67: 5227–5242.

Doyle, H.A., Underwood, R. and Polack, H., 1966. Seismic velocities from explosions off the central coast of New South Wales. Geol. Soc. Australia, in press.

Drake, C.L., Ewing, M. and Sutton, G.H., 1959. Continental margins and geosynclines: The east coast of North America north of Cape Hatteras. In: L.H. Ahrens, F. Press, K. Rankama and S.K. Runcorn (Editors), Physics and Chemistry of the Earth. Pergamon, London, 3: 110–198.

Elsasser, W.M., 1963. Early history of the earth. In: J. Geiss and E. Goldberg (Editors), Earth Science and Meteorites in honour of F.G. Houtermans. North-Holland, Amsterdam, pp.1–30.

Elsasser, W.M., 1966. Thermal structure of the upper mantle and convection. In: P.M. Hurley (Editor), Advances in Earth Science. M.I.T. Press, Cambridge, Mass., pp.461–502.

Engel, A.E.J., 1963. Geologic evolution of North America. Science, 140: 143–152.

Engel, A.E.J. and Engel, C.G., 1962. Progressive metamorphism and amphibolite, Northwest Adirondack Mountains, New York. In: A.E.J. Engel, H.L. James and B.F. Leonard (Editors), Petrologic Studies - Buddington Volume. Geol. Soc. Am., New York, N.Y., pp.37–82.

Engel, A.E.J., Engel, C.G. and Havens, R.G., 1965. Chemical characteristics of oceanic basalts and the upper mantle. Bull. Geol. Soc. Am., 76: 719–734.

Eskola, P., 1921. On the eclogites of Norway. Skrifter, Norske Videnskaps-Akad, Kristiania, I: Mat. Naturv. Kl., 1921: 1–128.

Everingham, I.B., 1965. The crustal structure of the southwest of Western Australia. Australia, Bur. Mineral Resources, Geol. Geophys., Record, 97: 1–6.

Fermor, L.L., 1913. Preliminary note on garnet as a geological barometer and on an infra-plutonic zone in the earth's crust. Records Geol. Surv. India, 43.

Fermor, L.L., 1914. The relationship of isostasy, earthquakes and vulcanicity to the earth's infra-plutonic shell. Geol. Mag., 51: 65–67.

Goldschmidt, V.M., 1922. Über die Massenverteilungen im Erdinneren, verglichen mit der Structur gewisser Meteoriten. Naturwissenschaften, 10: 918–920.

Green, D.H. and Lambert, I.B., 1965. Experimental crystallization of anhydrous granite at high pressures and temperatures. J. Geophys. Res., 70: 5259–5268.

Green, D.H. and Ringwood, A.E., 1966. An experimental investigation of the gabbro–eclogite transformation and some petrological applications. Geochim. Cosmochim. Acta, 71: 3589–3594.

Green, T., Ringwood, A.E. and Major, A., 1966. Friction effects and pressure calibration in a piston-cylinder high pressure–temperature apparatus. J. Geophys. Res., in press.

Gutenberg, B., 1955. Wave velocities in the earth's crust. Geol. Soc. Am., Spec. Papers, 62: 19–34.

Gutenberg, B., 1959. Physics of the Earth's Interior. Acad. Press, New York, N.Y., 240 pp.

Harris, P.G. and Rowell, J.A., 1960. Some geochemical aspects of the Mohorovičić discontinuity. J. Geophys. Res., 65: 2443–2459.

Heier, K.S. and Adams, J.A.S., 1965. Concentration of radioactive elements in deep crustal material. Geochim. Cosmochim. Acta, 29: 53–61.

Hess, K., 1962. History of ocean basins. In: A.E.J. Engel, H.L. James and B.F. Leonard (Editors), Petrologic Studies - Buddington Volume. Geol. Soc. Am., New York, N.Y., pp.599–620.

Holmes, A., 1926a. Contributions to the theory of magmatic cycles. Geol. Mag., 63: 306–329.

Holmes, A., 1926b. Structure of the continents. Nature, 118: 586–587.

Holmes, A., 1927. Some problems of physical geology and the earth's thermal history. Geol. Mag., 64: 263–278.

Howard, L.E. and Sass, J.H., 1964. Terrestrial heat flow in Australia. J. Geophys. Res., 69: 1617–1625.

Kennedy, G.C., 1956. Polymorphism in the felspars at high temperatures and pressures. Bull. Geol. Soc. Am., 67: 1711–1712 (abstract).

Kennedy, G.C., 1959. The origin of continents, mountain ranges, and ocean basins. Am. Sci., 47: 491–504.

Korzhinsky, D.S., 1937. Dependence of mineral stability on depth. Zap. Vses. Mineral. Obshestva, 66 (2): 369–396 (in Russian).

Kozlowski, K., 1958. On the eclogite-like rocks of Stary Gieraltow (east Sudeten). Bull. Acad. Polon. Sci., Ser. Sci. Chim., Geol., Geograph., 6 (11): 723–728.

Kushiro, I., 1965. Clinopyroxene solid solutions at high pressure: the join diopside-albite. Carnegie. Inst. Wash., Yearbook, 64: 112–120.

Kushiro, I. and Yoder, H.S., 1964. Experimental studies on the basalt–eclogite transformation. Carnegie Inst. Wash., Yearbook, 63: 108–114.

Kushiro, I. and Yoder, H.S., 1965. The reactions between forsterite and anorthite at high pressures. Carnegie Inst. Wash., Yearbook, 64: 89–94.

Lee, W., and Uyeda, S., 1965. Review of heat flow data. In: W. Lee (Editor), Terrestrial Heat Flow– Geophys. Monograph,6 (8): 87–190.

Lovering, J.F., 1958. The nature of the Mohorovičić discontinuity. Trans. Am. Geophys. Union, 39: 947–955.

Lovering, J.F., 1962. The evolution of the meteorites. In: C.B. Moore (Editor), Researches on Meteorites. Wiley, New York, N.Y., pp.179–198.

MacDonald, G.A. and Katsura, T., 1961. Variations in lava of 1959 eruption in Kilauea Iki. Pacific Sci., 5 (3): 358–369.

MacDonald, G.J.F., 1959. Chondrites and the chemical composition of the earth. In: P.H. Abelson (Editor), Researches in Geochemistry. Wiley, New York, N.Y., pp.476–494.

MacDonald, G.J.F., 1963. The deep structure of continents. Rev. Geophys., 1: 587–665.

MacDonald, G.J.F. and Ness, N.F., 1960. Stability of phase transitions in the earth. J. Geophys. Res., 65: 2173–2190.

Tectonophysics, 3 (5) (1966) 383–427

MacGregor, I.D., 1964. The reaction 4 enstatite + spinel = forsterite + pyrope. Carnegie Inst. Wash., Yearbook, 63: 156–157.

Morgan, W.J., 1965. Gravity anomalies and convection currents, 2. The Puerto Rico Trench and the Mid-Atlantic Rise. J. Geophys. Res., 70: 6189–6204.

Nakamura, Y. and Howell, B.F., 1964. Maine seismic experiment: frequency spectra of refraction arrivals and the nature of the Mohorovičić discontinuity. Bull. Seismol. Soc. Am., 54: 9–18.

Pakiser, L.C., 1965. The basalt–eclogite transformation and crustal structure in the western United States. U.S. Geol. Surv., Prof. Papers, 525-B: 1–8.

Phinney, R.A., 1964. Structure of the earth's crust from spectral behaviour of long-period body waves. J. Geophys. Res., 69: 2997–3018.

Ramberg, H., 1963. Experimental study of gravity tectonics by means of centrifuged models. Bull. Geol. Inst. Univ. Uppsala, 42 (1): 1–97.

Ringwood, A.E., 1962a. A model for the upper mantle, 1. J. Geophys. Res., 67: 857–867.

Ringwood, A.E., 1962b. A model for the upper mantle, 2. J. Geophys. Res., 67: 4473–4477.

Ringwood, A.E., 1966a. Mineralogy of the mantle. In: P.M. Hurley (Editor), Advances in Earth Science. M.I.T. Press, Cambridge, Mass., pp.357–399.

Ringwood, A.E., 1966b. The chemical composition and origin of the earth. In: P.M. Hurley (Editor), Advances in Earth Science. M.I.T. Press, Cambridge, Mass., pp.287–356.

Ringwood, A.E. and Green, D.H., 1964. Experimental investigations bearing on the nature of the Mohorovičić discontinuity. Nature, 201: 566–567.

Robertson, E.C., Birch, F. and MacDonald, G.J.F., 1957. Experimental determination of jadeite stability relations to 25,000 bar. Am. J. Sci., 255: 115–137.

Roy, R. and Tuttle, O.F., 1956. Investigations under hydrothermal conditions. In: L.H. Ahrens, K. Rankama and S.K. Runcorn (Editors), Physics and Chemistry of the Earth. Pergamon, London, 1: 138–180.

Rubey, W.W., 1951. Geologic history of sea water. Bull. Geol. Soc. Am.,62: 1111–1147.

Rubey, W.W., 1955. Development of the hydrosphere and atmosphere with special reference to the probable composition of the early atmosphere. Geol. Soc. Am., Spec. Papers, 62: 631–650.

Smulikowski, K., 1960. Petrographical notes on some eclogites of the East Sudetes. Bull. Acad. Pol. Sci., Ser. Sci. Geol. Geograph., 8 (1): 11–19.

Steinhart, J.S. and Meyer, R.P., 1961. Explosion studies of continental structure. Carnegie Inst. Wash., Publ., 622: 409 pp.

Stishov, S.M., 1963. The nature of the Mohorovičić discontinuity. Akad. Nauk S.S.S.R., Ser. Geofiz., 1963 (1): 42–48.

Sumner, J.S., 1954. Consequences of a polymorphic transition at the Mohorovičić discontinuity. Trans. Am. Geophys. Union, 35: 385 pp.

Talwani, M., Sutton, G.H. and Worzel, J.L., 1959. A crustal section across the Puerto Rico Trench. J. Geophys. Res., 64: 1545–1555.

Tatel, H.A. and Tuve, M.A., 1955. Seismic exploration of a continental crust. Geol. Soc. Am., Spec. Papers, 62: 35–50.

Thompson, G.A. and Talwani, M., 1964a. Crustal structure from Pacific Basin to central Nevada. J. Geophys. Res., 69: 4813–4837.

Thompson, G.A. and Talwani, M., 1964b. Geology of the crust and mantle, western United States. Science, 146: 1539–1549.

Tuve, M.A., Tatel, H.E. and Hart, P.J., 1954. Crustal structure from seismic exploration. J. Geophys. Res., 59: 415–422.

Van Bemmelen, R.W., 1964. The evolution of the Atlantic Mega-Undation. Tectonophysics, 1 (5): 385–430.

Von Herzen, R.P. and Uyeda, S., 1963. Heat flow through the eastern Pacific ocean floor. J. Geophys. Res., 68: 4219–4250.

Wetherill, G.W., 1961. Steady-state calculations bearing on geological implications of a phase-transition Mohorovičić discontinuity. J. Geophys. Res., 66: 2983–2993.

Wilson, J.T., 1954. The development and structure of the crust. In: G.P. Kuiper (Editor), The Earth as a Planet. Univ. Chicago Press, Chicago, Ill., pp.138–214.

Woollard, G.P., 1962. The relation of gravity anomalies to surface elevation, crustal structure and geology. Dept. Geol., Univ. Wisconsin Res. Rept., 62-9 (III): 73.

Worzel, J.L. and Shurbet, G.L., 1955. Gravity interpretations from standard oceanic and continental sections. Geol. Soc. Am., Spec. Papers, 62: 87–100.

Wyllie, P.J., 1963. The nature of the Mohŏroviči̇́c̄ discontinuity. A compromise. J. Geophys. Res., 68: 4611–4619.

Yoder, H.S. and Tilley, C.E., 1962. Origin of basalt magmas: an experimental study of natural and synthetic rock systems. J. Petrol., 3: 342–532.

Tectonophysics, 3 (5) (1966) 383–427

10

Reprinted from *Phys. Earth Planet. Interiors*, **3**, 405–421 (1970)

CLASSIFICATION OF ECLOGITES IN TERMS OF PHYSICAL CONDITIONS OF THEIR ORIGIN

SHOHEI BANNO

Department of Earth Sciences, Kanazawa University, Kanazawa, Japan

The apparent Fe–Mg distribution coefficient between garnet and clinopyroxene K' is defined by

$$K' = \left(\frac{X_{Fe}}{X_{Mg}}\right)^{ga} \bigg/ \left(\frac{X_{Fe}}{X_{Mg}}\right)^{cpx}.$$

The effects of pressure, temperature, and chemistry of the rocks on K' were examined, using formula volumes of end members of these minerals, and the K' values of natural eclogites. It is shown that K' increases with increasing pressure, and decreases with increasing temperature, and that the effect of chemistry on K' is usually not large.

It is shown that the eclogite types defined by the geological mode of occurrences are also characterized by particular range of K' values, and that the relative temperatures of eclogite crystallization as estimated by K' and by ordinary petrological considerations are usually in harmony with each other. Further, K' can be applied to distinguish the difference in temperatures among the eclogites of some types. The crystallization temperature of eclogites increases in the following order, which is shown by localities: Colombia, Ural and Guatemala, California, Alps and Japan, Bavaria and Spain, Norway, East Sudetes, and granulite facies eclogites. Eclogite inclusions in basalt and in kimberlite represents highest temperature but the former represents higher pressure than the later.

The crystallization temperature of some eclogites which are not included above are also discussed.

1. Introduction

In recent years, the genesis of eclogite, garnet-clino-pyroxene rock with basaltic composition, has been discussed by many authors in relation to the status of upper mantle materials. The argument that eclogite is a high pressure modification of basalt may be traced back at least to GRUBENMANN (1904), and classical papers by ESKOLA (1921) and GOLDSCHMIDT (1922) established the basis for later approaches. Recent contributions to this problem may be represented by the papers by GREEN and RINGWOOD (1967) and RINGWOOD and GREEN (1966), who, based on experimental work on the basalt to eclogite transformation, have extensively discussed the petrological, geophysical and tectonophysical aspects of this problem.

Petrologically, it is considered that there are several types of eclogites, each representing a particular mode of occurrence and probably representing a particular field of temperature and pressure of formation. Therefore, there are "eclogites and eclogites" (COLEMAN *et al.*, 1965): some eclogites are crustal basic metamorphics and the others are mantle materials. Not all eclogites came from the mantle.

Geologically, the following mode of occurrences of eclogites or basaltic garnet-clinopyroxene rocks are knwon:

1) Eclogite inclusions in kimberlite;
2) Eclogite inclusions in alkali basalt;
3) Eclogite or pyrope-diopside rock inclusions in peridotite;
4) Granulite facies eclogite or garnet-clinopyroxene granulite;
5) Amphibolite facies eclogite;
6) Low temperature eclogite or eclogite in glaucophanitic metamorphic terranes.

BORG (1956), SMULIKOWSKI (1964), BANNO (1964), WHITE (1964), COLEMAN *et al.* (1965) and others have shown that each of, or groups of, these eclogite types are characterized by particular compositional range of garnet or clinopyroxene solid solutions. SOBOLEV (1964), COLEMAN *et al.* (1965), BANNO and MATSUI (1965) and ESSENE and FYFE (1967) have shown that the pattern of the distribution of elements, mainly Fe and Mg, between garnet and clinopyroxene varies from type to type, and that this offers a more sound basis of the petrological classification of eclogites than the compositional range of individual minerals. This view is

405

also the basic standpoint of the present paper, in which the author examines the extent over which the element distribution between garnet and clinopyroxene could be applied as a measure of physical conditions, and examines the possibility of classifying some of the eclogite types in more detail. Throughout this paper, the values of temperature and pressure are not referred to quantitatively. It is the author's opinion that if the equilibrium relations at two temperatures are experimentally determined, the temperature of eclogite crystallization is uniquely determined by the use of the distribution relations to be discussed in this paper.

It is helpful to define the nomenclature to be used. The term eclogite is used in a wide sense, and hence it denotes garnet-clinopyroxene rock with more or less basaltic composition, but the term basaltic is used rather vaguely. FORBES (1965) has mentioned that many, if not all, eclogites are not basaltic in chemical composition. According to this nomenclature, almanine-salite rock in the granulite facies, and pyrope-diopside rock enclosed in peridotite are called eclogite, along with typical eclogite containing pyrope-rich garnet and omphacite. Therefore, we have to accept the existence of eclogites which do not belong to the eclogite facies. The definition of the eclogite facies follows that given by O'HARA (1960), i.e., the eclogite facies is characterized by the assemblage kyanite + garnet + omphacite (or in a wide sense Ca-rich clinopyroxene), or by quartz + garnet + omphacite.

2. Some basic concepts of distribution relations

2.1. *Definition of distribution coefficient*

In later sections, the classification of eclogites will be discussed mainly in terms of the distribution coefficient between garnet and clinopyroxene, and hence this coefficient is defined first.

The apparent distribution coefficient of A and B cations of the same valency between phases α and β, $K'^{\alpha \cdot \beta}_{A \cdot B}$ or simply K' is defined by

$$K'^{\alpha \cdot \beta}_{A \cdot B} = \left(\frac{X_A}{X_B}\right)^{\alpha} \Big/ \left(\frac{X_A}{X_B}\right)^{\beta}, \qquad (1)$$

where X_A^{α} etc. denote the mole fraction of A in phase α etc. This coefficient corresponds to the following exchange reaction:

$$AY^{\alpha} + BZ^{\beta} = AZ^{\beta} + BY^{\alpha}, \qquad (2)$$

or simply expressed by

$$A^{\alpha} + B^{\beta} = A^{\beta} + B^{\alpha}, \qquad (3)$$

where AY, BZ etc. denote the components of solid solutions.

If the solid solutions α and β are ideal, K' defined by eq. (1) is equal to the thermodynamic distribution constant $K^{\alpha \cdot \beta}_{A \cdot B}$ or simply K, as defined by

$$K^{\alpha \cdot \beta}_{A \cdot B} = \left(\frac{a_A}{a_B}\right)^{\alpha} \Big/ \left(\frac{a_A}{a_B}\right)^{\beta} = \exp\left(\frac{\Delta G}{RT}\right),$$

where a_A^{α} etc. denote the activities of AY in phase α etc., and ΔG denotes the difference in the free energies between the right- and left-hand sides of eq. (2).

In dealing with the distribution relations of trace elements such as Mn and rare earth elements (REE), the distribution coefficients are normalized to Mg for Mn, and to Sm for REE. The distribution relations of trace elements between garnet and clinopyroxene are not based on Nernst's distribution law, but on the exchange reaction such as shown in eq. (2).

Generally speaking, the apparent distribution coefficient K' is a function of temperature, pressure and the chemical composition of the system.

2.2. *Effect of pressure on the distribution coefficient*

The effect of pressure on the distribution constant K as defined by eq. (4) is obtained as

$$\frac{\partial \ln K}{\partial P} = \frac{\partial}{\partial P}\left(\frac{\Delta G}{RT}\right) = \frac{\Delta V}{RT}. \qquad (5)$$

Therefore, with a crude approximation that ΔV is independent of pressure, the pressure coefficient of K is obtained as follows:

$$\ln K = \frac{\Delta V}{RT}(P - P_0) + \ln K_0, \qquad (6)$$

$$K = K_0 \exp \frac{\Delta V(P - P_0)}{RT}, \qquad (7)$$

where K and K_0 are the distribution constants at pressure $P = P$ and $P = P_0$ at given temperature, respectively.

The relationship between the thermodynamic distribution constant K and the apparent distribution coefficient K' as defined by eq. (1) is given as

$$\ln K' = \ln K - \ln \left\{ \left(\frac{\gamma_A}{\gamma_B} \right)^{\alpha} \middle/ \left(\frac{\gamma_A}{\gamma_B} \right)^{\beta} \right\}, \qquad (8)$$

where γ_A^{α} etc. denote the activity coefficient of AY in phase α etc. The term containing the activity coefficient in eq. (8) is composition dependent, and hence the pressure coefficient of K' is not easily calculated. We have, however, the following relation:

$$\frac{\partial \ln \gamma_A}{\partial P} = \frac{V_{AY} - V_{AY}^{\circ}}{RT}, \qquad (9)$$

where V_{AY} and V_{AY}° denote partial formula volume of AY in phase α and formula of pure AY, respectively. If there is an additivity of formula volume, the activity coefficient is independent of pressure. For neso- and ino-silicate solid solutions, the additivity of formula volume has been examined in detail on several series such as diopside-Ca-Tschermakite (CLARK et al., 1962), olivine and orthopyroxenes for Mg–Fe, Mg–Co and Mg–Ni substitutions (MATSUI and SYONO, 1968; MATSUI et al., 1968), diopside–jadeite series (KUSHIRO, personal communication) and others. Most of these solid solutions have slight volume of mixing, but detailed studies on diopside–hedenbergite, and pyrope–almandine series, with which we are concerned here, have not yet been obtainable to the author. If the volume of mixing is ignored, we have a crude approximation that the pressure affects K' only through the change of K by pressure.

Let us examine the pressure effect on K' in a simplified system where the phase α is ideal and the phase β is non-ideal solid solution. The apparent distribution coefficient K' at pressures P_1 and P_2 at the same temperature are denoted by K_1' and K_2', respectively, and similarly the subscripts 1 and 2 denote the quantities under pressures P_1 and P_2, respectively. We have then

$$\ln K_1'^{!\}} = \ln K_1 + \ln \left(\frac{\gamma_A}{\gamma_B} \right)^{\beta}_1 \qquad (10)$$

and

$$\ln K_2' = \ln K_2 + \ln \left(\frac{\gamma_A}{\gamma_B} \right)^{\beta}_2. \qquad (11)$$

From eqs. (10) and (11), we have

$$\ln \frac{K_1'}{K_2'} = \ln \frac{K_1}{K_2} + \ln \left\{ \left(\frac{\gamma_A}{\gamma_B} \right)^{\beta}_1 \middle/ \left(\frac{\gamma_A}{\gamma_B} \right)^{\beta}_2 \right\}. \qquad (12)$$

If we fix the composition of the phase β, γ_A^{β} and γ_B^{β} are constant under the crude approximation, so that we have

$$\ln \frac{K_1'}{K_2'} = \ln \frac{K_1}{K_2} = \frac{\Delta V}{RT} (P_1 - P_2). \qquad (13)$$

Therefore, the difference in K' between rocks formed under different pressures is best examined by comparing the values of K' for the fixed composition of the non-ideal phase, or of the more non-ideally looking phase, because by doing so the pressure coefficient of K' is the same or similar to that of K.

In this paper, we are mainly concerned with the Fe–Mg distribution coefficient between garnet and clinopyroxene K'^{ga-cpx}_{Fe-Mg}, which corresponds to the following exchange reaction:

$$\begin{array}{cc} \text{diopside} & \text{almandine} \\ \text{CaMgSi}_2\text{O}_6 + \tfrac{1}{3}\text{Fe}_3\text{Al}_2\text{Si}_3\text{O}_{12} = \\ 66.10 & 38.43 \end{array}$$

$$\begin{array}{cc} \text{hedenbergite} & \text{pyrope} \\ = \text{CaFeSi}_2\text{O}_6 + \tfrac{1}{3}\text{Mg}_3\text{Al}_2\text{Si}_3\text{O}_{12} \\ 68.10 & 37.76 \end{array} \qquad (14)$$

$$\text{(formula volume in cm}^3\text{)}$$
$$\Delta V = 1.33 \text{ cm}^3.$$

The formula volumes given above were taken from the compilation by ROBIE et al. (1966), but the formula volume of hedenbergite given in their table 5–2 appears unreasonably small, so that its formula volume was calculated from the cell constants listed in their table 5–1.

The apparent Fe–Mg distribution coefficient between garnet and clinopyroxene, K'^{ga-cpx}_{Fe-Mg}, is defined as follows, but in many cases, it will be denoted only as K':

$$K'^{ga-cpx}_{Fe-Mg} = \left(\frac{X_{Fe}}{X_{Mg}} \right)^{ga} \middle/ \left(\frac{X_{Fe}}{X_{Mg}} \right)^{cpx}. \qquad (15)$$

With increasing pressure, the reaction (14) proceeds from the right- to left-hand sides, so that K increases with increasing pressure. The pressure coefficient of K is shown in fig. 1, in which we see that the pressure coefficient is close to unity in dealing with crustal rocks, but it is not so when we deal with the rocks formed within the mantle. The pressure coefficient of K' may not deviate much from that of K, if we accept MUEL-

SHOHEI BANNO

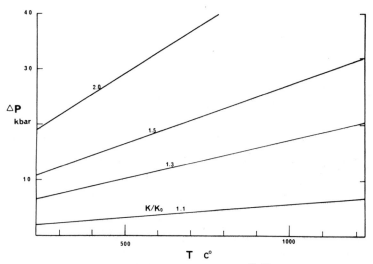

Fig. 1. The pressure coefficient for $K'^{\text{ga--cpx}}_{\text{Fe--Mg}}$.

LER's (1962) idea that diopside–hedenbergite series is a nearly ideal solid solution.

2.3. *Effect of temperature on distribution coefficient*

The effect of temperature on the distribution coefficient K is given by

$$\frac{\partial \ln K}{\partial T} = \frac{-\Delta H}{RT^2}. \tag{16}$$

Assuming that at sufficiently high temperatures, ΔC_p of the exchange reaction is negligible, we have

$$\ln K = \frac{\Delta H_0}{RT} - \frac{\Delta S_0}{R}, \tag{17}$$

where the subcript 0 refers to the quantities at $T = T_0$, which is sufficiently high.

It follows that with increasing temperature K approaches to a certain constant, probably close to, but not equal to unity. This obvious conclusion is sometimes mistaken for that K approaches to unity with increasing temperature.

The effect of temperature on K' is seen from eq. (17), too. The temperature effect on the activity coefficient is shown as follows, assuming that the excess specific heat C_p^E is negligible at high temperature,

$$\frac{\partial \ln \gamma}{\partial T} = \frac{H^E}{RT^2}, \quad \ln \gamma = \frac{H^E}{RT} - \frac{S^E}{R}, \tag{18}$$

where the superscript E denotes the excess thermodynamic quantities of mixing. The activity coefficient for a given composition approaches to a certain constant as the temperature increases, and if S^E/R is small, the non-ideality decreases with increasing temperature.

For the equilibrium between garnet and clinopyroxene, we have no reliable thermodynamic data for calculating ΔG, so that the effect of temperature on K and K' can only be obtained by comparing the values for natural eclogites of distinct mode of occurrences. To determine the sense of $\partial K/\partial T$, the values of K' of three distinct types of eclogites are compared in fig. 2, in which the atomic Fe^{2+}/Mg ratios of garnet are plotted against those of the associated clinopyroxene. The three representative types are California low temperature eclogites described by COLEMAN *et al.* (1965), Norwegian amphibolite facies eclogites (metabasites) as compiled by GREEN (in press), and the eclogite inclusions in kimberlite at the Robert Victor mine described by KUSHIRO and AOKI (1968). Geological observations suggest that these rocks were formed under more or less similar physical conditions as the associated non-eclogitic rocks, and that the temperature

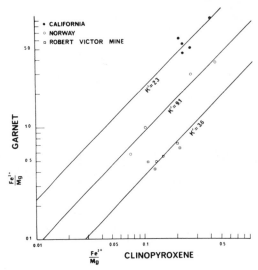

Fig. 2. The Fe–Mg distribution between garnet and clinopyroxene for three distinctive eclogite types: low temperature eclogite, amphibolite facies terrane eclogite, and eclogites inclusions in kimberlite.

emphasized as supporting the adequacy of a crude approximation that both these minerals are nearly ideal. We have, however, several reasons to doubt their ideality. The substitution of Mg by Ca in garnet may not be ideal, as their ionic radii are distinctly different from each other and the only common rock-forming mineral which forms continuously a solid solution between the Ca- and Mg-end members is garnet under high pressure (SOBOLEV et al., 1965, 1968).

Clinopyroxene solid solution may not be ideal for the diopside–Ca–Tschermakite series, as there is an excess volume of mixing. The orthopyroxene component necessarily results in the non-ideality of clinopyroxene as it breaks one of the conditions of the ideal solid solution, that the substituting cations should occupy the energetically equivalent site. The contents of Ca–Tschermakite and orthopyroxene components are, however, subordinate in ordinary eclogitic clinopyroxenes, and hence they may be ignored in crude treatments. The diopside–jadeite series has an excess volume of mixing (KUSHIRO, private communication) and an ordered phase is formed between them (CLARK and PAPIKE, 1968). The activity of the jadeite component in the diopside–jadeite series as calculated from the phase equilibrium diagram proposed by KUSHIRO (1965) shows a positive deviation from Raoult's law. The diopside–hedenbergite series was considered by MUELLER (1962) to be nearly ideal, on the basis that the Fe–Mg distribution between clinopyroxene and actinolite is explained as the equilibrium between two ideal solutions.

The existence of non-ideality in various solid solutions series of clinopyroxene, however, does not necessarily lead to a very pessimistic view, because in a ternary solid solution (A, B, C)X, in which the series (A, B)X is ideal, the (A, B, C)X series behaves as if ideal (A, B,C)X solid solution if the concentration of CX is sufficiently low (MATSUI and BANNO, 1968).

It follows that we need not worry too much about the effect of minor components. We have to be careful with the Fe/(Fe + Mg) ratio of minerals, the grossular content of garnet and the jadeite content of clinopyroxene as the possible major sources of the non-ideality.

In the absence of adequate thermodynamic data, the dependence of K' on the chemistry of rocks has to be examined using the data on natural eclogites. For this purpose, we need a set of isofacial eclogites covering a

of formation increases in the same order as mentioned above. The effect of pressure is negligible in comparing the low temperature and amphibolite facies terrane eclogites, as both are crustal rocks. The distinctly high pressure of the crystallization of the inclusions affects and increases K'. Therefore, it is concluded that the apparent Fe–Mg distribution coefficient between garnet and clinopyroxene decreases with increasing temperature.

2.4. Effect of chemistry on distribution coefficient

We have as yet no direct measurement of the activities in garnet and clinopyroxene solid solutions, so that the dependence of K' on the chemistry of the rocks is not clear.

Both garnet and clinopyroxene have only one structural site for Fe, Mg, Mn etc., provided that the clinopyroxene is low in orthopyroxene component, then the non-ideality due to the non-equivalence of lattice sites for the substituting cations (MATSUI and BANNO, 1965, 1968; BANNO and MATSUI, in press) need not be taken into account. The Fe–Mg distribution in pairs involving garnet or clinopyroxene is often approximated by that between the ideal solid solutions, and this can be

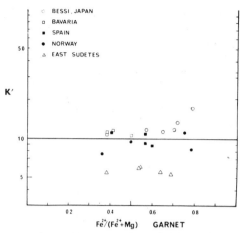

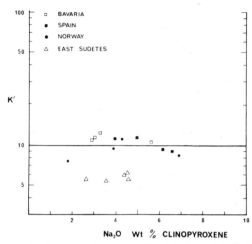

Fig. 3.   The relationship between K' and the Fe/(Fe + Mg) ratio of garnet. Data are from the amphibolite facies terrane eclogites, and the low temperature eclogites of the Bessi area, Japan.

Fig. 4.   The relationship between K' and the Na_2O content of clinopyroxene. The data are from the amphibolite facies terrane eclogites.

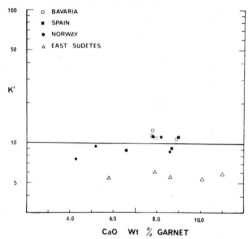

Fig. 5.   The relationship between K' and the CaO content of garnet. The data are from the amphibolite facies terrane eclogites.

wide range of chemical composition, but such data are as yet not available to the author. The best data now available are those by GREEN (in press) who described two eclogites from one and the same eclogite lense and obtained two K' values, 7.6 and 11.1, respectively, for the strictly isophysical rocks. Fig. 3 shows the plots of the values of K' of the amphibolite facies terrane eclogites against the Fe/(Fe + Mg) ratio of garnet. Included in the figure are the eclogites from Norway (ESKOLA, 1921; GREEN, in press; K' by Green for both data), East Sudetes (SMULIKOWSKI, 1967), Bavaria (YODER and TILLEY, 1962; BANNO, 1967a), and Spain (VOGEL, 1967), along with the eclogites of the Sanbagawa metamorphic terrane to be referred later. In the text, only ferrous iron is taken into consideration, and thus Fe refers always to ferrous iron. The dependence of K' on the Fe/(Fe + Mg) ratio of garnet cannot be denied for Green's Norwegian data, but it cannot be detected for other eclogites. Similar plots of K' against the Na_2O content of clinopyroxene and the CaO content of garnet as shown in figs. 4 and 5 are not conclusive as to the dependence of K' on these parameters, but rather suggest that K' is nearly independent of these parameters. The absence of clear compositional dependence of K' of the eclogites other than the Green's data might well be due to the fact that the data are from isolated

localities and the rocks may not have crystallized strictly under the same physical conditions.

Therefore, it may be concluded that the value of K' is constant within about 20% error or less. The 20% error in K' is not too unsatisfactory because K' is sensitive to temperature and the minerals of eclogites are often heterogeneous. The Fe content of the garnet of the bronzite eclogite described by MATSUI et al. (1966) varies by 20% as revealed by the electron probe study. Heterogeneity of garnet and clinopyroxene of eclogitic rocks have been described by PHILIPSBORN (1930), ESSENE and FYFE (1967), GREEN et al. (1968) and others. Further the determination of FeO by conventional wet chemical analysis is sometimes very difficult on eclogite minerals. A preliminary study of FeO analysis in diopside of the Higasiakaisi eclogite by Mössbauer spectroscopy suggests that the FeO content has been underestimated by the factor of 20% (MATSUI et al., in preparation).

The arguments on the apparent Fe–Mg distribution coefficient between garnet and clinopyroxene, K', are summarized as follows:

1) K' increases with increasing pressure, but the pressure effect is negligible in comparing the crustal eclogites;

2) K' decreases with increasing temperature;

3) There may be the compositional dependence of K', and care must be taken for the effect of the Fe–Mg substitution, the jadeite and orthopyroxene contents of clinopyroxene, and the grossular content of garnet as the possible sources of non-ideality, but in ordinary eclogites, the dependence of K' on chemistry is not large.

In the following, it is considered that the eclogites were formed under more or less similar physical conditions as the associated rocks. The eclogites associated with metamorphic rocks are metabasites belonging to the same or similar metamorphic facies as the enclosing rocks, and the eclogite inclusions are cumulates from basaltic or kimberlitic magmas or the solidified magma, which may have suffered metamorphism at lower temperature than the solidus temperature of the magmas. This assumption does not conflict with the observation that many eclogite lenses have tectonic contact with schists. A more detailed summary of the mode of occurrences of metamorphic eclogites were given elsewhere (BANNO, 1966).

3. Subdivision of and mutual relations among eclogite types

3.1. General statement

In the previous section, it was shown that there is a distinct difference of K' values between three distinctive types of eclogites.

It was pointed out by SMULIKOWSKI (1964), BANNO (1964) and COLEMAN et al. (1965) that the compositional range of garnet is different among some eclogite types, and that the pyrope content increases with increasing temperature. The difficulty of defining the limit of the pyrope content for each eclogite type was noticed, and the reason for this was explained by the consideration of the element distribution as given by COLEMAN et al. (1965) and BANNO and MATSUI (1965). Two methods of analysis of the distribution relations of elements have been proposed, one of which is based on the intersection of tie lines on a Ca–Mg–Fe ternary plot of garnet and clinopyroxene compositions and the other is based on the apparent distribution coefficient. Each of them has its own merit, but it is the author's opinion that the one which possesses the theoretical basis at least for an idealized system, i.e., for the equilibrium of ideal garnet and clinopyroxene solid solutions in our case, is preferable. Fig. 6 illustrates the tie lines for the

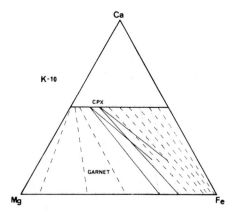

Fig. 6. The Ca–Mg–Fe plot of hypothetical garnet–clinopyroxene equilibrium. Clinopyroxene is ideal for Fe–Mg substitution, and garnet is ideal for Ca–Mg–Fe. The tie lines intersect with each other even if the CaO content of the system differs.

equilibrium of hypothetical ideal grossular-pyrope-almandine, and diopside-hedenbergite solid solutions with $K = 10$. It is seen in the figure that the tie lines intersect with each other even among the pairs being formed under the same physical conditions. This is due to the fact that the Fe–Mg distribution under consideration is not of a ternary but of a quaternary system. Therefore, in this paper, the phase equilibrium relations are discussed mainly in terms of the apparent distribution coefficient between coexisting garnet and clinopyroxene.

3.2. *Subdivision of low temperature eclogites*

The eclogites included in this type occur mainly in glaucophanitic metamorphic terranes. This type was called ophiolitic eclogite by SMULIKOWSKI (1964) and group "C" by COLEMAN *et al.* (1965). The term ophiolitic is rather vague, and it is not used here.

The low temperature eclogites have been described from the following localities:

1) Urals: Lawsonite and glaucophane are stably associated (CHESNOKOV, 1960).

2) Colombia: Boulders in a tertiary conglomerate. Lawsonite is considered to have been stable with garnet and clinopyroxene, but it is now changed to zoisite. One of the specimens described contains jadeite-rich pyroxene+quartz assemblage (GREEN *et al.*, 1968).

3) Guatemala: A boulder in a low temperature metamorphic terrane. Lawsonite is associated (MCBIRNEY *et al.*, 1967).

4) California: Glaucophane is stably associated. COLEMAN *et al.* (1965) considered that the associated lawsonite was formed later than the eclogite minerals, but ESSENE and FYFE (1967) considered it to be stably associated with them. The assemblage jadeite+quartz and aragonite are unstable. The chemical data are from COLEMAN *et al.* (1965).

5) New Caledonia: The detail of the mode of occurences is not known, but chemical data were described by COLEMAN *et al.* (1965).

6) Western Alps: According to BEARTH (1966), glaucophane schists are associated, but lawsonite is unstable. The chemical data are from VAN DER PLAS (1959) and BEARTH (1965).

7) Bessi area, Sanbagawa metamorphic belt, Japan: Eclogites occur as layers in an epidote amphibolite mass, and their occurrences appear to be restricted to the neighbourhood of the Higasiakaisi peridotite mass which contains Almklovdalen type eclogites in itself. The chemical data are listed in table 1. The eclogites enclosed in peridotite of glaucophanitic terrane are not included in this type.

The plots of K' values against the Fe/(Fe+Mg) ratios of garnet are shown in fig. 7. In the figure, three groups of the low temperature eclogites are distinguished: the

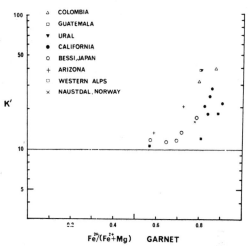

Fig. 7.   The relationship between K' and Fe/(Fe+Mg) ratio of garnet for low temperature eclogites.

first group is composed of the Ural, Colombia and Guatemala eclogites, the second group is the Californian ones, and the third group includes the Alpine and Japanese eclogites.

The clinopyroxenes of the low temperature eclogites are rich in Na_2O, as first suggested by BORG (1956). One of the Colombia pyroxenes contains 68% jadeite and 14% acmite. The effect of Na_2O content of the pyroxene on K' is examined in fig. 8, in which the values of K' are plotted against the Na_2O content of the pyroxene. There is no clear compositional dependence of K'. The CaO content of garnet generally increases with increasing FeO content, so that it is difficult to separate its effect from that of FeO. It is considered that the observed difference in K' among three groups of the low temperature eclogites reflects the difference in the physical conditions among them.

The Ural, Colombia and Guatemala eclogites are of the lawsonite–glaucophane schist facies, and the Alpine and Japanese ones are of the lawsonite-free glaucophane schist facies and of the albite–epidote amphibolite facies, respectively, and hence the temperature of metamorphism is higher in the latter than in the former. This view is in good agreement with the conclusion obtained from the comparison of K', that the former group has higher values than the latter. The metamorphic facies of the Californian eclogites are not

TABLE 1

The chemical compositions of garnet and clinopyroxene from the Bessi area
(The eclogites enclosed in the Higasiakasi peridotite mass are not included)

	1		2		3		4		5	
	Gar	Cpx	Gar	Cpx	Gar	Cpx	Gar	Cpx	Gar	Cpx
SiO_2	37.15	55.18	39.91	52.9						
TiO_2	0.26	0.38	0.12	0.6						
Al_2O_3	21.08	9.56	20.88	11.7						
Fe_2O_3	1.49	5.14	1.83	6.3		5.04		4.38		4.64
FeO	26.04	2.93	20.26	1.4	20.31*	2.56	24.65*	2.53	22.47*	2.09
MnO	0.87	0.01	0.65	0.02						
MgO	4.23	8.40	8.75	7.2	4.98	7.34	5.34	7.51	6.80	7.12
CaO	8.49	11.26	6.76	12.3						
Na_2O	<0.1	6.52	0.08	5.48						
K_2O	<0.1	<0.01	0.07	0.37						
H_2O^+	0.74	0.43	0.71	1.4						
H_2O^-	0.05	0.05	0.15	0.0						
P_2O_5	0.04	0.03	0.13							
Total	100.44	99.89	100.30	99.67						
K'	17.6		11.9		11.7		13.7		11.3	
$Fe^{2+}/(Fe^{2+}+Mg)$ in garnet		0.78		0.57		0.70		0.72		0.65

* Total Fe as FeO
1. SBD122 Boulder at the Hodono valley ⎫
2. SB56081103 Gongen shrine ⎬ The corrected analyses given by BANNO (1964).
3. ⎫
4. ⎬ Boulders at Hodono valley.
5. ⎭
Analysts. 1, 2: H. Haramura, 3: Y. Hirano, 4, 5: Y. Oki.

clear, as the presence of lawsonite is interpreted differently by different authors. Petrologically, they are certainly of lower temperature than the Japanese eclo-

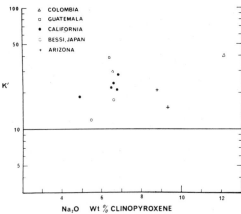

Fig. 8. The relationship between K' and the Na_2O content of clinopyroxene for low temperature eclogites.

gites, and of higher temperature than the Colombia ones, which accompany the jadeite + quartz assemblage. Judging from the element distribution, it is more plausible to consider that the Californian eclogites represent intermediate temperatures between the other two groups, rather than to consider that the difference in the K' values between the Californian and other eclogites is simply due to the poor accuracy of the apparent distribution coefficient to be used as a geological thermometer. Therefore, it is considered that K' can be used to subdivide the eclogite type.

We have three more occurrences of possible low temperature eclogites. BINNS (1967) described an eclogite from Naustdal, Norway, which forms a lens in albite-epidote amphibolite facies terrane. The K' value is 16, which lies within the range for the low temperature eclogite type, and it is distinctly higher than those of other Norwegian eclogites. Geological and phase equilibrium considerations seem to favor the view that the Naustdal eclogite is a metabasite formed under the same or similar physical conditions as the enclosing schists.

SPRY (1958) described an eclogite from Tasmania. The associated metamorphic rocks are of the albite-epidote amphibolite facies. The K' value of this eclogite is 8, which is typical of the amphibolite facies terrane eclogites. More mineralogical data are needed for this eclogite occurrence, but it is not unreasonable to consider that it belongs to an upper albite-epidote amphibolite facies, because the K' values of the eclogites of the albite-epidote amphibolite and amphibolite facies may overlap with each other, for reasons to be discussed later.

O'HARA and MERCY (1966) described the mineralogy of two eclogites found in breccia pipes in Arizona and New Mexico and considered them to be related to basalt or kimberlite activity. BANNO (1967b) mentioned that the extremely low CaO content of the Arizona eclogite garnet is due to the very high Na_2O content of the associated clinopyroxene, and hence it does not indicate physical conditions. He and GREEN et al. (1968) mentioned that the distribution coefficients of the Arizona eclogite are similar to those of glaucophanitic

metamorphic terranes. In fig. 8, it is seen that the high Na_2O content of the pyroxene cannot be the sole reason of high K'. GREEN et al. (1968) also mentioned the occurrence of lawsonite-bearing eclogitic rocks in the same pipes (WATSON, 1962), which indicates the presence of low temperature and high pressure metamorphic rocks within the crust of this area. In fig. 9, the distribution of rare earth elements between garnet and clinopyroxene of various eclogites is shown. The factors controlling the distribution of REE are not well known, but the figure shows that the distribution pattern for the Arizona eclogite is distinctly different from those of eclogites that occur as amphibolite facies metamorphics, and the inclusions in basalt. For these reasons, the author favors the view that the Arizona eclogites are crustal rocks belonging to the low temperature eclogite type.

3.3. Subdivision of amphibolite facies terrane eclogites

The eclogites of this type form lenticular masses in amphibolite facies terranes. They are typical eclogite,

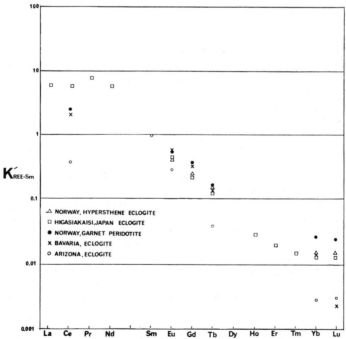

Fig. 9.   The rare earth distribution pattern for various eclogites. The apparent distribution coefficient is normalized to Sm.

historically and in phase equilibrium relations. The eclogites enclosed in peridotite of these terranes are excluded again. The known localities of this eclogite type include Fichtelgebirge in Bavaria, Basa Gneiss region in SW Norway, Glenelg in Scotland, Saualpe, Greenland, the Śnieżnik Mts in East Sudetes, Cabo Ortegal in Spain and others, and sufficient mineralogical data for analysing the distribution relations are obtainable from Bavaria, Norway, East Sudetes and Spain. The plots of K' values of these eclogites against the $Fe/(Fe + Mg)$ ratio of garnet, the Na_2O content of pyroxene, and the CaO content of garnet are shown in figs. 3, 4 and 5, by which it was concluded that the composition dependence of K' is small.

A distinct difference in K' is seen between the eclogites of the East Sudetes, and of the others, thereby suggesting that the Sudetes eclogites were formed at higher temperature than the others. The East Sudetes eclogites have the critical mineral assemblages of the eclogite facies, i.e., kyanite + garnet + omphacite, and quartz + garnet + omphacite, but the described specimens are from scattered localities so that it is not certain to what extent they are isophysical. In discussing the jadeite to Ca-Tschermakite ratio of the eclogitic clinopyroxenes, BANNO and YAMASAKI (in preparation) concluded that the Sudetes eclogites belong to the eclogite facies rather than to the granulite facies. They also suggested that the Sudetes eclogites represent higher temperatures or lower pressures than the typical amphibolite facies terrane ones. Therefore, the conclusions deduced from both K' and compositional range of clinopyroxene are in harmony and suggest that the East Sudetes eclogites are of higher temperature than the ordinary amphibolite facies terrane eclogites. According to SMULIKOWSKI (1967), the associated metamorphic rocks in this area are migmatite and amphibolite.

The subdivision of other eclogites of this type is difficult, but plots of K' values in figs. 3, 4 and 5, as well as the average K' values of each of terranes as shown in table 2 suggest that the Bavarian and Spanish eclogites are of slightly lower temperature than the Norwegian ones, though this is not conclusive.

3.4. Comparison of low temperature and amphibolite facies terrane eclogites

Inspection of figs. 3 and 7, in which the plots of K' values against the $Fe/(Fe + Mg)$ ratio of garnet for the

low temperature and amphibolite facies terrane eclogites are shown, reveals the fact that the minimum K' of the low temperature eclogites i.e., for Alpine and Japanese ones and the maximum value of K' of the others are rather similar to each other. The overlapping of the K' values between the albite-epidote amphibolite and amphibolite facies eclogites is, however, not against the validity of using K' as a geological thermometer. The pressure-temperature relationships between these two metamorphic facies are shown in fig. 10, which is a

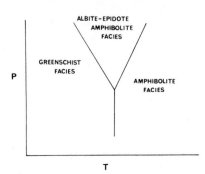

Fig. 10. A schematic diagram showing the pressure–temperature relations among the green schist, albite-epidote amphibolite and amphibolite facies.

qualitative quotation of the diagram given by MIYA-SHIRO (1961). Most low temperature eclogites occur in glaucophanitic metamorphic terranes, while those of the amphibolite facies terrane probably suffered the kyanite-sillimanite type metamorphism. The pressure of metamorphism is therefore higher in the former than in the latter, so that the temperature of an upper albite-epidote amphibolite facies could well be higher than that of a lower amphibolite facies. As the effect of pressure on K' is negligible in dealing with the crustal eclogites, the overlapping of K' values between these two facies is generally expected. This argument may give some suggestions as to the origin of the problematical Tasmanian eclogite.

3.5. Granulite facies eclogites and their relations to the amphibolite facies eclogites

The garnet-clinopyroxene assemblage is stable in the granulite facies, too. In Mg-rich rocks, this assemblage is stable only in SiO_2-undersaturated rocks, while in a Fe-rich environment, it is stable even in SiO_2-saturated

SHOHEI BANNO

TABLE 2

The values of K' for different garnet–clinopyroxene pairs

Type and locality	Number of samples	K'	References	Remarks
Amphibolite facies eclogites				
Bavaria	4	11.3	1, 2	
Cabo Ortegal, Spain	4	10.3	3	
Norway	4	9.1	4, 5	1
Glenelg, Scotland	2	8.0	1, 6	
Śnieżnik Mts, East Sudetes	5	5.6	7	
Granulite facies eclogites				
Varberg	8	6.3	8	2
Other areas	14	7.3	9, 10	
Eclogite inclusion in basalt				
Hawaii	3	2.5	1, 11	
Others	2	2.7	12, 13	
Eclogite inclusion in kimberlite				
Robert Victor mine	6	3.7	14	
Zagadochnaya	5	10.9	15	3
Basutoland	3	4.9	16	
Garnet peridotite inclusions in kimberlite				
	4	3.2	5, 16, 17	
Garnet peridotite and eclogite enclosed in peridotite				
Norway	8	5.7	4, 5, 17	4
Higasiakaisi, Japan	3	8.3	This paper	5
Czechoslovakia	3	2.6	18, 19	

Remarks

1) K' as calculated assuming that some Fe_2O_3 of garnet is actually FeO (GREEN, in press).
2) Probe analysis. Reciprocal of K_D given by SAXENA (1968).
3) CaO of garnet is very high.
4) K' neglecting Fe_2O_3.
5) Table 3 of this paper.

Key for references

1. YODER and TILLEY (1962)
3. VOGEL (1967)
5. GREEN (in press)
7. SMULIKOWSKI (1967)
9. SOBOLEV (1964)
11. KUNO (in press)
13. GIROD (1967)
15. SOBOLEV et al. (1968)
17. O'HARA and MERCY (1963)
19. MIKHAIROV and ROVSHA (1966)

2. BANNO (1967)
4. ESKOLA (1921)
6. O'HARA (1960)
8. SAXENA (1968)
10. WARNAARS (1967)
12. DICKEY (1968)
14. KUSHIRO and AOKI (1968)
16. NIXON et al. (1963)
18. FIALA (1966)

ones (O'HARA, 1960; BANNO, 1966; GREEN and RINGWOOD, 1967). SOBOLEV (1964) and ESSENE and FYFE (1967) have shown that the average of K' of the granulite facies eclogites is lower than that of the amphibolite facies terrane eclogites. The plots of K' values of granulite facies eclogites against the Fe/(Fe + Mg) ratio of associated garnet are shown in fig. 11 in which the plots for the Norwegian and East Sudetes eclogites are also shown for comparison. The data on the granulite facies rocks are from SOBOLEV (1964) and WARNAARS (1967). In the figure, it is seen that the K' values of the granulite facies eclogites are lower than those of Nor-

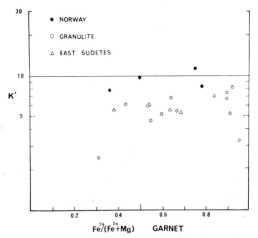

Fig. 11. The relationship between K' and $Fe/(Fe+Mg)$ ratios of the granulite facies eclogites. The data for Norwegian and East Sudetes eclogites are also shown for comparison.

wegian ones, which are the representatives of the amphibolite facies eclogites, but the values of the East Sudetes eclogites are similar to the granulite facies ones. The fact that the granulite facies eclogites generally have lower K' values than the ordinary amphibolite facies eclogites is in harmony with petrological considerations, that the former represents higher temperature than the latter. It has been suggested that the contents of jadeite and Ca-Tschermakite differ between the granulite and amphibolite facies eclogites (WHITE, 1964), but the compositional dependence of K' is negligible for the amphibolite facies eclogites. The agreement of the conclusions by mineral facial considerations and that by K' further supports the adequacy of using K' as a geological thermometer.

The East Sudetes eclogites, however, have similar K' values to the average granulites. They are associated with migmatite, probably of the amphibolite facies, and this is contradictory to the previous view that the granulite facies represent higher temperatures than the amphibolite facies, so that a tentative explanation to this controversy is given below.

The amphibolite-granulite facies boundary is usually defined by the appearance of orthopyroxene in basic metamorphic rocks. The mineral assemblages of pelitic metamorphic rocks are similar to each other between an upper amphibolite and a lower granulite or the

hornblende granulite facies. Under high pressures, and within the stability field of the quartz + garnet + clinopyroxene assemblage, the appearance of orthopyroxene in basic metamorphic rocks has to take place at much higher temperatures than under lower pressures, because the following reaction proceeds from the left- to right-hand sides with increasing pressure:

$$CaAl_2Si_2O_8 + 4MgSiO_3 =$$

$$\begin{array}{cc} \text{anorthite} & \text{enstatite} \\ 100.7 & 4 \times 31.5 \end{array}$$

$$= Mg_3Al_2Si_3O_{12} + CaMgSi_2O_6 + SiO_2 \quad (19)$$

$$\begin{array}{ccc} \text{pyrope} & \text{diopside} & \text{quartz} \\ 113.3 & 66.1 & 23.7 \end{array}$$

$$\text{(formula volume cm}^3\text{)}$$

$$\Delta V = -23.6 \text{ cm}^3$$

where the reaction is expressed in terms of Fe-free end members.

A schematic representation of the pressure-temperature relationships between the amphibolite and granulite facies is shown in fig. 12. In field A of the figure,

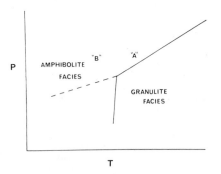

Fig. 12. A schematic diagram showing the pressure-temperature relations between the amphibolite and granulite facies. The boundary between field "A" of the amphibolite facies and the granulite facies is the univariant curve of reaction (19).

eclogite is stable, and orthopyroxene does not coexist with basic plagioclase. The mineral assemblages of pelitic metamorphic rocks are similar between the amphibolite, and the hornblende granulite facies, i.e., the lower granulite facies, and thus they are similar between the fields "A" and "B". The mineral assemblages of basic metamorphic rocks are also similar between the fields A and B, as the appearance of orthopyroxene is depressed by reaction (19) and then the amphibolites between these fields are not easily distinguished. Therefore, the rocks metamorphosed in the field "A" may well be classified as belonging to the amphibolite facies. If we consider that the metamorphic rocks of the East Sudetes belong to field "A", the overlapping of K' values between the Sudetes and granulite facies eclogites is reasonably explained. It is emphasized, however, here that the above discussions are qualitative, and do not propose nor assume that the eclogites are formed under the same water fugacity as the associated metamorphics.

3.6. *Eclogite inclusions in basalt and in kimberlite*

The phase equilibrium relations of the eclogite inclusions in basalt were discussed by LOVERING and WHITE (lecture in this symposium), so they are only briefly mentioned here. The most striking feature of the eclogite inclusions in basalt is the fact that their K' values are lower than those of the eclogite inclusions in kimberlite. The eclogite inclusions in basalt are considered to have been formed at the uppermost mantle or the lowermost crust, probably at the former, while some eclogite inclusions in kimberlite contain diamond (WILLIAMS, 1932; SOBOLEV *et al.*, 1965), and are undoubtedly formed under very high pressures. The shallow origin of the inclusions in basalt is also supported by the presence of plagioclase in some specimens (LOVERING and WHITE, 1964). The melting point of eclogite under dry conditions increases with increasing pressure, so that the inclusions in basalt are considered to have crystallized at lower temperatures than those in kimberlite, but this is contradictory with the fact that the K' values are generally lower in the former than in the latter. This controversy may be avoided by assuming that the magmas are far from being dry, and the water in magmas affected greatly the crystallization temperature of the inclusions, or by assuming that the differences in the chemistry of the host magma as well as that of eclogite itself were so large that the temperature of crystallization has little connection with the depth of the formation and then the inclusions in basalt actually represent higher temperature than the others. Another and more plausible explanation to this controversy is, however, that the pressure affected K'. It was shown in a previous discussion that 20 kb difference in pressure at 1000 °C affects K' by the factor of

1.5, the ratio of K' values between two eclogite types. Judging from the experimental determination of eclogite crystallization as reported by GREEN and RINGWOOD (1967), 20 kb of pressure difference between two types of eclogite inclusions is not an unreasonable estimate.

The dependence of K' on the chemistry of rocks was considered generally to decrease with increasing temperature. However, the ideality of the equilibrium between garnet and clinopyroxene at high temperatures may be affected by the high concentration of orthopyroxene component in clinopyroxene, which necessarily results in the increase of non-ideality. Therefore, a detailed comparison of high temperature eclogites being under consideration in terms of K' is not so reliable as that for the metamorphic eclogites, and only a rough comparison has meaning. It appears that both these inclusions crystallized under more or less similar temperatures to each other.

3.7. *Eclogite enclosed in peridotite and garnet peridotite*

The discussion based on the apparent distribution coefficient of Fe and Mg between garnet and clinopyroxene can be applied in discussing the genesis of eclogite, or pyrope-diopside rock, enclosed in peridotite, and of garnet peridotite. The average K' values of garnet peridotites, which occur as inclusions in kimberlite and as the intrusive mass in metamorphic terranes, as well as those of eclogites enclosed in peridotite, i.e., Almklovdalen type eclogites, are shown in table 2. The data on the garnet peridotite inclusions in kimberlite are rather scanty, but available data show that the K' values are higher for the intrusives than for the inclusions, suggesting that the latter, probably mantle materials, are of higher temperatures than the crustal intrusives. This conclusion is in harmony with that given by O'HARA and MERCY (1963), who discussed this problem on the basis of the compositional range of pyroxenes.

Eclogites consisting of pyrope and diopside are often found to be enclosed in peridotite and garnet peridotite intruded into metamorphic terranes. The known localities include SW Norway, Bohemian massif, Spain and Higasiakaisi, Japan; the analyses for the last are listed in table 3.

The genesis of this eclogite type has not been discussed in detail. Their bulk chemical compositions are not

TABLE 3

Chemical compositions of eclogite, Higasiakaisi, Japan.
(Fe-rich eclogites are not included in this paper, as the consanguinity between Mg-rich and Fe-rich eclogites are in doubt).

	1		2		3	
	Gar	Cpx	Gar	Cpx	Gar	Cpx
SiO_2	39.68	51.86	40.46	53.38	41.20	54.73
TiO_2	0.98	0.20	0.23	0.10	0.13	0.08
Al_2O_3	22.05	1.02	22.72	1.43	22.50	0.84
Fe_2O_3	4.10	0.99	0.49	0.66	0.98	0.95
FeO	11.33	1.73	13.12	2.01	14.03	1.66
MnO	1.57	0.02	0.62	0.07	0.37	0.03
MgO	13.63	17.09	16.28	17.71	14.60	17.19
CaO	6.75	25.24	6.42	24.57	6.33	24.41
Na_2O	0.19	0.08	<0.02	0.37	tr.	0.52
K_2O	0.04	0.06	<0.02	<0.02	tr.	tr.
H_2O^+	0.28	2.13	0.10	0.05	0.00	0.10
H_2O^-			0.00	0.00	0.00	0.00
P_2O_5			0.05	0.12		
Cr_2O_3			0.12	0.08	0.30	0.059
Total	100.60	100.42	100.61	100.55	100.44	100.569

1. MIYASHIRO and SEKI (1958).
2. BANNO and YOSHINO (1965).
3. BANNO and YOSHINO (1965) for garnet. Diopside: new analysis.
Analyst: H. Haramura.

strictly basaltic, and very low Na_2O and high CaO contents are noteworthy. The normative mineral assemblage is olivine + plagioclase + diopside (+ hypersthene), with colour index more than 50. The hypothesis that they crystallized from basaltic magma within the crust is rejected from the experimental data on the basalt to eclogite transformation, and high pressure solidus phases of basaltic magmas. The possible mechanism for their genesis includes the metamorphism of olivine eucrite in peridotite, or the intrusion of upper mantle materials in essentially solid state.

The K' values of these eclogites are also shown in table 2. They range from 6 to 9, and are within the range of the amphibolite facies terrane eclogites. The plots of the K' values against the $Fe/(Fe+Mg)$ ratio of garnet of this eclogite type are shown in fig. 13, together with the plots of the K' values of the associated garnet peridotite, and of metamorphic eclogites in the same area. The Norwegian and Japanese data are used in the diagram, as they are the only areas which have mineralogical data being capable of being examined by phase equilibrium principles. Inspection of the figure reveals the fact that the values of K' are different between Norwegian and Japanese occurrences, and this difference is accompanied by the difference in the K'

values of associated metamorphic eclogites. In Norway, where both eclogite types have comparatively low K' values, the host peridotite is intruded into the amphibolite facies terrane, while in Japan, where prevailing metamorphic rocks are of the albite-epidote amphibolite facies, both eclogite types have comparatively high K' values.

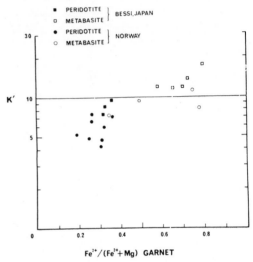

Fig. 13. The relationship between K' and $Fe/(Fe+Mg)$ ratio of garnet for eclogites in peridotite, and garnet peridotite. For comparison, K' for the metamorphic eclogites of the same area is shown.

In this connection, it is worthy of note that in Spain, pyrope-diopside eclogites occur in peridotite intruded into an amphibolite facies terrane, where metamorphic eclogite occurs, while spinel peridotite, which represents lower pressure than the garnet peridotite, occurs in the granulite facies terrane (VOGEL, 1967; WARNAARS, 1967). However, this parallelism between the peridotite types and the metamorphic facies of the enclosing rocks is not necessary, as we have garnet peridotite intrusives in granulite terrane in Bohemia (DUDEK and KOPECKY, 1966).

It is as yet not clear if the similarity of K' between the eclogites enclosed in peridotite and in neighbouring metamorphic rocks represents the compositional dependence of K' among isofacial rocks, or if it is due to the difference in temperatures between them. If the temperatures are considered different, the eclogites in peridotite represent higher temperatures than those in associated metamorphic rocks, and this is geologically reasonable.

It appears that the association of two eclogite types in the same area, and the parallelism of K' between them are not fortuitous, and some genetical connection may exist between them. A possible interpretation of this is that the temperature-pressure conditions of a geosyncline and the underlying upper mantle during regional metamorphism are intimately related with each other, i.e., the upper mantle underneath the glaucophanitic metamorphic terrane was of lower temperature than that underneath the amphibolite facies metamorphic terranes. Another possible interpretation is the metamorphic recrystallization of olivine eucrite and dry basalt under more or less similar depth of a geosyncline, at slightly different temperatures.

4. Distribution of trace elements

BANNO and MATSUI (1965) demonstrated that the apparent Mn–Fe distribution coefficient, K'_{Mn-Fe} between garnet and clinopyroxene varies systematically with K'_{Mg-Fe}. Table 4 shows the examples of the Mn distribution in some representative eclogite types. The values of K' are normalized to Mg, instead of Fe of the previous paper, because the determination of FeO and Fe_2O_3 of eclogite minerals are sometimes questioned because of extreme difficulty in dissolving them in acid. The problem in dealing with the trace element distribution lies in the unnecessarily rounded figures of the Mn concentration of clinopyroxene. For low temperature eclogites, in which a remarkable preferential concentration of Mn into garnet takes place, the published MnO contents of clinopyroxenes are far from being satisfactory.

It was shown in the previous section that K'_{Fe-Mg} is nearly composition independent. If this is accepted it is expected that K'_{Mn-Mg}, K'_{Co-Mg} and K'_{Ni-Mg} are composition independent as well (cf. section 2.4).

A detailed consideration of the trace element distribution in eclogite minerals will be discussed elsewhere.

5. Concluding remarks

In the foregoing discussions it was shown that K'_{Fe-Mg} can be used as a geological thermometer to distinguish various eclogite types, and to subdivide some of the eclogite types. The temperature of crystallization of eclogites is considered to increase in the following order:

1) Ural, Colombia and Guatemala; ⎫ Low
2) California;   ⎬ temperature
3) Alps and Japan;   ⎭ eclogites
4) Bavaria and Spain;
5) Norway;
6) East Sudetes and granulites. The former represents higher pressures than the latter at more or less similar temperatures;
7) Inclusions in basalt and in kimberlite.

TABLE 4

Apparent Mn–Mg distribution coefficient for various eclogites

Locality	K'_{Fe-Mg}	K'_{Mn-Mg}	References
California	28	35	21
Spain	10.3	12	3
Norway	9.1 (7.6)*	8.5*	5, 22
East Sudetes	5.6	14	7, 23
		10.5**	
Inclusions in kimberlite	3.6	3.8	14
Inclusions in basalt	2.6	3.7	11, 12
Inclusions in peridotite			
Higasiakaisi	8.3	15	24
			This paper
Norway	5.7	8	5, 17, 25

Numbers in the last column are the source of data as given in table 2. The additional references are:
21. COLEMAN et al. (1965)   22. MATSUI et al. (1966)
23. BAKUN-CZUBAROW (1968)   24. MATSUI (unpublished)
25. GREEN (unpublished)
* K' by (22).   ** K' by (23).

Geological and petrological implication of determining the relative temperature of eclogite crystallization will be discussed elsewhere, as it needs detailed consideration on the relationships between the mineralogy of eclogites and associated rocks. The coexistence of eclogite and basic schists in many metamorphic terranes cannot be explained, if we accept the classical assumption that the chemical potential of water during regional metamorphism was the same or similar within a mineral zone of particular metamorphic terranes. It is also worthy to mention that most of the low temperature eclogites occur in glaucophanitic metamorphic terranes, and no eclogite has been described from low pressure regional metamorphic terranes. This may require the revision of GREEN and RINGWOOD's (1967) suggestion that the eclogite mineral assemblage may be stable in basic rocks even at very low pressures, if the temperature is low.

Acknowledgements

The author is deeply indebted to Drs. D. H. GREEN and Y. MATSUI for helpful discussions on this problem and the permission to quote their unpublished data. He is also indebted to Prof. H. KUNO, Drs. I. KUSHIRO, Y. OKI, and H. HIGUCHI for permission to use their unpublished data, and to Prof. M. YAMASAKI for the critical reading of the manuscript. The grant from the Australian National University for the visiting appointment from 1965–1967 is also acknowledged.

References

BAKUN-CZUBAROW, N. (1968) Arch. Mineral 28, 243.
BANNO, S. (1964) J. Fac. Sci. Univ. Tokyo, Sec. II 15, 203.
BANNO, S. (1966) Japan. J. Geol. Geography, Trans. 37, 105.
BANNO, S. (1967a) Neues Jahrb. Mineral. Monatsh., 116.
BANNO, S. (1967b) Earth Planet. Sci. Letters 2, 249.
BANNO, S. (in press) Korzhinskii volume, Moscow.
BANNO, S. and Y. MATSUI (1965) Proc. Japan Acad. 41, 716.
BANNO, S. and G. YOSHINO (1965) Upper Mantle Symposium, New Dehli, 150.
BEARTH, P. (1965) Schweiz. Mineral. Petrog. Mitt. 45, 179.
BEARTH, P. (1966) Schweiz. Mineral. Petrog. Mitt. 46, 13.
BINNS, R. A. (1967) J. Petrol. 8, 349.
BORG, I. W. (1956) Bull. Geol. Soc. Am. 67, 1563.
CHESNOKOV, B. V. (1960) Intern. Geol. Rev. 2, 936.
CLARK, J. R. and J. J. PAPIKE (1968) Am. Mineralogist 53, 840.
CLARK, S. P., J. F. SCHAIRER and J. DE NEUFVILLE (1962) Yearbook 1961–62 Geophys. Lab. Carnegie Inst. Wash., 59.
COLEMAN, R. G., D. E. LEE, L. B. BEATTY and W. W. BRANNOCK (1965) Bull. Geol. Soc. Am. 76, 483.
DICKEY, J. S., JR. (1968) Am. Mineralogist 53, 1304.
DUDEK, A. and P. KOPECKY (1966) Kristallinikum 4, 7.
ESKOLA, P. (1921) Oslo Vidensk. Skr. Mat.-Naturw. Kl. No. 8.
ESSENE, E. and W. S. FYFE (1967) Contrib. Miner. Petrol. 15, 1.
FIALA, J. (1966) Kristallinikum 4, 31.
FORBES, R. B. (1965) J. Geophys. Res. 70, 1515.
GIROD, M. (1967) Bull. Soc. Franç. Minéral. Crist. 90, 202.
GOLDSCHMIDT, V. M. (1922) Naturw. 42, 1.
GREEN, D. H. (1966) Earth Planet. Sci. Letters 1, 414.
GREEN, D. H. (in press) Korzhinskii volume, Moscow.
GREEN, D. H. and A. E. RINGWOOD (1967) Geochim. Cosmochim. Acta 31, 767.

GREEN, D. H., A. J. P. LOCKWOOD and E. C. KISS (1968) Am. Mineralogist 53, 1320.
GRUBENMANN, U. (1904) Die Kristallinschiefer I (Borntraeger, Berlin).
KUSHIRO, I. (1965) Yearbook 64–65 Geophys. Lab. Carnegie Inst. Wash., 112.
KUSHIRO, I. and K. AOKI (1968) Am. Mineralogist 53, 1347.
KUNO, H. Geol. Soc. Am., Mem., in press.
LOVERING, J. F. and A. J. R. WHITE (1964) J. Petrol. 5, 195.
MATSUI, Y. and S. BANNO (1965) Proc. Japan Acad. 41, 461.
MATSUI, Y. and S. BANNO (1968) Kagaku-no-Ryoiki 156, 256 (in Japanese).
MATSUI, Y., S. BANNO and I. HERNES, Norsk Ged. Tidsskr 46, 364.
MATSUI, Y. and Y. SYONO (1968) Geochem. J. 2, 51.
MATSUI, Y., Y. SYONO, S. AKIMOTO and K. KITAYAMA (1968) Geochem. J. 2, 61.
MIYASHIRO, A. (1961) J. Petrol. 2, 277.
MIYASHIRO, A. and Y. SEKI (1958) Japan. J. Geol. Geography, Trans. 29, 199.
MUELLER, R. F. (1962) Geochim. Cosmochim. Acta 26, 581.
NIXON, P. H., O. VON KNORRING and J. M. ROOKE (1963) Am. Mineralogist 48, 1090.
O'HARA, M. J. (1960) Geol. Mag. 97, 145.
O'HARA, M. J. and E. L. P. MERCY (1963) Trans. Roy. Soc. Edinburgh 65, 251.
O'HARA, M. J. and E. L. P. MERCY (1966) Am. Mineralogist 51, 336.
PHILIPSBORN, H. VON (1930) Chem. Erde 5, 200.
RINGWOOD, A. E. and D. H. GREEN (1966) Tectonophysics 3, 383.
ROBIE, R. A., P. M. BETHKE, M. S. TOULMIN and J. L. EDWARDS (1966) in: S. P. Clark, ed., Handbook of physical constants (Geol. Soc. Am., Mem. No. 97) p. 30.
SAXENA, S. K. (1968) Am. Mineralogist 53, 2018.
SMULIKOWSKI, K. (1964) Bull. Acad. Pol. Sci. 12, 27.
SMULIKOWSKI, K. (1967) Geol. Sudetica 3, 7.
SOBOLEV, N. V. (1964) Paragenetic types of garnet (in Russian, Nauka, Moscow) p. 218.
SOBOLEV, N. V. and I. K. KUZNETZOVA (1966) Dokl. Acad. Nauk SSSR 167, 1365 (in Russian).
SOBOLEV, N. V., N. I. ZYUZIN and I. K. KUZNETZOVA (1966) Dokl. Acad. Nauk SSSR 167, 902 (in Russian).
SOBOLEV, N. V., I. K. KUZNETZOVA and N. I. ZYUZIN (1968) J. Petrol. 9, 253.
SPRY, A. H. (1963) Mineral. Mag. 33, 589.
VAN DER PLAS, L. (1959) Leidse Geol. Mededel. 24, 415.
VOGEL, D. E. (1967) Leidse Geol. Mededel. 40, 121.
WARNAARS, F. W. (1967) Ph.D. Thesis (Univ. Leiden).
WATSON (1960) Bull. Geol. Soc. Am. 71, 2082.
WHITE, A. J. R. (1964) Am. Mineralogist 49, 883.
WILLIAMS, A. F. (1932) The genesis of diamond (Benn, London) p. 636.
YODER, H. S., JR. and C. E. TILLEY (1962) J. Petrol. 3, 342.

Erratum

On p. 254, the entry "Watson (1960)" should read "Watson, K. D. (1960)."

11

Reprinted from *Verh. Geol. Bundesanstalt, Vienna,* **2,** 257–265 (June 1971)

On the Significance of the Eclogite Facies in Alpine Metamorphism

N. FRY

W. S. FYFE

Zusammenfassung

Tieftemperatur-Eklogite sind charakteristische Gesteine, die bei Krustenabsenkungen unter Kontinentalrändern gebildet werden. Der Grad ihrer Bildung könnte die Geschwindigkeit der Krustenbewegung widerspiegeln. Eklogite sind Metamorphoseprodukte basaltischer Gesteine unter lokalen Bedingungen, wenn der Druck der fluiden Phase viel geringer ist, als der Belastungsdruck. Sie bilden sich entweder unter Reaktionen in festem Zustand oder an einem Intergranularfilm mit einem sehr niedrigen Wasser-Partial-Druck, möglicherweise unter Beteiligung gesättigter Lösungen im System NaCl-KCl-H₂O. Bei Eklogitbildung aus submarinen Laven ist zunächst eine Dehydrierung des basaltischen Materials notwendig. Wahrscheinlich erfolgt dies unter Bildung von Hydraten wie Lawsonit in der Glaukophan-Lawsonit-Fazies. Fortschreitende Metamorphose der Naß-Trocken-Vergesellschaftung muß zu einer Verringerung der Menge des Eklogits führen und zu einem ständigen Anwachsen des Verhältnisses von Amphibolit zu Eklogit.

Abstract

Low temperature eclogites are characteristic rocks formed during plate descent beneath continental margins. The degree of their formation may reflect velocity of plate motion. Eclogites are the product of metamorphism of basaltic rocks under local conditions where fluid pressures are much less than load pressures. Their formation involves either solid state reactions or reations in fluid films with a very low partial pressure of water, possibly saturated solutions in the system NaCl-KCl-H₂O. Where submarine lavas are involved a preliminary step to eclogite formation must involve dehydration of the basaltic material, probably by formation of hydrates such as lawsonite in the glaucophane-lawsonite facies. Progressive metamorphism of the wet-dry assemblage must lead to reduction in the amount of eclogite and a steady increase in the ratio of amphibolite to eclogite.

Introduction

EsKOLA (1921) introduced the eclogite facies for rocks of basic composition and a dominant mineralogy of pyroxene (jadeite-diopside) and garnet (almandine-pyrope). GOLDSCHMIDT (1922) suggested that eclogite could be a major constituent of upper mantle of basaltic composition, and recently, particularly

---

Address: N. FRY und W. S. FYFE, Department of Geology, The University, Manchester M 13 9 PL, England.

257

with the development of ideas of plate tectonics, there has been much interest in the question whether crustal eclogites represent tectonically emplaced mantle or metamorphosed crustal material. Alternatively DOBRETSOV & SOBOLEV (1970) suggest that eclogites are intruded at depth during metamorphism. BEARTH (1959) has clearly demonstrated that some Alpine eclogites are metamorphosed submarine effusives and new oxygen isotope data supports the extension of this thesis to other occurrences. Workers on eclogites have been impressed by their association with rocks of glaucophane-lawsonite, greenschist and amphibolite facies, and COLEMAN et al. (1965) have suggested that the eclogite facies should be abolished.

As with many problems in modern petrology, a clearer understanding of the physico-chemical conditions involved would remove some of the difficulties in understanding transitions between eclogites and rocks of different facies. Up to the present time much experimental data has involved basalt-eclogite-amphibolite transitions near $1000°$ C. or above, so that the extrapolation required to cover the metamorphic conditions of recent mobile belts, with temperatures of perhaps $300—500°$ C., is very great. The state of confusion which has arisen is clearly illustrated by the different conclusions on the eclogite facies in two of the most recent texts of metamorphic petrology (WINKLER, 1967; TURNER, 1968).

YODER & TILLEY (1962) have shown that at temperatures near melting and with excess water, eclogites are unstable relative to amphibolites or basalts at pressures below 10 kb. GREEN & RINGWOOD (1967) studied the basalt-eclogite transition at high temperatures ($> 1000°$ C.) and proposed a phase boundary as in fig. 1. This suggests that, given favourable kinetic factors, eclogites could form in zeolite to greenschist facies conditions. ITO & KENNEDY (1970) have suggested a different boundary which projects into the high-pressure, low-temperature conditions of the glaucophane-lawsonite facies. In fact, their boundary is very close to that for the reaction:

$$albite \rightarrow jadeite + quartz.$$

Present field observations tend to support the ITO-KENNEDY boundary.

The other major difficulty involves the influence of water pressure on eclogite stability. This was discussed by ESSENE & FYFE (1967) and later by FRY & FYFE (1969) who showed by thermodynamic arguments based on YODER & TILLEY's data that if water is in excess amphibolite is more stable than eclogite under any reasonable crustal conditions. This is confirmed by experimental work of ESSENE, HENSON & GREEN (1970). FRY & FYFE show that an eclogite facies, in the sense of a P-T field in which eclogite is the most stable assemblage for basic rocks of normal water content, must lie in upper mantle conditions. Otherwise, eclogites represent dry, low P_{H_2O} metamorphism inside the P-T fieds of other accepted facies. Equilibrium apparent between eclogite minerals and assemblages of, say, glaucophane-schist or amphibolite facies is real, the mineral proportions reflecting the water content at any given site. In this context an additional point should be stressed. For most rock compositions amphibole bearing assemblages are stable to lower water pressures than other hydrous assemblages. Thus, not only are assemblages in equilibrium with eclogite likely to be amphibole bearing, but reactions which would produce eclogite by dehydration such as:

258

peridotite + amphibolite → serpentine + eclogite

are not possible. This is strictly correct only when both assemblages are at the same pressure and temperature, that is, under isobaric isothermal conditions.

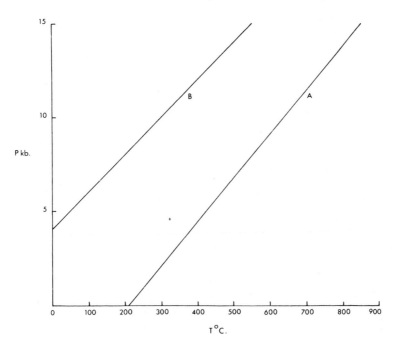

Fig. 1: Field boundary between eclogite and basalt:
A as proposed by GREEN and RINGWOOD (1967),
B as proposed by ITO and KENNEDY (1970).

Low and Medium Grade Metamorphism

Several authors have pointed out changes in eclogite mineral composition with changing grade (see BANNO, 1970). At lower temperatures garnets tend increasingly to almandine composition, so much so that it is doubtful whether at very low grade a bimineralic omphacite-garnet rock can have a basaltic bulk composition.

This hypothesis is difficult to verify because most minerals known to occur with omphacite and garnet at low grade are hydrates. So it may be that where, for example, eclogite and glaucophane schist coexist, their combined bulk composition is basaltic though neither this eclogite nor this glaucophane schist has itself a basaltic composition. Two points about such eclogites should be noted. Firstly, they do not disprove the possibility of an anhydrous, bimineralic eclogite of basaltic composition at the same P-T conditions. Secondly, the presence of hydrous minerals in a rock is not itself evidence of excess water or high P_{H_2O}. On the contrary, where eclogite is present the hydrous phases are

259

likely to be at their low P_{H_2O} limit of stability, thus providing a buffered P_{H_2O} of very low value, with P_{H_2O} increasing with temperature.

At low grades important restrictions are placed on the precursors of eclogite. BRYHNI et al. (1970) have suggested that in certain high grade metamorphic regimes, partial melting may give conditions appropriate to eclogite formation: $P_{load} = P_{fluid} = P_{melt} \gg P_{H_2O}$. At low to moderate grades such a process is not applicable. In combination with the statement above, that dehydration to eclogite by another solid assemblage is impossible, this leads to an important conclusion. It appears that at low grades eclogite can only form from another anhydrous assemblage, which must usually mean an igneous basaltic assemblage.

In general, hydrous assemblages contain more water at lower metamorphic grades, and with prograde reactions release water. Where eclogite is present the following situation exists:

1. For a given overall water content, the lower the grade, the greater is the bulkratio of eclogitic to hydrous material. For a given proportion of eclogite, the lower the grade, the greater is the overall water content. Thus basalts which metamorphosed at low grades may give a mixture of eclogite and a hydrous assemblage, may at higher grades give no eclogite at all.

2. Prograde reactions releasing water from hydrates enable some eclogite to be hydrated. So, with increasing grade, progressively less unaltered eclogite will remain. It should be stressed that partial alteration of eclogite to amphibolite or greenschist is not indicative of retrograde metamorphism. On the contrary, it could be typical of prograde metamorphism.

Mechanism of Basalt Dehydration

As common amphibolites of basaltic composition contain about 2% water, to totally or partially form an eclogite assemblage requires that the water content of a rock is of the order of 1% or less. There is probably little difficulty in achieving such low water contents in gabbros if these are buried and suffer only slight fracturing during loading or if such rocks are introduced during deep metamorphism. The transformation of gabbroic rocks to eclogites in a glaucophane schist facies terrain necessitates very small diffusion coefficients for water; diffusion coefficients of the same order of magnitude as those in solids themselves.

It is now certain that some of the precursors to eclogites are submarine volcanics. In many cases these rocks have spilitic affinities and the work of VOGEL & GARLICK (1970) has clearly shown that some eclogites have suffered extensive oxygen isotope exchange with light meteoric waters. It is difficult to see how submarine effusives could be buried with less than 2% water. If they are to form eclogites, they must first be dehydrated and then rendered impermeable to water.

Submarine volcanics, particularly pillow lavas, will have a significant glass content and will be mixed with muds, carbonates and cherts and other fine-grained reactive materials. There are thus four low strength materials capable of facile deformation. Excess water will be expelled by compaction and the

260

residue will be taken up by the formation of hydrates. Such hydrates will tend to nucleate and grow most easily at reactive sites such as provided by silica gel, fine muds and glass. Thus a glassy pillow rim may be the site of low grade mineral formation while the more crystalline core may remain intact. At the same time as water moves to reactive sites the elements may also tend to move, particularly if residual fluids are saline. Potassium, an element often very low in eclogites, may migrate to clay mineral sites to generate mica.

It is perhaps reasonable to suggest that if a large fraction of dry basalt is to survive, burial and deformation should be rapid. Otherwise it is difficult to see how, in a submarine environment, it is possible to avoid a degree of hydration sufficient to preclude eclogite formation. Motion of the same order as that observed on some continental margins (10 cms./year) would be suitable. In such circumstances rocks will remain cold and hence reaction rates will be even more sensitive to fine rock structures. Thus hydrate formation at sites at a distance from dry basalt is a preliminary process in the formation of eclogite from basic lava complexes.

Lawsonite Precursors

In rocks of the Täschtal (Zermatt) a conspicuous feature of some meta-lavas is the presence of white pseudomorphs which in shape resemble lawsonite porphyroblasts in glaucophane schists (Figs. 2 and 3). These pseudomorphs are apparently widely distributed (BEARTH, pers. comm.) within the Inner Zone of the Pennine Alps (BEARTH, 1966) in which Täschtal lies. In Täschtal they consist of aggregates or intergrowths of zoisite or clinozoisite, usually with small amounts of mica, but more rarely kyanite. Chemically, replacement of lawsonite by an epidote mineral must release alumina and water, which could lead to mica metasomatism, by alkali fixation at these sites.

Lawsonite in eclogite has been recorded, together with zoisite alteration or pseudomorph formation, by WATSON (see WYLLIE, 1967, p. 263) and DIXON (1969). GREEN et al. (1968) have interpreted pseudomorphs of clinozoisite and paragonite as after lawsonite in a Colombian eclogite from which fresh lawsonite is absent. The authors believe that such an interpretation applies to pseudomorph-bearing eclogites and glaucophane-garnet rocks of the Pennine Alps.

This, together with the association of eclogites and glaucophane-lawsonite rocks in California, Urals, etc. leads the authors to believe that frequently the formation of eclogite from basalt occurs at temperatures of perhaps 300 to 400° C. where pressures are appropriate to glaucophane-lawsonite formation (5—10 kb.). With increasing temperature these hydrates break down and as progressive metamorphism takes place the amount of eclogite is reduced in the heterogeneous rock assemblage.

As an example of the role played by hydrates in eclogitic metamorphism lawsonite is superb. It is distinctive in composition, and is recognizeable both as a fresh mineral and when pseudomorphed because of its porphyroblastic habit. Its stability field of pressure and temperature is not only known, but

261

Fig. 2. Lawsonite porphyroblasts in chloritic glaucophane schist from the Franciscan of California (× 1).

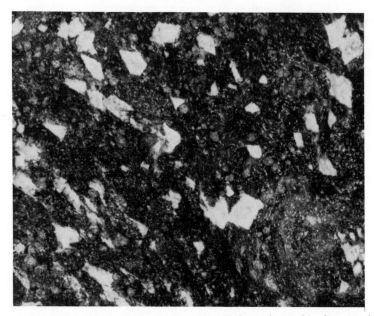

Fig. 3. Pseudomorphs of clinozoisite in garnetiferous glaucophane schist from Täschtal, Switzerland (× 1).

262

restricted to unusual conditions (hence its use as a determinative mineral of the glaucophane-lawsonite facies). Lawsonite is also one of the most water rich metamorphic minerals. Its breakdown in eclogite therefore leads to alteration of a large quantity of eclogitic minerals, perhaps in some cases destroying them completely by prograde metamorphism in a chemically closed system.

Mechanism of Eclogite Formation

From what has been said above, the simplest method of forming eclogites is to transform dry basalts or gabbros to eclogites in the solid state. If solid state reactions of this type occur, we would expect that the textures observed would be those resulting from the shortest possible diffusion paths. A plagioclase-pyroxene rock would be converted to a garnet-omphacite rock by the shortest possible exchange distances. We would anticipate that there would be a simple textural relationship between parent material and product. This is often seen, particularly in coronitic gabbros (see, for example, MILLER, 1970). It is of note that GREEN & RINGWOOD (1967) and ITO & KENNEDY (1970) were able to study the solid state reaction over a considerable temperature range (800—1300° C.) in times which vary by less than a power of ten. This suggests quite a low activation energy for the process and given geological times reaction as low as 300° C. might be possible. There are no data yet available to substantiate this point.

But there are other eclogites in the Alps whose textures have all the appearance of normal hydrothermal growth, for example large radiating groups of omphacite or mineral layering and non-random development of minerals. Such textures if not widespread could also be inherited from special igneous textures but they are often very normal in terms of wet metamorphic crystallization. Is it possible that fluids with low P_{H_2O} are present; fluids whose partial pressure of water is lower than that in equilibrium with amphiboles appropriate to the P-T regime?

From the submarine volcanic nature of some rocks transformed to eclogite, mentioned above, it is not unreasonable to suppose that salts in the NaCl-KCl system are likely to be trapped in such rocks. The presence of saline materials in fluid inclusions in rocks appears to be very common. At very high load pressures most hydrate forming reactions will not reduce P_{H_2O} below that of saturated salt solutions. Hence, such solutions may remain in the rocks, possibly as films on grain boundaries. They could be critical in catalyzing the basalt-eclogite reaction, as well as creating some of the observed textures.

The critical question to answer is whether or not the partial pressure of water in equilibrium with solid NaCl + KCl is lower than that in equilibrium with eclogite-amphibolite assemblages. Thermodynamic calculations we have performed using the data for tremolite stability indicate that it is a "touch and go" situation. At temperatures above 600° C. saline fluids are likely, below they might exist. All the data needed to perform an exact calculation are not available but the results are interesting. There could be a very strong case for searching for fluid inclusions in the minerals of low temperature eclogites. It is perhaps worth noting that the presence of such salt-solution films could be ubiquitous in

263

metamorphic rocks, particularly in rocks containing phases such as micas, chlorites, etc. Such films could have profound effects on the mechanical properties of metamorphic rocks.

The Significance of the Eclogite Facies

There is little doubt that eclogites can form from dry rocks within the P-T fields where wet rocks of otherwise identical composition would be glaucophane-lawsonite, greenschist or amphibolite rocks. We already recognize the necessity to distinguish silica deficient or excess assemblages. Clearly we must do the same with water, possibly recognizing degrees of deficiency.

As for the eclogite facies, it may still be too soon to restrict this facies to the upper mantle. If facies are to be treated as mutually exclusive P-T fields, each defined by stability of a different mineral assemblage in rocks with excess water, then clearly the eclogite facies must be consigned to mantle conditions. In this sense it does not exist as a metamorphic facies.

However, within certain types of metamorphic terrain eclogites have wide-spread distribution, though each outcrop may be small. Eclogites are easily recognized and (pace GREEN & RINGWOOD, 1967) represent, these authors believe, elevated pressures. They may be sufficiently indicative of certain plate tectonic phenomena, and resulting types of metamorphic terrains, to justify retaining the eclogite facies as a high pressure facies defined by dry rocks. Presence or absence of eclogite may be recording the velocity of plate motion. But in the end nomenclature is of little significance as long as it serves and does not restrict our developing understanding of metamorphism.

A k n o w l e d g e m e n t. The writers wish to express their deep gratitude to Prof. P. BEARTH for introducing them to the Zermatt area and freely providing from his store of knowledge, while they naturally accept responsibility for any misconceptions in this paper.

References

BANNO, S. (1970): Classification of eclogites in terms of physical conditions of their origin: in, Phase transformations and the Earth's interior. Special volume of Physics of the Earth and Planetary Interiors. p. 405—421. North Holland publishing Comp. Amsterdam.

BEARTH, P. (1959): Über Eklogite, glaukophanschiefer und metamorphe Pillow-laven. SMPM, 39, 267—286, Zürich.

BEARTH, P. (1966): Zur mineralfaziellen Stellung der Glaukophangesteine der Westalpen. SMPM, 46, 13—23, Zürich.

BRYHNI, I., GREEN, D. H., HEIER, K. S., & FYFE, W. S. (1970): On the occurrence of eclogite in Western Norway. Contr. Mineral. and Petrol. 26, 12—19, Springer-Verlag.

COLEMAN, R. G., LEE, D. E., BEATTY, L. B., & BRANNOCK, W. W. (1965): Eclogites and eclogites: their differences and similarities. Bull. Geol. Soc. Am., 76, 483—508, New York.

DIXON, J. E. (1969): Unpublished Ph. D. thesis, Cambridge University.

DOBRETSOV, N. L., & SOBOLEV, N. V. (1970): Eclogites from metamorphic complexes of the U. S. S. R.: in, Phase transformations and the Earth's interior. Special volume of Physics of the Earth and Planetary Interiors. p. 462—470.

264

ESKOLA, P. (1921): On the eclogites of Norway. Videnskap. Skrift. I Mat.-Naturv. Kl., 1. Bd., Nr. 8, 1—118, Kristiania.

ESSENE, E. J., & FYFE, W. S. (1967): Omphacite in Californian metamorphic rocks. Contr. Mineral. and Petrol. *15*, 1—23, Springer-Verlag.

ESSENE, E. J., HENSEN, B. J., & GREEN, D. H. (1970): Experimental study of amphibolite and eclogite stability: in, Phase transformations and the Earth's interior. Special volume of Physics of the Earth and Planetary Interiors. p. 378—384.

FRY, N., & FYFE, W. S. (1969): Eclogites and water pressure. Contr. Mineral. and Petrol. *24*, 1—6, Springer-Verlag.

GOLDSCHMIDT, V. M. (1922): Über die Massenverteilung im Erdinnern, verglichen mit der Struktur gewisser Meteoriten. Naturwissenschaften, 42, S. 1—3, Berlin.

GREEN, D. H., & RINGWOOD, A. E. (1967): An experimental investigation of the gabbro to eclogite transformation and its petrological implications. Geochim. Cosmochim. Acta, *31*, 767—833, London.

GREEN, D. H., LOCKWOOD, J. P., & KISS, E. (1968): Eclogite and almandine-jadeite-quartz rock from the Guajiva Peninsula, Columbia, South America. Amer. Min. *53*, 1320—1335, Menasha.

ITO, K., & KENNEDY, G. C. (1970): The basalt-eclogite transition and the structure of the upper mantle. Abstracts Geol. Soc. Am. Annual meeting, 586, Boulder.

MILLER, C. (1970): Petrology of some eclogites and metagrabbros of the Ötztal Alps, Tirol, Austria. Contr. Mineral. and Petrol., *28*, 42—56, Springer-Verlag.

TURNER, F. J. (1968): Metamorphic Petrology. McGraw-Hill Book Co., New York etc.

VOGEL, D. E., & GARLICK, G. D. (1970): Oxygen-isotope ratios in metamorphic eclogites. Contr. Mineral. and Petrol., *28*, 183—191, Springer-Verlag.

WINKLER, H. G. F. (1967): Die Genese der Metamorphen Gesteine. 2nd. Edit. Springer, Berlin.

WYLLIE, P. J. (1967): Ultramafic and related rocks. John Wiley and Sons Inc., New York etc.

YODER, H. S., & TILLEY, C. E. (1962): Origin of basalt magmas. J. Petrology, *3*, 346—521, Oxford.

265

Editor's Comments on Papers 12 Through 17

Ophiolite Suite

The association of radiolarian cherts and/or deep-water carbonate strata with spilitic pillow lavas, gabbros, and serpentinized peridotites has been recognized for many years in the Alps (see, e.g., Steinmann, 1905, 1927), but this assemblage of rock types is characteristic of many mountain belts (see, e.g., Benson, 1926; Hess, 1938, 1939, 1955; Peive, 1973). The significance of these units as fragments of oceanic crust + mantle underpinnings, surmounted by deep-sea sediments, has been stressed by Hess (1962, 1965) and Dewey and Bird (1970, 1971), among others. The "alpine"-type peridotites in general have been emplaced in a relatively cold condition, hence do not produce high-temperature aureoles in the surrounding country rocks (for an exception, however, see Williams and Smyth, 1973). Because of the refractory nature of ophiolites and alpine peridotites relative to sialic crust, cold emplacement argues persuasively for a solid-state tectonic process rather than magmatic intrusion.

One of the first clear postulations that alpine-type peridotites represent mantle material was that by de Roever (Paper 12). In this publication de Roever described the association of the ultramafic masses with amphibolites and lower-grade schists and suggested that the latter represent the overlying basaltic oceanic crust dragged upward with the mantle peridotite as it was tectonically emplaced in the orogenic belt. The fabric of these serpentinized peridotites also suggests a solid-state, or metamorphic, origin.

Papers 13 and 14, by Thayer and by Medaris and Dott, respectively, provide mineralogic, petrologic, chemical, and tectonic data on the material interpreted as constituting the basal portions of the oceanic crust (layer 3) and the mantle material directly beneath the Mohorovicic discontinuity. As undepleted asthenosphere (e.g., pyrolite) rises toward the surface beneath a mid-oceanic ridge, greater degrees of partial melting of the relatively fusible constituents occur, providing a gravity-driven phase separation of basaltic melt which congeals as overburden resting upon the underlying, now depleted, lithospheric residue. The separation is probably imperfect, and to the extent that slow spreading rates allow for the development and persistence of magma chambers within the basal oceanic crustal levels along the spreading center, crystal fractionation will result in peridotitic crystal cumulates as well as gabbros, diabases, and alkalic, silicic differentiates. Thus we expect a mid-oceanic ridge to contain a broad range of ultramafic + mafic rock types, somewhat interlayered, with minor amounts of intermediate and felsic units as well. High-temperature, high-pressure peridotite parageneses such as those described by Medaris and Dott are inherited from material brought up from deep within the upper mantle, whereas the high-temperature, low-pressure gabbroic associations discussed by Thayer reflect near-surface equilibria. Relatively high-temperature, low-pressure metamorphism of the oceanic crust would be expected to be initiated during the short time interval in which the material lay adjacent to the spreading center.

Examples of ophiolite suites from California and Cyprus have been described in detail by Bailey, Blake, and Jones (Paper 15) and by Moores and Vine (Paper 16), respectively. In the occurrences in the California Coast Ranges, there is a general downward succession from radiolarian chert to keratophyric and basaltic pillow lavas, to diabase, gabbro, and leucogabbro, and finally to serpentinized harzburgite. Not all units are present in each section, and nowhere does the stratigraphic column approach the minimal thickness of 3–5 km that is appropriate to oceanic crust(layers 2 and 3); therefore, tectonic thinning due to faulting and/or penecontemporaneous erosion must be called upon to account for the relatively thin column of ophiolitic rocks. The Troodos Massif in Cyprus is thicker and consists of a somewhat similar sequence of units. However, a sheeted dike complex, such as probably characterizes certain spreading centers, is very fully developed, in contrast to the dike-poor California localities. Many of these hypabyssal intrusions are "half-dikes," that is, they possess chilled contacts only on one side; this relationship suggests that successive magma injections intruded along the centers of preexisting dikes, effectively splitting them in two, probably in response to continuing divergent plate motion at the ridge crest.

The last ophiolite paper chosen is that by Coleman (Paper 17). This work dealt with the mode of emplacement of alpine-type peridotite complexes. In it, Coleman called attention to the fact that ophiolites are situated at present or inferred ancient convergent lithospheric plate boundaries, where they seem to have been tectonically transported into juxtaposition with continental crust. They are recognized as representing the thrusting of oceanic crust over a continental crust-capped lithospheric slab. This phenomenon appears to take place in at least two different ways: (1) a continental crust-capped plate begins to descend beneath an oceanic crust-capped stable slab (subduction), then buoyantly returns toward the surface; or (2) a thin wedge of a subducting, oceanic crust-capped plate decouples from the leading edge of the downward curving plate and is thrust over the stable, continental crust-capped slab (this process was termed "obduction" by Coleman, and was referred to as "flake tectonics" by Oxburgh, 1972). In either case, relatively weakly metamorphosed ophiolite is thrust into a continental margin terrane that has experienced a long and complicated petrotectonic history.

265

References

Benson, W. N. (1926) The tectonic conditions accompanying the intrusion of basic and ultrabasic igneous rocks: *U.S. Nat. Acad. Sci. Mem.,* **1,** 1–90.

Dewey, J. F., and Bird, J. M. (1970) Mountain belts and the new global tectonics: *Jour. Geophys. Res.,* **75,** 2625–2647.

Dewey, J. F., and Bird, J. M. (1971) Origin and emplacement of the ophiolite suite: Appalachian ophiolites in Newfoundland: *Jour. Geophys. Res.,* **76,** 3179–3206.

Hess, H. H. (1938) A primary peridotite magma: *Amer. Jour. Sci.,* 4th ser., **35,** 321–344.

Hess, H. H. (1939) Island arcs, gravity anomalies and serpentinite intrusions: *Internat. Geol. Congr., Moscow,* Rept. 17, **2,** 263–283.

Hess, H. H. (1955) Serpentines, orogeny and epeirogeny: p. 391–408, in *Geol. Soc. America Spec. Paper* **62,** Poldevaart, ed., 762 p.

Hess H. H. (1962) History of ocean basins: p. 599–620, in *Geol. Soc. America, Buddington Vol.,* A. E. Engel, H. L. James, and B. F. Leonard, eds., 660 p.

Hess, H. H. (1965) Mid-oceanic ridges and tectonics of the sea-floor: p. 317–332, in *Submarine Geology and Geophysics,* W. F. Whittard and R. Bradshaw, eds., Butterworth, London, 464 p.

Oxburgh, E. R. (1972) Flake tectonics and continental collision: *Nature,* **239,** 202–204.

Peive, A. V., ed. (1973) *Folded Belts and Ophiolites,* Parts I, II, and III: Acad. Sci. USSR, Moscow, 532 p.

Steinmann, G. (1905) Geologische Beobachtungen in den Alpen, 2, Die Scherdt'sche Überfaltungstheorie und die geologische Bedeutung der Tiefseeabsetze und der ophiolithischen massargesteine: *Ber. Nat. Ges. Reiburg,* i, **16,** 44–65.

Steinmann, G. (1927) Die ophiolitischen Zonen in den Mediterranean Kettengebirgen: *Rept. Int. Geol. Congr., 14th, Madrid,* **2,** 638–667.

Williams, H., and Smyth, W. R. (1973) Metamorphic aureoles beneath ophiolite suites and alpine peridotites: tectonic implications with West Newfoundland examples: *Amer. Jour. Sci.,* **273,** 594–621.

12

Reprinted from *Geologischen Rundschau*, **46**(1), 137–146 (1957)

SIND DIE ALPINOTYPEN PERIDOTITMASSEN VIELLEICHT TEKTONISCH VERFRACHTETE BRUCHSTÜCKE DER PERIDOTITSCHALE?

Von W. P. DE ROEVER, *Leiden*

Zusammenfassung

Das Fehlen intensiver Mylonitisierung in vielen Olivingesteinen und die Häufigkeit ungestörter Maschenstrukturen in Serpentinen weisen darauf hin, daß Fließen in kristallinem Zustand und durch Wasserdampf erleichterte Gleitung von Kristallen keine allgemeingültige Erklärung für die Platznahme der alpinotypen Ultramafite darbieten.

Die hier erörterte Deutung der alpinotypen Ultramafitmassen als tektonisch verfrachtete Bruchstücke der Peridotitschale würde eine einfache Erklärung ergeben für mehrere Phänomene, die sonst schwer zu verstehen sind; es ist aber noch zu überprüfen, ob sie tektonisch möglich ist. Die vorliegende Arbeit hat also einen vorläufigen Charakter und beabsichtigt nur, diese Deutung zur Erwägung zu empfehlen.

Nach der hier erörterten Deutung erfolgte die Erstarrung der alpinotypen Peridotite vielleicht schon während einer sehr frühen Periode der Bildung des Erdmantels. Die in Peridotiten vieler Fundorte aufgefundene Gefügeregelung wäre aber möglicherweise auf nachträgliche Metamorphose in der Peridotitschale zurückzuführen: die alpinotypen Peridotite würden also vielleicht als metamorphe Gesteine zu deuten sein. Der Kissenlavenvulkanismus der STEINMANN-Trinität wäre eine normale Begleiterscheinung der tektonischen Platznahme der Peridotitmassen; das betreffende Magma würde der Peridotitschale entstammen und während der Bewegung der oberen Teile dieser Schale hochgepreßt worden sein.

Die hier erörterte Deutung der alpinotypen Ultramafitmassen würde eine einfache Erklärung darbieten für das augenscheinliche Fehlen sicherer Kontaktmetamorphose, für den immer wieder zu beobachtenden tektonischen Charakter der Ultramafitkontakte, für das übliche Fehlen von Gängen von alpinotypen Ultramafiten in nichtultramafischen Gesteinen und für die häufige Vergesellschaftung der Ultramafite mit offenbar der Unterlage der Geosynklinale entstammenden Massen von Amphiboliten und untergeordneten anderen kristallinen Schiefern. Solche amphibolitreichen Gesteinsmassen wären vielleicht als mitgerissene Bruchstücke der Bedeckung der Peridotitschale zu deuten; es wäre also möglich, daß derartige Amphibolite usw. wenigstens einen Teil der sog. Basaltschale aufbauen.

Es ist für die hier erörterte Hypothese entscheidend, ob sie tektonisch möglich ist. Es ist schon längst bekannt, daß Überschiebungsbahnen oft von Peridotiten oder Serpentinen markiert sind. Nach den Angaben in der bezüglichen Literatur wäre es aber nicht möglich, für alle alpinotypen Ultramafitmassen eine weite Verfrachtung durch Überschiebung, Auspressung und Abgleitung anzunehmen. Die hier erörterte Hypothese hat weiter den Nachteil, daß sie für mehrere Gebiete einen sehr großen Zusammenschub postuliert, größer als man gewöhnlich annimmt. Es gibt jedoch noch soviel Unstimmigkeit über die Tektonik vieler peridotit- und serpentinführender Gebiete, daß es berechtigt erscheint, die hier erörterte Hypothese bei künftigen Ultramafitstudien zur Erwägung zu empfehlen.

BOWEN und TUTTLE (1949) kamen in ihrer Arbeit über das System MgO—SiO_2—H_2O zu der Schlußfolgerung: "There seems no escape from the conclusion that ultramafics can be intruded only in the solid state." Die „Intrusion" würde durch Fließen in kristallinem Zustand und durch die infolge Anwesenheit von Wasserdampf erleichterte Gleitung von Kristallen ermöglicht werden. Als allgemeingültige Erklärung der Platznahme alpinotyper Peridotite und Serpentine wurde diese Auffassung aber z. B. von HIESSLEITNER (1951/52), WILKINSON (1953) und H. H. HESS (1955) abgelehnt, u. a. wegen des Auftretens einer inneren Zonengliederung in vielen Ultramafitmassen, wegen des Fehlens intensiver Mylonitisierung und Kataklase in vielen Olivingesteinen und wegen der Häufigkeit ungestörter Maschenstrukturen in Serpentinen; alle diese drei Autoren befürworteten die Existenz eines ultramafischen Magmas.

Wichtige Argumente gegen die Entstehung der alpinotypen Ultramafite aus basaltischem Magma sind u. a. die Seltenheit von Übergangsgesteinen, wie z. B. feldspathaltigen Peridotiten, und die Anwesenheit von großen Peridotitmassen in Gebieten, in denen die zu erwartenden viel größeren Massen von komplementären Differentiaten fehlen [1]) (HESS 1938).

Da auch eine metasomatische Entstehungsweise dieser Gesteine m. E. als allgemeingültige Erklärung scheitert, würde man HESS (1955) beipflichten können: "So the problem remains unsolved. Some vital piece of evidence is still missing."

Obgleich sowohl HIESSLEITNER als HESS die Meinung geäußert haben, daß tektonische Verfrachtung bei der Platznahme von ultramafischen Gesteinskörpern eine gewisse Rolle spielen kann, scheint jedoch die Möglichkeit noch nicht genügend berücksichtigt zu sein, daß die alpinotypen Peridotitmassen bloß tektonisch verfrachtete Bruchstücke der Peridotitschale seien [2]), deren Platznahme nichts mit magmatischer Intrusion zu schaffen hat; daß also ein sogenannter alpiner Peridotit vielleicht schon in einer sehr frühen, möglicherweise präarchäischen Periode der Bildung des Erdmantels aus einem ultramafischen Urmagma erstarrte und während der alpinen Orogenese nur tektonisch, in vollständig festem Zustand, verfrachtet würde [3]). Solch eine Verfrachtung würde nur selten mit wichtigen Internbewegungen verknüpft sein, z. B. unter Bildung von Peridotitmyloniten oder Serpentinschiefern, aber meistens starre, von Verschiebungen begrenzte Blöcke und Linsen betreffen: die Gesteine der obersten Teile der Peridotitschale wären also nach dieser Hypothese z. T. in unveränderter Form der direkten Beobachtung zugänglich.

Die in Peridotiten vieler Fundorte aufgefundene Gefügeregelung wäre möglicherweise auf Regionalmetamorphose in großen Tiefen oder auf sonstige Bewegungen in der Peridotitschale zurückzuführen; dasselbe würde für den Lagen- und Schichtenbau der alpinotypen Peridotite zu-

---

1) Man würde also in alpinotypen Gebieten nur ausnahmsweise aus basaltischem Magma gebildete peridotitische Kristallisationsdifferentiate auffinden können; von solchen Peridotiten, die u. a. gabbroide Abkühlungsrinden, deutliche Kontaktmetamorphose des Nebengesteins und allmähliche Übergänge in größere Gabbromassen zeigen würden, wird in dieser Arbeit nicht weiter die Rede sein.

2) Siehe auch Symposium Metamorphism and Orogenesis, General discussion, Geologie en Mijnbouw Bd. 18, S. 145 (1956).

3) KOSSMAT (1937, S. 323) rechnete mit der Möglichkeit, daß die Ultramafite „als eingeschobene, zwar hoch erhitzte, nicht aber flüssige Masse den Ergüssen der Kissenlaven und verwandter Magmen folgten" und „daß die Peridotite nach Art des Dunits zur Zeit ihrer Intrusion großenteils in kristallinem Zustand sind, wahrscheinlich vergesellschaftet mit einer komplexen Schmelzlösung, deren Pressung die Bewegung entlang von Translationsflächen ermöglicht".

138

treffen können. Die alpinotypen Peridotite würden also vielleicht als metamorphe Gesteine zu deuten sein.

Die Bildung von Serpentin fand nach HESS (1954, 1955) möglicherweise schon z. T. in der Peridotitschale statt.

Die hier erörterte Deutung der alpinotypen Ultramafitmassen als tektonisch verfrachtete Bruchstücke der Peridotitschale würde eine einfache Erklärung darbieten für mehrere Phänomene, die sonst schwer zu verstehen sind, nämlich für das augenscheinliche Fehlen sicherer Kontaktmetamorphose, für das übliche Fehlen von Gängen von alpinotypen Ultramafiten in nichtultramafischen Gesteinen, für den immer wieder zu beobachtenden tektonischen Charakter der Ultramafitkontakte und für die häufige Vergesellschaftung der Ultramafite mit offenbar der Unterlage der Geosynklinale entstammenden Massen von Amphiboliten und untergeordneten anderen kristallinen Schiefern; es wird davon in den folgenden Seiten noch näher die Rede sein. Es ist aber noch zu überprüfen, ob diese Deutung der alpinotypen Ultramafitmassen tektonisch möglich ist. Die vorliegende Arbeit kann also nur einen vorläufigen Charakter haben und beabsichtigt nur, diese Deutung zur Erwägung zu empfehlen.

Die hier zur Diskussion gestellte Hypothese stützt sich z. T. auf Beobachtungen derselben Art wie diejenigen, die den peridotiterfahrenen HIESSLEITNER veranlaßten, die üblichen Altersdeutungen der Balkanperidotite abzulehnen und ein größeres Alter dieser Gesteine zu befürworten; sie ist gewissermaßen eine Erweiterung seiner Auffassung, daß alle Berührungen der Balkan- und Alpen-Peridotite mit mesozoischen und jüngeren Gesteinsserien — soweit dieselben nicht transgressiv über Serpentin greifen — ausgesprochen tektonische Kontakte sind (1951/52, S. 586). HIESSLEITNER schloß auf ein paläozoisches Alter der Balkanperidotite, da diese Gesteine s. E. eine weniger hochgradige Metamorphose zeigen als das vorpaläozoische Hochkristallin desselben Gebietes. Er berücksichtigte jedoch nur die Neubildung von späten Mineralien; die Peridotite selbst wurden als magmatische Intrusivgesteine gedeutet. Es wäre aber m. E. durchaus nicht unmöglich, daß diese Gesteine — wegen der anderswo in vielen Peridotiten aufgefundenen Gefügeregelung des Olivins — als metamorphe Gesteine, und zwar als hochgradig metamorphe Gesteine zu deuten sind; ein größeres Alter der Balkanperidotite würde deshalb gar nicht ausgeschlossen sein.

Die hier erörterte Hypothese muß im Licht der rezenten Untersuchungsergebnisse gesehen werden, daß peridotitisches Material schon in einer Tiefe von wenigen Kilometern unter dem Ozeanboden auftritt (siehe auch die Bemerkungen TILLEYS in Proc. Geol. Soc. London **1521**, 1955, S. 49). HESS (1955) hat schon darauf hingewiesen, daß diese Verhältnisse eine Platznahme der Peridotite in festem Zustand viel annehmbarer machen. Wenn man z. B. das von RAITT, FISHER und MASON (1955) aus seismischen und magnetischen Daten abgeleitete Strukturprofil des Tongatrogs (S. 253, Abb. 9) besieht, würde man sich vorstellen können, daß die Verhältnisse in vielen Gebieten späterer intensiver Orogenese vor der Platznahme der ultramafischen Gesteine ungefähr so aussahen, mit seich-

139

ten Peridotitteufen, speziell an einer Seite des Haupttrogs. Ein Zusammenschub eines solchen asymmetrischen Gebildes würde leicht zu einer einseitigen Überschiebung oder Durchspießung der obersten Teile der Peridotitschale über bzw. durch die aufliegenden Gesteine führen können. Nach weiterer Entwicklung des Faltengebirges würden dann speziell die von der Seite des Troges herstammenden höheren Decken reich an Peridotiten sein, was in der Tat zutrifft: nach HESS (1955, S. 395) kommen die größten ultramafischen Gesteinsmassen in den obersten Teilen der ursprünglichen Gebirgsstrukturen vor. Während der Entwicklung des Faltengebirges würde es auch leicht zu einer Aufteilung der Peridotitmassen in Linsen, also zu einer Unterbrechung ihrer Verbindung mit der eigentlichen Peridotitschale, kommen können.

Der Kissenlavenvulkanismus der STEINMANN-Trinität wäre eine normale Begleiterscheinung des hier erörterten Prozesses; das betreffende Magma würde der Peridotitschale entstammen und während der Bewegung der oberen Teile dieser Schale hochgepreßt worden sein.

Die hier zur Diskussion gestellte Deutung der alpinotypen Ultramafitmassen als tektonisch verfrachtete Bruchstücke der Peridotitschale würde eine einfache Erklärung darbieten für das übliche Fehlen von Gängen von alpinotypen Ultramafiten in nichtultramafischen Gesteinen (siehe unten) sowie für das augenscheinliche Fehlen sicherer Kontaktmetamorphose. Was das letztere anbelangt, würde man doch bei einer magmatischen Platznahme der alpinotypen Peridotite um die Ultramafitkörper herum eine deutliche, sogar sehr hochgradige Kontaktmetamorphose erwarten; solch eine deutliche Kontaktmetamorphose fehlt jedoch. HIESSLEITNER z. B. sprach schon von den „oft mehr unsicher als sicher deutbaren Anzeichen von Kontaktmetamorphose". Es scheint durchaus nicht unmöglich, daß alle bisher als Kontaktphänomene der alpinotypen Ultramafite geschilderten Mineralneubildungen einer anderen Ursache zugeschrieben werden müssen, z. B. Reaktion zwischen Ultramafit und Nebengestein während späterer Metamorphose oder Aufstieg von Lösungen den Kontakten entlang (siehe auch HIESSLEITNER 1951/52, S. 437—446).

BAILEY und McCALLIEN (1953) zogen aus dem Fehlen von Kontaktmetamorphose in direkt auf Serpentin lagernden Schichten die Schlußfolgerung, daß die Serpentinmassen keine Intrusionen, sondern Effusionen darstellen. Diese Beobachtung braucht aber nicht auf einen effusiven Ursprung der Serpentinmassen hinzuweisen, sondern ist auch mit der hier erörterten Hypothese im Einklang.

Bei einer magmatischen Platznahme der alpinotypen Ultramafite würde man neben tektonischen Kontakten auch häufig ungestörte Intrusivkontakte beobachten können. Die Berührungen der alpinotypen Ultramafite mit anderen Gesteinen — soweit dieselben nicht transgressiv überlagernde Schichten sind oder jüngere Intrusionen darstellen — scheinen aber immer wieder tektonischen Charakters zu sein. HIESSLEITNER z. B. hat betont, daß „mit völlig ungestörten Peridotitkontakten auf Balkan und in Kleinasien nirgends zu rechnen ist, überall gibt es Anzeichen tektonischer Beanspruchung und Bewegung, besonders an den Rändern" (1951/52, S. 421).

140

Auch dieser immer wieder zu beobachtende tektonische Charakter der alpinotypen Ultramafitkontakte ist mit Hilfe der hier erörterten Hypothese in einfacher Weise zu erklären.

Ein Argument für die Möglichkeit der hier gegebenen Deutung ist weiter der häufigen Vergesellschaftung der alpinotypen Ultramafite mit offenbar der Unterlage der Geosynklinale entstammenden Amphiboliten und anderen kristallinen Schiefern zu entnehmen. Auch in den obersten Teilen der ursprünglichen Gebirgsstrukturen, in einer sonst wenig oder gar nicht metamorphen Umgebung, werden die Ultramafite sehr oft von offenbar weit verfrachteten, größeren oder kleineren Massen von Amphiboliten und untergeordneten anderen kristallinen Schiefern begleitet [4]. Diese Häufigkeit der Vergesellschaftung von alpinotypen Ultramafiten mit offenbar der Unterlage der Geosynklinale entstammenden Gesteinen ist m. E. ein wichtiges Argument für die Möglichkeit einer Deutung der Ultramafitmassen als tektonisch verfrachtete Bruchstücke der wahrscheinlich nur wenig tieferen Peridotitschale. Die Häufigkeit dieser Vergesellschaftung wäre übrigens bei einer magmatischen Platznahme der Ultramafite nur verständlich, wenn die Intrusionen ursprünglich sehr oft kristalline Schiefer als Nebengestein hätten, also fast auf kristalline Teile der Unterlage der Geosynklinale beschränkt wären; auch die die Peridotite begleitenden Amphibolite und sonstigen kristallinen Schiefer weisen aber keine sichere Kontaktmetamorphose auf, so daß eine intrusive Platznahme der Ultramafitmassen, auch in diesem Licht gesehen, wenig wahrscheinlich erscheint.

Eine solche Vergesellschaftung von Peridotiten und Serpentinen mit Amphiboliten und untergeordneten anderen kristallinen Schiefern ist z. B. in Timor, Celebes und der benachbarten Insel Kabaena zu beobachten (de Roever 1940, 1950, 1953); sie tritt hier offenbar in Decken auf, die wenig oder gar nicht metamorphe Gesteinsserien überlagern (siehe auch Brouwer 1942, S. 377—381). Dieselbe Vergesellschaftung von alpinotypen Ultramafiten mit Amphiboliten und anderen kristallinen Schiefern ist aus sehr vielen anderen Gebieten der Erde bekannt, so daß hier wohl auf ein Literaturverzeichnis verzichtet werden kann. Hiessleitner hat diese Gesteinsvergesellschaftung eingehend besprochen (1951/52, speziell S. 510 bis 516) und sagt darüber u. a.: „Bei den feldgeologischen Aufnahmen von Peridotitgebieten ist immer wieder die Feststellung zu machen, daß Amphibolite in Peridotiten, namentlich auch Amphibolit als Grenzgestein von Peridotit angetroffen werden neben frischen, dynamometamorph nicht oder kaum betroffenen Gabbrogesteinen. Dabei stehen magmatische Verbundenheit von Gabbro mit Peridotit, aber auch die Abhängigkeit der Amphibolitzone von der Peridotitmasse außer Zweifel. Wie sehr auch die Grenzbereiche der Peridotitmassive tektonischer Bewegung und Durchstörung ausgesetzt sind, so klären doch keine Beobachtungen auf, wie zwischen Gabbro und Amphibolit ein so bedeutender metamorpher Hiatus

---

[4] Wenn die Ultramafite in einer stärker metamorphen Umgebung vorkommen, kann man in ihrer Nähe auch Amphibolite finden, die durch Metamorphose aus den die Ultramafite begleitenden Gabbros und Vulkaniten hervorgegangen sind; solche Amphibolite werden hier natürlich nicht gemeint.

141

sich herausbilden konnte, falls etwa der Amphibolit als orthometamorpher Gabbro oder Pyroxenit gedeutet werden sollte (S. 339—340).... Fast kaum ein Peridotitgebiet des Balkans, das nicht über kürzeren oder längeren Grenzverlauf seinen Amphibolitgürtel aufweist, ob nun kristalline oder halbkristalline, karbonatische oder nichtkarbonatische Hüllgesteine herantreten (S. 437)... Die häufig wiederkehrende Niveaustellung der Peridotitgesteine an der Grenze Hochkristallin zu Minderkristallin bzw. Paläozoikum ... ist um so bemerkenswerter, als auch in den Ostalpen ähnliche Beobachtungen gelten, beispielsweise für die steirischen Serpentine, für die Serpentine des Rhätikon (F. ANGEL), des Engadin (H. P. CORNELIUS), ja schließlich auch für die Hohen Tauern. Diese so häufig festzustellende Position der Serpentine kann jedenfalls nicht Anlaß geben, auf der Vorstellung von jüngerem, nachmesozoischem Intrusionsalter zu beharren, eher kommt darin ein variszisches Intrusionsniveau zum Ausdruck" (S. 471). Mit der letzten Auffassung HIESSLEITNERS ist der Verfasser nur so weit einverstanden, daß die häufige Vergesellschaftung von Ultramafiten mit offenbar der Unterlage der Geosynklinale entstammenden Amphiboliten und anderen kristallinen Schiefern in der Tat darauf hinweist, daß auch die Ultramafite der Unterlage der Geosynklinale entstammen; sie braucht aber nicht ein paläozoisches Alter der Peridotite anzudeuten, sondern ist auch mit einem größeren Alter derselben im Einklang.

Es kann noch darauf hingewiesen werden, daß die hier besprochenen Amphibolite in der Tat richtige kristalline Schiefer sind, deren Struktur nicht während der Erstarrung eines Magmas gebildet wurde, wie z. B. durch ihre Vergesellschaftung auf Timor und Kabaena mit untergeordneten Granat-Piemontit-Quarziten, Granatglimmerschiefern, granatführenden Gneisen, staurolith- und disthenführenden Gesteinen, regionalmetamorphen kristallinen Kalken und anderen hochgradigen kristallinen Schiefern angedeutet wird. Es sagt auch HIESSLEITNER: „In kristallinem Grundgebirge, das Peridotite einhüllt, aber auch in solchem, wo keine Peridotitkörper zur Beobachtung gelangen, treten Amphibolite in Typen auf, die, soweit heute die Feststellungen vorangekommen sind, sich von echten Grenzamphiboliten oder Amphiboliteinschlüssen in Peridotit weder chemisch noch strukturell zu unterscheiden brauchen" (S. 514).

Die in dieser Arbeit besprochenen Amphibolite sind auch nicht durch Metamorphose aus den die Ultramafite begleitenden Gabbros usw. entstanden, weil es zwischen diesen Amphiboliten und den metamorphen Gabbros mindestens sehr oft einen bedeutenden Metamorphosehiatus gibt.

Der Häufigkeit der Amphibolite unter den die Peridotite unmittelbar begleitenden kristallinen Schiefern kann noch eine spezielle Bedeutung zugeschrieben werden. Diese kristallinen Schiefer würden nämlich nach der hier erörterten Hypothese die Bedeckung der Peridotitschale bilden, die von Bruchstücken der Peridotitschale überschoben oder durchspießt wurde, unter Mitschleppung größerer oder kleinerer Schiefermassen. Es würde also mit der Möglichkeit gerechnet werden können, daß wenigstens ein Teil der sogenannten Basaltschale von derartigen kristallinen Schiefern aufgebaut wird, was bei einer hauptsächlichen Zusammensetzung der-

142

selben aus Amphiboliten hinsichtlich der physischen Konstanten durchaus möglich scheint. Die die Peridotite unmittelbar begleitenden kristallinen Schiefer würden nach der hier erörterten Hypothese wahrscheinlich den ältesten Gesteinen der Erdkruste einzureihen sein.

Die u. a. von ERNST (1935) beschriebene Gefügeregelung der Ultramafite wäre vielleicht z. T. von der von diesen kristallinen Schiefern erlittenen Regionalmetamorphose hervorgerufen, z. T. auf sonstige Bewegungen in der Peridotitschale zurückzuführen.

Was das übliche Fehlen von Gängen von alpinotypen Ultramafiten in nichtultramafischen Gesteinen anbelangt, muß erstens darauf hingewiesen werden, daß bei einer Entstehung der alpinotypen Peridotite als Glieder der Peridotitschale die Verfestigung dieser Gesteine aus einem ultramafischen Magma durchaus nicht unmöglich scheint. Peridotitgänge in ultramafischem Nebengestein würden also vielleicht als magmatische Gänge zu deuten sein.

Außer Zweifel erscheint die Intrusionsnatur des „Peridotit"-Ganges, dessen dem sedimentären Nebengestein entstammende Kokseinschlüsse von SOSMAN (1938) zur Bestimmung der Intrusionstemperatur untersucht wurden. Es handelt sich hier aber nach der Beschreibung von KEMP und ROSS (1907) um ein biotitreiches Ganggestein mit viel Perowskit und viel sekundärem Kalzit und Dolomit, das nach einer chemischen Analyse einen hohen Gehalt an CaO und TiO_2 aufweist (siehe auch BENSON 1927, S. 58—59). Dieses Gestein ist also kein alpinotyper Peridotit. HESS (1955, S. 394) schrieb über solche Gesteine: "Alnoites, kimberlites, mica peridotites, and related rock types can hardly be confused with alpine ultramafics because they differ conspicuously in mineralogy, chemistry and tectonic setting. Considering the superfluity of names with which petrology is burdened it is most unfortunate that some name other than *mica peridotite* is not used for this dike rock. Gross errors have probably resulted from applying conclusions drawn from facts related to mica peridotites to alpine peridotites and vice versa. If mica peridotite had been called humptydumptyite, these probably would not have arisen."

Das übliche Fehlen von Gängen von alpinotypen Ultramafiten in nichtultramafischen Gesteinen kann noch mit einer Aussage HIESSLEITNERS illustriert werden: „Die Peridotitmassive kommen in geschlossenen Massen hoch, soviel wie nie verlieren sich diese Eruptionsvorgänge in Bildung von Apophysen und seitlichem Ganggeäder, wie etwa Granite, deren Gangdurchbrüche oft gehäuft in Nähe der Hauptmasse, entfernt davon abklingen. Peridotitisch injizierte Schiefer gibt es nicht. Was da von ‚Serpentinapophysen' und ‚Serpentingängen', die Hüllgesteine der Balkanperidotite durchbrechend, beschrieben wurde, hat sich noch immer als tektonischer Schubkeil oder, wie beim Serpentingang von Lojane, als längliche Scholle älteren Serpentins, von jüngerem Granit hochgefördert, erwiesen" (S. 422).

Es kann weiter darauf aufmerksam gemacht werden, daß eventuelle Gänge von Serpentin oder teilweise serpentinisiertem Peridotit in nichtultramafischen Gesteinen vielleicht auch als Folge etwaiger Volumenvermeh-

143

rung bei der Serpentinisierung größerer Peridotitmassen gebildet werden können — gewissermaßen als großzügiges Analogon des von FLETT (1912) aus Troktolithen des Lizarddistriktes beschriebenen Phänomens, daß serpentinisierte Olivinkristalle in dem umgebenden Feldspalt von zahlreichen radialen serpentingefüllten Expansionsrissen umgeben werden. Rein hydrothermale Prozesse würden auch eine Bildung von Serpentingängen bewirken können.

Es muß nach der hier erörterten Hypothese damit gerechnet werden, daß Gabbrogänge in Peridotitmassen vielleicht auch schon als fertiges Gebilde der Peridotitschale entstammen. Dies trifft vielleicht auch zu für nicht von ultramafischen Gesteinen umgebene Gabbromassen, die keine Kontakthöfe zeigen, aber gilt natürlich nicht für die die Peridotite und Gabbros begleitenden Vulkanite; nur diese letzteren wären also unzweifelhaft als Initialmagmatite zu deuten (siehe auch HIESSLEITNER 1951/52, S. 470). Wie bereits erwähnt, wäre die Entstehung dieser Vulkanite nach der hier erörterten Hypothese eine normale Begleiterscheinung der tektonischen Platznahme der Peridotite; die zwei Gesteinsgruppen wären aber gar nicht von gleichem Alter. Ihre häufige Vergesellschaftung würde also zu Fehlschlüssen geführt haben, wie auch schon von HIESSLEITNER betont wurde: „Daß die Ultrabasite a priori, auch dort, wo keine zwingenden Gründe der Feldgeologie vorliegen, mit den Diabasen („Ophiten') zusammen zu den „Ophiolithen' zusammengezogen werden und dadurch Attribute erhalten, die eigentlich nur den Diabasen zukommen, wie Hochsteigen in jüngere Schichten, gangförmiger Intrusionsmechanismus, Verbindung mit Radiolaritgesteinen usw. ist geeignet, in der Beurteilung der Grüngesteinsphänomene gewisse Vorurteile zu bringen" (1951/52, S. 434—435; siehe auch HESS 1955, S. 393).

Die auffällige Beschränkung von Chromerzen auf ultramafische Gesteine wäre übrigens nach der in dieser Arbeit gegebenen Hypothese sehr verständlich; die Chromitlagerstätten würden als fertige Gebilde der Peridotitschale entstammen.

Es ist für die hier erörterte Hypothese entscheidend, ob sie tektonisch möglich ist. Es ist schon längst bekannt, daß Überschiebungsbahnen oft von Peridotiten oder Serpentinen markiert sind. Nach den Angaben in der bezüglichen Literatur wäre es aber nicht möglich, für alle alpinotypen Ultramafitmassen eine weite tektonische Verfrachtung anzunehmen. Es gibt jedoch noch so viel Unstimmigkeit über die Tektonik vieler peridotit- und serpentinführender Gebiete, daß es berechtigt erscheint, die hier erörterte Hypothese zur Erwägung in künftigen Ultramafitstudien zu empfehlen. Es wäre z. B. wichtig, festzustellen, ob es überhaupt alpinotype Gebiete gibt, für die man die Möglichkeit einer weiten tektonischen Verfrachtung der Ultramafite m i t S i c h e r h e i t ablehnen kann [5]; dabei soll übrigens nicht nur an Überschiebung gedacht werden, sondern auch an

---

[5] DUBERTRET (1955) lehnte eine weite tektonische Verfrachtung der sehr mächtigen Gabbro- und Serpentinmassen von Nordwestsyrien und vom angrenzenden Hatay (Türkei) ab und befürwortete für diese Gesteine eine lakkolithartige magmatische Platznahme unter einer Abkühlungsrinde von Kissenlaven. Er erwähnte jedoch keine Kontaktmetamorphose im Liegenden; auch das Vorkommen von Amphiboliten wurde nicht in befriedigender Weise erklärt (S. 173). DUBERTRETS Auffassung mag deshalb weiterer Überprüfung bedürfen.

144

Auspressung in der folgenden, von HESS (1955, S. 402) beschriebenen Weise: "Slickensided, slippery shear zones form easily in serpentine. Once formed, movement takes place readily along them. Thus solid serpentine bodies may move into overlying sediments in much the same way that a watermelon seed moves when squeezed between one's fingers." Abgleitung kann natürlich auch stattgefunden haben.

Auch die überaus großen Peridotit- und Serpentinmassen von Celebes und Kuba würden nach der hier erörterten Hypothese als tektonisch verfrachtete Bruchstücke der Peridotitschale zu deuten sein; diese Hypothese würde also für solche Gebiete einen sehr großen Zusammenschub postulieren, größer als man gewöhnlich für sonstige Faltengebiete annimmt. Für Celebes (und für die benachbarte Insel Kabaena) wird die Deckennatur der bezüglichen Ultramafitkomplexe angedeutet durch ihre Vergesellschaftung mit kristallinen Schiefern und ihre Überlagerung auf Gesteinsserien, die jünger als die Metamorphose dieser kristallinen Schiefer sind (DE ROEVER 1953, 1956). Auf Kuba wurde von THAYER und GUILD (1947) an einigen Stellen tektonische Überlagerung von Serpentin auf jüngeren Gesteinen beobachtet, während Bewegungsphänomene in den Ultramafiten als Begleiterscheinungen großer Überschiebungen gedeutet wurden.

HIESSLEITNER hat sich gegen Platznahme in festem Zustand aller Peridotitmassen der Balkanhalbinsel ausgesprochen, z. B. in folgender Weise: „Dort aber, wo größere geschlossene basische Massen vorliegen, wie in dem so oft herangezogenen Raduschabezirk, aber auch in allen anderen Peridotitmassen, deren Innenbau genauer untersucht wurde, enthüllt das geologische und petrographische Einzelstudium einen gliederbaren magmatischen Stockwerksbau, *der häufig auch mit der endgültigen Lage des Eruptivs in seiner Hülle Zusammenhänge besitzt — zumindest jenes des gemeinsamen Lotes mit der Hülle zur Zeit der Intrusion* (Kursivierung vom Verfasser dieser Arbeit) — was alles nur aus schmelzflüssigem Gehaben der Peridotitmasse zum Zeitpunkt der Raumnahme hergeleitet werden kann" (1951/52, S. 423—424). Es ist nach der in dieser Arbeit erörterten Hypothese durchaus nicht unmöglich, daß die hier gemeinte innere Zonengliederung der Peridotitmassive während der Verfestigung aus einem ultramafischen Magma entstanden ist; die kursivierte Stelle würde jedoch Verfestigung nach der Platznahme andeuten, die intrusiven Charakters wäre. Es scheint aber gar nicht ausgeschlossen, daß sich speziell in größeren Peridotitmassen eine schon in der Peridotitschale entstandene Schichtung erhalten hat, die die Hauptbegrenzung von linsenförmigen Bruchstücken dieser Schale bestimmte und deshalb jetzt parallel zu den Kontakten verlaufen würde. Diese Schichtung könnte dann aber auch als eine metamorphe Erscheinung gedeutet werden.

Es kann zum Schluß noch darauf aufmerksam gemacht werden, daß die hier zur Diskussion gestellte Hypothese nicht nur mit einer Anzahl wichtiger Beobachtungen von HESS, HIESSLEITNER, WILKINSON, und BAILEY und McCALLIEN im Einklang ist, sondern auch den Resultaten der Experimente von BOWEN und TUTTLE entspricht und den Grundgedanken BOWENS bestätigen würde, daß die Platznahme von Peridotiten nicht als

Folge von Intrusion eines peridotitischen Magmas gedeutet werden kann.

Am Ende dieser Arbeit möchte ich den Herren Prof. E. BEDERKE, Dr. C. G. EGELER, Dr. E. KÜNDIG, H. KONING und A. C. TOBI meinen herzlichen Dank aussprechen für ihre kritischen Bemerkungen beim Durchlesen des Manuskripts; es sei aber hervorgehoben, daß nur der Autor selbst für den Text verantwortlich ist.

Geologisches und Mineralogisches Institut der Universität Leiden,
April 1957.

Literaturhinweise

BAILEY, E. B., und McCALLIEN, W. J. (1953): Serpentine lavas, the Ankara Mélange and the Anatolian Thrust. Trans. Roy. Soc. Edinb. **62**, S. 403—442. — BENSON, W. N. (1927): The tectonic conditions accompanying the intrusion of basic and ultrabasic igneous rocks. Mem. Nat. Acad. Sci. **19**—1, S. 1—90. — BOWEN, N. L., und TUTTLE, O. F. (1949): The system $MgO—SiO_2—H_2O$. Bull. Geol. Soc. America **60**, S. 439—460. — BROUWER, H. A. (1942): Summary of the geological results of the expedition. Geol. Exp. of the Univ. of Amsterdam to the Lesser Sunda Islands, Bd. 4, S. 345—402, Amsterdam. — DUBERTRET, L. (1955): Géologie des roches vertes du Nord-Ouest de la Syrie et du Hatay (Turquie). Notes et Mém. sur le Moyen-Orient, Mus. National d'Hist. Naturelle, Paris, Bd. **6**, S. 5—224. — ERNST, Th. (1935): Olivinknollen der Basalte als Bruchstücke alter Olivinfelse. Nachr. Ges. Wiss. Göttingen, Math.-Phys. Kl., N. F., Fachgr. IV, Bd. **1**, S. 147—154. — FLETT, J. S., und HILL, J. B. (1912): The geology of the Lizard and Meneage. Mem. Geol. Surv. Engl. and Wales, Expl. Sheet 359. — HESS, H. H. (1938): A primary peridotite magma. Amer. Journ. Sci. **235**, S. 321—344. – (1954): Geological hypotheses and the earth's crust under the oceans. Proc. Roy. Soc. London, Ser. A, **222**, S. 341—348. – (1955): Serpentines, orogeny, and epeirogeny. Geol. Soc. America Spec. Paper **62**, S. 391—408. — HIESSLEITNER, G. (1951/52): Serpentin- u. Chromerz-Geologie d. Balkanhalbinsel u. eines Teiles v. Kleinasien. Jb. Geol. B.-A. Wien, Sonderbd. **1**. — KEMP, J. F., u. ROSS, J. G. (1907): A peridotite dike in the Coal Measures of southwestern Pennsylvania. Ann. N. Y. Acad. Sci. **17**, S. 509—518. — KOSSMAT, F. (1937): Der ophiolithische Magmagürtel in den Kettengebirgen des mediterranen Systems. Sitzungsber. Preuß. Akad. Wiss., Phys.-math. Kl., **24**, S. 308—325. — RAITT, R. W., FISHER, R. L., und MASON, R. G. (1955): Tonga Trench. Geol. Soc. America Spec. Paper **62**, S. 237—254. — DE ROEVER, W. P. (1940): Geological investigations in the southwestern Moetis region (Netherlands Timor). Geol. Exp. of the Univ. of Amsterdam to the Lesser Sunda Islands, Bd. **2**, S. 97—344, Amsterdam. – (1950): Preliminary notes on glaucophane-bearing and other crystalline schists from South East Celebes, and on the origin of glaucophane-bearing rocks. Proc. Kon. Nederl. Akad. Wetensch. **53**, S. 1455—1465. – (1953): Tectonic conclusions from the distribution of the metamorphic facies in the island of Kabaena, near Celebes. Proc. 7th Pacific Sci. Congr. (New Zealand 1949), Bd. **2**, S. 71—81. – (1956): Some additional data on the crystalline schists of the Rumbia and Mendoke Mountains, South East Celebes. Verh. Kon. Nederl. Geol. Mijnbouwk. Gen., Geol. Ser. **16**, S. 385—393. — SOSMAN, R. B. (1938): Evidence on the intrusion-temperature of peridotites. Amer. Journ. Sci. **235** A, S. 353—359. — THAYER, T. P., und GUILD, P. W. (1947): Thrust faults and related structures in eastern Cuba. Trans. Amer. Geoph. Union **28**, S. 919—930. — WILKINSON, J. F. G. (1953): Some aspects of the Alpine-type serpentinites of Queensland. Geol. Mag. **90**, S. 305—321.

146

13

Reprinted from *Geol. Soc. America Bull.*, **80**, 1515–1522 (Aug. 1969)

T. P. THAYER   *U.S. Geological Survey, Washington, D. C.*

Peridotite-Gabbro Complexes as Keys to Petrology of Mid-Oceanic Ridges

Abstract: Two suites of olivine-rich ultramafic and feldspathic rocks appear to be present in the Mid-Atlantic Ridge: one which seems to have alkalic affinities, and one similar to the chromitite-bearing alpine peridotite-gabbro complexes. The similarities of rocks in the two environments—continental and oceanic—imply that much about the petrology of mid-oceanic ridges may be learned from studies of continental complexes, and that silicic rocks have been formed in the mantle. Although gabbros in St. Paul Rocks and similar rocks at Tinaquillo, Venezuela, and Lizard, England, have been interpreted as not comagmatic with intimately associated peridotite by some petrologists, evidence to the contrary at Lizard is discussed. Association of fresh gneissic gabbro, some containing quartz, with talcose serpentinite, amphibole schist, quartz diorite and epidotic but unsheared basalts along the Mid-Atlantic Ridge is believed to indicate presence of alpine-type rocks that occur normally in eugeosynclinal belts.

Gabbro, described as partly interlayered with peridotite by gravitational differentiation, forms major parts of three widely separated ultramafic complexes which have been interpreted as slices of oceanic crust and upper mantle: the Troodos massif in Cyprus, the Bowutu Mountains in Papua, and the Camagüey complex in central Cuba. If, as Dietz has suggested, peridotite and related rocks in eugeosynclines represent fragments of ocean rind formed along mid-oceanic ridges and moved laterally by ocean-floor spreading, gabbro must be an essential constituent of the upper mantle. This could account for many geophysical anomalies, but would complicate some postulated mechanisms involved in ocean-floor spreading.

CONTENTS

INTRODUCTION

Despite the widespread interest, very little direct evidence is available on the petrology of plutonic rocks under mid-oceanic ridges and its relation to the hypothesis of ocean-floor spreading. Explanations of variations in physical properties of the upper mantle are restricted by theories as to what kinds of rocks should be present. The hypothesis of serpentinization of mantle peridotite under oceanic crust (Hess, 1962) is supported by samples from scattered dredge hauls, and is attractive because it provides a simple explanation of several problems. For example, it meshes neatly with the hypothesis that basaltic magma and olivine-rich peridotite form syngenetically by partial melting of garnet peridotite.

Geophysical data, however, indicate complexity in the upper mantle, especially under mid-oceanic ridges (Talwani and others, 1965). Complexity is implied also by widespread association of magnesium-rich gabbro with chromitite-bearing alpine-type dunite and peridotite which appear to have been formed by gravitational segregation in the mantle, or under conditions believed to obtain in the mantle (Thayer, 1967 and in press). Direct evidence on the composition and structure of the upper mantle from dredging and drilling will be very scanty for many years, so there is a great premium on making the most of the petrologic information we do get. A case in point involves gabbro dredged by the *Atlantis*, and described by Quon and Ehlers (1963); they interpreted the gabbro and gabbroic anorthosite as "derived by crystallization from the basaltic magma and . . . not related to the peridotite" brought up in the same dredge haul (p. 7). Examination of thin sections kindly loaned by Professor Ehlers confirmed the fact that the gabbros are fresh and gneissic, but also revealed other evidence of the presence of a distinctive suite of alpine plutonic rocks (Thayer, 1967). More recent descriptions of St. Paul Rocks by Melson and others (1967) indicate that more than one peridotite-gabbro association may be present in the Mid-Atlantic Ridge. The presence of plutonic feldspathic rocks in both suites seems highly significant.

The purposes of this paper are (1) to call attention to some recent descriptions of peridotite and feldspathic rocks in widely scattered localities that are believed by the authors to bear on the petrology of the upper mantle, and (2) to present interpretations which are at variance with some of the others. In presenting the different interpretations, some factual points of contention involved in resolution of the differences are discussed.

ACKNOWLEDGMENTS

C. B. Raleigh, W. B. Joyner, and E. D. Jackson, all of the U.S. Geological Survey, and W. G. Melson, U.S. National Museum, contributed substantially to this paper by discussions and review of the manuscript. Comments by I. G. Gass on the parts pertaining to Cyprus helped to eliminate some ambiguities.

AMPHIBOLE-BEARING PERIDOTITE AND GABBRO IN ST. PAUL ROCKS, TINAQUILLO, AND LIZARD

At St. Paul Rocks (1° N. lat.; 29°15′ W. long.) on the Mid-Atlantic Ridge, Melson and others (1967) have identified three seemingly dissimilar kinds of mylonitic rocks: olivine-rich, chromite-bearing, two-pyroxene peridotite (lherzolite), with or without pargasite and blue spinel; mylonite characterized by abundant augen of brown hornblende and lesser amounts of plagioclase, olivine, clinopyroxene, and other minerals (hornblende mylonite); and mylonite composed of clinopyroxene, plagioclase, scapolite, and subordinate brown hornblende (pyroxene gabbro). The hornblende-free peridotite mylonite is low in alkalis, Fe, TiO_2, and P_2O_5, like alpine peridotite (Thayer, 1960, 1967). The hornblende mylonite, in contrast, is markedly alkalic, rich in TiO_2 and P_2O_5, and critically undersaturated in SiO_2. The pyroxene gabbro is very rich in diopside and also rather rich in TiO_2; with increase in brown hornblende, it grades into the hornblende mylonite (W. G. Melson, 1968, oral commun.). The three kinds of rocks are interlayered, interlaminated, and intergraded in a manner "consistent with movement of a relatively hot (but solid) plastic rock mass through the suboceanic mantle, and incorporation and shearing out of a variety of unrelated rocks during ascent" (Melson and others, 1967, p. 1534).

The mylonites in St. Paul Rocks are remarkably similar to ultramafic complexes at Tinaquillo, Venezuela, and Lizard, England (Melson and others, 1967, p. 1533). At all three places, olivine-rich peridotite (lherzolite) is interlayered with rocks of gabbroic composition; alkalic amphibole is an important if not conspicuous constituent; blue or green spinel is present; and all the rocks are strongly sheared or mylonitized (Melson and others, 1967; MacKenzie, 1960; Green, 1964). Interpretations of the relations of the gabbroic rocks, however, could hardly differ more. Green, MacKenzie, and Melson and others deny, in essence, any close magmatic ties between the peridotite and interlayered gabbroic rocks, whereas Thayer and Brown (1961) and Flett (1946) have called attention to or described features indicating comagmatic relationship.

As Melson and others have observed (1967, p. 1533), the pargasite augen in the peridotite mylonite at St. Paul Rocks "are similar to the dominant amphiboles of the banded parts of the spinel peridotite bodies" at Lizard and Tinaquillo (Table 1). Some of the amphiboles at St. Paul Rocks, especially the brown hornblende, are regarded by Melson (1967, p. 1533) as mantle-derived, but MacKenzie (1960) and Green (1964) believe that the amphiboles at Tinaquillo and Lizard, respectively, were formed by reaction at a late stage, during or right after emplacement of the peridotite. Flett (1946, p. 51), however, described the Lizard peridotite as having been intruded very closely in time with basic rocks which have been reduced to flaser gneiss and amphibole schist (the Traboe schist) by regional metamorphism. Flett's interpretation agrees with observations by Fox and Teall (1893).

Comagmatic relationships between gabbro and peridotite at Lizard are strongly indicated by the textures and lack of chilling in gabbro dikes that cut peridotite, described by Flett (1946, p. 84) in great detail. The large dike of coarse troctolite at Coverack (Flett, 1946, p. 81) crosses the foliation in the peridotite, but during a visit in 1968, the author and others found that in places the troctolite clearly has been foliated with the peridotite and the two rocks are mixed magmatically along gradational boundaries. Green's analyses (1964, p. 157), furthermore, indicate that clinopyroxene in the troctolite differs from that in "primary assemblage" pyroxenite only by minor amounts of iron and titanium, and may contain more chromium. Relict textures and intimate interlayering of Traboe schist with peridotite at Pol Cornick and between Porthallow and Portkerris Point are typical of transition zones between major gabbro and peridotite units in many alpine complexes (Thayer, 1963b, 1967). The field relations show that Flett and Fox and Teall were right, I believe, on two essential points with which Green disagreed, namely: comagmatic relations of the peridotite with gabbro[1] which has been metamorphosed to form the Traboe schist, and extensive alteration of both rocks during regional metamorphism.

---

[1] The gabbro dikes which cut, but are not chilled against, peridotite are believed to be syngenetic with the gabbroic rocks from which the Traboe schist was derived. As Flett has indicated, the gabbro of St. Keverne and Crousa Downs appears to be younger; it probably is of a different type. My correlation of the Crousa Downs gabbro with the Traboe schist (Thayer, 1967, p. 236) I now believe to have been incorrect.

TABLE 1. COMPOSITIONS OF AMPHIBOLES FROM ST. PAUL ROCKS; TINAQUILLO, VENEZUELA; AND THE LIZARD AREA, ENGLAND

	1	2	3	4
SiO_2	43–47	43.61	46.19	44.35
Al_2O_3	10–12	15.06	12.83	15.23
TiO_2	0.2–0.5	1.15	.90	.86
Fe as FeO	3.6–7.8	6.57	4.03	5.28
MgO	17–20	16.53	18.71	17.26
CaO	11–13	11.80	12.45	12.14
Na_2O	2–3.3	2.78	1.87	2.45
K_2O	0.4–0.9	.11	.27	.32

(1) Pargasite from peridotite in St. Paul Rocks (Melson and others, 1967, p. 1533).

(2) Amphibole from amphibole-rich layer in peridotite, Tinaquillo, Venezuela (MacKenzie, 1960, p. 307).

(3) Colorless amphibole from "recrystallized hydrous" facies of peridotite, Lizard, England (Green, 1964, p. 168).

(4) Colorless amphibole from peridotite in contact with "granulite" at Pol Cornick, Lizard, England (Green, 1964, p. 168).

PERIDOTITE, GABBRO, AND DIORITIC ROCKS FROM MID-OCEANIC RIDGES

Plutonic rocks dredged from the Mid-Atlantic Ridge between 30° N. and 36° N. lat. by the *Atlantis* in 1948–1949 include altered derivatives of olivine-rich peridotite, gneissic gabbro, and quartz diorite (Table 2). These are associated with fresh and altered basalts, and the absence of other rocks, except at Station 78, indicates they are not ice-rafted erratics. Shand (1949) described two specimens of gabbro, and Quon and Ehlers (1963) later found that the collections also include schists derived from serpentinite; granite; basalt (some containing quartz, chlorite, and epidote in vesicles); and one specimen of quartz-epidote rock. Although some of the serpentinites contain relict olivine, enstatite, and diopside (Quon and Ehlers, 1963, p. 4), most of them are thoroughly altered to felted antigorite or serpentine in which talc locally is abundant, to talc schist, and also to amphibole schist (Pl. 1, fig. 3). The gabbro ranges from olivine-bearing and anorthositic to a quartz-bearing variety that was found at Station 78, where biotite-hornblende quartz diorite was also dredged up. Quon and Ehlers (1963, p. 5) found 6 percent amphibole plus some chlorite and biotite in one specimen of gabbro, and I identified another gabbro partly altered to albite, epidote,

TABLE 2. MID-ATLANTIC RIDGE LOCATIONS OF ALPINE-TYPE PLUTONIC AND ALTERED VOLCANIC ROCKS
THAT WERE DREDGED BY THE ATLANTIS

Station	N. Latitude	W. Longitude	Depth (fathoms)	Rock type
5	30°00'	42°10'	2100–2500	Olivine gabbro, one specimen (Shand, 1949).
6	30°06'	42°08'	800	Serpentinite; gabbro; sedimentary breccia containing pieces of talc, chlorite, and amphibole schists; and metavolcanic rocks.
10	31°55'	40°26'	1450–1500	Serpentinite (Quon and Ehlers, 1963).
20 (20A)	30°04'	42°16'	2250	Serpentinite, fresh and albitized gabbro, quartz-epidote rock.
21	30°08'	43°37'	2300	Talcose serpentinite, amphibole schist.
78	35°59'	47°10'	2250	Gabbro, quartz diorite, granite, epidotic basalt, limestone, and calcareous sandstone.

Modified *after* Quon and Ehlers (1963)

chlorite, and quartz. All the gabbro is strongly deformed, but most is unaltered (Pl. 1, figs. 1, 2). Epidotic amygdaloidal basalts, in contrast, appear undeformed (Pl. 1, fig. 4).

Dredging in the vicinity of the Mid-Atlantic Ridge at about 45° N. confirms the presence, at surface or at depth, of similar plutonic and metamorphic rocks. Aumento (1969, p. 257) states that:

"Two elongated, block faulted sea mounts yielded, in association with basalts and serpentinites, large quantities of in situ cummingtonite-bearing amphibolites. These occurrences are a further indication that considerable metamorphism . . . has taken place on the Mid-Atlantic Ridge."

Large crystals of olivine, plagioclase, and chromite in basalts dredged by the R.R.S. *Discovery II* at 45°50' N. led Muir and Tilley (1964) to infer the presence of layered troctolite at depth:

"In about one-fifth of the basalts, and in the olivine-enriched dolerite, large xenocrysts of very magnesian olivine ($Fo_{90.5}$), highly calcic plagioclase ($An_{88.2}$), and chrome spinel are found. In the basalts they may be distinguished from the true microphenocrysts by their grain size, different chemical composition, and different habit. . . . Rarely they are found as very small xenoliths composed of all three minerals. It is notable that no pyroxenes have been found as xenocrysts." (Muir and Tilley, 1964, p. 414).

The implications of the xenocrysts were clearly recognized by Muir and Tilley (p. 431):

"Of much greater significance is the widespread occurrence in both the basalts and dolerites of crystals of highly magnesian olivine, picotite, and calcic bytownite. Among plutonic rocks, assemblages of minerals with these compositions (although usually with a more iron-rich olivine,

Fa_{12-16}) are found only in troctolites from the lower zones of layered intrusions such as those of Rhum. The presence of glide lamellae in the olivines would also be consistent with such a source. A possible genetic relationship between the xenocrysts and the host basalts is suggested by the absence of pyroxene as a xenocryst mineral and by its comparatively late appearance in the crystallization of the basalts. On this evidence it is suggested that well-developed layered gabbroic intrusions should occur at quite moderate depths below the basalts forming the crust of the rift. The mechanism by which the xenocrysts were derived is less clear. . . ."

Similar "xenocrysts" were found in basalts at about 22° N. by Melson and others (1968, p. 5928), but the olivines lack glide lamellae. Regardless of disagreement as to whether the xenocrysts may be cognate or not, they are interpreted by both groups of authors as evidence of gabbroic gravitational differentiates under the Mid-Atlantic Ridge.

Along the Carlsberg Ridge in the Indian Ocean, Matthews and others (1965) found spilitic basalt associated with peridotite altered completely to talc, talc-chlorite, and chlorite-actinolite rocks. Gabbro dredged at a nearby station was described as crushed and strongly banded; the banding is formed by variations in proportions of andesine, sheafed green hornblende, chlorite, and clinozoisite.

Association of gneissic gabbro (Pl. 1, figs. 1 and 2) with olivine-rich peridotite, diorite, and more salic rocks, and undeformed spilitic basalts (Pl. 1, fig. 4) is characteristic of alpine mafic (ophiolitic) complexes (Thayer, 1967, p. 232). The presence of these rocks together in the Mid-Atlantic and Carlsberg Ridges, and the kind of alteration some of them have undergone, introduce the possibility of the existence of a complete suite of alpine-type intrusive rocks, from dunite to albite granite (Thayer, 1963a, 1967), in these and other mid-oceanic ridges.

The occurrence, in mid-oceanic ridges as in continental areas, of peridotite and gneissic gabbro with dioritic or subsilicic rocks which appear to be closely related, has two corollaries: (1) petrologic relations of rocks in the mid-oceanic ridges may be studied in continental analogs; (2) *comagmatic* dioritic and more salic rocks associated with alpine mafic complexes in continental areas (Thayer, 1963a, 1967) may have been formed in the mantle, because such rocks in mid-oceanic ridges can have no other source. Although the similarities of the rocks imply a common origin, much more information is needed to elucidate the fundamental differences in interpretations of their relationships. Just as rocks at Lizard and Tinaquillo appear to present the best opportunities to learn more about the peridotite, gabbro, alkalic amphiboles, and mylonite displayed in St. Paul Rocks, the Troodos massif in Cyprus may present an unusual opportunity to study the relationships between basaltic lavas and alpine plutonic rocks that appear to have formed in an ancient oceanic environment.

THE TROODOS MASSIF IN CYPRUS: PART OF A TETHYAN MID-OCEAN RIDGE?

The Troodos massif consists of two principal groups of rocks: Mesozoic tholeiitic pillow lavas and related dikes, and the Troodos Plutonic Complex (Bear, 1966; Gass, 1967, 1968; Wilson and Ingham, 1959). The volcanic rocks are divided into three units, the Sheeted Intrusive Complex, and Lower and Upper Pillow Lavas. The volcanic rocks are unusual in two respects. Despite a total thickness of 3 to 5 km of lavas (Gass, 1968), sedimentary rocks have been found only between the topmost flows of the Upper Pillow Lavas. Basaltic dikes between septa of pillow lava constitute at least 90 percent of the Sheeted Intrusive Complex, and Gass (1968, p. 42) calculates that about 48,000 dikes crop out in it over an east-west cross-strike section more than 100 km long.

The Troodos Plutonic Complex was intruded into the lowest unit (the Sheeted Intrusive Complex) of the volcanic sequence. The Complex includes rocks that grade from olivine-rich harzburgite and dunite containing chromitite, through gabbro, to granophyric trondhjemite

or albite granite. The rocks are distributed rudely concentrically with the harzburgite and dunite at the center and silicic rocks at the margins. The peridotite and olivine gabbro have been described as alpine (Bear, 1966, p. 28) or closely comparable to alpine (Gass, 1967, p. 126), but their structural and genetic relations to pyroxene gabbro and more silicic rocks are debated (Bear, 1966; Gass, 1967, p. 126; Thayer, 1967, p. 230). Regardless of theories of petrogenesis, however, Bear's (1966, p. 30) descriptions of the more silicic rocks in the Troodos Plutonic Complex apply equally well to rocks in the Canyon Mountain Complex in Oregon (Thayer and Himmelberg, 1968) which typifies alpine peridotite-gabbro complexes (Thayer, 1967).

In discussing the genesis of the plutonic rocks and their tectonic setting, Gass (1967) said:

"In composition, and in the nature of the main rock types present, the Troodos plutonic complex is closely comparable to Thayer's 'Alpine' type complexes. This . . . indicates a similar parent magma, but not necessarily a similar tectonic setting. . . . All evidence points to the upper mantle as the source of the Troodos Plutonic Complex magma." (p. 126)

"It is evident from the north-south structures within the massif that east-west tensional stress was dominant throughout its evolution. Although no direct evidence is available, the structure, particularly of the Sheeted Intrusive Complex, is of the type that could be formed on the crest of a mid-ocean ridge. . . . So, it is possible that the Troodos massif represents a volcanic edifice formed on the median ridge of Tethys. . . ." (p. 133)

In 1968 (p. 42) Gass amplified his concepts as follows:

"If the Sheeted Intrusive Complex of the massif represents oceanic layer 2, then the Troodos Plutonic Complex may represent layer 3. The crudely stratiform arrangement ranging from central ultramafic varieties through gabbros to marginal silicic members is the type of structure that could be formed in areas of excessive heat flow, such as the crest of mid-ocean ridges. . . . In the case of the Troodos massif, the mantle material would have to be fused sufficiently to allow gravity fractionation to . . . produce a gross compositional layering. If the Troodos massif does lie athwart the mid-Tethyan ridge, then the most obvious place to draw the axis of this ridge would be N/S through the Troodos Plutonic Complex."

Gass' hypothesis that the rocks of the Troodos Plutonic Complex were formed by gravity fractionation in the mantle is similar to ideas I have advanced to account for the genesis of the same kinds of alpine-type rocks in eugeosynclines (Thayer, 1967, in press). This concept postulates genesis of layered complexes consisting of peridotite, gabbro, and more silicic rocks at or near the top of the oceanic mantle, and later emplacement into crustal rocks. Although gravity data indicate that the structures of Cyprus and the Mid-Atlantic Ridge are very different, the assumption of similar densities to explain the anomalies in the two areas implies presence of the same kinds of rocks. Cyprus is the locus of a major gravity high ranging from 100 to 250 mgals above the surrounding region (Gass, 1968, p. 40), whereas a gravity low of comparable magnitude is indicated by Talwani and others (1965, p. 349) over the Mid-Atlantic Ridge. Talwani and others presented three models which satisfied the gravity data with densities of 3.05 to 3.2 g/cm^3 to depths of 25 to 40 km below the oceanic crust, across a belt 1500 km wide along the Mid-Atlantic Ridge. Gass (1968, p. 40) postulates a mass under Cyprus, 110 km wide by 240 km long, that ranges from 11 to 32 km thick at different places and "has a density of at least 3.3 g/cm^3." In an earlier paper, however, Gass and Masson-Smith (1963) indicated that a minimum density of 3.18 g/cm^3 for a mass about 11 km thick would account for the Troodos anomaly. The high positive anomaly is explained (Gass, 1968, p. 41) by thrusting of an undeformed slice of volcanic and upper mantle material over sialic crust during the Tertiary Alpine orogeny.

Gass' and Masson-Smith's earlier assumption of a density of 3.18 g/cm^3 for the Troodos block seems more probable to me than 3.3 g/cm^3, for at least two reasons. The Plutonic Complex includes a large volume of feldspathic rocks whose lower limits are unknown. Although Gass and his predecessors believe the rocks follow a stratiform pattern of distribution, the prevalence of steep banding and steep contacts between gabbro and peridotite suggest more complex structure and distribution (Thayer, 1967, p. 230). The 3.3 g/cm^3 figure implies solid fresh peridotite, but Gass (1967, p. 128) ascribes a large negative anomaly over Mount Olympus to complete serpentinization of harzburgite and dunite to a depth of 6.5 miles. If the peridotite is serpentinized to such a depth in one place, is it likely to be fresh everywhere else? The lower density figure would represent about a 70:30 mixture of peridotite and feldspathic rocks, and allow for some serpentinization.

The models proposed by Talwani and others (1965, p. 348) across the Mid-Atlantic Ridge postulate rocks of density 3.05 gm/cm^3 at a depth of 25 to 30 km, or of density 3.15–3.2 g/cm^3 to depths of 35 km or more. Hess (1962, p. 605) has reasoned that serpentinization probably does not extend more than 5 km below the sea floor. It seems more consistent, therefore, to suppose the low anomalies found across the Mid-Atlantic Ridge are largely caused by the presence of feldspathic rocks similar to those exposed in the Troodos massif, or to those postulated by Muir and Tilley.

MAGNESIUM-RICH GABBRO UNDER OCEANIC BASALT FLOORS

The most widely accepted version of the ocean-floor spreading hypothesis (Girdler, 1965; Hess, 1962) implies that some characteristic associations of volcanic and plutonic rocks should be found both in eugeosynclines and in mid-ocean ridges, despite the postulated tectonic differences in the two environments. If the rocks that formed along and under a mid-ocean ridge move laterally and then down under continental margins, like a conveyor belt (Isacks and others, 1968), some original rock relations should survive and be preserved in eugeosynclines in what Dietz (1963) has called ocean rind fragments. Comparison of the Troodos massif, on a postulated ancient mid-ocean ridge, with rocks interpreted as oceanic at continental margins in Papua and in Cuba, does, in fact, reveal many similarities.

In eastern Papua, Davies (1968) interprets the Bowutu Mountains as a slice of oceanic crust and upper mantle rocks thrust over continental sialic crust, in a manner directly comparable to Gass' interpretation of the Troodos massif. A 4- to 5-km thick layer of magnesium-rich gabbro lies between peridotite and a thick cover of Cretaceous basalts. The gabbro is described as generally intruding peridotite, "but at a few localities the two are interlayered in accumulative sequences" (Davies, 1968, p. 211). The peridotites are mostly olivine-rich harzburgite and dunite which are fresh except along contacts and shear zones. Small masses of diorite and quartz diorite intrude the gabbro near contacts with volcanic rocks.

Relict cumulus chromite, plagioclase, and olivine have recently been described in the Camagüey district in Cuba (Thayer, in press), where Hess (1964) interprets the peridotite as part of "layer 3." Previously, Flint and others (1948) and Thayer (1942) had described intimate structural and petrologic relationships which are of major economic importance, between peridotite, chromite deposits, and gabbroic rocks.

We can now point to three presently continental-type areas where rocks interpreted as oceanic crust and upper mantle are exposed: the Troodos massif in Cyprus, the Bowutu Mountains in Papua, and eastern Cuba. In all three areas, large volumes of magnesium-rich gabbro are intimately associated with olivine-rich peridotite, and gravitational differentiation has been cited to explain some features of the rocks. It would appear that some of the geophysical phenomena attributed to serpentinization of peridotite under ocean basins more probably are caused by presence of gabbro. The consistency of petrologic relations in the three widely separated areas implies that the concept of ocean basins as floored by tholeiitic basalt lying directly on partly serpentinized peridotite is greatly oversimplified.

CONCLUSIONS

In conclusion, it appears that two suites of feldspathic plutonic rocks which are associated with olivine-rich peridotite in the Mid-Atlantic Ridge have analogs on the continents. Fresh mylonites in St. Paul Rocks, which contain various amounts of subsilicic alkalic hornblende, are similar to hornblende-bearing mylonitic peridotite and probably comagmatic gabbroic rocks near Tinaquillo, Venezuela, and Lizard, England. Altered peridotite, gneissic gabbro, and quartz diorite dredged from the Mid-Atlantic Ridge, 30° to 36° north of St. Paul Rocks, closely resemble the various members of alpine mafic complexes which are widely distributed in land areas. The Troodos massif in Cyprus offers a possibly unique opportunity to study the relations between basaltic lavas and plutonic rocks which appear to have evolved under oceanic conditions, yet are very much like alpine mafic (ophiolitic) complexes which occur in eugeosynclinal environments.

Petrologic relations between peridotite and obviously comagmatic magnesium-rich gabbro are similar in 3 widely separated complexes which are interpreted as uplifted ocean floor and upper mantle. Widespread distribution of feldspathic rocks with peridotite in the upper part of the mantle would not be inconsistent with the hypothesis of ocean-floor spreading, but would complicate some of the postulated mechanisms involved.

REFERENCES CITED

Aumento, F., 1969, Geological investigations, Mid-Atlantic Ridge: Canada Geol. Survey, Activities Rept., Paper 69-1, pt. A., p. 253–257.

Bear, L. M., 1966, The evolution and petrogenesis of the Troodos Complex: Cyprus Geol. Survey Ann. Rept. 1965, p. 26–38.

Davies, H. L., 1968, Papuan ultramafic belt: 23d Internat. Geol. Cong., Prague, Rept., sec. 1, p. 209–220.

Dietz, R. S., 1963, Alpine serpentines as oceanic rind fragments: Geol. Soc. America Bull., v. 74, p. 947–952.

Flett, J. S., 1946, Geology of the Lizard and Meneage: Great Britain Geol. Survey Mem., 208 p.

Flint, D. E., de Albear, J. F., Guild, P. W., 1948, Geology and chromite deposits of the Camagüey district, Camagüey Province, Cuba: U.S. Geol. Survey Bull. 954-B, p. 39–63.

Fox, H., and Teall, J. J. H., 1893, Notes on some coast sections of the Lizard: Geol. Soc. London Quart. Jour., v. 49, p. 199–210.

Gass, I. G., 1967, The ultrabasic volcanic assemblage of the Troodos massif, Cyprus, p. 121–134, in Wyllie, P. J., Editor, Ultramafic and related rocks: New York, John Wiley & Sons, 464 p.

—— 1968, Is the Troodos massif of Cyprus a fragment of Mesozoic ocean floor?: Nature, v. 220, p. 39–42.

Gass, I. G., and Masson-Smith, D., 1963, The geology and gravity anomalies of the Troodos Massif, Cyprus: Royal Soc. London Philos. Trans., ser. A, v. 255, p. 417–467.

Girdler, R. W., 1965, The formation of new oceanic crust, p. 123–136, in A symposium on continental drift: Royal Soc. London Philos. Trans., ser. A, v. 258, 323 p.

Green, D. H., 1964, The petrogenesis of the high-temperature peridotite intrusion in the Lizard area, Cornwall: Jour. Petrology, v. 5, p. 134–188.

Hess, H. H., 1962, History of ocean basins, p. 599–620, *in* Petrologic studies (Buddington Volume): Geol. Soc. America, 660 p.

—— 1964, The oceanic crust, the upper mantle and the Mayaguez serpentinized peridotite, p. 169–175, *in* Burk, C. A., *Editor*, A study of serpentinite: Natl. Acad. Sci.–Nat. Research Council Pub. 1188, 175 p.

Isacks, B., Oliver, J., and Sykes, L. R., 1968, Seismology and the new global tectonics: Jour. Geophys. Research, v. 73, p. 5855–5899.

MacKenzie, D. B., 1960, High-temperature alpine-type peridotite from Venezuela: Geol. Soc. America Bull., v. 71, p. 303–318.

Matthews, D. H., Vine, F. J., and Cann, F. J., 1965, Geology of an area of the Carlsberg Ridge, Indian Ocean: Geol. Soc. America Bull., v. 76, p. 675–682.

Melson, W. G., Jarosewitch, E., Bowen, V. T., and Thompson. G.. 1967, St. Peter and St. Paul Rocks: a high-temperature, mantle-derived intrusion: Science, v. 155, p. 1532–1535.

Melson, W. G., Thompson, G., and Van Andel, T. H., 1968, Volcanism and metamorphism in the Mid-Atlantic Ridge, 22° N. latitude: Jour. Geophys. Research, v. 73, p. 5925–5941.

Muir, I. D., and Tilley, C. E., 1964, Basalts from the northern part of the rift zone of the Mid-Atlantic Ridge: Jour. Petrology, v. 5, p. 409–434.

Quon, S. H., and Ehlers, E. G., 1963, Rocks of northern part of Mid-Atlantic Ridge: Geol. Soc. America Bull., v. 74, p. 1–8.

Shand, S. J., 1949, Rocks of the Mid-Atlantic Ridge: Jour. Geology, v. 57, p. 89–92.

Talwani, M., Le Pichon, X., and Ewing, M., 1965, Crustal structure of the mid-ocean ridges: 2. Computed model from gravity and seismic refraction data: Jour. Geophys. Research, v. 70, p. 341–357.

Thayer, T. P., 1942, Chrome resources of Cuba: U.S. Geol. Survey Bull. 935-A, p. 1–74.

—— 1960, Some critical differences between alpine-type and stratiform peridotite-gabbro complexes: 21st. Internat. Geol. Cong., Copenhagen, Repts., pt. 13, p. 247–259.

—— 1963a, The Canyon Mountain Complex, Oregon, and the alpine mafic magma stem: U.S. Geol. Survey Prof. Paper 475-C, p. C82–C85.

—— 1963b, Flow layering in alpine peridotite-gabbro complexes: Mineral Soc. America Spec. Paper 1, p. 55–61.

—— 1967, Chemical and structural relations of ultramafic and feldspathic rocks in alpine intrusive complexes, p. 222–239, *in* Wyllie, P. J., *Editor*, Ultramafic and related rocks: New York, John Wiley & Sons, 464 p.

——1969, Gravity differentiation and magmatic re-emplacement of podiform chromite deposits, *in* Wilson, H. D. B., *Editor*, Symposium on magmatic deposits: Econ. Geology mon. 4, p. 132–146.

Thayer, T. P., and Brown, C. E., 1961, Is the Tinaquillo, Venezuela, "pseudogabbro" metamorphic or igneous?: Geol. Soc. America Bull., v. 72, p. 1565–1570.

Thayer, T. P., and Himmelberg, G. R., 1968, Rock succession in the alpine-type mafic complex at Canyon Mountain, Oregon: 23rd Internat. Geol. Cong., Prague, Rept., sec. 1, p. 175–186.

Wilson, R. A. M., and Ingham, F. T., 1959, The geology of the Xeros-Troodos area with an account of the mineral resources: Cyprus Geol. Survey Mem. 1, 184 p.

Manuscript Received by The Society December 2, 1968
Revised Manuscript Received March 3, 1969
Publication Authorized by The Director, U.S. Geological Survey

Erratum

On p. 284, the tenth line from the top should read "Matthews, D. H., Vine, F. J., and Cann, J.R. .."

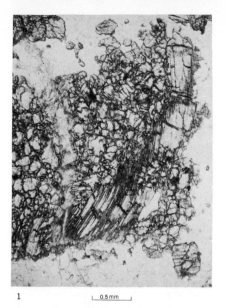

1   ⌊ 0.5 mm ⌋

2   ⌊ 0.2 mm ⌋

Figure 1. Photomicrograph of granulitic norite containing deformed bronzite in matrix of clear granular plagioclase (around three sides of field) and pyroxene with some interstitial amphibole. Uralitic amphibole occurs along veinlet near left edge of field. *Atlantis* station 6, thin-section 6–16; plane light.

Figure 2. Photomicrograph of gneissic anorthosite containing crushed feldspar grain 1.3 mm long in mosaic of fresh feldspar of similar composition (labradorite?). *Atlantis* station 6, thin-section 6-2; crossed polarizers.

3   ⌊ 0.2 mm ⌋

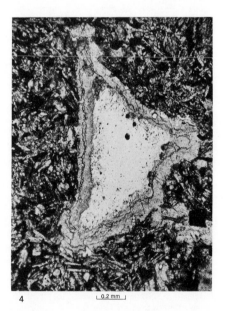

4   ⌊ 0.2 mm ⌋

Figure 3. Photomicrograph of amphibole schist. *Atlantis* station 21, thin-section 21-4, crossed polarizers.

Figure 4. Photomicrograph of metabasalt containing amygdule filled with epidote (outer layer), chlorite (middle layer), and quartz (center). *Atlantis* station 78, thin-section 78-19, plane light.

PHOTOMICROGRAPHS OF ROCKS FROM THE MID-ATLANTIC RIDGE

THAYER, PLATE 1

Geological Society of America Bulletin, v. 80, no. 8

14

Reprinted from *Science*, **169**, 971–974 (Sept. 4, 1970)

Mantle-Derived Peridotites in Southwestern Oregon: Relation to Plate Tectonics

L. G. MEDARIS, JR., and R. H. DOTT, JR.

Abstract. A group of peridotites in southwestern Oregon contains high-pressure mineral assemblages reflecting recrystallization at high temperatures (1100° to 1200°C) over a range of pressure decreasing from 19 to 5 kilobars. It is proposed that the peridotites represent upper-mantle material brought from depth along the ancestral Gorda–Juan de Fuca ridge system, transported eastward by the spreading Gorda lithosphere plate, and then emplaced by thrust-faulting in the western margin of the Cordillera during late Mesozoic time.

The origin of ultramafic rocks in orogenic belts has long been a subject of debate (*1, 2*). Despite careful field and petrologic studies of these enigmatic rocks in recent years, a unifying theme has failed to emerge, and instead, several different modes of origin now seem plausible (*2*). According to models of plate tectonics, ultramafic rocks of the alpine type may represent fragments of the oceanic crust and upper mantle transported by spreading lithosphere plates and then tectonically emplaced in orogenic belts (*3*). In the course of a comparative study of selected ultramafic rocks in northwestern California and southwestern Oregon, we discovered that a group of peridotites located in coastal southwestern Oregon contains high-pressure mineral assemblages, and the petrologic data are consistent with the idea that these assemblages are derived from the upper mantle.

Southwestern Oregon is situated on the western margin of the Cordilleran mobile belt, in proximity to the seismically active Gorda–Juan de Fuca ridge system and its associated transform faults (*4, 5*) (Fig. 1). Late Cenozoic underthrusting of the American lithosphere plate by the Gorda plate in this region is indicated by the presence of magnetic anomaly number 3 (5 million years old) beneath the continental slope and by deformation of late Cenozoic, possible trench sediments at the base of the continental slope (*6*). The late Cenozoic Cascade volcanoes appear to represent a volcanic arc complementary to the inferred trench, although an active Benioff zone defined by deep-focus earthquakes is not known at the present time (*4, 5*).

As in California (*7*), Mesozoic underthrusting in Oregon and Washington is inferred from mélange terranes associated with volcanic rocks and serpentinite sheets. During the culmination of late Mesozoic mountain-building, sedimentary and volcanic rocks were deformed and extensively thrust-faulted, metamorphosed, and intruded by igneous plutons (*8–10*). The entire Mesozoic complex, including the thrust plates, was overlapped by Eocene strata. Then all these rocks and thrust plates were cut by late Cenozoic vertical faults trending north-northwest, presumed to be related to the San Andreas system (*9, 11*) (Fig. 1).

That the region offshore from northwestern California and southwestern Oregon is unusual in showing structural patterns related to spreading and underthrusting as well as faulting of the San Andreas type was suggested by Tobin and Sykes (*5*) and confirmed by Silver

(*6*). Spreading ceased south of the Mendocino fracture zone after the East Pacific Rise was destroyed there by collision with the American lithosphere plate (*4, 5*), but it has continued farther north. As the rise disappeared farther south, the San Andreas transform system became fully developed and similar faults were propagated northward along the western edge of the American plate into Oregon, as evidenced by onshore geology (*9*) and earthquake first motions (*12*). An unusually complex pattern of contemporaneous strike-slip faulting and underthrusting has resulted at the edge of the continental (American) plate.

Peridotite masses, serpentinized to varying degrees, characterize southwestern Oregon. The four of interest (Fig. 1) are in fault contact with surrounding rocks, and metasomatic reaction zones (rodingites) commonly are present along contacts, an indication of the relatively low temperatures of final emplacement (*13*). The Vondergreen Hill and Carpenterville peridotites, located near the coast (Fig. 1, Nos. ① and ②), have been intensely sheared. The Carpenterville mass consists of large, randomly oriented blocks of peridotite, pyroxenite, serpentinite, gabbro, and Jurassic sedimentary and volcanic rocks, all intimately sheared together in a matrix of crushed material, whereas the Vondergreen Hill mass consists solely of peridotite blocks in a sheared serpentinite matrix. Farther inland, the

Table 1. Chemical analyses of peridotites, recalculated to 100 percent (by weight) on a water-free basis (total Fe as FeO). Sample 1, average value for six peridotites (range of values is given in parentheses) from southwestern Oregon: four from Vondergreen Hill and one each from Carpenterville and Signal Butte; sample 2, lherzolite, Central Indian Ridge, Indian Ocean (*16*); sample 3, St. Paul's Rocks (WD 55), Atlantic Ocean (*18*); sample 4, Lizard, England, average (*19*); sample 5, pyrolite (*20*).

Oxide	Sample				
	1	2	3	4	5
SiO₂	44.1 (43.4–44.7)	43.69	44.63	44.77	45.18
Al₂O₃	4.0 (3.6–4.2)	3.13	4.10	4.16	3.54
Cr₂O₃	0.4 (0.4–0.5)	0.55	0.46	0.40	0.43
FeO	9.1 (8.7–9.8)	7.34	7.91	8.21	8.45
MgO	40.0 (38.7–41.2)	40.30	39.12	39.22	37.50
CaO	2.4 (1.9–2.7)	4.28	2.87	2.42	3.08
Na₂O		0.23	0.32	0.22	0.57
K₂O		.02	.07	.05	.13
TiO₂		.01	.12	.19	.71
MnO		.07	.13	.11	.14
P₂O₅			.02	.01	.06
NiO		0.38	.25	.24	.20
CoO					.01
Total	100.00	100.00	100.00	100.00	100.00

286

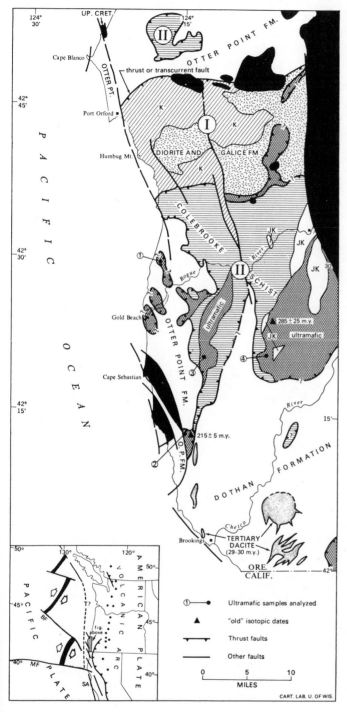

Fig. 1. Generalized geologic map of the coastal region of southwestern Oregon showing four ultramafic rock localities and two isotopic date localities discussed in text [adapted from (*8–10, 14*); see (*25*)]. Inset map shows location of region studied with respect to Gorda–Juan de Fuca ridge system and associated transform faults, the Cascade volcanic arc, and San Andreas fault system; *T?* designates site of inferred filled trench along margin between American and Gorda (shaded) lithosphere plates. Sawtooth symbol denotes major late Mesozoic thrust-fault boundary between Coast Range geologic province and the Cordilleran orogen [adapted from (*4*)].

Signal Butte and Snow Camp peridotites are much larger and less sheared, exhibiting semiconcordant contacts with surrounding strata and extending north-south for many miles (Fig. 1, Nos. ③ and ④). Both bodies are comprised of massive peridotite with minor serpentinite, cut by small gabbroic dikes. Dioritic stocks also occur within the Snow Camp peridotite. In the past, all four bodies were interpreted as narrow, steeply dipping intrusive rocks sheared and serpentinized during post-intrusive faulting (*9, 14*). Recently, however, Coleman (*10*) has proposed the plausible hypothesis that most of the ultramafic bodies are associated wih low-dipping late Mesozoic thrust sheets modified along late Cenozoic vertical fault zones (Fig. 1). At least one, and possibly two, major sheets are indicated, and Coleman believes (*10*) that associated peridotites and schists (block II of Fig. 1) may represent a series of several additional thrust sheets. Where serpentinization was thorough, the ultramafic masses appear to have risen as diapirs or tectonic intrusions through more dense surrounding rocks to further complicate the region.

We have analyzed six samples of peridotite, four from Vondergreen Hill and one each from Carpenterville and Signal Butte, for Si, Fe, Mg, Ca, Cr, and Al by electron probe techniques (*15*). Except for a slight enrichment in total Fe, the peridotites closely resemble in chemistry peridotites from mid-ocean ridges (*16–18*), high-temperature peridotites (*19*), and the hypothetical mantle material, pyrolite (*20*) (Table 1).

Mineralogically, specimens from all four peridotites are the same, since each contains forsterite, enstatite, diopside, and accessory spinel. Although serpentinization is appreciable in many

287

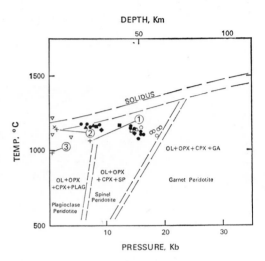

DEPTH, Km

Fig. 2. Conditions of recrystallization for peridotites of southwestern Oregon, based on O'Hara's pyroxene grid for natural aluminous four-phase peridotites (*22*). ○, ●, Vondergreen Hill; □, ■, Carpenterville; △, ▲, Signal Butte (open symbols, augen or coarse-grained pyroxene; solid symbols, matrix pyroxene); ♦, Snow Camp; ①, Lizard, England (primary mineral assemblage); ②, Horoman, Japan (primary); ③, Miyamori, Japan (primary); +, recrystallized assemblages for ①, ②, and ③; ▽, New Zealand peridotites; ✕, Bushveld peridotite.

samples, producing mesh and bastite textures, sufficient amounts of anhydrous silicates are preserved from which to deciper the original textures. The Vondergreen Hill peridotite is texturally distinct from the other three, in that it contains strained augen (≦ 10 mm) of ensatite, diopside, and forsterite in a fine-grained matrix of forsterite, enstatite, diopside, and spinel. The other peridotites have inequigranular textures, in which large, anhedral pyroxenes (≦ 6 mm) are set in a medium-grained, allotriomorphic granular matrix of forsterite, two pyroxenes, and spinel. Nonuniform extinction and straining are evident in silicates from all four localities, but these features are most conspicuous in the Vondergreen Hill peridotite. Exsolution lamellae of diopside are prominent in coarse-grained enstatite, but lamellae are absent or only poorly developed in coarse-grained diopside, and absent in matrix pyroxenes. Spinel is anhedral, pale brownish-yellow to brown, rimmed by magnetite, and interstitial to matrix silicates in all specimens.

Microprobe analyses (*21*) indicate that the silicates are relatively uniform with respect to the chemistry of the major elements. Values (in mole percent) are: Forsterite: forsterite, 90 to 91; enstatite: wollastonite, 1 to 3; enstatite, 88 to 90; ferrosilite, 9 to 10; diopside: wollastonite, 47 to 49; enstatite, 48 to 49; ferrosilite, 4 to 5. The Al_2O_3 contents of the pyroxenes are exceptionally high (up to 5.60 percent for enstatite, 6.88 percent for diopside),

but variable within a given specimen, ranging, for example, from 3.36 to 5.15 percent for enstatite and from 4.97 to 6.43 percent for diopside in a sample of Vondergreen Hill peridotite. Invariably, the coarse-grained pyroxenes contain more Al_2O_3 than associated matrix pyroxenes do. Spinels in the Vondergreen Hill, Carpenterville, and Signal Butte peridotites are unusually rich in $MgAl_2O_4$ [$Mg/(Mg + Fe^{2+})$ = 0.74 to 0.81; $Cr/(Cr + Al)$ = 0.12 to 0.21], whereas spinel in the Snow Camp mass contains slightly more iron and chromium [$Mg/(Mg + Fe^{2+})$ = 0.69 to 0.75; $Cr/(Cr + Al)$ = 0.30 to 0.34].

Conditions of recrystallization for the peridotites of southwestern Oregon have been estimated by means of O'Hara's pyroxene grid (*22*), based on the composition of clinopyroxene in a four-phase assemblage consisting of clinopyroxene (CPX), orthopyroxene (OPX), olivine (OL), and an aluminous phase, either plagioclase, spinel, or garnet (Fig. 2). As the plots for coarse-grained and matrix clinopyroxenes in Fig. 2 indicate, the peridotites were derived from depths of 50 to 60 km, and recrystallization took place at high and perhaps slightly increasing temperatures on the order of 1100° to 1200°C as the peridotites were brought to shallower depths. The prevalence of disequilibrium assemblages is demonstrated by different Al_2O_3 and Cr_2O_3 contents, and thus different inferred temperatures and pressures of recrystallization, for individual clinopyroxene grains in a given peridotite specimen.

Despite this, however, partitioning of Mg and Fe^{2+} among forsterite, enstatite, diopside, and spinel indicates uniformly high temperatures of recrystallization (*23*). The apparent equilibrium distribution of Mg and Fe^{2+} in these disequilibrium assemblages is due to recrystallization within a restricted temperature interval but over a wide range of pressure, combined with the relative insensitivity of $Mg\text{-}Fe^{2+}$ partitioning to changes in pressure. If recrystallization of the anhydrous phases were related to tectonic emplacement of the peridotites in the continental crust, one would expect much lower temperatures of recrystallization, particularly because blueschist and greenschist facies metamorphic rocks (Colebrooke Schist) are widely developed in this region, and rodingites occur at peridotite contracts. Clearly, the episode of high-temperature recrystallization recorded in these peridotites is earlier than, and unrelated to, their ultimate emplacement in the continental crust.

We propose that this group of peridotites represents fragments of the upper mantle originally brought from depth beneath the ancestral Gorda–Juan de Fuca ridge system. During this initial stage of development, high-temperature recrystallization occurred, which resulted in the breakdown of alumina-rich pyroxenes with the concomitant formation of spinel and less-aluminous matrix pyroxenes. Partial melting and generation of gabbros or basalts may have taken place at depths of 20 to 25

288

km, as suggested by the occurrence of gabbroic dikes in the peridotites and by the approach of the recrystallization trend toward the solidus in Fig. 2. The peridotites and associated rocks were then carried eastward by the spreading Gorda plate, eventually to be emplaced by thrust-faulting in the western margin of the Cordillera. Low-temperature serpentinization and formation of rodingites probably first occurred at shallow depths beneath the ridge, continued during eastward transport, culminated during tectonic emplacement, and persisted through Cenozoic time; perhaps these processes are continuing today (*24*).

The sequence of tectonic events summarized above provides an explanation for what were previously thought to be anomalous isotopic dates. Many K-Ar dates indicate that the late Mesozoic regional metamorphism and dioritic plutonism in southwestern Oregon and northwestern California occurred chiefly between 130 and 145 million years ago (*8, 9*), but dioritic and gabbroic masses within the Carpenterville and Snow Camp peridotites yielded K-Ar dates on amphiboles of 215 ± 5 and 285 ± 25 million years, respectively (*9*). The mineralogic evidence from the peridotites is consistent with the previously suggested hypothesis that several large blocks of old diorite and gabbro have been carried up within the mantle peridotite masses, either with their isotopic clocks left intact, or with excess argon having been incorporated into their minerals in a high-pressure environment.

L. G. Medaris, Jr.
R. H. Dott, Jr.

Department of Geology and Geophysics, University of Wisconsin, Madison 53706

References and Notes

1. H. H. Hess, *Geol. Soc. Amer. Spec. Pap. 62* (1955), p. 391; N. L. Bowen and O. F. Tuttle, *Bull. Geol. Soc. Amer.* **60**, 439 (1949).
2. P. J. Wyllie, in *Ultramafic and Related Rocks*, P. J. Wyllie, Ed. (Wiley, New York, 1968), p. 407; *Tectonophysics* **7**, 437 (1969).
3. R. S. Dietz, *Bull. Geol. Soc. Amer.* **74**, 947 (1963); B. Isacks, J. Oliver, L. R. Sykes, *J. Geophys. Res.* **73**, 5855 (1968).
4. W. J. Morgan, *J. Geophys. Res.* **73**, 1959 (1968).
5. D. G. Tobin and L. R. Sykes, *ibid.*, p. 3821.
6. E. A. Silver, *Science* **166**, 1265 (1969); thesis, University of California, San Diego (1969).
7. W. Hamilton, *Bull. Geol. Soc. Amer.* **80**, 2479 (1969); W. G. Ernst, *J. Geophys. Res.* **75**, 886 (1970); B. M. Page, *Bull. Geol. Soc. Amer.* **81**, 667 (1970).
8. W. P. Irwin, *U.S. Geol. Surv. Prof. Pap. 501-C* (1964), p. C1; M. C. Blake, W. P. Irwin, R. G. Coleman, *U.S. Geol. Surv. Prof. Pap. 575-C* (1967), p. C1; J. Suppe, *Bull. Geol. Soc. Amer.* **80**, 135 (1969); R. Lent, thesis, University of Oregon (1969).
9. R. H. Dott, Jr., *J. Geophys. Res.* **70**, 4687 (1965).
10. R. G. Coleman, Abstract, Cordilleran Section, annual meeting of the Geological Society of America, Eugene, Oregon (1969), p. 12; ———, personal communication.
11. R. H. Dott, Jr., *Science* **166**, 874 (1969).
12. T. V. McEvilly, *Nature* **220**, 901 (1968).
13. R. G. Coleman, *U.S. Geol. Surv. Bull. 1247* (1967).
14. J. G. Koch, *Bull. Amer. Ass. Petrol. Geol.* **56**, 25 (1966).
15. B. L. Gulson and J. F. Lovering, *Geochim. Cosmochim. Acta* **32**, 119 (1968).
16. C. G. Engel and R. L. Fisher, *Science* **166**, 1136 (1969).
17. W. G. Melson *et al.*, *ibid.* **155**, 1532 (1967); E. Bonatti, *Nature* **219**, 363 (1968).
18. H. H. Hess, in *A Study of Serpentinite near Mayaguez, Puerto Rico*, C. A. Burk, Ed. [*Nat. Acad. Sci. Nat. Res. Counc. Publ. 1188* (1964), p. 172].
19. D. H. Green, *J. Petrol.* **5**, 134 (1964).
20. A. E. Ringwood, in *The Earth's Crust and Upper Mantle*, P. J. Hart, Ed. [American Geophysical Union (*Geophysical Monograph 13*), Washington, D.C., 1969], p. 1.
21. Polished thin sections were prepared for five peridotite specimens from Vondergreen Hill, two each from Carpenterville and Signal Butte, and one from Snow Camp. In the mineral analysis we used an electron probe (Applied Research Labs) and followed the procedures recommended by A. E. Bence and A. L. Albee [*J. Geol.* **76**, 382 (1968)]. Reduction of probe data was performed on a computer (Univac 1108), with the aid of a program written by D. Gast.
22. M. J. O'Hara has established a provisional petrogenetic grid for the system, CaO, (Mg, Fe^{2+})O, $(Al,Cr,Fe^{3+})_2O_3$, SiO_2, on the basis of experimental studies. Two sets of intersecting curves, related to the CaO and R_2O_3 (where R is a trivalent metal) contents of clin. pyroxene, permit the determination of temperature and pressure conditions from the chemical composition of clinopyroxene in a four-phase assemblage [M. J. O'Hara, in *Ultramafic and Related Rocks*, P. J. Wyllie, Ed. (Wiley, New York, 1968), p. 383].
23. L. G. Medaris, in preparation; E. D. Jackson, in *Magmatic Ore Deposits*, H. D. B. Wilson, Ed. [*Econ. Geol. Monogr. 4* (1969), p. 41]; R. Kretz, *J. Geol.* **71**, 773 (1963).
24. I. Barnes, V. C. LaMarche. Jr., G. Himmelberg, *Science* **156**, 830 (1967).
25. Mesozoic ocean floor underthrusting apparently culminated in late Cretaceous time with thrusting of at least one and probably more large plates (for example, Fig. 1, block II) over late Jurassic Otter Point and Dothan formations, which are similar to the Franciscan complex of California. Eocene (and possibly latest Cretaceous) strata then were deposited. Finally, vertical faults trending northwest and dacite intrusions formed in post-Eocene time. Block I contains Jurassic greenschist and diorite basement unconformably overlain by Lower Cretaceous strata (Fig. 1, K), all of which represent Klamath Province rocks apparently transported relatively at least 25 miles (40 km) from the east either by thrusting or by transcurrent faulting. Block II (Fig. 1) contains schists transitional between blueschist and greenschist facies, ultramafic masses, and unmetamorphosed uppermost Jurassic(?) and Lower Cretaceous strata (*JK*), all of unknown location prior to thrusting. The ultramafic masses are inferred to have been transported eastward from the Gorda ridge and structurally injected into the Colebrooke Schist by underthrusting. Subsequently the entire complex was emplaced in its present position during culmination of the Mesozoic episode of spreading.
26. Supported by grants from the Wisconsin Alumni Research Foundation. We are indebted to R. G. Coleman for permission to include his thrust interpretations in Fig. 1 and for his constructive criticism during the evolution of this paper. We thank C. Craddock and C. V. Guidotti for reading the manuscript, and E. D. Glover for providing invaluable assistance in the operation of the electron probe.

6 August 1969; revised 27 April 1970

15

Reprinted from *U.S. Geol. Survey Prof. Paper 700-C*, 70–81 (1970)

ON-LAND MESOZOIC OCEANIC CRUST IN CALIFORNIA COAST RANGES

By EDGAR H. BAILEY, M. C. BLAKE, JR.,
and DAVID L. JONES, Menlo Park, Calif.

Abstract.—The basal mudstones of the Upper Jurassic to Upper Cretaceous Great Valley sequence rest despositionally on a typical ophiolite ultramafic-mafic succession of igneous rocks. The ophiolite succession from top downward typically consists of chert; keratophyric to basaltic lavas; diabase, gabbro, or norite; and serpentinized peridotite, although not all parts are present everywhere. The volcanic rocks have an average thickness of 3,000 feet (900 m), and the serpentine may be as much as 5,000 feet (1,500 m) thick above a basal thrust fault. Comparison of occurrence, lithology, and thickness with the present in situ sea floor indicates that the ophiolite is the exposed Mesozoic oceanic crust on which sedimentary rocks of the Great Valley sequence were deposited. Coeval eugeosynclinal rocks of the Franciscan assemblage have been dragged below the rocks of the Great Valley sequence by sea-floor spreading. A great thrust fault, herein named the Coast Range thrust, separates the Franciscan and Great Valley sequence. Serpentine immediately above the thrust, previously thought to have been intruded into the fault zone, is the basal part of the Mesozoic oceanic crust lying beneath the Great Valley sedimentary rocks and thus was present before thrusting commenced.

The California Coast Ranges and adjacent Great Valley contain two coeval Upper Jurassic to Upper Cretaceous sequences, both possibly as much as 50,000 feet (15,000 meters) thick (see fig. 1). The western unit is the eugeosynclinal Franciscan assemblage of Bailey and others (1964), consisting of graywacke, shale, mafic volcanic rock, chert, limestone, and metamorphic rocks of zeolite and blueschist facies. The eastern unit is the Great Valley sequence, which consists predominantly of graywacke and shale with some conglomerate. This sequence was deposited in an area lying continentward from the site of accumulation of the eugeosynclinal Franciscan rocks, and it was referred to as miogeosynclinal by Bailey and others (1964) and as shelf and slope facies by Irwin (1964). Although its base is not exposed, the eugeosynclinal Franciscan assemblage has

been regarded as having been deposited in a deep ocean environment on oceanic crust. In contrast, geologists have tended to regard the Great Valley sequence as having been deposited on continental crust because Cretaceous clastic strata of this sequence in the northern and eastern parts of the valley rest depositionally on the metamorphic and granitic rocks of the Klamath Mountains and Sierra Nevada. As is described herein, however, the Jurassic part of the Great Valley sequence, exposed west of the valley, rests depositionally upon an accumulation of generally mafic volcanic rocks that in turn rest on serpentinized ultramafic rocks (Bailey and Blake, 1969). This basal ultramafic-mafic sequence is similar in lithology to other composite igneous piles elsewhere referred to as ophiolite, a term we use throughout the paper in a descriptive sense without genetic implication. Because of its resemblance to present in situ oceanic crust, we believe that the ophiolite is the Mesozoic oceanic crust on which the Great Valley sediments were deposited.

The significance of the ultramafic-mafic sequence as oceanic crust has not been generally recognized; and even though these rocks are widespread and fairly well exposed in many places in the California Coast Ranges, they remain little studied. We, as well as others, are only beginning to study in detail their occurrence, structure, petrography, and chemistry. This preliminary paper, therefore, leans heavily on the available descriptions of other geologists, most of whom had no idea that they were studying or writing about ancient oceanic crustal material. However, focusing attention on these rocks seems worthwhile, especially as recognition of the character of the serpentine has an important bearing on the geologic history of a major part of the Pacific margin, now being widely discussed under the impetus of concepts of sea-floor spreading and new global tectonics.

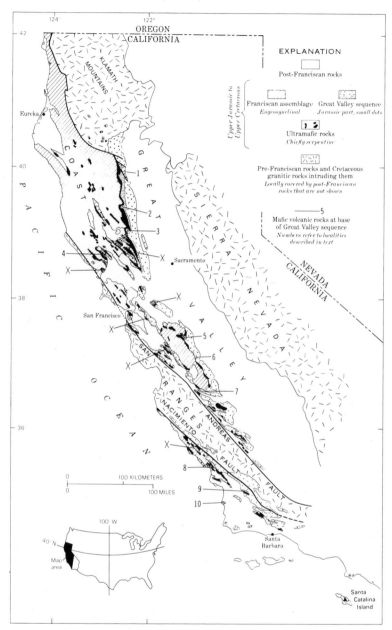

FIGURE 1.—Map of western California, showing the distribution of the coeval Great Valley sequence and Franciscan assemblage and the serpentine that in most places lies between these major units. Numbers indicate localities, described in text, where mafic volcanic rocks underlie Upper Jurassic sedimentary rocks of the Great Valley sequence; other similar localities not described are indicated by X's.

DISTRIBUTION OF THE OPHIOLITE

More than a dozen localities where Jurassic sedimentary rocks of the Great Valley sequence rest directly upon volcanic rock, which in turn lies on serpentine, are shown by numbers and X's in figure 1. Also shown is the distribution of the younger rocks of the Great Valley sequence, the coeval Franciscan assemblage, and the sheet of serpentine separating the two units. The localities where the ophiolitic succession is known to underlie Jurassic sedimentary rocks of the Great Valley sequence are widely distributed and coextensive with the outcrop area of the Jurassic strata. Furthermore, near the north end of the Great Valley, the volcanic rocks and serpentine both terminate abruptly at the fault that limits the northern extent of the Jurassic sedimentary strata, as they would if they formed the basal beds of the sedimentary sequence.

DESCRIPTION OF LOCALITIES

Columnar sections showing the succession of rocks from the lowest exposure of serpentine upward through the mafic igneous rocks and overlying sedimentary strata at 10 localities are shown on figure 2. We have examined some of these localities and modified published geologic descriptions; for localities not visited, we have liberally reinterpreted the published reports and maps in light of the ophiolite or oceanic crust concept. Such preliminary reinterpretations are necessary at this time because the relations of one kind of igneous rock to another commonly are not stated, and from available reports we could not be certain, for example, whether a gabbro occurs above or below a diabase. In such instances, we have stacked the mafic igneous rocks in the most probable order. Similarly, on some maps the original sequence in the ophiolite has been obscured by cover or by real or postulated faults. Additional work can be expected to modify the columns as drawn, but probably only to a limited degree. Most thicknesses shown are averages based on sections through mapped areas, and again, because of lack of data on dips, are in part "best guesses." Despite these limitations, the section proved to be surprisingly uniform with respect to thickness, kind, and succession of rocks, especially in the ophiolite sequences. The areas represented by the columns are discussed below in geographic order from north to south as shown on figure 1, and from left to right as shown on figure 2.

Paskenta

West of Paskenta, a succession of at least 10,000 feet (3,000 m) of Jurassic mudstone with some sandstone and conglomerate (1, figs. 1 and 2) lies depositionally on mafic igneous breccias that lie on gabbro, banded gab-bro, pyroxenite, and serpentinized peridotite and dunite. The kinds of ophiolitic rocks are not discriminated on any published geologic map, although they extend from the foothills west of Paskenta westward for a distance of 3 miles (5 km) and to an altitude almost 4,000 feet (1,200 m) above the valley. Most of this section is basalt occurring as breccia or pillow lava, but about one-quarter is ultramafic rock, chiefly serpentinized peridotite. As the rocks stand nearly vertical, the apparent thickness is nearly 15,000 feet (4,600 m), but because of repetition of ultramafic-mafic successions, we believe the main section is repeated at least once by faulting.

Of special interest in this area is an uncontestable depositional succession, extending up from ultramafic rock through mafic volcanic rock to mudstone, exposed in the channel of the South Fork of Elder Creek. The contact of the Jurassic mudstone on volcanic basaltic breccia is shown on figure 3, and the contact of the breccia on pyroxenite on figure 4. The breccia immediately above the massive pyroxenite contains 6-inch cobbles of pyroxenite along with pieces of other coarse-grained mafic and ultramafic rocks, such as gabbro, quartz gabbro, and diorite. The total thickness of the ophiolite sequence here is only a few hundreds of feet, but because mudstone is again exposed in the canyon west of the ultramafic rock, the lower part of the ophiolite is believed to be faulted off. Farther west, beneath the mudstone, is more ophiolite containing considerable gabbro, part of which is segregated into anorthosite and pyroxenite layers; mafic breccias with coarse-grained pieces cemented by fine-grained igneous material of similar composition; and pyroxenite, peridotite, and dunite that are almost completely serpentinized.

Stonyford

Within the Stonyford quadrangle (2, figs. 1 and 2), which has been studied by Brown (1964), the Upper Jurassic portion of the Great Valley sequence consists of about 10,000 feet (3,000 m) of siltstone, sandstone, and conglomerate. The lower half is largely tuffaceous siltstone containing interbeds of basaltic sandstone and tuff. These sedimentary rocks in turn rest on a volcanic section, as much as 3,000 feet (1,000 m) thick, that is made up chiefly of mafic pillow lavas and breccias but includes numerous thin diabase sills as well as altered tuffs and radiolarian chert lenses. Below this unit is a wedge of peridotite and serpentine having a thickness of about 5,000 feet (1,500 m). In Brown's interpretation, a major thrust fault, the Stony Creek fault zone, occurs between serpentine and Jurassic siltstone in the eastern part of the map area, and between serpentine and mafic volcanic rocks farther west. Although minor faults undoubtedly occur at the positions shown on

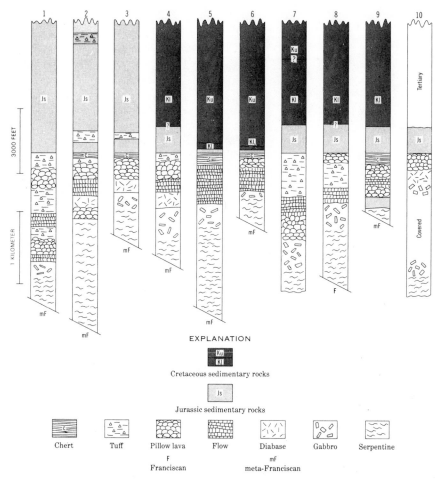

EXPLANATION

Cretaceous sedimentary rocks

Jurassic sedimentary rocks

| Chert | Tuff | Pillow lava | Flow | Diabase | Gabbro | Serpentine |

F
Franciscan

mF
meta-Franciscan

FIGURE 2.—Columnar sections showing details of the ophiolite succession at the numbered localities on figure 1.

Brown's map, the regional relations described in this paper suggest that the main zone of thrusting separating the Great Valley sequence from the underlying deformed and metamorphosed Franciscan rocks lies at the base of the serpentine, not at the top.

Wilbur Springs

The Wilbur Springs area (3, figs. 1 and 2) has been studied by several geologists, including Taliaferro (1943), Lawton (1956), Rich (1968), and Moisseeff (1966, 1968). Taliaferro (1943, p. 196–197, 210) described basalt, agglomerate, and chert as being present in the lower part of the Jurassic strata of the Great Valley sequence and, noting their similarity to rocks in the Franciscan, believed that there was a gradational contact between the Franciscan and Great Valley units. He described a section going upward from slaty Franciscan rocks as consisting of 1,200 feet (380 m) of serpentine overlain by 2,000 feet (600 m) of pillow basalts with interbeds of red chert and black shale, intruded by autobrecciated diabase. Moisseeff, who mapped the area in detail, indicates that the rocks above the serpentine range from tachylite to coarse gabbro and diabase and include soda-rich and hornblende-bearing varieties.

FIGURE. 3.—Upper Jurassic shale and minor graywacke of the Knoxville Formation lying depositionally on mafic volcanic breccia exposed in the canyon of the South Fork of Elder Creek, Tehama County, Calif.

FIGURE. 4.—Igneous breccia containing blocks of pyroxenite lying depositionally on massive pyroxenite exposed in canyon of South Fork of Elder Creek. Finger points to contact.

Basalts are most abundant and generally have pillow structure; coarse-grained igneous rocks are mainly restricted to the lower part. Prehnite and pumpellyite are found in the diabase near the basal serpentine. Sedimentary rocks are "scarce," except for beds of radiolarian chert. Higher in the section the effusive rocks are overlain by tuff, graywacke, and shale. Moisseeff (1966) initially considered the section to be ophiolite, and clearly states (p. 15) "the serpentine unit belongs to the Lower Knoxville Formation and is therefore a part of the Great Valley sequence." He later (Moiseyev (Moisseeff), 1968, p. 170) seems to have had some reservations regarding this conclusion, for he states, "The serpentinized ultramafic mass that has been intruded between Franciscan and Great Valley assemblages * * *.", though he still regards the basalt flows, tuffs, and radiolarian cherts as the base of the Great Valley sequence.

Healdsburg

The Healdsburg quadrangle, which is 30 miles (50 km) west of the western edge of the Great Valley, has been mapped and reported on by Gealey (1951). It contains near its northwestern corner (4, figs. 1 and 2) an unusually complete, though poorly exposed, ophiolite succession beneath unfossiliferous black shales of the Great Valley sequence that he regarded as Jurassic in age. Above a tabular mass of sheared serpentine, possibly 1,000 feet (300 m) thick, is a mafic "sill" about 2,000 feet (600 m) thick with gabbro at the base and diabase at the top. Locally the sill is layered with olivine- or pyroxene-rich bands. Overlying the sill, and

barely distinguishable in some places, is basalt that locally has pillow structure, and above the basalt is volcanic breccia and agglomerate. At least part of the rocks referred to as basalt we have found to be keratophyre and very siliceous quartz keratophyre. Gealey suggests that the gabbro, olivine gabbro, enstatite rock, and serpentinized periodite were all differentiates from a single sill. Here we are still concerned with the problems of relations between various ultramafic and mafic rocks, and whether or not they are intrusive or extrusive—problems that are typical of ophiolites in other areas.

At another locality, 15 miles (24 km) to the east near Mount Saint Helena, Bezore (1969) has studied ophiolite at the base of the Great Valley sequence. Here, above a tectonic contact with Franciscan rock, is a succession of 2,000 feet (600 m) of serpentinized harzburgite cut by dikes of pyroxenite and olivine gabbro, overlain by 1,200 feet (370 m) of mixed serpentinized dunite and gabbro that is overlain by 500 feet (160 m) of massive olivine-free gabbro, in turn capped by 1,000 feet (300 m) of diabase breccia. The overlying black shales and conglomerates have not yielded fossils but are believed to be of Late Jurassic age. Bezore suggests that this ophiolite on which the sedimentary rocks of the Great Valley sequence were deposited might be oceanic crust.

Mount Boardman

The ultramafic rocks of the Mount Boardman area (5, figs. 1 and 2) were described by Hawkes and others (1942), and a larger area was mapped in detail by Maddock (1964). Here beneath Jurassic shale of the Great Valley sequence is a typical ophiolite succession, but Maddock mapped major faults between some of the units. Beneath the Jurassic shale is his Lotta Creek Tuff Member, a unit 900 feet (275 m) thick consisting of mafic or keratophyric material with increasingly abundant siliceous shales or impure chert near the top. The tuff lies on a pile of keratophyre and quartz keratophyre flows 1,500 feet (450 m) thick with no sedimentary interbeds. Beneath the keratophyre, though everywhere mapped as separated by a fault, is hornblende gabbro cut by aplite, or perhaps trondjemite, dikes containing secondary prehnite. The gabbro in one area is in the center of a synclinal tabular mass of ultramafic rock, which we believe it overlies but which Maddock has separated by a fault. The sill-like ultramafic sheet is about 4,000 feet (1,200 m) thick and is largely serpentinized peridotite, although some dunite is present locally both near its base and top. The sheet is banded in places and contains segregations of chromite. It was mapped as intrusive into the Franciscan rocks by Maddock (1964), but its lower contact is shown as a fault by Hawkes and others (1942). We regard the surface below the serpentine as the major thrust fault that separates the Franciscan and Great Valley units.

Quinto Creek

There is no published large-scale map of the exposures at the base of the Great Valley sequence along Quinto Creek (6, figs. 1 and 2), but the San Jose sheet of the 1:250,000 geologic map of California depicts the distribution of rock types reasonably well, although it is incorrect on age assignments. Fossil-bearing black shale only a few hundred feet thick in this locality is definitely of Jurassic age. It dips steeply eastward and is underlain by a few tens of feet of tuff, which is considerably silicified and has been correlated with the Lotta Creek Tuff Member of the Mount Boardman area. The tuff lies depositionally on light-colored pillow lavas that are probably keratophyre, and these lavas are underlain by a volcanic succession, more than 1,000 feet (300 m) thick, containing basalt, diabase, and, locally, breccias cut by deep-green volcanic glass. Underlying the volcanic rock is a few hundred feet of mixed coarse-grained mafic rocks including quartz gabbro, leucogabbro, hornblende gabbro, and hornblendite. The coarse-grained rocks overlie about 1,000 feet (300 m) of serpentinized peridotite with some dunite. The exact thickness of the ultramafic and mafic units is unknown, as are their contact relations, but there is nothing to suggest that this is not a typical ophiolite succession.

380–189 O—70——6

Vallecitos

The Vallecitos area (7, figs. 1 and 2) has been mapped by Enos (1963, 1965), who described a succession of Jurassic and Lower Cretaceous sedimentary rocks of the Great Valley sequence lying depositionally on a pile of volcanic rocks 3,500 feet (1,100 m) thick. However, he assigned the volcanic rocks to the Franciscan Formation, and emphasized that the relations demonstrated the pre-Portlandian age of these Franciscan rocks. We believe that the volcanic pile, which contains no sedimentary rocks, is not Franciscan but a part of the ophiolitic succession on which the Great Valley sequence was deposited.

As described by Enos, the thick volcanic pile consists of andesite tuff and tuff breccia, underlying flow layered andesite porphyry, keratophyre, and quartz keratophyre, and in the lower part subolivine basalt with well-developed pillow structure. Locally the volcanic rocks lie upon hornblende gabbro, which in turn seems to be above norite. The norite is in contact with serpentinized periodotite that in at least part of the area overlies Franciscan jadeitic metagraywacke and glaucophane schist. Enos showed no fault below the serpentine and considered it as intrusive into the Franciscan. To interpret this area as ophiolite beneath the Great Valley sequence requires drastic reinterpretation of Enos' data. However, all parts of the ultramafic-mafic sequence are shown in the typical ophiolite order on his geologic map, though partly covered by younger rocks so that the stacking is not immediately obvious.

Black Mountain

The Black Mountain area (8, figs. 1 and 2), about half way between Morro Bay and Atascadero, has been mapped in detail by Fairbanks (1904). The oldest sedimentary rocks of the Great Valley sequence are black shales, which he named the Toro Formation and listed as Cretaceous in age, but the shales are now known to include rocks of Late Jurassic age. These sedimentary rocks are underlain by 2,800 feet (850 m) of volcanic rock, termed Cuesta Diabase, which is in turn underlain by 3,500 feet (1,100 m) of serpentinized peridotite and some dunite. The sedimentary strata, which are more than 3,000 feet (900 m) thick, occupy the central part of a long syncline; the mafic and ultramafic rocks are found on both limbs, as is well shown on Fairbanks' cross sections. He regarded the Cuesta Diabase as an intrusive sill, although he mentions that it is generally amygdaloidal and has friction breccia or tuffaceous facies at its top. Other workers (Taliaferro, 1944, p. 545) have reported pillows and pillow breccias, indicating that at least part of the Cuesta is extrusive. Fairbanks' "diabase" shows great variation, and apparently includes keratophyre and quartz keratophyre as well as gabbro. Gabbro and norite are also mentioned as occurring along the edges of the serpentinized peridotite.

Stanley Mountain area

The Stanley Mountain area (*9*, figs. 1 and 2), 15 miles (25 km) northeast of Santa Maria, has been known for many years a place where well-bedded Jurassic shale is interlayered with mafic volcanic rocks and chert. Taliaferro (1943) presented a map showing these relations and cited the area as one where the Franciscan Formation graded into the Jurassic (Knoxville) part of the Great Valley sequence, which included abundant volcanic rocks in its lower part. A different interpretation was supplied by Easton and Imlay (1955), who used the data afforded by fossils found in the shale, which they assigned to the Franciscan Formation, to prove the Jurassic age of this part of the Franciscan. Recently Brown (1968) mapped the area in detail, and he concluded that the fossiliferous Jurassic shale and volcanic rocks were a part of the Great Valley sequence that had been tectonically superposed over much more deformed and metamorphosed Franciscan rocks.

The ophiolite sequence has at its top interbedded red chert and black shale 400 feet (120 m) thick lying conformably beneath graywacke. The chert and shale unit lies depositionally on a pile of mafic volcanic rocks 1,500 feet (460 m) thick consisting chiefly of flows, pillow lavas, and tuffs. Most of the lava is described as spilitic andesite; the remainder is referred to as basalt. In the upper and lower parts of the pile, the lavas commonly show pillow structure, and throughout the pile most of them are vesicular. Amygdules contain epidote and quartz, and in some varieties both minerals also occur as metamorphic replacements that make up a major part of the rock. Pumpellyite is present also as a metamorphic mineral, and in one tuff layer near the base, prehnite is abundant. The chert and underlying volcanic pile have features typical of an ophiolite sequence, but beneath the volcanic rocks is black shale with fossiliferous limestone nodules in place of the usual ultramafic rocks. This occurrence of shale below the volcanic part of a presumed ophiolite sequence is the only such occurrence that we know of in the California Coast Ranges. As shown on Brown's map, both volcanic rocks and the shale believed to underlie it have attitudes strongly divergent from the intervening contact, suggesting that the contact might be a thrust fault that has brought the volcanic rocks over the shale. Beneath the lower shale is the major thrust that carries the Great Valley sequence over metamorphosed Franciscan rocks, but only locally is a little serpentine found along it.

Point Sal

The geology of Point Sal (*10*, figs. 1 and 2), about 30 miles south of San Luis Obispo, was mapped by Fairbanks (1896). In this area, upper Jurassic shale of the Great Valley sequence lies upon volcanic rocks, which are shown on the 1:250,000 San Luis Obispo sheet of the geologic map of California as belonging to the Franciscan Formation, although no sedimentary rocks are included. Just below the shales is nearly 1,000 feet (300 m) of pillow basalt containing amygdules of epidote and quartz. Cutting the basalt locally are dikes of andesite and diabase. Beneath the basalt is a mass of diabase grading downward to gabbro that is at least 1,000 feet (300 m) thick, but the base is concealed by overlapping Tertiary strata. About a mile to the south, gabbro again crops out in a belt that trends southeastward along the coast. At this northwestern end, this gabbro is uniform in texture and composition, but farther to the southeast it becomes regularly banded with layers of diorite and hornblende gabbro. Still farther southeast, and presumably lower in the sequence, is olivine pyroxenite, anorthosite, hypersthene gabbro, serpentinized peridotite, and serpentinized dunite. Mafic dikes of various kinds transect the layered mass, and the sequence of rock types may be even more complex than indicated herein. Nevertheless, within this area an ultramafic-mafic succession generally similar to the ophiolite found elsewhere in the Coast Ranges appears to underlie the oldest rocks of the Great Valley sequence. Because of the good exposure provided by the sea cliffs, it is an area where future work might be rewarding.

SUMMARY OF OCCURRENCES OF OPHIOLITE

The ophiolite lying beneath the Jurassic strata of the Great Valley sequence seems remarkably similar from place to place. The lowest part everywhere, except perhaps in the Stanley Mountain area, is a completely hydrated serpentine, and most of the parent rock was a pyroxene-bearing peridotite. As its base is everywhere a fault, its original thickness is unknown, but the thickness of the portion present generally does not exceed 5,000 feet (1,500 m) except where repeated by imbrication. Dunite, where present, is not consistently confined to the base, and pyroxenite is widespread but occurs only in small amounts, usually as thin layers or dikes. The upper part of this zone commonly gives way to a feldspathic banded or layered complex consisting of norite, gabbro, hornblende gabbro, anorthosite, hornblendite, and, rarely, trondjemite. Overlying these coarse-grained rocks everywhere is a layer of mafic, or locally even silicic, volcanic rock that ranges in thickness from 2,000 to 5,000 feet (600–1,500 m) and averages about 3,000 feet (900 m). Its lower part is diabasic or basaltic, and may show pillow structure. Locally, prehnite, epidote, and quartz occur in amygdules and as alteration products in the lavas. Higher in the succession, keratophyre or quartz keratophyre is locally dominant. The upper rocks tend to be breccias or tuffs. At the top of the volcanic pile chert is common in some areas as a result of silification of the tuffs, or more locally as rhythmically layered radiolarian cherts possibly precipitated directly as silica. Intrusions, especially dikes, are commonly noted in the mafic volcanic succession but are rarely described as occurring in the underlying ultramafic rocks. Exceptions are found in the Mount Boardman area, where gabbro dikes intrude the upper part of the serpentinized peridotite (Maddock, 1964; Himmelberg and Coleman, 1968), and perhaps locally in the lower ultramafic mass of Point Sal (Fairbanks, 1896).

COMPARISON OF MESOZOIC OPHIOLITE WITH MODERN OCEANIC CRUST

Because of its mode of occurrence, there is little doubt that the ophiolite described here is the Mesozoic oceanic crust on which the Great Valley sequence was deposited. Several authors have suggested that ophiolite sequences elsewhere in the world represent ancient oceanic crust (Hess, 1962; Dietz, 1963; Davies, 1968; Gass, 1968; Thayer, 1969; Varne and others, 1969). Their view is based on comparison of ophiolites with present oceanic crust, but unfortunately, knowledge of the characteristics of this crust is obtained mainly through interpretation of geophysical data supported by scattered dredging and drilling. According to widely spaced seismic data, the average oceanic structure consists of ultramafic rock (mantle) overlain by: (1) an "oceanic" layer about 4.8 km thick of gabbro or serpentine; (2) an intermediate layer about 1.7 to 2.0 km thick probably of basalt and consolidated sediment; and (3) an upper layer, 0.2–0.3 km thick, of unconsolidated sediments (Dietz, 1961; Raitt, 1963). The existence of the upper layer of consolidated sediments and underlying basalt has been well documented by the JOIDES deep drilling program (California Univ., Scripps Inst. Oceanography, 1969). Dredge hauls have provided data on the deeper structure. For example, peridotite was dredged from the lower slopes, gabbros and greenschists from intermediate levels, and basalt from the upper slopes of the deep fracture zones that intersect the Mid-Atlantic Ridge (Bonatti, 1968). Distribution appears to be similar along the rifted crest of the Mid-Indian Ocean Ridge (Engel and Fisher, 1969), where tholeiitic basalt overlies coarse-grained gabbro, anorthosite, and lherzolite. The lithology and thickness of the layers in the present oceanic crust appear to be grossly similar to those of the ophiolite at the base of the Great Valley sequence. This similarity supports our conclusion that these rocks do indeed represent the ancient oceanic crust upon which the Late Jurassic sediments were deposited.

TECTONICS RESPONSIBLE FOR EXPOSURE OF OCEANIC CRUST

The depositional and tectonic activities responsible for the onland exposure of Mesozoic oceanic crust in the California Coast Ranges have recently been discussed by Bailey and Blake (1969) and by Hamilton (1969). The present distribution of the major Mesozoic units is shown on figure 5, and our concept of the sedimentation and tectonism leading to this distribution is illustrated, on figure 6, by a series of sequential cross sections drawn northeastward through the northern Coast Ranges from a point on the coast about 70 miles north of San Francisco. These cross sections show that in the late Mesozoic in this area the Franciscan and Great Valley sequence rocks were being deposited in parallel basins, and from mid-Cretaceous onward were being simultaneously deformed by underthrusting due to seafloor spreading. Superposition of the two units was brought about by movement along the great thrust fault, whose position is nearly everywhere marked by serpentine. Beneath the thrust, the Franciscan rocks were sheared and converted to blueschists in an inverted sequence, in which higher grade jadeite- or lawsonite-bearing metagraywacke near the fault grades downward to mildly altered pumpellyite-bearing graywacke. The distribution of this great thrust through the Coast Ranges and the metamorphic zones in the Franciscan are shown on figure 5. Prior to its recognition throughout the Coast Ranges, different parts of the thrust had received several different names; for simplicity, we recommend that it be referred to everywhere as the Coast Range thrust. The cross sections making up figure 6 make it clear that the ultramafic-mafic succession is a part of the Mesozoic oceanic crust that has been brought into view mainly because it has overridden the Franciscan rocks along the Coast Range thrust.

SIGNIFICANCE OF THE ON-LAND MESOZOIC OCEANIC CRUST

Recognition that the ultramafic-mafic rock at the base of the Great Valley sedimentary sequence is oceanic crust is significant for several reasons. For one, it provides an unusual opportunity for geologists to study on-land exposures of oceanic crust. At the present time much effort is being put forth to learn more about the oceanic crust through dredging, deep-sea drilling, and

PETROLOGY AND MINERALOGY

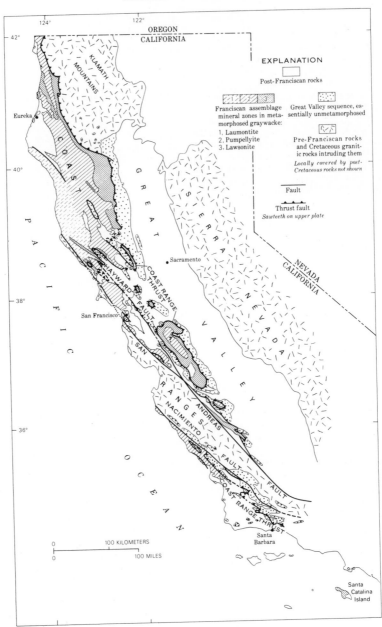

FIGURE 5.—Map of western California, showing the position of the Coast Range thrust beneath the rocks of the Great Valley sequence and the inverted metamorphic zones in the Franciscan rocks below the thrust.

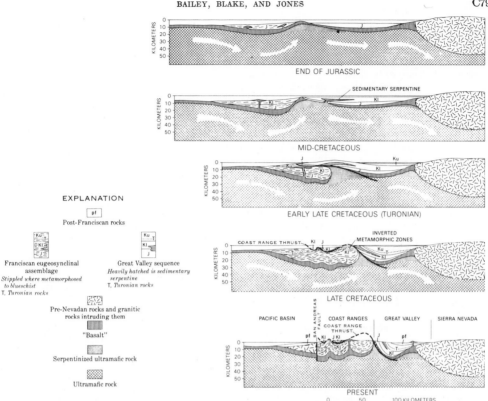

EXPLANATION

pf
Post-Franciscan rocks

Franciscan eugeosynclinal
assemblage
*Stippled where metamorphosed
to blueschist*
T, *Turonian rocks*

Great Valley sequence
*Heavily hatched is sedimentary
serpentine*
T, *Turonian rocks*

Pre-Nevadan rocks and granitic
rocks intruding them

"Basalt"

Serpentinized ultramafic rock

Ultramafic rock

FIGURE 6.—Sequential sections through the Coast Ranges northeastward from a point on the coast 70 miles north of San Francisco. No attempt has been made to show the contemporaneous tectonism and intrusion in the Sierra Nevada block at right edge of sections or to show the complex structure within the Franciscan assemblage. Symbols: J, Jurassic; Kl, Lower Cretaceous; Ku, Upper Cretaceous.

remote sensing in the ocean basins. A few hours spent in examing the on-land exposures can yield more about the lithologic character and sequence than can weeks of dredging, although of course one cannot assume that the crust of the vast ocean basins will prove to be everywhere like the sample provided by examination of a limited on-land area.

Recognition of the character of the ultramafic-mafic rocks is also significant in interpreting the history of the Coast Ranges, especially as it leads to a new concept regarding the age, origin, and emplacement of the serpentine at the base of the Great Valley sequence. As was noted by Taliaferro (1943) and emphasized by Irwin (1964), the serpentine makes up a great sheet that intervenes nearly everywhere between the rocks of the Great Valley sequence and the underlying Franciscan rocks (see fig. 1). Prior to its dissection by faulting and

erosion, this sheet of sheared serpentine extended many thousands of square miles, for at least the length of the Great Valley and from its western edge to the Pacific coast. It marks the position of the extensive Coast Range thrust, and the serpentine has been regarded by most geologists as having been injected into the thrust as a cold intrusion squeezed up from the upper mantle. As the serpentine was thought to have intruded along the plane of the fault, between its two walls, geologists were faced with the problem of where to show the fault on a geologic map. Virtually all geologists somehow decided that the serpentine was more like the Franciscan assemblage than the Great Valley sequence, and consequently drew the fault at the base of the Great Valley sequence so that the serpentine would be in the Franciscan side. The recognition of the serpentine as a part of the ophiolite beneath the Great Valley clastic

299

strata requires that the thrust fault be placed below it, and that the serpentine be in existence prior to faulting and to deposition of the Great Valley rocks. The depositional contacts on the serpentine, together with the occurrence of ultramafic fragments in overlying volcanic breccia and in the Jurassic strata lying on the ophiolite, proves that the serpentine antedates the sedimentation and faulting.

The total extent of the ultramafic sheet and overlying volcanic and sedimentary rocks that formed the upper plate of the thrust is not known because of dissection by later faulting and erosion. It is at least 15,000 square miles, and this is a minimum for the area that was initially oceanic crust and has now been welded to the continent. This figure, however, includes only rocks in the upper plate, and because the Franciscan rocks in the lower plate were also deposited on oceanic crust still farther from the continental margin, the total width of material added to the continent is at least 150 miles (250 km). Underthrusting has telescoped this new material with the result that the effective continental growth along this part of the Pacific margin during late Mesozoic time has been reduced to about 100 miles (160 km).

REFERENCES

Bailey, E. H., and Blake, M. C., 1969, Tectonic development of western California during the late mesozoic: Geotektonika, pt. 3, p. 17–30; pt. 4, p. 24–34.

Bailey, E. H., Irwin, W. P., and Jones, D. L., 1964, Franciscan and related rocks, and their significance in the geology of western California: California Div. Mines and Geology Bull. 183, 177 p.

Bezore, S. P., 1969, The Mount Saint Helena ultramafic-mafic complex of the northern California Coast Ranges [abs.]: Geol. Soc. America Abstracts with Programs 1969, pt. 3, Cordilleran Sec., Eugene, Oreg., 1969, p. 5–6.

Bonatti, Enrico, 1968, Ultramafic rocks from the Mid-Atlantic Ridge: Nature, v. 219, p. 363–364.

Brown, J. A., Jr., 1968, Thrust contact between Franciscan group and Great Valley sequence northeast of Santa Maria, California: Univ. Southern California, Los Angeles, Calif., Ph. D. thesis (geology), 236 p.

Brown, R. D., Jr., 1964, Geologic map of the Stonyford quadrangle, Glenn, Colusa, and Lake Counties, California: U.S. Geol. Survey Mineral Inv. Field Studies Map MF–279, scale 1 : 48,000.

California University, Scripps Institution of Oceanography, 1969, Initial reports of the Deep Sea Drilling project, prepared for the National Science Foundation, National Ocean Sediment Coring Program, v. 1 : 672 p.

Davies, H. L., 1968, Papuan ultramafic belt, in Upper mantle (Geological processes): Internat. Geol. Cong., 23d, Prague, 1968, Repts., v. 1, Proc. Sec. 1, p. 209–220.

Dietz, R. S., 1961, Continent and ocean basin evolution by spreading of the sea floor: Nature, v. 190, no. 4779, p. 854–857.

——— 1963, Alpine serpentinites as oceanic rind fragments: Geol. Soc. America Bull., v. 74, no. 7, p. 947–952.

Easton, W. H., and Imlay, R. W., 1955, Upper Jurassic fossil localities in Franciscan and Knoxville Formations in southern California: Am. Assoc. Petroleum Geologists Bull., v. 39, no. 11, p. 2336–2340.

Engel, C. G., and Fisher, R. L., 1969, Lherzolite, anorthosite, gabbro, and basalt dredged from the Mid-Indian Ocean Ridge: Science, v. 166, p. 1136–1141.

Enos, Paul, 1963, Jurassic age of Franciscan Formation south of Panoche Pass, California: Am. Assoc. Petroleum Geologists Bull., v. 47, no. 1, p. 158–163.

——— 1965, Geology of the Western Vallecitos syncline, San Benito County, California: California Div. Mines and Geology, Map sheet 5, scale 1 :31,680.

Fairbanks, H. W., 1896, The geology of Point Sal: California Univ. Dept. Geology Bull., v. 2, no. 1, p. 1–92.

Fairbanks, H. W., 1904, Description of the San Luis quadrangle [California]: U.S. Geol. Survey Geol. Atlas, Folio 101, 14 p.

Gass, I. G., 1968, Is the Troodos Massif of Cyprus a fragment of Mesozoic ocean floor?: Nature, v. 220, no. 5162, p. 39–42.

Gealey, W. K., 1951, Geology of the Healdsburg quadrangle, California: California Div. Mines Bull. 161, 50 p.

Hamilton, Warren, 1969, Mesozoic California and the underflow of Pacific mantle: Geol. Soc. America Bull., v. 80, p. 2409–2429.

Hawkes, H. E., Jr., Wells, F. G., and Wheeler, D. P., Jr., 1942, Chromite and quicksilver deposits of the Del Puerto area, Stanislaus County, California: U.S. Geol. Survey Bull. 936–D, p. 79–110.

Hess, H. H., 1962, History of ocean basins, in Petrologic studies: Geol. Soc. America, Buddington volume, p. 599–620.

Himmelberg, G. R., and Coleman, R. G., 1968, Chemistry of primary minerals and rocks from the Red Mountain–Del Puerto ultramafic mass, California, in Geological Survey Research 1968: U.S. Geol. Survey Prof. Paper 600–C, p. C18–C26.

Irwin, W. P., 1964, Late Mesozoic orogenies in the ultramafic belts of northwestern California and southwestern Oregon, in Geological Survey Research 1964: U. S. Geol. Survey Prof. Paper 501–C, p. C1–C9.

Lawton, J. E., 1956, Geology of the north half of the Morgan Valley quadrangle and the south half of the Wilbur Springs quadrangle [California]: Stanford Univ., Ph. D. thesis, 259 p.

Maddock, M. E., 1964, Geology of the Mount Boardman quadrangle, Santa Clara and Stanislaus Counties, California: California Div. Mines and Geology Map Sheet 3, scale 1 : 62,500.

Moisseeff, A. N., 1966, The geology and the geochemistry of the Wilbur Springs quicksilver district, Colusa and Lake Counties, California: Stanford Univ., Ph. D. thesis, 214 p.

Moiseyev (Moisseeff), A. N., 1968, The Wilbur Springs quicksilver district (California) example of a study of hydrothermal processes by combining field geology and theoretical geochemistry: Econ. Geology, v. 63, no. 2, p. 169–181.

Raitt, R. W., 1963, The crustal rocks, in The sea, v. 3: London, Interscience Publishers, p. 85–102.

Rich, E. I., 1968, Geology of the Wilbur Springs quadrangle, Colusa and Lake Counties, California: Stanford Univ., Ph. D. thesis, 101 p.

Taliaferro, N. L., 1943, Franciscan-Knoxville problem: Am. Assoc. Petroleum Geologists Bull., v. 27, no. 2, p. 109–219.

——— 1944, Cretaceous and Paleocene of Santa Lucia Range, California: Am. Assoc. Petroleum Geologists, v. 28, no. 4, p. 449–521.

Thayer, T. P., 1969, Peridotite-gabbro complexes as keys to petrology of mid-oceanic ridges : Geol. Soc. America Bull., v. 80, no. 8, p. 1515–1522.

Varne, Richard, Gee, R. D., and Quilty, P. G. J., 1969, Macquarie Island and the cause of oceanic linear magnetic anomalies : Science, v. 1966, p. 230–233.

16

Reprinted from *Phil. Trans. Royal Soc. London*, **A268**, 443–466 (1971)

The Troodos Massif, Cyprus and other ophiolites as oceanic crust: evaluation and implications

By E. M. Moores

Department of Geology, University of California, Davis, U.S.A.

and F. J. Vine*

Department of Geological and Geophysical Sciences, Princeton University, U.S.A.

[Plate 5 and 6]

Many Alpine ophiolite complexes characteristically display a pseudostratiform sequence of ultramafics, gabbro, diabase, pillow lava and deep-sea sediments. These masses resemble the known rock suite from the ocean floor. They are either fragments of old oceanic crust and mantle caught up in deformed belts, or results of diapiric emplacement of partly molten mantle material on or near the sea bottom. Such complexes are widespread in the Tethyan mountain system and have been recognized also from the circum-Pacific region. The Troodos Massif, Cyprus, consists of a pseudostratiform mass of harzburgite, dunite, pyroxenite, gabbro, quartz diorite, diabase and pillow lava arranged in a dome-like manner. The diabase forms a remarkable dyke swarm, trending mostly north–south in which 100 km of extension is indicated over 100 km of exposure. Such a feature suggests formation by sea-floor spreading. Layering of pyroxenite, harzburgite and dunite generally is perpendicular to subhorizontal rock unit contacts. The harzburgite and dunite are tectonites and probably represent uppermost mantle. Pyroxenite, gabbro, quartz diorite and diabase may represent the products of partial fusion of mantle material or of fractional crystallization of such partial fusion products. Chemical compositions of mafic intrusive and extrusive rocks do not fit well with oceanic tholeiite compositions, but resemble greenstones and associated rocks recently reported from the oceans.

The massif probably formed about an old Tethyan ridge. Some pillow lavas may be crust added after the main spreading episode. A fault zone active during emplacement of the lower units of the complex may represent a fossil transform fault. Complex chilled margins in the dyke swarms and mutually contradictory cross-cutting relations between dykes and plutonic mafic rock suggest formation of ocean crust by multiple intrusion of small portions of liquid. Uneven top surface of the dyke swarm and some conjugate dyke systems suggest independently varying rates of magma supply and extension.

Other Tethyan ophiolites, particularly in Greece and Italy, exhibit internal structure parallel to, rather than perpendicular to, major rock units, and some show much less diversity in mafic rock type. If these masses are fragments of ocean floor and mantle, such differences in internal structure may be due to differences in spreading processes—perhaps differences in spreading rate.

Introduction

'...in Cyprus there is apparently no sign of a floor of country rocks occurring at the base of the Troodos plutonic rocks. The very high positive gravity anomaly... existing in the Troodos area is an indication that, at least in southern Cyprus, the granitic layer of the earth's crust is missing. The sialic crust...appears to have moved aside under great tensional stress while the numerous dykes of the diabase and pillow lava series were intruded' (Wilson 1959, p. 126).

'The dykes, as the lava flows accumulated, were intruded into progressively higher levels in the volcanic rocks. Thus low down in the volcanic series there is an overwhelming preponderance of intrusive material and towards the top, lavas become increasingly abundant' (Wilson 1959, p. 75).

Ever since Steinmann (1906, 1926) drew attention to the association of serpentinite, pillow lavas, and chert in Alpine ophiolite complexes, they have been a subject of controversy. Much

---

* Present address: School of Environmental Science, University of East Anglia, Norwich.

of this controversy has been engendered by exposure of these rocks in the Alps where they occur primarily in the Pennide zone (Vuagnat 1963; Burri & Niggli 1945), and where they have been subjected to extensive subsequent metamorphism and tectonism. Outside of the Alps proper, however, ophiolites are recognized widely throughout the Tethyan region (Maxwell & Azzaroli 1963; Trümpy 1960; Gansser 1959; Brunn 1960; Aubouin 1965), and have been reported as well from the circum-Pacific region, e.g. California (Steinmann 1906; Bezore 1969; Hsu 1969), Japan (Miyashiro 1966), Papua (Davies 1969), Macquarie Island (Varne, Gee & Quilty 1969). Where least deformed, these masses characteristically display a consistent sequence, from bottom to top, of magnesian ultramafics, gabbro, diabase, extrusive lava, and deep-sea sediments. Though most masses are allochthonous and many inverted, nearly all preserve the above sequence, regardless of how many zones are preserved.

Three hypotheses for the origin of Alpine ophiolites are currently under consideration, as follows:

(1) That they represent slices of oceanic crust and mantle (Hess 1965; Vine & Hess 1970; deRoever 1957; Moores 1969; Vuagnat 1963).

(2) That they represent diapiric emplacement of partially fused mantle material (Maxwell 1969, 1970; Moores 1969).

(3) That they represent the intrusion into the crust and stratiform crystallization of mafic or ultramafic magma (Smith 1958; Brunn 1956, 1960; Aubouin 1965; Dubertret 1955).

Of the three hypotheses, we favour the first one in the light of presently available evidence. Arguments for and against each hypothesis are complex and involve consideration of individual situations and local details of stratigraphy and regional structure, which are beyond the scope of this article. However, if the first hypothesis is accepted, then the study of ophiolites should reveal a great deal of information about the petrology and structure of oceanic crust.

In this article we present a progress report of a study of a selected ophiolite complex, the Troodos Massif, Cyprus, in an attempt to unravel the processes of oceanic crustal evolution. The Troodos Complex was chosen for this study because of the long-recognized positive gravity anomaly found over the island (Gass & Masson-Smith 1963), and Gass & Masson-Smith's (1963) hypothesis that the Troodos represents an upthrust slice of oceanic crust and mantle. Furthermore, Wilson's work (1959) demonstrated the presence of an apparently unique feature, the 'sheeted' or diabase dyke complex, which seems to imply a mode of formation amounting to sea-floor spreading. Also the Troodos Complex is relatively undeformed and unmetamorphosed; much of it is well mapped; and logistics are easy. It therefore appeared that a restudy would be feasible within the limited amount of time available. Our study would not have been possible without the previous work of the members of the Cyprus Geological Survey Department (Wilson 1959; Bear 1960; Carr & Bear 1960; Gass 1960; Bear & Morel 1960; Bagnall 1960, 1964; Gass & Masson-Smith 1963; Bear 1966; Pantazis 1967, etc.). The observations of Bear & Wilson on the Sheeted Complex and the Plutonic rocks, and of Gass on the Pillow Lava relations were especially valuable. These observations were made before the development of the sea-floor spreading hypothesis by Hess (1962) and Vine & Matthews (1963), but they fit it perfectly. It is clear from the above quotations from Wilson's memoir, for example, that he was conceptually very near to the idea of sea-floor spreading, but did not relate it to the development of the ocean basins.

General description of the Troodos Complex

For the purposes of discussion, the Troodos Complex can be divided into three broad units (Wilson 1959; Gass & Masson-Smith 1963; Gass 1967, 1968): Pillow Lava, Sheeted Complex, and Plutonic Complex. It must be emphasized that this division is for descriptive purposes only, and that gradations exist between all units (see figure 1).

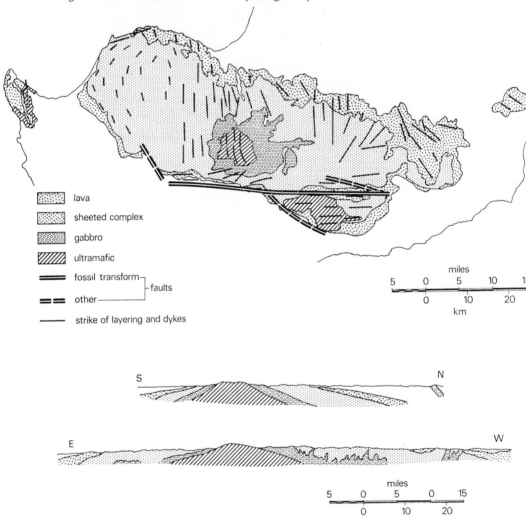

FIGURE 1. Map and cross-sections of Troodos Complex, modified after Gass & Masson-Smith (1963) and Bear (1965). The Mt Olympus or Troodos ultramafic area is north of the Fossil Transform Fault; the Limassol Forest area borders the Fossil Transform Fault on the south. The western inlier is the Akamas Peninsula, the eastern one is Troulli.

The Plutonic Complex consists of two principal exposures, the Mount Olympus or Troodos area proper, and the Limassol Forest area. These rocks include harzburgite, dunite with accessory chromitite, olivine pyroxenite, gabbro, uralitized gabbro, and albitized quartz diorite.

The ultramafic rocks are partly to completely serpentinized. In addition, some remobilization of serpentinite has occurred to the east of Troodos and in the Limassol Forest area. This complex is overlain gradationally by the Sheeted Complex, a massive dyke swarm consisting of 90 to 100 % dykes intruded into a screen of pillow lava in the upper part, and quartz diorite, rare extrusive volcanics, and gabbro in the lower part. The dykes consist of epidosite, keratophyre, and saussuritized dolerite. The pillow lava screens consist primarily of keratophyre, quartz porphyry and epidosite. This complex grades by decreasing abundance of dykes into the pillow lavas above. The Lower Pillow Lavas contain up to 50 % dykes, and are of andesitic basalt, keratophyre and quartz andesite. The Upper Pillow Lavas contain basalt, olivine basalt, and ultrabasic pillow lavas and a few dykes.

Figure 1 also summarizes the local trends of the dyke structure. Generally the dykes trend north–south. In three areas, significant departures from this regional strike are present: in the northwestern part of the island where the dykes strike about N 60° E; in the eastern part of the exposure of the complex, where the dykes strike northwest and northeast, in addition to north–south, in apparently converging trends; and in the south, where marked east–west trends are found associated with east-trending faults which were active during emplacement of the lower units of the complex. At one level within the Sheeted Complex there are 100 % dykes implying, therefore, 100 km of extension in 100 km of exposure. The only possible mechanism which has been proposed to account for such extension is sea-floor spreading.

Our interpretation is that the Troodos Massif as a whole represents a slice of oceanic crust and uppermost mantle, that the harzburgite and dunite of the ultramafic rocks represent depleted mantle, the olivine pyroxenite and gabbro represent intrusive or cumulate magmatic material, that the gabbro and Lower Sheeted Complex (or Diabase) represent seismic layer three of the oceanic crust, and that the upper Sheeted Complex (basal group), Lower Pillow Lavas, and Upper Pillow Lavas represent layer two. Part of the Upper Pillow Lavas represent material added after the main crustal formation.

Upper Pillow Lavas

The Upper Pillow Lavas are discontinuously exposed around the margin of the massif. They consist of pillowed flows and sparse to abundant extrusive breccia, relatively free from intrusives (Wilson 1959; Pantazis 1967; Bagnall 1960). Particularly in the southern portion of the complex, abundant breccias are interbedded with radiolarian manganiferous shales. Most pillow lavas contain olivine phenocrysts, commonly altered to chocolate-brown calcite and are altered to propylitized and zeolite-facies assemblages. Some are quite fresh, however, especially in the northeastern part of the massif (Gass 1960, 1958) and the southern marginal exposures (Wilson 1959; Bear 1960). Figures 2 and 3, plate 5, show typical exposures of pillow lavas and breccia, respectively. Figure 4, plate 5, shows a thin section of a typical olivine basalt. Table 1 presents new and previously published analyses of rocks from the Upper Pillow Lavas. Though variations are present, the analyses show a relatively low SiO_2 and low to high alkalis. Plotted in an alkalis–silica diagram (figure 5), these rocks show a tendency towards the alkaline side, though no mineralogy on pyroxenes has yet been done to test the validity of this trend. However, there is a general compositional consistency between altered and unaltered rocks. The Upper Pillow Lavas apparently overstep onto rocks of the Sheeted Complex in the north-central portion of the massif (Wilson 1959; Carr & Bear 1960; Gass 1960), but apparently no break can be found in the northeast (Gass 1960). In the south, however, Upper Pillow Lavas overlie with marked

FIGURE 2. Exposure of Upper Pillow Lavas. Note intrusive-free exposure, hyaloclastic matrices between pillows, veining of calcite and analcite. Northwest margin of the massif.

FIGURE 3. Photo of breccia in Upper Pillow Lava, near southeastern margin of massif.

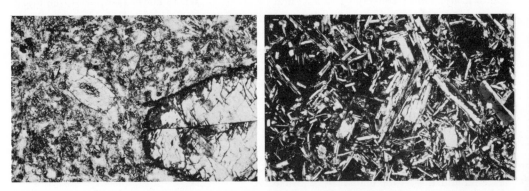

FIGURE 4. Photomicrograph of Upper Pillow Lava rock. Phenocrysts are calcite replacements of olivine in microcrystalline to glassy altered groundmass. Plain light, max. dimension 1.6 mm.

FIGURE 6. Photomicrograph of dyke intrusive into Lower Pillow Lavas. Plagioclase and augite microphenocrysts in a seriate groundmass of plagioclase and alteration products. Crossed Nicols, max. dimension 3.2 mm.

FIGURE 7. Stream exposure of Lower Pillow Lavas, near contact with Sheeted Complex. Note composite dykes intruding flat-lying pillow lavas. Area below distinctive pillow structure is brecciated pillow lavas intruded by numerous small irregular sills.

(Facing p. 446)

Moores & Vine

Phil. Trans. Roy. Soc. Lond. A, volume 268, plate 6

FIGURE 11. Photo of Sheeted Diabase showing typical aspect. Max. width of photo approx. 1.5 km.

FIGURE 15. Photomicrograph of olivine gabbro, labradorite, olivine, augite, and iron ore. Crossed Nicols, max. dimension 3.2 mm.

FIGURE 16. Thin section of cumulate-textured olivine pyroxenite. Subhedral olivine, anhedral, twinned, poikilitic augite. Crossed Nicols, max. dimension 3.2 mm.

FIGURE 17. Foliated harzburgite. Fabric marked by planar orientation of olivine grains and enstatite layers. Troodos ultramafic area.

FIGURE 18. Isoclinal fold in dunite and chromitite. Troodos ultramafic area.

TABLE 1. CHEMICAL ANALYSES, UPPER PILLOW LAVAS

	(a)	(b)	(c)	(d)	(e)	(f)	(g)	(h)	(i)	(j)	(k)	(l)	(m)	(n)	(o)
SiO_2	45.73	43.00	42.38	43.39	45.04	49.86	47.60	47.43	42.11	48.00	46.5	45.5	46.4	50.8	49.7
Al_2O_3	15.42	4.64	4.97	12.83	7.13	11.92	11.78	14.81	6.59	14.75	14.4	12.1	14.6	15.2	15.6
Fe_2O_3	5.68	2.42	3.15	4.90	3.13	3.35	2.96	6.19	3.62	6.33	9.2	8.5	7.9	8.2	7.8
FeO	2.20	6.47	5.25	2.10	5.43	4.94	6.70	2.49	5.38	1.61	—	—	—	—	—
MgO	8.91	33.45	31.72	8.06	26.16	10.17	14.08	3.44	30.00	7.12	8.2	7.1	10.8	9.4	8.9
CaO	5.84	3.99	4.66	11.84	5.56	9.65	7.91	14.32	3.39	8.50	11.7	11.7	7.8	7.8	2.3
Na_2O	2.63	0.25	0.40	1.04	0.77	1.70	1.58	2.43	0.25	0.82	4.2	1.7	4.2	1.8	0.08
K_2O	1.90	0.05	0.12	3.25	0.06	0.37	0.75	1.43	0.17	4.85	0.04	2.1	0.8	2.7	0.09
H_2O+	3.81	3.83	4.18	3.16	3.81	4.49	3.62	1.49	6.05	3.33	—	—	—	—	—
H_2O-	5.92	1.22	1.52	3.84	1.89	3.07	1.44	0.68	1.89	2.53	—	—	—	—	—
CO_2	1.12	—	0.86	4.85	0.07	—	0.04	4.41	—	2.19	—	—	—	—	—
TiO_2	0.73	0.18	0.28	0.49	0.36	0.53	1.34	0.64	0.23	0.34	0.8	0.38	0.64	0.54	0.7
P_2O_5	0.09	—	0.05	0.08	0.12	0.06	0.02	0.09	0.05	0.06	—	—	—	—	—
MnO	0.17	0.15	0.14	0.11	0.14	0.14	0.15	0.10	0.16	0.08	0.21	0.08	0.12	0.08	0.15
Cr_2O_3	0.04	0.51	0.39	—	0.32	0.11	0.09	—	0.28	0.02	—	—	—	—	—
NiO	—	—	—	—	—	—	0.05	—	0.12	S 0.05	—	—	—	—	—
	100.19	100.16	100.06	99.94	99.99	100.36	100.11	99.95	100.29	100.58					

Remarks:

(a) Basalt, Kambia Village: Bear (1960), table III, Analysis 1370, p. 83.

(b) Ultrabasic pillow lava, 1.2 km SW of Margi: Gass (1960), table III, Analysis 1374, p. 83.

(c) Ultrabasic lava, 1.2 km west of Margi: Bear (1961), Analysis 1, p. 11.

(d) Mugearite, Alikos River, 800 m west of Margi: Bear (1961), Analysis 2, p. 11.

(e) Ultrabasic lava, 2 km south of Margi: Bear (1961), Analysis 3, p. 11.

(f) Limburgite, 800 m west of Margi: Bear (1961), Analysis 5, p. 11.

(g) Olivine norite intrusive, east of Agrokipia: Bear (1960), table IV, Analysis 1564, p. 77.

(h) Quartz gabbro intrusive, east of Agrokipia: Bear (1960), table IV, Analysis 1566, p. 77.

(i) Peridotite plug, 800 m SW of Margi: Gass (1960), table III, Analysis 1374, p. 83.

(j) Trachybasalt dyke, 1.6 km east of Vouni: Wilson (1959), table I, Analysis 684, p. 69.

(k) Altered olivine basalt, near Troulli: new XRF analysis, sample 19 D.

(l) Altered dolerite dyke, near Kellaki: new XRF analysis, sample 23F.

(m) Altered olivine basalt, near Malounda: new XRF analysis, sample 30 A.

(n) Altered dolerite dyke, near Malounda: new XRF analysis, sample 30B.

(o) Basalt, core from drill hole, near Koutraphas, North-central margin of Complex: new XRF analysis, sample UPL-K.

unconformity faulted exposures of Sheeted Complex and Lower Pillow Lavas. These relations taken together indicate that a slight to substantial unconformity separates Lower and Upper Pillow Lavas.

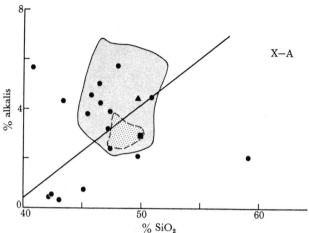

FIGURE 5. Alkalis–silica diagram for Upper Pillow Lava rocks. Heavy line is Hawaii alkalic-tholeiitic division line, as defined by MacDonald & Katsura (1964); fine-stippled area is field of analyses from mid-Atlantic Ridge at 45° N (Aumento & Loncarevic 1969; Aumento 1968). Heavy pattern field of analyses reported by Muir & Tilley (1964, 1966). Mid-Atlantic diorite (− A) is from Aumento (1969). Average spilite (▲) from Poldervaart (1955), average oceanic tholeiite (■) from Engel *et al.* (1965). ●, flows and intrusives.

Lower Pillow Lavas

Lower Pillow Lavas characteristically display plagioclase and pyroxene phenocrysts in a groundmass of altered plagioclase and iron ore (see figure 6, plate 5). Albitization and celadonite alteration are common, as are quartz and chalcedony amygdules. Characteristic also is the presence of 10 to 50 % dykes. showing chilled margins and multiple intrusive nature (see figure 7, plate 5). Chilled margins are also common in pillows, in contrast to similar exposures in the Upper Pillow Lavas. The contact between the Lower Pillow Lavas and the Basal Group (the upper unit of the Sheeted Complex) is marked by intrusion of sills, a sharp increase of dyke density in some areas (Gass 1960), gradual increase in other areas (Bear 1960; Wilson 1959), and local relief of up to 2 km (Gass 1960. pl. 1). The sharp increase in dyke density and relief on the contact in the eastern part of the area has prompted Gass (1960, 1967, 1968) to postulate an unconformity between Lower Pillow Lavas and Basal Group. In view of the transitional nature of the contact elsewhere, however, and the lack of an erosional surface and intercalated sediments, we prefer to interpret this apparent disconformity as resulting from relative variations in spreading rate against magma supply as discussed below.

Chemically, as well as petrographically, the Lower Pillow Lavas differ somewhat from the Uppers. Table 2 presents available chemical analyses of rocks from the Lower Pillow Lavas. It will be seen that they are often characterized by high silica, low potash, and variable amounts of other elements. There is, however, a consistent grouping of K_2O contents around 0.25 % as shown in figure 8, in contrast to the greater spread of Upper Pillow Lava values. Figure 9 shows a plot of Lower Pillow Lava rocks on a silica–alkalis diagram. Compared with the Upper Pillow Lavas (figure 6) most rocks are lower in alkalis, higher in SiO_2 and show a general increase of alkalis with increasing SiO_2 content.

TABLE 2. CHEMICAL ANALYSES, LOWER PILLOW LAVAS

	(a)	(b)	(c)	(d)	(e)	(f)	(g)	(h)	(i)	(j)	(k)	(l)	(m)	(n)	(o)	(p)	(q)	(r)	(s)	(t)	(u)	(v)	(w)	(x)
SiO_2	51.71	65.22	50.67	52.01	66.18	63.32	46.6	47.8	51.6	53.2	52.8	56.1	69.9	50.9	59.6	41.0	53.6	53.8	63.2	74.3	48.4	55.7	49.7	48.0
Al_2O_3	14.70	12.71	15.61	14.77	12.78	15.68	15.2	15.9	15.4	15.3	14.4	14.4	13.0	15.2	13.6	15.1	15.8	14.6	13.1	10.5	17.6	15.2	16.6	15.2
Fe_2O_3	1.86	2.84	8.15	4.32	1.37	7.37	10.9	10.9	10.9	8.5	8.4	11.5	5.5	13.5	10.5	8.4	10.3	11.1	8.4	3.6	13.2	10.5	13.5	13.2
FeO	6.34	4.72	2.67	5.83	3.52	—	—	—	—	—	—	—	—	—	—	—	—	—	—	—	—	—	—	—
MgO	7.55	0.64	2.21	6.54	0.72	1.39	6.6	8.4	7.5	4.8	7.6	5.8	1.8	5.8	5.1	6.3	4.1	4.4	4.5	2.5	3.4	4.4	3.5	6.0
CaO	10.74	3.81	11.21	4.00	2.98	5.04	9.8	9.2	8.7	11.2	9.9	5.7	4.8	8.6	5.6	11.0	9.3	9.6	4.9	2.0	6.8	7.8	8.4	6.9
Na_2O	1.88	4.58	3.05	2.08	4.03	2.55	2.2	1.5	2.3	2.2	1.8	3.0	4.0	2.6	3.0	5.4	2.8	2.4	3.6	4.5	3.4	3.4	3.3	3.1
K_2O	0.25	0.60	0.36	2.48	1.60	1.69	0.20	0.50	0.37	0.22	0.19	0.20	0.22	0.16	0.19	1.0	0.57	0.60	0.20	0.22	1.69	0.20	1.05	0.34
H_2O+	3.70	3.63	0.81	3.86	4.41	—	—	—	—	—	—	—	—	—	—	—	—	—	—	—	—	—	—	—
H_2O-	0.82	0.39	1.31	2.23	2.39	1.65	—	—	—	—	—	—	—	—	—	—	—	—	—	—	—	—	—	—
CO_2	—	0.04	2.06	0.44	—	—	—	—	—	—	—	—	—	—	—	—	—	—	—	—	—	—	—	—
TiO_2	0.48	0.55	1.01	1.01	0.34	1.15	0.98	0.86	0.84	0.56	0.52	1.15	1.12	1.32	1.45	0.52	1.0	0.87	0.66	0.29	1.58	1.36	1.30	1.47
P_2O_5	0.06	0.19	0.11	0.10	0.09	0.03	0.25	—	—	—	—	—	—	—	—	—	—	—	—	—	—	—	—	—
MnO	0.15	0.18	0.35	0.27	0.13	—	—	0.10	0.11	0.10	0.11	0.10	0.10	0.16	0.16	0.24	0.16	0.14	0.07	0.07	0.16	0.14	0.16	0.24
Cr_2O_3	—	—	S 1.17	—	—	—	—	—	—	—	—	—	—	—	—	—	—	—	—	—	—	—	—	—
	100.24	100.10	100.75	99.94	100.54	99.87																		

Remarks:

(a) Augitite dyke, south of Kambia: Bear (1960), table III, Analysis 1767, p. 58.
(b) Dacite glass, Xyliatos: Carr & Bear (1960), Analysis 1063, p. 35.
(c) Quartz basalt dyke, 1.6 km SW of Skouriotissa: Wilson (1959), table I, Analysis 188, p. 69.
(d) Andesite, Mitsero: Bear (1960), table III, Analysis 1569, p. 58.
(e) Glassy dacite, Kokkinoyia: Bear (1960), table III, Analysis 1265, p. 58.
(f) Andesite, 2.4 km south of Kambia: Bear (1960), table III, Analysis 1623, p. 58.
(g) Vesicular andesite, south of Troulli: new XRF analysis, sample 21D.
(h) Altered basalt, Malounda: new XRF analysis, sample 30E.
(i) Interior of altered basalt pillow, Malounda: new XRF analysis, sample 30G.
(j) Altered basalt, base of flow, Malounda: new XRF analysis, sample 30J.
(k) Altered basalt dyke margin, Malounda: new XRF analysis, sample 30F.
(l) Altered basaltic dyke, Klirou: new XRF analysis, sample 32M.
(m) Altered glassy dyke, Klirou: new XRF analysis, sample 32P.
(n) Altered basaltic pillow margin, Klirou: new XRF analysis, sample 32N.
(o) Altered basaltic dyke: new XRF analysis, sample 32R.
(p) Fresh pillow lava, NW of Mathiati: new XRF analysis, sample 8A.
(q) Altered pillow lava, near Karpedhes: new XRF analysis, sample 12R.
(r) Altered basalt sill, SE of Mathiati: new XRF analysis, sample 17P.
(s) Altered quartz dolerite dyke, SE of Mathiati: new XRF analysis, sample 17S.
(t) Altered dacite, Troulli: new XRF analysis, sample 19B.
(u) Altered pillow lava, Karpedhes: new XRF analysis, sample 12C.
(v) Altered basaltic intrusive, Karpedhes: new XRF analysis, sample 12Y.
(w) Altered dolerite, Troulli: new XRF analysis, sample 20C.
(x) Drill core near Analiondas, altered basalt: new XRF analysis, sample AN 270.

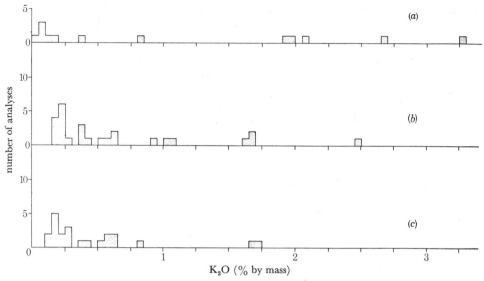

FIGURE 8. Plot of K$_2$O content against number of analyses for (a) Upper Pillow Lava, (b) Lower Pillow Lava and (c) Sheeted Complex.

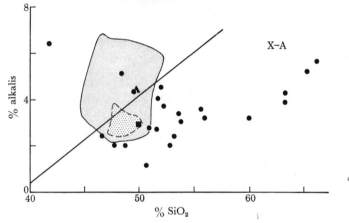

FIGURE 9. Silica-alkalis plot for flows and intrusives of Lower Pillow Lavas. See figure 5 for explanation of symbols.

Sheeted Complex

The Sheeted Intrusive Complex has been subdivided by the Cyprus Geological Survey Department into two units depending upon the type of screens present. The Basal Group consists of 90 to 100 % dykes in screens of pillow lava. Figure 10, adapted from Wilson, shows several typical sections through this unit. Pillow lavas of this unit include greenstones, keratophyres, and andesites, similar to those of the Lower Pillow Lavas, but which have been subjected to more pronounced greenschist metamorphism. Dykes in this unit include epidosites, keratophyres and greenstones. The distinction between this unit and the underlying diabase has been made on the basis of the presence or absence of pillow lava screens. Consequently the contacts are subjective and approximate.

Rocks assigned to the Diabase include dyke swarms composed of 100 % dykes, many of which show little or no chilled margins. In addition, much of the massif mapped as Diabase includes diabase intruded into screens of gabbro, quartz-diorite, and unpillowed amygdaloidal, porphyritic (presumably extrusive) rock. Figure 11, plate 6, shows a typical exposure of sheeted Diabase. Particularly apparent is the vertically dipping 'grain' extending the width of the photograph.

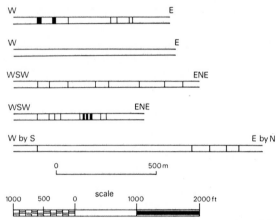

FIGURE 10. Representative cross-sections of Basal Group exposures, adapted from Wilson (1959). Black vertical lines are screens of pillow lava. White areas are dyke complexes.

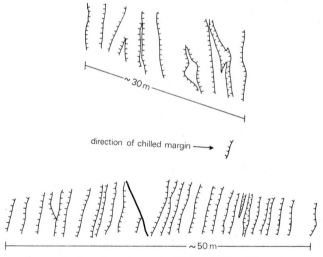

FIGURE 12. Representative cross-sections of Sheeted Complex.

The multiple intrusive nature of the dykes is illustrated in figures 7 and 11. In figure 7 the 'ribs' standing in relief represent individual chilled margins of a single dyke. Apparently the process has been that one dyke is intruded and has its margins chilled against the wall rock. The next dyke then intrudes up the middle of the previous one, and in turn forms chilled margins, then the next one repeats, and so on. This process apparently has been operative in the Sheeted Complex as well, but much more extensively, so that in a single exposure one can rarely see the two sides of a multiple intrusive dyke. Figure 12 shows two sketches of road cuts illustrating the preponderance of one-sided chilled margins of such an exposure. These relations indicate that the main process of formation was one of multiple intrusion of small volumes of magma, and that enough time was available for the previous rock to cool sufficiently for the succeeding pulse to be chilled.

Although most dykes are oriented so that their original dip must have been vertical, in some places conjugate sets of cross-cutting dykes are present within the same exposure. We interpret these conjugate dykes to indicate variations in extension against magma injection, as discussed below. Some kilometre-sized areas of consistent moderate to steep dip are adjacent to similarly sized areas of opposite dip. These areas of opposing dip may represent conjugate sets on a larger scale or tilted blocks originally of the same dip.

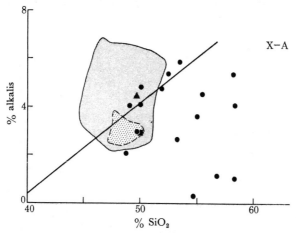

FIGURE 13. Silica–alkalis plot for rocks of Sheeted Complex. See figure 5 for explanation of symbols.

Though the strike of the intrusives in a given local area is generally uniform, in a few places sharp changes of almost 90° are present; this is in addition to the more regional changes present in the northwestern, southern, and eastern portions of the massif. Remanent magnetization directions suggest that these changes in strike are a function of original attitude at time of intrusion, rather than any case of tectonic rotation as implied by Bagnall (1964).

Available chemical analyses of rocks from the Sheeted Complex (table 3) display similarities with the Lower Pillow Lavas in generally moderate to high SiO_2 and low K_2O (see also figure 8). The silica–alkalis plot of these rocks (figure 13) places them generally in the same areas as the Lower Pillow Lavas (see figure 9).

The contact of the Diabase with the underlying Plutonic rocks is complex. Although much of the lower Diabase consists of diabase dykes intrusive into screens of gabbro and quartz diorite, intrusive contacts of gabbro and quartz diorite into Diabase are also present. Also, much rock mapped as gabbro and quartz diorite actually has a considerable number of dykes. Hence the contact between the plutonic and diabasic rocks is irregularly gradational and apparently represents a zone of multiple intrusion.

TABLE 3. CHEMICAL ANALYSES, SHEETED COMPLEX

	(a)	(b)	(c)	(d)	(e)	(f)	(g)	(h)	(i)	(j)	(k)	(l)	(m)	(n)	(o)	(p)	(q)	(r)	(s)	(t)
SiO_2	51.78	58.36	50.07	53.40	78.27	77.04	49.10	54.98	58.31	54.70	52.67	56.36	49.84	58.2	53.2	53.4	49.5	74.6	48.7	51.6
Al_2O_3	14.10	14.08	15.54	15.40	11.43	10.95	15.60	14.17	13.73	14.86	15.28	16.92	17.87	14.4	16.0	15.3	14.9	11.8	15.5	15.2
Fe_2O_3	3.88	5.92	2.08	10.46 }	1.44	2.14	2.08	7.14	3.78	5.88	3.68	11.68	9.98	9.6	9.1	10.1	11.8	4.4	9.3	10.4
FeO	5.15	1.80	8.06	}	0.14	0.79	5.04	4.95	5.53	4.21	6.02	—	6.48	4.6	5.2	7.2	8.4	3.7	6.1	7.2
MgO	8.07	2.91	8.78	5.12	0.57	0.70	10.42	4.69	4.63	4.67	5.51	3.68	6.48	4.6	5.2	7.2	8.4	3.7	6.1	7.2
CaO	7.23	12.74	4.50	7.37	0.92	1.86	7.58	4.51	5.34	10.89	6.80	7.45	9.35	4.1	11.7	3.4	8.7	2.3	13.3	6.1
Na_2O	4.50	0.75	4.43	3.87	4.70	4.63	2.38	1.83	3.45	0.16	4.62	1.96	2.50	5.0	2.0	5.6	3.6	4.8	1.6	4.2
K_2O	0.23	0.25	0.35	0.60	0.53	0.18	1.65	1.73	0.59	0.13	0.82	0.18	0.42	0.78	0.63	0.20	0.17	0.16	0.44	0.61
H_2O^+	2.73	1.32	4.14	2.24	1.39	0.79	4.22	2.93	2.14	2.60	2.83 }	1.30	2.94	—	—	—	—	—	—	—
H_2O^-	1.27	0.79	0.96	—	0.53	0.26	1.64	1.99	0.90	1.10	0.62 }	—	—	—	—	—	—	—	—	—
Co_2	0.14	—	0.26	—	—	—	—	—	—	0.17	0.12	—	—	—	—	—	—	—	—	—
TiO_2	0.54	0.58	0.48	0.95	0.25	0.20	0.30	1.11	1.30	0.68	0.93	0.35	0.40	1.02	0.54	0.53	1.07	0.21	0.55	0.66
P_2O_5	0.08	0.03	0.09	—	0.05	0.05	0.04	0.10	0.11	0.06	0.07	—	—	—	—	—	—	—	—	—
Cr_2O_3	—	—	—	—	—	—	—	—	—	—	—	—	—	—	—	—	—	—	—	—
MnO	0.26	0.20	0.20	0.12	0.03	0.05	0.18	0.18	0.25	0.26	0.16	0.16	0.23	0.10	0.13	0.14	0.2	0.07	0.13	0.19
NiO	—	—	—	—	—	—	—	—	—	—	—	—	—	—	—	—	—	—	—	—
F	—	—	0.07	—	—	—	—	S 0.06	—	S 0.06	—	—	—	—	—	—	—	—	—	—
	99.96	99.73	100.01	99.83	100.25	99.64	100.23	100.37	100.06	100.03	100.13	100.04	100.01							

Remarks:

(a) Andesite pillow lava, Basal Group, 2.4 km north of Pharmakas: Bear (1960), table II, Analysis 1582, p. 51.

(b) Epidosite pillow lava, Basal Group, 2.4 km SW of Kalokhorio: Bear (1960), table II, Analysis 1583, p. 51.

(c) Greenstone (chlorite-actinolite) pillow lava, Basal Group, Kalokhorio: Bear (1960), table II, Analysis 1790, p. 71.

(d) Microdiorite intrusive: Bear (1960), table II, Analysis 1775, p. 51.

(e) Quartz-albite microporphyry. 1.6 km NE or Pyrga: Gass (1960), table III, Analysis 1366, p. 83.

(f) Quartz-albite porphyry, 800 m north of Psevda: Gass (1960), table III, Analysis 1381, p. 83.

(g) Serpentinized microgabbro, 1.6 km north of Mosphiloti: Gass (1960), table III, Analysis 1379, p. 83.

(h) Quartz diabase dyke, Basal Group, 1.6 km SW of Kato Vlaso: Wilson (1959), table I, Analysis 149, p. 69.

(i) Quartz diabase, Kykko: Wilson (1969), table I, Analysis 662, p. 69.

(j) Epidosite, Lefka-Pedoulas Road: Wilson (1959), table IV, Analysis 667, p. 98.

(k) Microdiorite, Gourrie: Bear & Morel (1960), table II, Analysis 2808, p. 30.

(l) Microgabbro, Ayios Theodoros: Bear & Morel (1960), table II, Analysis 2871, p. 30.

(m) Microgabbro, Ayios Theodoros: Bear & Morel (1960), table II, Analysis 2872, p. 30.

(n) Altered dolerite, north of Kellaki: new XRF analysis, sample 25C.

(o) Altered dolerite, south of Troulli: new XRF analysis, sample 21F.

(p) Altered dolerite, south of Melini: new XRF analysis, sample 29E.

(q) Altered dolerite, south of Melini: new XRF analysis, sample 29A.

(r) Altered dolerite, Basal Group, west of Lefkara: new XRF analysis, sample 33A.

(s) Altered dolerite, south of Troulli: new XRF analysis, sample 21B.

(t) Altered microgabbro, south of Lefka: new XRF analysis, sample 120B.

Plutonic rocks

Rocks of the Plutonic Complex include 'granophyre' or quartz diorite, uralitized gabbro, norite, olivine gabbro, troctolite, poikilitic olivine pyroxenite, dunite with accessory chromitite and harzburgite or 'enstatite olivinite' (see figure 14).

Quartz diorite is present as screens in Diabase, as 'sheeted' and 'unsheeted' bodies intrusive into Diabase, and as residual segregations in gabbro. Most of these rocks are medium to coarse grained, containing abundant quartz, zoned plagioclase, and chlorite alteration after mafics. Chemical analyses of these rocks exhibit very high SiO_2 and low K_2O contents (see table 4). Table 4 also includes an analysis of a Mid-Atlantic quartz diorite (Aumento 1969) for comparison. Zoned feldspars have andesine cores and albitic rims.

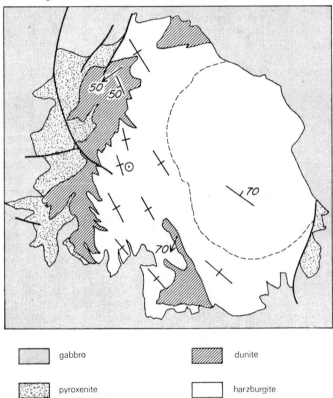

	gabbro		dunite
	pyroxenite		harzburgite

FIGURE 14. Generalized map of Troodos ultramafic area, showing attitudes of foliation and lineations. Dashed line in harzburgite outlines remobilized serpentinite plug.

Gabbroic rocks are found in two main areas—extensively surrounding the Troodos ultramafic mass, and in limited outcrops in the Limassol Forest area. In the former area, uralitized gabbro, norite, olivine gabbro, and troctolite are present. Uralitized gabbro is gradational into both quartz diorite and norite, and includes rocks in which amphibole has clearly altered from pyroxene, as well as rocks where it apparently represents a primary phase. The former have heavily altered calcic plagioclases surrounded by albitic rims, the latter have fresh calcic plagioclases with albitic rims. Norite contains two pyroxenes and plagioclase, and commonly grades into olivine-bearing rocks. Textures tend to be anhedral granular with reaction rims of pyroxene

TABLE 4. QUARTZ DIORITES, ANORTHOSITES, AND GABBROS

	(a)	(b)	(c)	(d)	(e)	(f)	(g)	(h)	(i)	(j)	(k)	(l)	(m)	(n)
SiO_2	73.06	76.58	73.31	76.85	64.54	46.07	49.10	54.76	51.20	49.9	48.16	45.56	39.92	56.43
Al_2O_3	12.30	11.61	13.40	12.36	14.03	22.21	18.20	15.31	15.80	5.2	9.35	29.90	26.00	26.10
Fe_2O_3	2.32	2.06	5.59	0.76	4.12	0.87	1.18	2.06	6.84	6.8	9.58	2.00	1.59	0.51
FeO	2.23	0.54	—	0.50	3.38	2.74	4.23	5.81	—	—	—	—	—	0.63
MgO	0.65	0.60	0.71	0.63	6.16	8.82	10.56	7.01	8.40	19.8	17.18	4.90	5.00	0.92
CaO	3.56	4.10	1.35	1.84	3.52	17.48	15.64	10.33	11.73	16.3	13.60	17.07	19.47	8.34
Na_2O	3.90	3.36	5.00	5.60	5.40	0.68	0.38	2.43	3.81	0.4	0.50	0.25	0.15	6.36
K_2O	0.18	0.18	tr	0.25	0.58	0.11	0.08	0.24	0.43	0.01	tr	tr	0.21	0.07
H_2O^+	1.09	0.97	0.54	0.54	0.52	1.03	0.77	1.49	—	—	1.41	1.50	7.50	0.23
H_2O^-	0.52	0.31	—	0.41	1.29	0.23	0.15	0.29	—	—	—	—	—	0.01
CO_2	—	—	—	0.02	0.07	—	—	—	—	—	—	—	—	—
TiO_2	0.28	0.19	0.10	0.29	0.92	0.08	0.12	0.55	0.29	0.19	tr	tr	0.35	0.18
P_2O_5	0.07	0.10	—	0.03	0.22	0.05	0.06	0.07	—	—	—	—	—	tr
S	—	—	—	—	—	0.09	—	—	—	—	—	—	—	—
Cr_2O_3	—	—	—	—	—	0.04	—	—	—	—	—	tr	tr	—
MnO	0.05	0.02	0.10	0.01	0.10	0.08	0.13	0.16	0.12	0.12	0.27	tr	tr	0.01
	100.21	100.62	100.10	100.09	99.85	100.45	100.60	100.51	—	—	100.05	100.08	100.29	99.79

Remarks:

(a) Granophyric hornblende-trondhjemite, SE Troodos–Limassol–Saittas Road: Wilson (1959), table IV, Analysis 1157, p. 98.

(b) Granophyric epidote trondjemite, Lefka–Pedoulas Road: Wilson (1959), table IV, Analysis 1153, p. 98.

(c) Quartz porphyry host rock, Zoopiyi: Bear & Morel (1960), table I, Analysis 2825, p. 28.

(d) Trondjemite, Gourrie: Bear (1960), table IV, Analysis 1577, p. 77.

(e) Tonalite, Gourrie: Bear (1960), table IV, Analysis 1576, p. 77.

(f) Gabbro-Platres end of old road to Troodos: Wilson (1959), table III, Analysis 124, p. 90.

(g) Hypersthene-gabbro, near milepost 39, Kakopetria–Troodos Road: Wilson (1959), table III, Analysis 1150, p. 90.

(h) Uralite gabbro, near milepost 47, Lefka–Pedhoulas Road: Wilson (1959), table III, Analysis 1158, p. 90.

(i) Gabbro, Limassol Forest area: Pantazis (1967), table V, Analysis 8612, p. 89.

(j) Olivine gabbro, Limassol Forest area, about 3.2 km south of Kellaki: new XRF Analysis, sample 24P.

(k) Olivine gabbro, east of Khandria: Bear & Morel (1960), table III, Analysis 2870, p. 41.

(l) Fine grained anorthosite, east of Louvaras: Bear & Morel (1960), table III, Analysis 2873, p. 41.

(m) Coarse-grained anorthosite, N.E of Apsiou: Bear & Morel (1960), table III, Analysis 2874, p. 41.

(n) Anorthosite, Central Indian ridge: Engel & Fisher (1969), table 3, p. 1138.

and/or amphibole around olivine (see figure 15, plate 6). Generally minerals are quite fresh, and little or no zoning is present in the plagioclase.

In the Limassol Forest area, gabbros and associated mafic rocks are not abundant, but occur as layered to massive rocks apparently overlying serpentinized peridotite. Subsequent remobilization of the latter, however, makes it difficult to determine the exact original relationships. Noteworthy in this area are the presence of three small bodies of anorthosite within the gabbro (Bear 1960, pp. 40–42). Bear describes an intimate interlayering of anorthosite and gabbro and uncommon dyke-like anorthosite masses. He also reports the presence of protoclastic texture. Table 4 presents an analysis of two of these rocks, together with one from the Indian Ocean (Engel & Fisher 1969) for comparison. Such a comparison is of doubtful validity, however, because the chemical composition strongly depends on modal percentages of minerals present.

TABLE 5. CHEMICAL ANALYSES, ULTRAMAFIC ROCKS

	(a)	(b)	(c)	(d)	(e)	(f)
SiO_2	46.27	39.54	33.72	37.68	42.93	38.59
Al_2O_3	2.31	5.15	0.51	0.56	4.43	1.97
Fe_2O_3	2.74	3.48	4.06	4.80	4.13	8.51
FeO	3.76	4.39	3.21	2.88	5.24	2.00
MgO	29.24	32.71	41.82	39.32	28.30	33.24
CaO	12.12	4.06	0.15	0.54	7.30	1.28
Na_2O	0.15	0.14	—	—	0.15	0.04
K_2O	0.07	0.09	—	—	0.05	0.05
H_2O^+	2.84	8.81	14.97	12.65	6.24	13.27
H_2O^-	0.28	0.85	0.93	0.89	—	—
Co_2	—	—	0.59	0.36	0.10	0.10
TiO_2	0.06	0.03	—	—	0.11	0.07
P_2O_5	0.13	0.20	0.04	0.04	0.03	0.03
S	—	0.18	0.02	0.08	—	—
Cr_2O_3	0.44	0.36	0.11	0.24	—	—
MnO	0.13	0.14	0.11	0.11	0.13	0.15
NiO	0.08	0.12	0.08	0.25	—	—
	100.62	100.25	100.32	100.40	99.16	99.30

Remarks:

(a) Olivine pyroxenite, 400 m SE of Army Leave Camp, Troodos: Wilson (1959), table II, Analysis 1155, p. 85.
(b) Olivine-rich pyroxenite, Trooditissa, Platres-Prodomos Road: Wilson (1959), table II, Analysis 1154.
(c) Dunite, 1.6 km west of Troodos Village: Wilson (1959), table II, Analysis 628.
(d) Harzburgite, 1.6 km west of Troodos Village: Wilson (1959), table II, Analysis 625.
(e) Harzburgite, Limassol Forest area: Pantazis (1967), table V, Analysis 8611, p. 89.
(f) Serpentinite, Limassol Forest area: Pantazis (1967), table V, Analysis 8608, p. 89.

Phase layering in the gabbro is common, especially in pyroxene and olivine gabbros. Some is clearly cumulate in origin; some shows features characteristic of flow layering (Thayer 1963) but may be deformed cumulate layering.

The gabbroic rocks grade downward by increasing concentration of mafic minerals into a unit of poikilitic olivine pyroxenite, pyroxenite, and felspathic olivine pyroxenite, designated by Wilson (1959) the harzburgite-wehrlite and the pyroxenite peridotite groups. Prominent layering in this unit is marked by varying proportions and grain size of pyroxene. Olivine is present as sub- to euhedral grains poikilitically included in pyroxene (see figure 16, plate 6). The abundance of olivine ranges gradationally from 20 to 70 % in different layers, and clinopyroxene is predominant only near the gabbro.

The lowermost rocks of the complex consist of dunite and harzburgite, the latter the 'enstatite olivinite' of the Cyprus Survey geologists. These rocks contain 80 to 100 % olivine with inter-layered enstatite. Rocks are strongly foliated, the latter marked by preferred orientation olivine grains as well as centimetre scale monomineralic enstatite layers (figure 17, plate 6). These two units contrast strongly with the pyroxenite-olivine pyroxenite unit in texture and in overall outcrop appearance. Chromite is present as accessory grains and as layered concentrations within dunite bodies. Both interlayered dunite and harzburgite and chromite concentrations within dunite display isoclinal similar folds, the axes of which plunge down the dip of the layering (see figure 18, plate 6).

Table 5 shows available chemical analyses of dunite, harzburgite and olivine pyroxenite. Available mineralogic data, mostly optical (Böttcher 1969) suggests that the magnesium content in dunite and harzburgite is approximately 90 to 94 %, and 82 to 87 % in the olivine pyroxenite. In addition, it is clear from the analyses that there is a tendency toward iron enrichment in olivine pyroxenite relative to the harzburgite.

TABLE 6. COMPARISON OF CHEMICAL RANGES FROM TROODOS ROCKS, AND
OTHER ALTERED VOLCANIC SEQUENCES

	Melson & Van Andel (1966) (a)		Smith (1968) (b)	Troodos (c)
	average fresh basalt	range of differences between greenstone and average basalt	range of values	range of values
SiO_2	49.38	-1.20 to $+2.32$	38.17 to 65.31	41.0 to 78.3
Al_2O_3	16.43	-0.13 to -1.73	12.18 to 18.27	4.64 to 17.6
Fe_2O_3	2.02	$+0.03$ to $+1.89$	2.13 to 9.93	1.58 to 13.5
FeO	6.98	-3.06 to $+0.33$	0.72 to 15.03	
MgO	8.34	-1.99 to $+1.13$	0.20 to 5.99	0.57 to 33.45
CaO	11.26	-4.82 to $+1.30$	3.54 to 22.40	0.92 to 14.32
Na_2O	2.74	$+0.19$ to $+1.74$	0.03 to 6.55	0.25 to 4.8
K_2O	0.28	-0.17 to -0.23	0.01 to 3.98	0.04 to 4.85
H_2O+	0.63	—	—	—
H_2O-	0.45	—	—	—
TiO_2	1.32	-0.41 to $+0.42$	0.42 to 1.63	0.12 to 1.34
P_2O_5	0.25	-0.07 to $+0.02$	0.24 to 1.50	0.11 to 0.19
MnO	0.15	—	0.06 to 0.28	0.03 to 0.35

Remarks:

(a) Data adapted from Melson & Van Andel (1966), table VII, p. 177.
(b) Data adapted from Smith (1968), table III, pp. 204–207.
(c) Data from tables 1–3, this paper.

COMPARISON WITH ROCKS FROM OCEANIC AREAS

The chemical variations of pillow lavas and associated intrusives are given above. Compared with basaltic and metabasaltic rocks dredged from the oceans, Troodos rocks display a similar range of SiO_2 content, alkalis and K_2O (see Melson & Van Andel 1966; Aumento 1968, 1969; Aumento & Loncarevic 1969; Aumento et al., this volume, p. 623; Muir & Tilley 1966, 1964). The degree of metamorphism and possible contamination in pillow lavas and Basal Group is reminiscent of that found by Smith (1968) in burial-metamorphosed basaltic flows in Australia. Table 6 shows the range of chemical variation found by Smith together with that found by Melson & Van Andel, as well as ranges for Troodos rocks. Such a comparison is not necessarily meaningful, but it is clear from Smith's work that low-grade metamorphism and metasomatism such as suffered by Troodos rocks, is a very complex chemical phenomenon, which is of major importance when comparing chemical data from altered volcanic piles. Consequently we feel that the chemical similarities or discrepancies between the altered Troodos and known fresh or altered oceanic rocks is neither an argument for nor against their origin at an oceanic ridge, but is an indication of a complex subsequent diagenetic and metamorphic history.

Another problem with the Troodos rocks when compared to oceanic rocks is the general lack, except for the Upper Pillow Lavas, of fresh or altered olivine phenocrysts or xenocrysts. There is no ready answer to this apparent discrepancy. Possibly some chlorite or celadonite alteration may be derived from olivine; some patches of secondary mafic minerals could be interpreted as having outlines resembling relict olivine crystals.

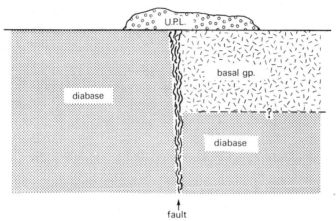

FIGURE 19. Cross-sectional sketch of relations on postulated Fossil Transform Fault. Sketch made looking eastward.

In summary, we feel that though the chemical and petrographic information available for the Troodos Complex differs somewhat from that for oceanic material, the effects of more comprehensive sampling of the various crustal units on Cyprus and of possibly differing post-magmatic chemical and mineralogic changes are not known with enough certainty to assess the significance of these discrepancies. On the other hand, the structural features of the Troodos Complex, in particular the dyke swarm, are readily compatible with origin by spreading at a ridge crest.

POSSIBLE FOSSIL TRANSFORM FAULT

The east–west trending fault zone interpreted as a fossil transform fault is shown on figure 1. The main characteristics of this feature are: (1) a broad, straight valley extending for 30 km or more; (2) a zone of brecciation as much as 100 to 200 m wide, along which Basal Group or Lower Pillow rocks are juxtaposed against Diabase; (3) a change in strike of dykes from mostly north–south to mostly east–west as one approaches the zone; (4) the presence of Upper Pillow Lavas in several places unconformably overlying faulted Lower Pillow Lavas, Basal Group and Diabase (see figure 19); (5) a zone of talus breccia with fragments of Diabase and some pillow lava, which is closely associated with the fault zone.

We interpret these features as resulting from transform movement during formation of the Basal Group, Lower Pillow Lava, and Diabase and prior to extrusion of the Upper Pillow Lavas. The presence of the east–west striking dykes in this region suggests intrusion along the fault zone.

Tectonic emplacement of the Troodos Massif

Subsequent to its formation about a postulated mid-Tethyan ridge, the Troodos Complex was caught up in the orogenic movements which gave rise to the Tehyan Mountain system. Gass & Masson-Smith (1963) and Bagnall (1964) have postulated thrusting over the African shield. The major tectonic emplacement was preCampanian in view of the widespread and generally undeformed Campanian chalks surrounding the mass. Minor subsequent late Tertiary movements have uplifted the massif and thrust it southward over the chalks. Possibly the main emplacement of the massif resulted from the collision of the African continent with a subduction zone dipping to the north (Temple & Zimmerman 1969; D. H. Roeder, personal communication). The Mamonia melange, which underlies the chalk sediments southwest of the Troodos Complex, may represent this fossil subduction zone or debris derived from it.

Magnetic properties

An indication of the direction and intensity of the natural remanent magnetization (n.r.m.) of over 900 hand specimens, from 150 localities within the Troodos Complex, has been obtained in the field with a portable fluxgate magnetometer (Doell & Cox 1967). In addition, a total of 200 oriented drill cores were obtained from 27 sites for further laboratory study. From previous experience with basic rocks it was felt that the direction of n.r.m. could be assessed in the field to within 20° of the true direction. This supposition was amply confirmed by the subsequent laboratory measurements of n.r.m. Moreover, in general, the directions of the remanent magnetic vectors for both mafic and ultramafic rock types change very little as a result of thermal or alternating field demagnetization treatment. Thus the n.r.m. directions obtained in the field can usually be taken to be indicative of the primary remanent magnetization. This feature of many of the rocks studied was clearly of great value in assessing preliminary results in the field, extending the area covered by magnetic sampling, and increasing the number of local or specific structural problems which could be investigated by the paleomagnetic method.

Obviously a primary objective of the project was to determine the distribution and extent of interfingering of normally and reversely magnetized material within the complex, in the hope of simultaneously confirming the validity of the Vine–Matthews hypothesis and providing compelling evidence that the massif was indeed formed by the process of sea-floor spreading. However, much to our chagrin no areas of convincingly or consistently 'reversed' material have been found, despite a fairly comprehensive reconnaissance survey of the main outcrop area and the two major inliers at Akamas and Troulli. If the complex represents a fragment of oceanic crust generated at a mid-ocean ridge crest, then the 110 km across strike extent of the outcrop area would represent between 11 and 5.5 Ma of spreading history if the ridge were spreading at a rate of 1 to 2 cm a^{-1} per ridge flank as suggested below. Such long epochs of consistently normal polarity of the Earth's magnetic field are thought not to occur during the late Cretaceous and Tertiary (Heirtzler et al. 1968) but may in the early and middle Cretaceous. Helsley & Steiner (1969) have suggested, on the basis of palaeomagnetic studies on land, that there are two such intervals, one of 20 and the other of 30 Ma duration, during this period. Although equivocal, it seems most probable, from the available evidence, that the Troodos Massif is middle Cretaceous in age. This conclusion is based on the Campanian age assigned to the faunal assemblage in the overlying radiolarian shales (Allen 1967) and preliminary potassium–argon age determinations (I. G. Gass, personal communication).

The pillow lavas and gabbros, which show little or no subsequent alteration, are thought to preserve their primary thermo-remanent magnetization. Clearly no 'reversal test' can be applied to the paleomagnetic vectors derived from these rocks but the results obtained are of considerable interest in that they yield a dip which is what one would expect for Cyprus during the Cretaceous, from extrapolation of African or European data (see, for example, Irving 1967), but a declination that is approximately due west. Such an azimuth for the palaeomagnetic vector is completely unexpected and most simply interpreted as indicating an anticlockwise rotation of the whole massif through approximately 90° since the time of its formation. This would imply formation at a ridge crest trending east–west, since the present predominant strike direction or 'grain' is north–south, and is more readily compatible with the assumed east–west elongation of the Tethys (Gass 1968).

The remobilized serpentinites yield randomly orientated and unstable remanent vectors, but the non-remobilized harzburgite and dunite units of the Plutonic Complex yield vectors which are stable and approximate to the present direction of the Earth's magnetic field. This stable remanence is presumably a chemical remanence acquired during serpentinization (cf. Saad 1969) and its direction implies that serpentinization is perhaps related to the late Tertiary uplift and emplacement of the massif. This observation and interpretation lends added credence to the correlation of these ultrabasics with submoho material since in an unserpentinized condition their densities would correspond to those inferred from the submoho velocity of 8.1 km s^{-1} for compressional seismic waves, i.e. 3.3 to 3.4 g cm^{-3} (Hess 1962; Talwani, Le Pichon & Ewing 1965). Many of these rocks, at present, have densities of 3.0 to 3.1 g cm^{-3} implying less than 50 % serpentinization. Thus we postulate that some ultramafic rocks dredged from the ocean floor might be derived from small, isolated pockets within the crust but most represent remobilized serpentinite which has been tectonically emplaced along fractures. The highly fractured nature of crust generated at more slowly spreading ridge crests, e.g. in the Atlantic and northwest Indian Oceans, may account for the fact that serpentinites are commonly dredged in these areas, but never in the Pacific.

The intensity of n.r.m. for the pillow lavas is typically of the order of 10^{-2} e.m.u. cm^{-3} and the Koenigsberger ratio (Q_n) $\sim$ 10, i.e. the n.r.m. intensity is on average ten times greater than the intensity of induced magnetization, as is the case for ocean floor basalts (e.g. Opdyke & Hekinian 1967). In contrast, the greenschist facies rocks (Basal Group and Diabase) have similar magnetic susceptibilities, i.e. intensities of induced magnetization $\sim$ 10^{-3} e.m.u. cm^{-3}, but Koenigsberger ratios of approximately 0.5. This drastic reduction in the intensity of the remanent magnetization is to be expected since the metamorphic process involves destruction of many of the primary iron–titanium oxide phases—titanium for example being taken up in newly formed sphene. This makes recovery of the primary remanent direction from these rocks very difficult and the directions of their remanent vectors are generally intermediate between those of the pillow lavas and gabbros and the present direction of the Earth's magnetic field. Presumably secondary viscous components of remanence are now comparable in intensity and stability to those of the residual primary remanence. The plutonics have natural remanent intensities of order 10^{-3} e.m.u. cm^{-3} and Koenigsberger ratios of approximately 1, the gabbros and diorites being more consistent in this respect than the ultramafics, which are highly variable.

Thus, pursuing the sea-floor analogue, the most potent sources of remanent magnetization contrasts within the oceanic crust are seen to reside in the upper part, the pillow lavas or seismic layer two, as suggested by Vine & Wilson (1965) and many others subsequently. Origin-ally it was thought that layer three might be less effective in this respect because it consists of gabbro or serpentinite, but it now seems probable, both from more recent dredge hauls and the analogy with Troodos, that the upper part of layer three (Diabase) and lower part of layer two (Basal Group) consist of greenschist facies dolerites equally impotent as regards con-tributing to the magnetic anomalies observed at or above sea-level.

More detailed documentation and consideration of the magnetic results was considered inappropriate to this meeting and will appear elsewhere.

DISCUSSION

Isoclinal folds, pronounced foliation, and preliminary petrofabric analysis of dunite and harzburgite suggest that these rocks are tectonites, rather than primary crystalline phases. We interpret these rocks as representing upper mantle material which has flowed in a solid state upward into an opening crack. The olivine pyroxenites, gabbros, Sheeted Complex and pillow lava rocks represent either the products of fractional fusion, segregated from parent mantle material during upward flow at a mid-ocean ridge, or fractional crystallization of such fusion products subsequent to their emplacement in the oceanic crust. Some layered gabbro-pyroxenite bodies clearly are of cumulate origin resulting from crystallization of small intrusive bodies of mafic magma. Some uralitized gabbro and quartz diorite are the residual liquids of such a process. Other masses of pyroxenite, gabbro, and quartz diorite show no such clear relation to one another, and may be partial fusion products of different compositions.

Figure 20 shows a schematic cross-section of the Troodos Complex. Several implications of this interpretation are as follows:

(1) The more olivine-rich Upper Pillow Lavas may have their source from the olivine pyroxenite zone near the Moho. Such a conclusion is supported by their more ultramafic com-position (see tables 1 and 5), and by the presence in the Sheeted Complex and Lower Pillow Lavas of small possibly pipe-like intrusions of olivine pyroxenite (see maps in Bear 1960; Gass 1960; and Wilson 1959).

(2) The observed field relations are best interpreted as the product of multiple intrusion and extrusion of small portions of liquid. Such a concept is supported by the abundant presence of chilled margins and multiply-intruded dyke complexes, the mutually contradictory relations between gabbro and diabase, with each intruding, and intruded by the other, and the presence of some apparently rootless dykes in the gabbro and diabase (L. M. Bear, oral communication, 1969). Such multiple intrusion and multiple chilling effects imply the emplacement of material, hence extension over a fairly broad zone, and a slow enough rate of spreading to allow for the production of chilling effects. This suggests that the ridge must have been fairly slow-spreading.

(3) The relief of the zone of maximum dyke formation—the top of the Sheeted Complex—and the presence of both vertical and conjugate dyke systems possibly are related to varying rates of magma supply and extension. Vertical dykes possibly suggest a role for fluid pressure and forceful injection of dyke material, causing the rock to fracture perpendicular to minimum compressive stress. On the other hand, if the rock is first extended by normal faulting and subsequently intruded, then conjugate systems of dykes may be formed. In addition, if the rate

of supply of magmatic material to the ridge axis is relatively large in relation to extension, then one might expect a relatively greater thickness of flows above and sill-like intrusions of gabbro below, a relatively thick Lower Pillow Lava and Basal Group (or Layer Two), and a relatively thin Sheeted Complex. If, on the other hand, the rate of creation of volume by extension gets slightly ahead of rate of supply of magmatic material to the ridge, one would expect a slightly thinner layer two crust, and a relatively greater proportion of sheeted material, perhaps as suggested for the East Pacific Rise by Menard (1967).

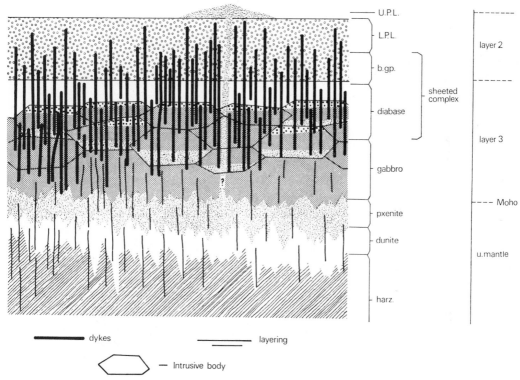

FIGURE 20. Interpretative cross-section of Troodos Complex. Large closed dots are quartz diorite-granophyre segregations in gabbro bodies. Open dots are pillow lava. Heavy black lines are dykes. Column to right is postulated correlation with seismic layers of the oceanic crust.

OTHER SHEETED COMPLEXES

It is now apparent that the Troodos Complex is not unique in exhibiting a sheeted diabase complex. Other complexes which have this or similar features are as follows:

(1) *Kizil Dagh, Hatay, Turkey.* Dubertret (1955) described a reconnaisance survey of this mass, and particularly referred to 'stratified beds' of dolerite, now dipping vertically. Recent reconnaisance of this exposure by Vuagnat & Çogolu (1968) and ourselves confirmed the presence of a sheeted complex.

(2) *Oman Mountains.* Reinhardt (1969) has summarized features of the Oman Mountains geosyncline in which he shows (his p. 5 and 27) subvertical Diabase dyke swarms, again suggesting similarities to Troodos.

(3) *Macquarie Island.* Varne *et al.* (1969) have described in the northern part of the island the presence of strip-like exposures of harzburgite, layered gabbro and sheeted Diabase intruded into a basaltic sequence.

In contrast to the above examples, however, most Diabases from ophiolite complexes have not been described as exhibiting sheeted structure, but as consisting of two main types: first, exemplified by the Vourinos Complex, northern Greece (see figures 21 and 22), where internal layering is parallel, rather than perpendicular, to major rock units; secondly, such as in the Italian occurrences, where no particular internal structure has been described (see figure 23). The Vourinos Complex (Moores 1969) displays a substantial development of phase and cryptic-layered pyroxenite, gabbro, and diorite, suggesting some intrusion and cumulate formation.

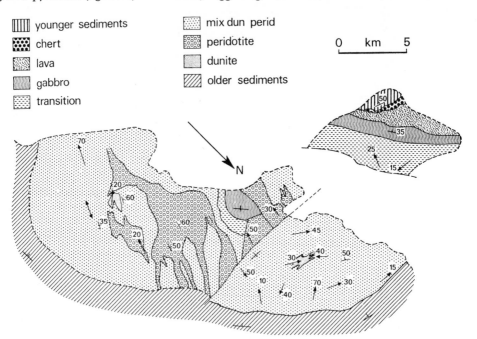

FIGURE 21. Map of Vourinos Complex, northern Greece, after Moores (1969). Attitudes are of foliation and minor fold axes.

Many Italian bodies (J. C. Maxwell 1969, 1970, and oral communication) contain peridotite or serpentinized peridotite, thin or absent gabbro, diabase without internal layering or dyke structure which grades non-descriptly into pillow lava or extrusive breccia. Though presently unexplained, these differences in internal structure apparently are primary and fundamental features. If these other ophiolites represent fragments of oceanic crust, then these differences in internal structure may represent differences in behaviour on spreading ridges. In speculating on a possible analogy with modern oceanic ridges, the difference in topography between slow and fast-spreading ridges (Menard 1964, 1967; Van Andel & Bowin, 1968) comes to mind. The generally prominent median valley and rugged topography of slow-spreading ridges such as the Mid-Atlantic and the northwest Indian Ocean Ridges contrast strongly with the smooth topography of the fast-spreading East Pacific Rise. The possible slow-spreading ridge origin of the Troodos Complex has been outlined above. Perhaps ophiolites without marked internal

42-2

structure represent crust and mantle formed as fast-spreading ridges, such as the East Pacific Rise, where all material would be injected in a narrower axial region (Vine 1968), and when injected, would move out without further addition, crystallizing as it moved. In this case, the upper part of the injected magma would crystallize immediately and display extrusive features, and the part immediately underlying it would crystallize more slowly and show gradation downward into coarser-grained intrusive material.

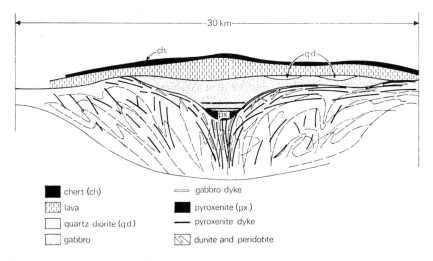

■ chert (ch)	⇒ gabbro dyke
▥ lava	■ pyroxenite (px.)
☐ quartz diorite (q.d.)	▬ pyroxenite dyke
☐ gabbro	◩ dunite and peridotite

FIGURE 22. Schematic cross-section of Vourinos Complex, after Moores (1969) showing complex structure of ultramafic rocks plunging towards the centre, pyroxenite and gabbro dykes concentrated towards the centre, and overlying mafic rocks.

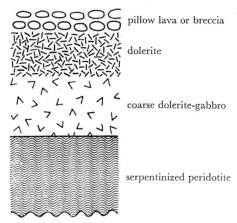

FIGURE 23. Schematic cross-section of Italian ophiolite complex, after J. C. Maxwell (personal communication).

The fact that successive intrusions in the Sheeted Complex and Lower Pillow Lava commonly dissect pre-existing dykes has been outlined above. This is an important relation, for it suggests that the intrusion process is not entirely random, at least on a short time scale, e.g. a few thousand years (4 m dykes intruded and repeatedly dissected every 200 years is equivalent to a half-spreading rate of 1 cm a^{-1}). Hence the process of emplacement invoked for fast-spreading ridges was operative at least locally on Troodos, and yet chilled contacts are developed on Troodos. The obvious difference, however, is the rate of emplacement of material. If the discrete volumes injected are the same in both cases, intrusives on fast-spreading ridges (e.g. East Pacific Rise) would be several times more frequent. Perhaps two effects are operative: (1) a narrower zone of injection over a long time span on a fast-spreading ridge, which would tend to make the central zone of formation hotter and narrower; and (2) a higher frequency of intrusion in fast-spreading ridges, which would tend to maintain higher temperatures at the ridge crest and prevent the formation of chilled contacts. Possibly the Vourinos Complex, which displays some differentiation and also parallel internal structure, represents a stage intermediate between the 'fast-spreading' Italian ophiolite and 'slow-spreading' Troodos models. The internal fabric of the ultramafic rocks also differs. On Troodos foliations of dunite and harzburgite are vertical, whereas in Vourinos they were originally horizontal away from the divergent and complexly domed centre. This difference implies that mantle layering on slow-spreading ridges tends to be vertical and becomes horizontal on fast-spreading ridges. Under this interpretation, the divergent structure present in the Vourinos Complex implies that it represents a piece of *the actual crest* of the former mid-Tethyan ridge. The zone of vertical dykes in the centre of the complex then would represent the zone of dyke intrusion at a ridge and the complexly domed ultramafics in the same area may represent fossil mantle diapirs.

The model outlined above is clearly speculative, but seems to be compatible with the known petrology and structure of Tethyan ophiolites. It provides a means of reconciling the three conflicting hypotheses for the origin of ophiolites outlined in the Introduction. If the model is correct, then it means that dyke intrusion, stratiform crystallization in magma chambers, and solid diapiric activity all occur at a ridge crest. Subsequently fragments of crust and uppermost mantle formed in this manner are thrust upon a continental margin. The conflicting interpretations of origin of ophiolites have resulted from study of individual occurrences in which these features have been variously preserved.

Throughout this study we drew on the experience and benefited from the advice and assistance of Dr L. M. Bear, formerly member and director of the Cyprus Geological Survey (1955–64) and of the U.N. Mission in Cyprus (1965–9). Especially valuable was his excellent grasp of the details of geology of the Troodos Massif, acquired over 14 years of enthusiastic observation.

We would also like to thank Dr Y. Haji Stavrinou, Director of the Cyprus Geological Survey, and many members of his staff, particularly Dr Theo Pantazis, for their hospitality, assistance and interest. We benefited from discussions with H. L. Davies, A. Gansser, I. G. Gass, H. H. Hess and J. C. Maxwell.

This study was supported by National Science Foundation Grants GA-1257 and GA-1395

REFERENCES (Moores & Vine)

Allen, C. G. 1967 *Ann. Rep. Geol. Surv. Dep. Cyprus for* 1966, p. 8.
Aubouin, J. 1965 *Geosynclines.* New York: Elsevier.
Aumento, F. 1968 *Can. J. Earth Sci.* **5**, 1–22.
Aumento, F. 1969 *Science, N.Y.* **165**, 1112–1113.
Aumento, F. & Loncarevic, B. D. 1969 *Can. J. Earth Sci.* **6**, 11–23.
Bagnall, P. S. 1960 *Geol. Surv. Cyprus Mem.* **5**, 1–116.
Bagnall, P. S. 1964 *J. Geol.* **72**, 327–345.
Bear, L. M. 1960 *Geol. Surv. Cyprus Mem.* **3**, 1–122.
Bear, L. M. 1965 *Geologic Map of Cyprus,* 1/250000. Cyprus: Geol. Surv. Dep.

Bear, L. M. 1966 *Ann. Rep. Geol. Surv. Dep. Cyprus for* 1965, pp. 26–37.
Bear, L. M. & Morel, S. W. 1960 *Geol. Surv. Cyprus Mem.* **7**, 1–88.
Bezone, S. P. 1969 *Geol. Soc. America, Abstracts with Programs for* 1969, **3**, 5–6.
Böttcher, W. 1969 *Neues Jb. Miner. Abh.* **110**, no. 2, 159–187.
Brunn, J. H. 1956 *Ann. Geol. Pays Hell.* **7**, 1–358.
Brunn, J. H. 1960 *Revue Géogr. phys. Géol. dyn.* **3**, 115–132.
Burri, C. & Niggli, P. 1945 *Vulkaninstitut Immanuel Friedländer Publikationen*, pp. 3–4.
Carr, J. M. & Bear, L. M. 1960 *Geol. Surv. Cyprus Mem.* **2**, 1–79.
Davies, H. L. 1969 *Int. Geol. Congr. 23rd Czech. Rep. Sect.* 1, *Proc.*, pp. 209–220.
DeRoever, W. P. 1957 *Geol. Rdsch.* **46**, 137–146.
Doell, R. R. & Cox, A. 1967 In *Developments in solid earth geophys.* **3**, pp. 159–162. New York: Elsevier.
Dubertret, L. 1955 *Notes at Memoires sur le Moyen Orient-Museé Natl. Hist. Paris* **6**, 5–224.
Engel, A. E. J., Engel, C. G. & Havens, R. G. 1965 *Bull. Geol. Soc. Am.* **76**, 719–734.
Engel, C. G. & Fisher, R. L. 1969 *Science, N.Y.* **166**, 1136.
Gansser, A. 1959 *Ec. Geol. Helv.* **52**, 659–680.
Gass, I. G. 1958 *Geol. Mag.* **95**, 241.
Gass, I. G. 1960 *Geol. Surv. Cyprus Mem.* **4**, 1–116.
Gass, I. G. 1967 In *Ultramafic and related rocks* (ed. P. J. Wyllie), pp. 121–134. New York: Wiley.
Gass, I. G. 1968 *Nature, Lond.* **220**, 39–42.
Gass, I. G. & Masson-Smith, D. 1963 *Phil. Trans. Roy. Soc. Lond.* A **255**, 417–467.
Heirtzler, J. R., Dickson, G. O., Herron, E. M., Pitman, W. C. & Le Pichon, X. 1968 *J. geophys. Res.* **73**, 2119–2136.
Helsley, C. E. & Steiner, M. B. 1969 *Earth Planet. Sci. Lett.* **5**, 325–332.
Hess, H. H. 1962 In *Petrologic studies: a volume to honor A. F. Buddington*, pp. 599–620. Geol. Soc. Am.
Hess, H. H. 1965 In *Submarine geology and geophysics, Colston Papers* **17** (ed. W. F. Whittard & R. Bradshaw), pp. 317–333. London: Butterworth.
Hsu, K. J. 1969 *Calif. Div. Mines Geol., Spec. Publ.* **35**.
Irving, E. 1967 In *Systematics Ass. Publ.* **7**, 59–76.
MacDonald, G. A. & Katsura, T. 1964 *J. Petrology* **5**, 82–133.
Maxwell, J. C. 1969 *Tectonophysics* **7**, 489–494.
Maxwell, J. C. 1970 In *The megatectonics of continents and oceans* (ed. H. Johnson). Rutgers University Press.
Maxwell, J. C. & Azzaroli, A. 1963 *Geol. Soc. Am. Spec. Paper* **73**, 203–204.
Melson, W. G. & Van Andel, Tj. H. 1966 *Marine Geol.* **4**, 165–186.
Menard, H. W. 1964 *Marine geology of the Pacific*, 271 pp. New York: McGraw-Hill.
Menard, H. W. 1967 *Science, N.Y.* **157**, 923–924.
Miyashiro, A. 1966 *Jap. J. Geol. Geogr. Trans.* **37**, 45–61.
Moores, E. M. 1969 *Geol. Soc. Am. Spec. Paper* **118**.
Muir, D. D. & Tilley, C. E. 1964 *J. Petrology* **5**, 409–434.
Muir, D. D. & Tilley, C. E. 1966 *J. Petrology* **7**, 193–201.
Opdyke, N. D. & Hekinian, R. 1967 *J. geophys. Res.* **72**, 2257–2260.
Pantazis, T. 1967 *Geol. Surv. Cyprus Mem.* **8**, 1–190.
Poldervaart, A. 1955 *Geol. Soc. Am. Spec. Paper* **62**, 119–144.
Reinhardt, B. M. 1969 *Schweiz. miner. petrogr. Mitt.* **49**, 1–30.
Saad, A. H. 1969 *J. geophys. Res.* **74**, 6507–6578.
Smith, C. H. 1958 *Geol. Surv. Can. Mem.* **290**, 132 pp.
Smith, R. E. 1968 *J. Petrology* **9**, 191–219.
Steinmann, G. 1906 *Freib. naturf. Gesell. Bericht* **16**, 18–66.
Steinmann, G. 1926 *14th Int. geol. Congr., Madrid, C.R.* **2**, 638–667.
Talwani, M., Le Pichon, X. & Ewing, M. 1965 *J. geophys. Res.* **70**, 341–352.
Temple, P. & Zimmerman, J. Jr. 1969 *Geol. Soc. Am., Abstracts with Programs for* 1969, **7**, 211–212.
Thayer, T. P. 1963 *Min. Soc. Am. Spec. Paper* **1**, 55–61.
Trümpy, R. 1960 *Bull. Geol. Soc. Am.* **71**, 843–908.
Van Andel, Tj. H. & Bowin, C. O. 1968 *J. geophys. Res.* **73**, 1279–99.
Varne, R., Gee, R. D. & Quilty, P. G. J. 1969 *Science, N.Y.* **166**, 230–232.
Vine, F. J. 1968 In *History of the Earth's crust* (ed. R. A. Phinney), pp. 73–89. Princeton: University Press.
Vine, F. J. & Hess, H. H. 1970 In *The sea* vol. 4 (ed. A. E. Maxwell, E. C. Bullard, E. Goldberg & J. L. Worzel). New York, London: Wiley–Interscience.
Vine, F. J. & Matthews, D. H. 1963 *Nature, Lond.* **199**, 947–49.
Vine, F. J. & Wilson, J. T. 1965 *Science, N.Y.* **150**, 485–489.
Vuagnat, H. 1963 *Geol. Rdsch.* **53**, 336–358.
Vuagnat, H. & Çogolu, E. 1968 *Soc. Phys. Hist. Natur. Geneve, C.R.* **2**, 210–216.
Wilson, R. A. M. 1959 *Geol. Surv. Cyprus Mem.* **1**, 1–135.

Erratum: On p. 327, the second line from the bottom should read: "Vuagnat, H., and Çogulu, E. . ."

17

Reprinted from *Jour. Geophys. Res.*, **76**(5), 1212–1222 (1971)

Plate Tectonic Emplacement of Upper Mantle Peridotites along Continental Edges

R. G. Coleman

U. S. Geological Survey, Menlo Park, California 94025

Recently developed ideas of global tectonics have provided a new framework within which to consider the origin of alpine-type peridotites. In plate theory, compressional zones associated with island arcs are considered to represent plate boundaries where oceanic lithosphere is subducted. The subduction zones are characterized by lithospheric underthrusting, andesitic volcanoes, and deep seismic activity that generally dips under the continental edge (the Benioff zone). The presence of large oceanic-mantle crustal slabs thrust over or into continental edges contemporaneously with blueschist metamorphism in New Caledonia and New Guinea establishes an important variant of plate tectonics in the zones of compression. The 'obduction' zones are characterized by a complete lack of volcanic activity and by high-pressure metamorphism. During formation, they can be represented by shallow seismic zones dipping oceanward. The common association of peridotites and blueschists in these orogenic belts may result from the initial stage of compressional impact (or orogeny) between an oceanic and a continental lithospheric plate. Disturbed zones combined with a lack of high-temperature contacts at boundaries between cold mantle-peridotite slabs and trench sediments provide geologic evidence of emplacement by obduction (tectonic overriding). Internal subsolidus plastic deformation of these peridotites can be attributed to deep-seated strain within the upper mantle during spreading. Serpentinites represent alteration developed during tectonic emplacement into wet sediments of the continental plate, which produces a less dense and plastic envelope that facilitates further tectonic movement in these compressional zones. Recognition of these peridotite-serpentinite-blueschist belts within exhumed subduction or obduction zones will allow delineation of ancient compressional impacts between moving lithospheric plates.

The origin of peridotites (alpine ultramafics) within orogenic zones of the earth has provoked nearly continuous debate among petrologists over many years [*Benson*, 1926; *Hess*, 1938, 1955a and b; *Steinmann*, 1905, 1927; *Thayer*, 1960; *Wyllie*, 1967; *Moores*, 1969; *Dubertret*, 1955]. The basis for much of the disagreement has been the paradox between the observed field petrologic evidence for generally cold intrusion and the higher temperatures expected for ultramafic intrusions derived from igneous melts [*Wyllie*, 1967, Table 12.1, p. 411]. Recent studies by experimental petrologists [*Green and Ringwood*, 1967; *Kushiro*, 1968; *MacGregor*, 1968] have shown that peridotite magmas require temperatures and pressures that greatly exceed those expectable within the continental crust and that formation of a peridotite magma is possible only within the mantle. The occurrence of peridotite in large tracts along modern

or ancient continental margins without high-temperature contact aureoles has led certain investigators to propose tectonic emplacement rather than igneous intrusion for these peridotites [*Benson*, 1926; *Moores*, 1969; *Hess*, 1965; *Dietz*, 1963; *Thayer*, 1969; *Coleman*, 1962, 1967; *deRoever*, 1957; *Davies*, 1968]. Assuming that tectonic emplacement is valid, there still is required a place for the magmatic development of these ultramafic rocks.

The purpose of this paper is to propose a mechanism of peridotite emplacement that overcomes the major petrogenetic problems and is consistent with both the new global tectonics and accumulated field observations related to the alpine ultramafics. New concepts of global tectonics, developed from independent confirmation that movement of rigid lithospheric plates can be related to present-day seismic activity and configuration of magnetic anomalies, have provided a completely new framework within which to consider the origin of alpine ultramafics [*McKenzie and Parker*, 1967; *Oliver et al.*, 1969; *Vine*, 1966; *Sykes*, 1967]. The largest known areas of alpine ultramafics are restricted to narrow geosutures or former continental boundaries. These ultramafic belts occupy structural positions that suggest development coincident with tectonic activity related to continental accretion [*Hess*, 1955a, 1965; *Dewey* and *Horsfield*, 1970]. *Hess* [1955a] believed that the ultramafic rocks were emplaced during the first great deformation of an orogenic event and that the traces of these belts marked the axes of ancient orogenic belts and island arcs. These ideas undoubtedly were prompted by the early observations of *Steinmann* [1927] and *Benson* [1926], both of whom were impressed by the association of ultramafic rocks (serpentines) with gabbros, diabase, and pillow lava (ophiolites) in orogenic zones. In the discussion to follow, the term ophiolite is retained as a convenient and traditional term to describe the mafic-ultramafic assemblages in orogenic zones; however, it is not used in the conceptual sense that has been applied by *Dubertret* [1955], *Brunn* [1960], *Maxwell and Azzaroli* [1962], and *Maxwell* [1969]. Recent work on the Vourinous and Troodos ophiolite complexes in the Mediterranean Tethyan belt has shown that the strong internal deformation of the peridotites, together with the lack of a basal chill zone and excessive proportion of ultramafic to mafic rocks, precludes any serious consideration that these ophiolites represent differentiated submarine extrusions of primary basalt liquid [*Moores*, 1969; *Moores and Vine*, 1971].

OCEANIC CRUST-MANTLE AND OPHIOLITES: COMPARISON

Recent intensive study of oceanic areas has provided a coherent picture of the possible nature of the oceanic crust [*Cann*, 1968; *Christensen*, 1970; *Melson et al.*, 1968; *Shor and Raitt*, 1969]. These studies represent a combination of geophysical and petrologic observations that have many internal consistencies. Seismic refraction studies have shown that the oceanic crust can be divided into three distinct layers and that each of these layers has a uniform seismic response when measured in various oceanic areas [*Shor and Raitt*, 1969; *Raitt*, 1963; *Hill*, 1957]. Layer 1 consists of sediments in various states of consolidation with an aver-

age thickness of 0.3 km and a seismic velocity (V_p) between 1.5 and 3.4 km/sec [*Shor and Raitt*, 1969]. Direct sampling by deep-sea drilling confirms these geophysical estimates [*Fischer et al.*, 1970]. Layer 2 is considered to be basalt that forms at oceanic ridges and has a magnetic character that produces linear magnetic anomalies of the ocean floors [*Vine and Matthews*, 1963; *Cann*, 1968; *Heirtzler and Le Pichon*, 1965]. The second layer has an average thickness of 1.4 km and a seismic velocity between 3.4 and 6.0 km/sec. Layer 3 has a uniform thickness of 4.7 km and a seismic velocity (V_p) of 6.8 km/sec. However, its exact petrologic nature is controversial [*Cann*, 1968; *Christensen*, 1970]. *Hess* [1962] suggested that perhaps the third layer consists of serpentinized peridotite, because seismic velocities of partially serpentinized peridotite could be matched to the 6.8-km/sec value observed in the third layer. Others, because of thermal problems and paucity of water for serpentinization, have raised serious objections to Hess' proposal for a serpentinized peridotite [*Christensen*, 1970; *Cann*, 1968; *Oxburgh and Turcotte*, 1968]. It has recently been suggested that the third layer is likelier to be amphibolite, gabbro, or a combination of gabbro, diabase, and their metamorphosed equivalents [*Christensen*, 1970; *Cann*, 1968]. Below the third layer, or Mohorovicic discontinuity, it is generally agreed that the mantle consists of peridotite, whose mineralogy depends on the amount of partial melting and its depth of crystallization below the Mohorovicic discontinuity.

Comparison of the oceanic crust sequence with the established stratigraphic sequences of the Tethyan belt ophiolites such as Vourinous [*Moores*, 1969; *Brunn*, 1956] and Cyprus [*Moores and Vine*, 1971; *Gass and Masson-Smith*, 1963] with other ophiolites from Papua [*Davies*, 1968], New Caledonia [*Deneufbourg*, 1969], and California [*Bailey et el.*, 1970], yields an obvious similarity (Figure 1). Another favorable comparison is that of the bulk composition of the ophiolite assemblage with the composition of oceanic basalts, gabbros, and peridotites. Characteristically, the basalts of the deep ocean areas are subalkaline tholeiites [*Engel et al.*, 1965; *Miyashiro et al.*, 1969; *Kay et al.*, 1970] strikingly similar to pillow lava, diabase, trondhjemite, and gabbro from ophiolite assemblages (Figure 2). The calc-alkaline trend

1214   R. G. COLEMAN

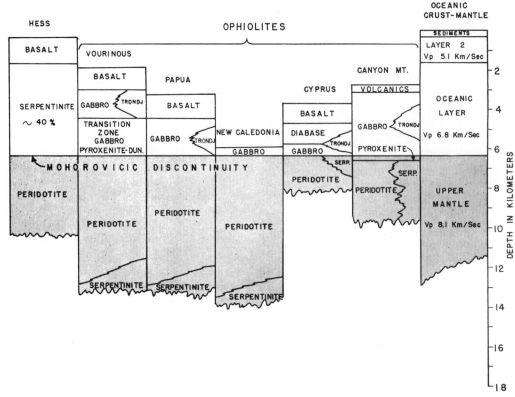

Fig. 1. Comparison of the stratigraphic thickness of igneous units from various ophiolite masses with the geophysical estimate of the oceanic crustal layers. Data from *Hess* [1962], Vourinous [*Moores*, 1969]; Papua [*Davies*, 1968]; New Caledonia [*Crenn*, 1953]; Cyprus [*Wilson*, 1959; *Moores and Vine*, 1971]; Canyon, Mt. Oregon [*Thayer and Himmelberg*, 1968]; oceanic crust and mantle [*Shor and Raitt*, 1969].

of the oceanic suite has been previously discussed by *Thayer* [1969] and *Moores* [1969]. The important aspect of the trend is its discontinuous nature and the general overlapping similarities between assemblages from the deep oceans and ophiolites. The ultramafic group consists primarily of harzburgite-dunite with minor plagioclase-bearing lherzolites, which characteristically display high-temperature subsolidus deformation [*Loney et al.*, 1971]. The high-calcium and alumina mafic gabbros common to the ophiolites have also been reported from deep-ocean dredge hauls [*Engel and Fisher*, 1969; *Miyashiro et al.*, 1970]. Many of the ophiolite gabbros and some ultramafic rocks exhibit cumulate textures that may have developed in shallow magma chambers that acted as feeders for the basalts extruded at ridge crests [*Moores and Vine*, 1971; *Davies*, 1969]. Basalts

in the ophiolite sequences are nearly identical with the subalkaline oceanic basalts; however, high total-iron contents in certain ophiolite and oceanic basalts suggests differentiation before extrusion [*Miyashiro et al.*, 1970]. The subordinate but persistent association of trondhjemite with ophiolites and oceanic basalts represents the end product of differentiation of a subalkaline basalt within the ridge system.

Partial melting of lherzolite, or the pyrolite of *Green and Ringwood* [1967], at depths less than 50 km could produce a subalkaline basalt liquid [*Green*, 1970]. Differentiation of some of this liquid in the upper levels of the ridge system could give rise to the plagioclase-rich gabbros often found associated with ophiolites and reported from oceanic ridges [*Miyashiro et al.*, 1970]. Depletion of the subalkaline basaltic liquid by precipitating predominantly plagioclase with olivine or pyroxene or both could produce the small amounts of trondhjemite-keratophyre often associated with ophiolites and oceanic crust. Residues developed by partial melting of primitive mantle lherzolites are now represented by the widespread depleted harzburgite-dunite masses that form the basement complex upon which the third layer or oceanic crust is formed.

Graphical expression of the oceanic ridge differentiation is shown on a plot of CaO versus Al_2O_3 (Figure 3). Partial melting (20–30%) of lherzolite produces a depleted harzburgite-dunite residue extremely low in calcium and aluminum and a subalkaline basalt liquid whose calcium and aluminum contents would be controlled by the amount of fractionation of this liquid at different pressures and temperatures before extrusion at the ridge axes. The gabbros unusually high in calcium and aluminum often associated with depleted harzburgites-dunites may then represent partially differentiated sub-alkaline basalt liquid rather than part of a continuous differentiated series from early ultramafics through gabbro, as has been suggested by *Thayer* [1969]. Until the volume relations between the various rock units are better known from geologic observations, the derivative nature of Figure 3 must be considered uncertain. The trends of igneous evolution suggested here are quite different from those characterizing large layered bodies such as the Bushveld or Stillwater complexes in that the initial liquid is developed by partial melting of ultramafics followed by differentiation at different levels (pressures) within the ridge system [*O'Hara*, 1968]. Many of the ophiolite gabbros and some of their associated peridotites exhibit cumulate layering that could be related to perched magma chambers produced in the rift zones. The depleted harzburgite-dunite basement complexes associated with ophiolites always exhibit strong deformational fabric superimposed on original compositional variations. Where these fabric elements have been studied, there is evidence of high-temperature subsolidus recrystallization [*Loney et al.*, 1971; *Ave'Lallemant*, 1967; *Den Tex*, 1969]. The deformational style may be related to the movement of these depleted harzburgites-dunites away from the axis of the ridge as part of the oceanic plate basement upon which the third layer is resting. The abundant

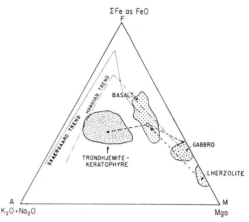

Fig. 2. FMA plot (in weight percent) of ophiolites and oceanic crustal rocks. Fields represent approximate and overlapping compositional variations of rock types from both occurrences. Circles in center of basalt field is the average given of oceanic basalts. Dashed lines indicate trends for the development of various rock types.

deformed harzburgites and dunites in ophiolitic zones that underlie the mafic rocks must represent the depleted mantle upon which the oceanic crust was constructed at active ridges and marks a major discontinuity in the oceanic crust.

Although the analytical data on ophiolites have not kept pace with the data now being developed on oceanic igneous rocks, there appear to be no major compositional inconsistencies between ophiolites and oceanic-ridge igneous rocks. In the following sections, ophiolite sequences within orogenic zones will be considered to have their igneous origins within the oceanic ridge crests and to have been modified by plate tectonic transport. Migration of the oceanic crust on depleted mantle rocks subjected these rocks to variable and irregular amounts of metamorphism before they reached the continental edges [*Melson et al.*, 1968].

General Distribution of Ophiolites and Blueschists in Ancient Geosutures

When ophiolite belts are plotted on a world map along with plate boundaries, there appears to be a very close correlation in space and time (Figure 4). Ancient linear ophiolite belts occupy geosutures that separate plates of quite different geologic age, structure, and sedimen-

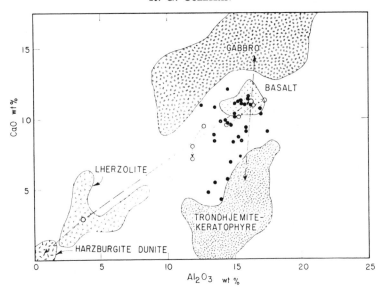

Fig. 3. CaO and Al₂O₃ plot of ophiolites and oceanic crustal rocks. Fields represent approximate and overlapping compositional variations of rock types from both occurrences. Solid dots represent ophiolite basalts; central open circle in basalt field, the average oceanic basalt. Open circle in lherzolite field represents the composition of pyrolite [*Green and Ringwood, 1967*]. Dashed lines indicate trends for the development of various rock types. Connected open circles show Skaergaard trend.

tary history. Structures within the suture zones in general are parallel to the strike of these zones. Another significant feature of the suture zones is the common occurrence of blueschist metamorphic belts [*Blake et al.*, 1969] parallel to, and intimately associated with, ophiolites. Where ages based on radiometric dating in these suture zones are available, it is found that the blueschist metamorphism is related in time to deformation within the suture, whereas the igneous age of the ophiolites is older than the enclosing sediments and their metamorphic equivalents [*Cogulu*, 1967; *Coleman*, 1967]. If we assume that these ancient exhumed geosutures represent the results of interaction between continental and oceanic plates as recently defined by seismic activity [*Barazangi and Dorman*, 1969], it may be possible to explain the long-standing problem of the duality of blueschists and ophiolites.

In the South Pacific, the youngest known areas of blueschist and ophiolite are exposed in New Caledonia and Papua, New Guinea [*Coleman*, 1967; *Davies*, 1968, 1969; *Lillie and Brothers*, 1970; *Lillie*, 1970]. A discussion of these young, little-complicated areas is given

in support of the concept that parts of the oceanic crust have been overthrust (obducted) onto thin continental edges. The Papuan ophiolite belt, consisting of harzburgite, dunite, gabbro, and basalt, extends 260 miles on the oceanic side of the island (Figure 5). According to *Davies* [1968, 1969], the ophiolites represent a slab of oceanic crust and mantle emplaced in Cretaceous or Eocene time by overthrusting (obduction) oceanic crust onto the continental crust. The ultramafic zone consists of peridotite that has both cumulate and deformational fabrics and grades upward into gabbro, which, in turn, is overlain by pillow lavas. Serpentinites are found as narrow zones along the base of the peridotites, and deformed sediments underlying the ophiolites have recrystallized sporadically to the blueschist facies [*Davies*, 1968, 1969]. Gravity data confirm the geologic assumption that the dense oceanic crust and mantle are slablike (Figure 6).

Large plates of ultramafic rock consisting primarily of dunite and harzburgite cover large areas in New Caledonia (Figure 6) [*Deneufbourg*, 1969]. Gravity data [*Crenn*, 1953] on the ultramafic masses indicate that, as in Papua,

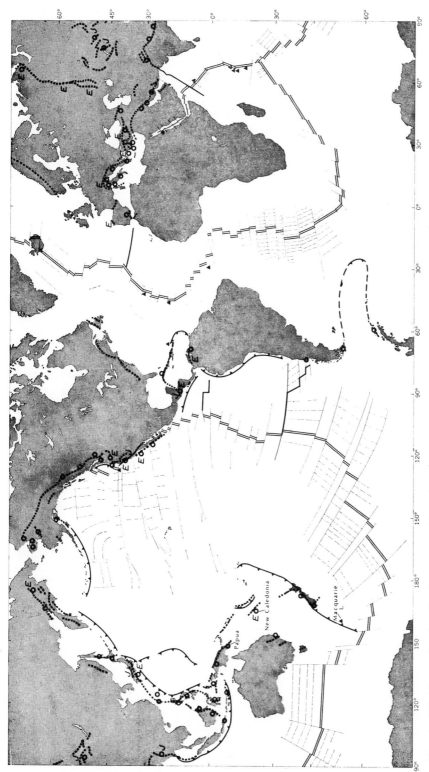

Fig. 4. World map showing distribution of ophiolite belts (solid squares) and associated blueschists (open circles) in orogenic zones and occurrences of crustal eclogites type C (*E*) and pure jadeite pods (*J*) in serpentinite. Heavy double line represents spreading ridge, heavy line with barbs represents active subduction zone, heavy and light solid lines represent transform faults and rises, dashed lines indicate approximate position of magnetic anomalies with respect to the ridges. Dredge hauls of ultramafic rocks from oceanic ridges are shown as solid triangles.

333

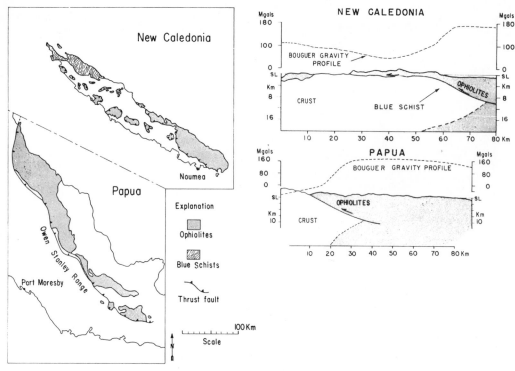

Fig. 5. Maps showing distribution of ophiolite masses in New Caledonia and Papua, New Guinea [modified after *Davies,* 1969, and *Lillie and Brothers,* 1970]. Generalized cross sections depict obduction of oceanic crust and mantle onto continental edges.

the ophiolites are slab shaped and obducted on top of Cretaceous and Eocene sedimentary rocks, which have been recrystallized to blues-chist facies. Obduction of the oceanic crust onto New Caledonia during Oligocene time postdates the Eocene and Cretaceous sediments [*Routhier,* 1953; *Avias,* 1967]. K-Ar ages on mica from the high-grade blueschists indicate that they were metamorphosed at the time the oceanic crust obducted the Pacific edge of New Caledonia [*Coleman,* 1967; *Brothers,* 1970]. Serpentinite is developed mainly along thrust contacts; the massive peridotite slabs show 10–20% serpentinization. Comparison of stratigraphy of Papua and New Caledonia ophiolites shows that these areas have a nearly equal thickness of peridotite (Figure 1), whereas Papua has a much more complete section of gabbro and basalt than New Caledonia. The basalt-gabbro part of the New Caledonia oceanic crust may have been removed by erosion as a result of uplift after emplacement. Obduction of the oceanic crust-mantle plate onto New Caledonia

and Papua explains juxtapositioning of blue-schists with oceanic crust-mantle (ophiolites). Lack of any high-temperature contact aurole and contemporaneous development of high-pressure, low-temperature blueschists strongly indicates that obduction of the oceanic crust plate onto thin margins of continental plates provides another mechanism to explain the development of blueschists associated with ophiolites.

Oceanic crust consumption, an integral part of the new global tectonics, is supported by seismic evidence and correlation with present-day volcanic activity [*Oliver et al.,* 1969]. When the oceanic plate impinges on the continental plate, it is believed to underthrust the continental plate; this zone of underthrusting is now referred to as the subduction zone (Figure 6). Development of volcanic arcs and calc-alkaline intrusives has been related to the consumption of oceanic crust [*Dickinson,* 1968], and *Ernst* [1970] has proposed that blueschist metamorphism results from down-buckling of trench

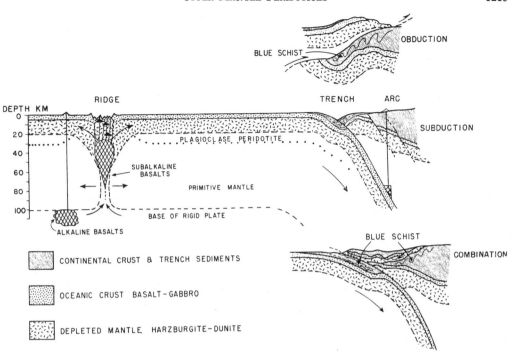

Fig. 6.   Conceptual model illustrating the development of oceanic crust at active ridges and its subduction or obduction or both at consuming plate margins.

sediments deep into the subduction zone. However, continuous plate consumption by subduction does not explain the occurrence of blueschists directly underneath oceanic crust slabs as reported from New Caledonia and New Guinea [*Coleman*, 1967; *Davies*, 1969; *Brothers*, 1970].

Obduction of the oceanic crust onto continental edges should be characterized by shallow seismic zones dipping away from the continental edge and by lack of volcanic arcs. Worldwide compilation of shallow seismic zones [*Barazangi and Dorman*, 1968] reveals that Macquarie ridge south of New Zealand may represent present-day obduction of oceanic crust, for the seismic activity here is shallow to intermediate. *Varne et al.* [1969] report that Macquarie Island represents a fragment of Pliocene oceanic crust in that it contains a belt of pillow lavas and layered gabbro masses cut by dike swarms associated with harzburgites. From present-day seismic evidence, it can be shown that the Tasman Sea oceanic crust is being consumed under the Macquarie ridge system [*McKenzie and Morgan*, 1969; *Hamilton and Evi-*

son, 1967]. To the north, the Macquarie ridge is truncated by the Alpine fault in New Zealand; farther north, the Alpine fault is terminated against the Kermadec trench system, where the Pacific oceanic crust is being consumed (Figure 4). This subduction zone is characterized by deep seismic activity and an active volcanic arc [*Cullen*, 1970]. In contrast, the Macquarie ridge system shows obduction of oceanic crust with a seismic zone dipping eastward away from the Tasman Sea [*Hamilton and Evison*, 1967] and no associated volcanic arc. Another example of present-day obduction of oceanic crust may be represented by the southern extension of the Marianas arch, where the western part of the west Caroline basin may represent oceanic crust obducting along a line between Palau Islands and Halmahea.

Focal mechanisms of recent intermediate-depth earthquakes in the eastern Mediterranean region suggest overthrusting of the Aegean Sea plates onto the Mediterranean Sea floor [*McKenzie*, 1970]. Similar overthrusting is detected from solutions of shallow focal mechanisms from the Caspian Sea to southern Iran [*Mc-Kenzie*, 1970]. The compressional nature of this Alpine zone has continued from Mesozoic time. Development of ophiolites and associated blueschists in the Tethyan belt has been documented in the literature [*Cogulu*, 1967; *Kaaden*, 1966; *Plas*, 1959]. It seems probable that the presently active overthrusting in the Mediterranean could be responsible for blueschist metamorphism under the advancing Eurasian plate. Continued movement would advance older oceanic crust on top of the younger underlying metamorphic rocks, such as that documented by *Cogulu* [1967] in Turkey. The great Oman ophiolite belt, which has stratigraphy and thickness comparable to the oceanic crust, is considered to have been emplaced sometime during the Late Cretaceous [*Reinhardt*, 1969]. Its obduction onto Maestrichian marine sediments by southwesterly overriding of the Eurasian plate onto the Arabian Peninsula fits with the regional tectonic synthesis of *Gansser* [1966] and the solutions of earthquake focal mechanisms [*McKenzie*, 1970].

Hess [1955a] originally suggested that emplacement of ultramafics initiated the early phases of orogeny. Modern geologic studies of ancient orogenic zones support this observation. Impact of lithospheric plates against one another initiates orogeny and obduction of oceanic crust-mantle slabs onto continental edges, providing mechanical force to emplace mantle ultramafics onto continental edges [*Dewey and Bird*, 1970]. The narrowness and the repetitive nature of ophiolite and blueschist metamorphic zones suggest that the tectonic process of obduction is much shorter in extent than the long-term oceanic plate consumption and volcanic arcs developed in subduction zones. It is clear, however, that both obduction and subduction of oceanic plates at continental margins have taken place and that the complexity of ancient continental margins reflects the fact that both mechanisms have been operative in the accretion of these margins.

Acknowledgments. The author has benefited from reviews by E. D. Jackson and Clark Blake, of the U.S. Geological Survey. All the information on Papua was provided by Hugh Davies, Bureau of Mineral Resources, Canberra, Australia, and the author is indebted to him for this, along with many fruitful discussions.

Publication authorized by the Director, U.S. Geological Survey.

REFERENCES

Avé Lallemant, H. G., Structural and petrofabric analysis of an alpine-type peridotite: The lherzolite of the French Pyrenees, *Leidse Geol. Mededel.*, *42*, 1–57, 1967.

Avias, J., Overthrust structure of the main ultrabasic New Caledonia massives, *Tectonophysics*, *4*, 531–542, 1967.

Bailey, E. H., M. C. Blake, Jr., and D. L. Jones, On-land Mesozoic oceanic crust in California Coast Ranges, *U.S. Geol. Survey Prof. Pap. 700-C*, 1970.

Barazangi, M., and J. Dorman, World seismicity maps compiled from ESSA Coast and Geodetic Service epicenter data, 1961–1967, *Bull. Seismol. Soc. Amer.*, *59*, 369–380, 1969.

Benson, W. N., The tectonic conditions accompanying the intrusion of basic and ultrabasic igneous rocks, *Mem. Nat. Acad. Sci.*, *19*, 1st mem., 90, 1926.

Blake, M. C., Jr., W. P. Irwin, and R. G. Coleman, Blueschist facies metamorphism related to regional thrust faulting, *Tectonophysics*, *8*, 237–246, 1969.

Brothers, R. N., Lawsonite-albite schists from northernmost New Caledonia, *Contrib. Mineral. Petrog.*, *25*(3), 185–202, 1970.

Brunn, J. H., Contribution à l'étude géologique du Pinde septentrional et d'une partie de la Macedonie occidentale; *Ann. Geol. Pays Hellen.*, *7*, 1–358, 1956.

Brunn, J. H., Mise en place et differenciation de l'association plutovolcanique du cortege ophiolitique, *Rev. Geogr. Phys. Geol. Dynam.*, *3*, 115–132, 1960.

Cann, J. R., Geological processes at mid-ocean ridge crests, *Geophys. J.*, *15*, 331–341, 1968.

Christensen, N. I., Composition and evolution of the oceanic crust, *Marine Geol.*, *8*, 139–154, 1970.

Cogulu, E., Etude pétrographique de la région de Mihaliccik (Turquie), *Bull. Suisse Mineral. Petrogr.*, *47*, 683–824, 1967.

Coleman, R. G., Metamorphic aragonite as evidence relating emplacement of ultramafic rocks to thrust faulting in New Zealand (abstract), *Trans. AGU*, *43*, 447, 1962.

Coleman, R. G., Low-temperature reaction zones and alpine ultramafic rocks of California, Oregon, and Washington, *U.S. Geol. Survey Bull. 1247*, 49, 1967.

Crenn, Y., Anomalies gravimetriques et magnetiques liées aux roches basiques de Nouvelle-Caledonie [Gravity and magnetic anomalies related to the basic (ultramafic) rocks of New Caledonia], *Ann. Geophys.*, *9*, 291–299, 1953.

Cullen, D. J., A tectonic analysis of the southwest Pacific, *New Zealand J. Geol. Geophys.*, *13*(1), 7–20, 1970.

Davies, H. L., Papuan ultramafic belt, *Intern. Geol. Cong., Prague, 1968, Rep., Proc. Sec. 1,* 209–220, 1968.

Davies, H. L., Peridotite-gabbro-basalt complex in eastern Papua: An overthrust plate of oceanic mantle and crust, Ph.D. thesis, Stanford Univ., Stanford, Calif., p. 88, 1969.

Deneufbourg, G., Observations sur la géologie du massif des péridotites du Sud de la Nouvelle-Calédonie, *Bull. Bureau de Recherches Géologiques et Minières, 4,* 27–55, 1969.

Den Tex, E., Origin of ultramafic rocks, their tectonic setting and history: A contribution to the discussion of the paper 'The origin of ultramafic and ultrabasic rocks' by P. J. Wyllie, *Tectonophysics, 7,* 457–488, 1969.

deRoever, W. P., de, Sind die alpinotypen Peridotit massen vielleicht tektonisch verfrachtete Bruchstücke der Peridotitschale, *Geol. Runds., 46,* 137–146, 1957.

Dewey, J. F., and J. M. Bird, Mountain belts and the new global tectonics, *J. Geophys, Res., 75,* 2625–2647, 1970.

Dewey, J. F., and Brenda Horsfield, Plate-tectonics, orogeny and continental growth, *Nature, 225,* 521–525, 1970.

Dickinson, W. R., Circum-Pacific andesite types, *J. Geophys. Res., 73,* 2261–2269, 1968.

Dietz, R. S., Alpine serpentines as oceanic rind fragments, *Bull. Geol. Soc. Amer., 74,* 947–952, 1963.

Dubertret, L., Géologie des roches vertes du Nord-Ouest de la Syrie et du Hatay (Turquie), *Mus. Nat. Hist. Nat. Notes Mem., 6,* 1–179, 1955.

Engel, A. E. J., C. G. Engel, and R. G. Havens, Chemical characteristics of oceanic basalts and the upper mantle, *Bull. Geol. Soc. Amer., 76,* 719–734, 1965.

Engel, C. G., and R. L. Fisher, Lherzolite, anorthosite, gabbro, and basalt dredged from the mid-Indian ocean ridge, *Science, 166,* 1136–1141, 1969.

Ernst, W. G., Tectonic contact between the Franciscan mélange and the Great Valley sequence: Crustal expression of a late Mesozoic Benioff zone, *J. Geophys. Res., 75,* 886–901, 1970.

Fischer, A. G., B. C. Heezen, B. C., R. E. Boyce, D. Bukry, R. G. Douglas, R. E. Garrison, S. A. King, V. Krasheninnikov, A. P. Lisitzin, and A. Dimm, Geological history of the western north Pacific, *Science, 168,* 1210–1214, 1970.

Gansser, A., The Indian Ocean and the Himalayas, a geologic interpretation, *Eclogae Geol. Helv., 59,* 1966.

Gass, I. G., The ultrabasic volcanic assemblage of the Troodos massif, Cyprus, in *Ultramafic and Related Rocks,* edited by P. J. Wyllie, pp. 121–134, Wiley, New York, 464 pp., 1967.

Gass, I. G., and D. Masson-Smith, The geology and gravity anomalies of the Troodos massif, Cyprus, *Phil. Trans. Roy Soc. London, A. 255,* 417–467, 1963.

Green, D. H., A review of experimental evidence on the origin of basaltic and nephelinitic magmas, *Phys. Earth Planet. Interiors, 3,* 221–235, 1970.

Green, D. H., and A. E. Ringwood, The stability fields of aluminous pyroxene peridotite and garnet peridotite and their relevance in upper mantle structure, *Earth Planet. Sci. Lett., 3,* 151–160, 1967.

Hamilton, R. M., and F. F. Evison, Earthquakes of intermediate depths in southwest New Zealand, *New Zealand J. Geol. Geophys., 10,* 1319–1329, 1967.

Heirtzler, J. R., and X. LePichon, Crustal structure of the mid-ocean ridges, 3, Magnetic anomalies over the mid-atlantic ridge, *J. Geophys. Res., 70,* 4013–4033, 1965.

Hess, H. H., A primary peridotite magma, *Amer. J. Sci., 35,* 321–344, 1938.

Hess, H. H., Serpentines, orogeny, and epeirogeny, *Geol. Soc. Amer. Spec. Pap. 62,* 391–407, 1955a.

Hess, H. H., The oceanic crust, *J. Marine Res., 14,* 423–439, 1955b.

Hess, H. H., History of ocean basins, in *Petrologic Studies,* Buddington volume, pp. 599–620, Geological Society of America, Boulder, Colo., 1962.

Hess, H. H., Mid-oceanic ridges and tectonics of the sea-floor, in *Submarine Geology and Geophysics, 17th Colston Research Synposium, Bristol, England,* edited by W. F. Whittard and R. Bradshaw, pp. 317–333, Butterworths, London, 1965.

Hill, M. N., Recent geophysical exploration of the ocean floor, *Progr. Phys. Chem. Earth, 2,* 129–163, 1957.

Kaaden, G. v.d., The significance and distribution of glaucophane rocks in Turkey, *Bull. Min. Res, Expl. Inst. Turkey, 67,* 36–67, 1966.

Kay, R., N. J. Hubbard, N. J. and P. W. Gast, Chemical characteristics and origin of oceanic ridge volcanic rocks, *J. Geophys. Res., 75,* 1585–1614, 1970.

Kushiro, I., Melting of a peridotite nodule at high pressures and high water pressures, *J. Geophys. Res., 73,* 6023–6028, 1968.

Lillie, A. R., The structural geology of lawsonite and glaucophane schists of the Ouegoa District, New Caledonia, *New Zealand J. Geol. Geophys., 13*(1), 72–116, 1970.

Lillie, A. R., and R. N. Brothers, The geology of New Caledonia, *New Zeland J. Geol. Geophys., 13*(1), 145–183, 1970.

Loney, R. A., G. R. Himmelberg, and R. G. Coleman, 1970, Structure and petrology of the alpine-type peridotite at Burro Mountain, California, U.S.A., *J. Petrol.,* in press, 1971.

MacGregor, I. D., Mafic and ultramafic inclusions as indicators of the depth of origin of basaltic magmas, *J. Geophys. Res., 73,* 3737–3745, 1968.

Maxwell, J. C., and A. Azzaroli, Submarine extrusion of ultramafic magma (abstract), *Geol. Soc. Amer. Spec. Pap. 73*, 203–204, 1962.

Maxwell, J. C., 'Alpine' mafic and ultramafic rocks—The ophiolite suite: A contribution to the discussion of the paper 'The origin of ultramafic and ultrabasic rocks' by P. J. Wyllie, *Tectonophysics, 7*, 489–494, 1969.

McKenzie, D. P. Plate tectonics of the Mediterranean region, *Nature, 226*, 239–243, 1970.

McKenzie, D. P. and W. J. Morgan, Evolution of triple junctions, *Nature, 224*, 125–133, 1969.

McKenzie, D. P. and R. L. Parker, The North Pacific: An example of tectonics on a sphere, *Nature, 216*, 1276–1290, 1967.

Melson, W. G., G. Thompson, and T. H. Van Andel, Volcanism and metamorphism in the mid-Atlantic ridge, 22°N latitude, *J. Geophys. Res., 73*, 5925–5941, 1968.

Miyashiro, A., F. Shido, and M. Ewing, Diversity and origin of abyssal tholeiite from the mid-Atlantic ridge near 24° and 30° north latitude, *Contrib. Mineral. Petrog., 23*(1), 38–52, 1969.

Miyashiro, A., F. Shido. and M. Ewing, Crystallization and differentiation in abyssal tholeiites and gabbros from mid-oceanic ridges, *Earth Planet. Sci. Lett., 7*, 361–365, 1970.

Moores, E. M.. Petrology and structure of the Vourinous ophiolitic complex Northern Greece. *Geol. Soc. Amer. Spec. Pap. 118*, 74, 1969.

Moores, E. M., and F. J. Vine, The Troodos massif, Cyprus and other ophiolites as oceanic crust, evaluation and implications, *Trans. Roy. Soc. London,* in press. 1971.

O'Hara, M. J.. Are ocean floor basalts primary magma?, *Nature, 220*, 683–686. 1968.

Oliver, J., L. Sykes. and B. Isacks, Seismology and the new global tectonics, *Tectonophysics, 7*, 527–541, 1969.

Oxburgh, E. R., and D. L. Turcotte, Mid-ocean ridges and geotherm distribution during mantle convections, *J. Geophys. Res., 73*, 2643–2661. 1968.

Plas, L. Van Der. Petrology of the northern Adula region, Switzerland, *Leid. Geol. Mededel., 24*, 415–602, 1959.

Raitt, R. W., The crustal rocks, in *The Sea,* vol. 3, edited by M. N. Hill, pp. 85–102, John Wiley, New York, 1963.

Reinhardt, B. M.. On the genesis and emplacement of ophiolites in the Oman Mountains geosyncline, *Schweiz. Mineral. Petrogr. Mitt., 49*, 1–30, 1969.

Routhier, P., Etude géologique du versant occidental de la Nouvelle-Calédonie entre le Cola de Boghen et la Pointe d'Arama, *Mem. Soc. Geol. France, 67*, 1–271, 1953.

Shor, G. G., Jr., and R. W. Raitt, Explosion seismic refraction studies of the crust and upper mantle in the Pacific and Indian Oceans, in *The Earth's Crust and Upper Mantle, Geophys. Monograph 13*, pp. 225–230, AGU, Washington, D. C., 1969.

Steinmann, G., Geologische Beobachtangen in den Alpen, 2, Die Schardt'sche uberfaltungstheorie und die geologische bedeutung der tiefseeabsatze und der ophiolithischen massargesteine, *Ber. Nat. Ges. Reiburg, i, 16*, 44–65, 1905.

Steinmann, G., Die ophiolithischen zonen in den mediterranean Kettengebirgen, *Rept. Int. Geol. Congr., 14th, Madrid, 2*, 638–667, 1927.

Sykes, L. R., Mechanism of earthquakes and nature of faulting on the mid-oceanic ridges, *J. Geophys. Res., 72*, 2131–2153, 1967.

Thayer, T. P., Some critical differences between alpine-type and stratiform peridotite-gabbro complexes, *Rept. Int. Geol. Cong., 21st, Copenhagen, 13*, 247–259, 1960.

Thayer, T. P., Peridotite-gabbro complexes as keys to petrology of mid-oceanic ridges, *Bull Geol. Soc. Amer., 80*, 1515–1522, 1969.

Thayer, T. P., and G. R. Himmelberg, Rock succession in the alpine-type mafic complex at Canyon Mountain, Oregon, *Rept. Int. Geol. Cong., 23rd, Prague, 1*, 175–186, 1968.

Varne, R., R. D. Gee, P. G. J. Quilty, Macquarie Island and the cause of oceanic linear magnetic anomalies, *Science, 166*, 230–232, 1969.

Vine, F. J., and D. H. Matthews, Magnetic anomalies over oceanic ridges, *Nature, 199*, 947–949, 1963.

Vine, F. J., Spreading of the ocean floor: New evidence, *Science, 154*, 1405–1415, 1966.

Wilson, R. A. M.. The geology of the Xeros-Troodos area, *Mem. Geol. Surv. Cyprus, 1*, 136. 1959.

Wyllie, P. J.. Review. in *Ultramafic and Related Rocks,* edited by P. J. Wyllie, pp. 403–416, Wiley, New York, 464 pp., 1967.

(Received August 25, 1970; revised October 26, 1970.)

Editor's Comments on Papers 18, 19, and 20

18 **de Roever:** *Overdruk van Tektonische Oorsprong of Diepe Metamorfose?*
English translation: *Overpressure of Tectonic Origin or Deep Metamorphism?*

19 **Gresens:** *Blueschist Alteration During Serpentinization*

20 **Ernst:** *Do Mineral Parageneses Reflect Unusually High-Pressure Conditions of Franciscan Metamorphism?*

Blueschists

For many years the origin of glaucophane schists and related rocks have been debated. Many petrologists have regarded these rocks as representatives of a distinct high-pressure, low-temperature metamorphic facies (Eskola, 1929, 1939; Miyashiro and Banno, 1958; Fyfe, Turner, and Verhoogen, 1958; Coleman and Lee, 1963; Ernst, 1963). This conclusion is based on many lines of evidence: (1) blueschists are denser than protoliths and—except for eclogites—they are more dense than other metamorphic rocks of similar bulk compositions; (2) many glaucophane schists are chemically indistinguishable from greenschists and albite-epidote amphibolites; (3) glaucophane schists are developed on a regional scale in petrotectonic environments generally interpreted as subduction-zone complexes; (4) blueschists and related rocks contain relatively high-pressure minerals such as jadeitic pyroxene, lawsonite, metamorphic aragonite, and glaucophane II.

Other workers (e.g., Wegmann, 1928; Suzuki, 1934, 1939; Taliaferro, 1943; Schürmann, 1953, 1956; Brothers, 1954; Fyfe and Zardini, 1967) have postulated that metastable crystallization or metasomatism can account for the distinctive mineralogies of blueschists. These authors have referred to the (1) unusual bulk compositions of some glaucophane schists; (2) irregular, patchy development of glaucophane schists apparently surrounded by unmetamorphosed rock, or purported occurrences of blueschists as aureoles surrounding serpentinites; (3) abundance of monomineralic veins and stringers in glaucophane schists; and (4) local interlayering of blueschists and greenschists, as supporting the view that unusual physical conditions are not required for the formation of blueschists and related rocks.

A controversy exists even among those authorities who invoke high pressures to account for the generation of these distinctive metamorphic rocks. Coleman and Lee (1962), Blake, Irwin, and Coleman (1967), and Brothers (1970), among others, have suggested that the operating pressures during recrystallization exceed lithostatic values (i.e., tectonic overpressures). This argument was pursued by de Roever (1967) in Paper 18, presented here as the original author's translation from the Dutch publication. Tectonic overpressures were called upon here, and by most other advocates of this mechanism, to provide the acknowledged high pressures of metamorphism in regions where the overlying column of rocks as now estimated was presumed to have been insufficient to provide the requisite total pressure. If, as is currently generally recognized, blueschists are confined to deeply subducted terranes, the reason for calling on tectonic stress buildup vanishes: the glaucophane schists and related rocks would have experienced a profound subduction in the high-pressure, low-temperature regime described in Paper 4, then would have decoupled from the downgoing lithospheric plate and buoyantly risen toward the surface, where they are seen today.

An example of the postulation that glaucophane schists and related rocks are generated metastably or metasomatically by the reaction of protoliths with a chemically active pore solution is presented in Paper 19 by Gresens. This author pointed to the common worldwide association of serpentinites and blueschists and invoked a genetic relationship. Briefly, the interaction of the ultramafic material and an aqueous solution was thought to provide fluids capable of promoting the production of blueschist-type minerals in the country rocks through metasomatism and/or metastable crystallization.

The various genetic hypotheses, including that of recrystallization at relatively high pressures and low temperatures, are set forth in Paper 20, by Ernst. Although the argument is presented for the Franciscan terrane of the California Coast Ranges, it can be extended to most, if not all, other blueschist belts. Fundamentally, the argument runs as follows. Rocks have insufficient strengths to support substantial tectonic over-pressures at geologically reasonable strain rates. Many blueschists have bulk chemistries essentially identical to rocks of the other low-grade metamorphic facies and to their protoliths; hence metasomatism cannot be responsible for the distinctive mineralogies. Observed mineral parageneses and oxygen isotopic data are compatible with a close approach to equilibrium at low temperatures and high pressures. Such physical conditions are characteristic of subduction zones, the exact petrotectonic environment in which blueschists are found. Thus all available data point to production of glaucophane schists and related rocks at relatively high pressures and low temperatures.

References

Blake, M. C., Jr., Irwin, W. P., and Coleman, R. G. (1967) Upside-down metamorphic zonation, blueschist facies, along a regional thrust in California and Oregon: *U.S. Geol. Survey Prof. Paper* **575-C,** 1–9.

Brothers, R. N. (1954) Glancophane schists from the North Berkely Hills, California: *Amer. Jour. Sci.,* **252,** 614–626.

Brothers, R. N. (1970) Lawsonite-albite schists from northernmost New Caledonia: *Contr. Mineral. Petrol.,* **25,** 185–202.

Coleman, R. G., and Lee, D. E. (1962) Metamorphic aragonite in the glaucophane schists of Cazadero, California: *Amer. Jour. Sci.,* **260,** 577–595.

Coleman, R. G., and Lee, D. E. (1963) Glaucophane-bearing metamorphic rock types of the Cazadero area, California: *Jour. Petrol.,* **4,** 260–301.

Ernst, W. G. (1963) Petrogenesis of glaucophane schists: *Jour. Petrol.,* **4,** 1–30.

Ernst, W. G. (1971) Do mineral parageneses reflect unusually high-pressure conditions of Franciscan metamorphism? *Amer. Jour. Sci.,* **270,** 81–108.

Eskola, P. (1929) Om Mineralfacies: *Geol. Fören. Stockh. Förh.,* **51,** 157–172.

Eskola, P. (1939) Die metamorphen Gesteine, p. 263–407, in *Die Entstehung der Gesteine* by T. F. W. Barth, C. W. Correns, and P. Eskola, Springer, Berlin, 422 p.

Fyfe, W. S., and Zardini, R. (1967) Metaconglomerate in the Franciscan Formation near Pacheco Pass, California: *Amer. Jour. Sci.,* **265,** 819–830.

Fyfe, W. S., Turner, F. J., and Verhoogen, J. (1958) Metamorphic reactions and metamorphic facies: *Geol. Soc. America Mem.*, **73**, 260 p.

Korzhinskii, D. S. (1959) *Physicochemical Basis of the Analysis of the Paragenesis of Minerals:* Consultants Bureau, New York, 142 p.

Miyashiro, A., and Banno, S. (1958) Nature of glaucophanitic metamorphism: *Amer. Jour. Sci.*, **256**, 97–110.

Schürmann, H. M. E. (1953) Beiträge zur Glaukophanfrage (2). *N. Jb. Min. Monatsschr.*, **85**, 303–394.

Schürmann, H. M. E. (1956) Beiträge zur Glaukophanfrage (3). *N. Jb. Min. Monatsschr.*, **89**, 41–85.

Suzuki, J. (1934) On some soda-pyroxene and amphibole-bearing quartz schists from Hokkaido: *J. Fac. Sci. Hokkaido Imp. Univ.* Ser. IV, **2**, 339–353.

Suzuki, J. (1939) A note on soda-amphiboles in crystalline schists from Hokkaido: *J. Fac. Sci. Hokkaido Imp. Univ.*, Ser. IV, **6**, 507–519.

Taliaferro, N. L. (1943) Franciscan-Knoxville problem: *Amer. Assoc. Petrol. Geol. Bull.*, **27**, 109–219.

Wegmann, C. E. (1928) Über das Bornitvorkommen von Saint-Véran, Hautes-Alpes: *Z. Prakt. Geol.*, **36**, 19–28, 36–43.

18

Reprinted from *Koninkl. Nederl. Akad. Wetensch.*,
Versl. Gew. Vergad. Afd. Natuurk., **76**(4), 69–74 (1967)

OVERDRUK VAN TEKTONISCHE OORSPRONG OF DIEPE METAMORFOSE?

DOOR

W. P. DE ROEVER

Van de drie grote groepen van gesteenten, die wel als speciaal onder hoge drukken gevormd werden beschouwd en waarvan de vormings-temperatuur dus naar verhouding niet zo hoog is, de eklogieten, de kya-niethoudende gesteenten en de gesteenten der glaukofaanschistfacies, worden er twee in dit opzicht nog omstreden. GREEN & RINGWOOD (1966) betoogden, dat de karakteristieke mineraalassociatie der eklogieten stabiel kan zijn in droge gesteenten van bazaltische samenstelling die gemetamorfoseerd zijn in de almandien-amfibolietfacies, de glaukofaan-schistfacies of een deel van de groenschistfacies. Eklogieten zullen hier derhalve verder buiten beschouwing worden gelaten. De tweede omstreden groep is die der kyaniethoudende gesteenten. Enige jaren geleden werd voor de vorming hiervan op grond van goed met elkander overeenkomende directe synthesen en reversibele omzettingen een druk van ongeveer 10 kb nodig geacht. Daarna kwam men vooral op grond van indirecte bepalingen en berekeningen, die overigens deels met elkander in strijd waren, tot veel lagere waarden. Volgens verscheidene onderzoekers zouden drukken van luttele kb voldoende zijn voor de vorming van kyaniet (zie bijv. NEWTON, 1966). Een drie maanden geleden gepubliceerde directe bepaling leverde echter weer ongeveer $6\frac{1}{2}$ kb op (ALTHAUS, 1967). Ook deze gegevens zijn nog te veel in strijd met elkander om er hier nader op voort te bouwen.

Ten slotte blijft er dan één groep gesteenten over, die der glauko-faanschistfacies, waarvoor minder tegenstrijdige resultaten werden ver-kregen en waarvoor bovendien de gegevens over verscheidene mineralen in dezelfde richting wijzen.

Voor glaukofaan, een Na-Mg-Fe-Al-silikaat van de amfiboolgroep, in de natuur praktisch altijd met een flink ijzergehalte, kan men voor de normale variëteiten op grond van de resultaten van W. G. ERNST (1963, zie ook 1964), die echter juist voor ijzerhoudende variëteiten van dit mineraal niet erg safe zijn, de volgende benaderende minimumdrukken berekenen: 4–$4\frac{1}{2}$ kb bij $200°$ C, 7–$7\frac{1}{2}$ kb bij $300°$ C en 10–$10\frac{1}{2}$ kb bij $400°$ C; deze cijfers gelden voor ' een glaukofaan die intermediair is tussen $Na_2Mg_3Al_2Si_8O_{22}(OH)_2$ en crossiet met 50% Fe i.p.v. Mg en Al. Voor deze crossiet zelf zijn de overeenkomstige waarden 3 kb hoger, voor ijzerloze glaukofaan 3 kb lager. Voor lawsoniet schatten CRAWFORD & FYFE (1965) — waarvan de laatste voor kyaniet juist relatief lage cijfers verkreeg — de volgende minimumdrukken: $4\frac{1}{2}$–5 kb bij 200° C, 5–$5\frac{1}{2}$ kb bij 300° C en $5\frac{1}{2}$–6 kb bij 400° C. Terzijde kan naar aanleiding van een vergelijking van deze beide groepen van cijfers worden opgemerkt, dat de invoering van een lawsoniet-glaukofaan-facies of een glaukofaanlawsonietschistfacies ter gedeeltelijke vervanging van ESKOLA's glaukofaanschistfacies blijkbaar heel weinig zin heeft, tenzij deze invoering bedoeld is om naast glaukofaanassociaties met of zonder lawsoniet ook lawsonietassociaties zonder glaukofaan te omvatten. De gegevens voor verdere karakteristieke mineralen van de glaukofaanschistfacies, nl. voor aragoniet en voor tezamen voorkomende jadeiet en kwarts berusten op thermodynamische gegevens zowel als redelijk goed daarmede overeenkomende directe waarnemingen van reversibele omzetting, ten dele met hydrostatische apparaten. Voor zuivere aragoniet kunnen we als benaderende minimumdrukken bij vorming als stabiele fase aanhouden $5\frac{1}{2}$–6 kb bij 200° C, 7–$7\frac{1}{2}$ kb bij 300° C en $8\frac{1}{2}$–9 kb bij 400° C; voor zuivere jadeiet naast kwarts 10 kb bij 200° C, 12 kb bij 300° C en 14 kb bij 400° C (zie bijv. CRAWFORD & FYFE, 1965); jadeiet bevat echter wat ijzer zodat voor natuurlijke jadeiet algemeen iets lagere waarden worden aangenomen. Als we voor de nog niet voldoende bekende metamorfosetemperatuur de meest vooraanstaande onderzoekers op dit gebied volgen, en een bedrag van ongeveer 200° C aanhouden, komen we voor de mineraalassociatie der glaukofaanschistfacies tot minimumdrukken van 4–5 kb, terwijl men algemeen aanneemt dat de druk minstens tot 8 kb oploopt.

Men kan zich nu de vraag stellen, of deze druk in essentie belastingdruk is, zoals door de meest vooraanstaande Amerikaanse onderzoekers op dit gebied wordt verondersteld. Zou dit het geval zijn, dan zou de vorming van deze mineralen hebben plaatsgehad op diepten tussen 15 en 30 km. Dit zou betekenen dat de geplooide sedimenten uit geosynklinale gebieden, waaruit later ketengebergten ontstaan, tot veel grotere diepten zouden zijn omlaaggedrukt dan tevoren werd aangenomen. En ook, dat op alle vindplaatsen van deze mineralen 15–30 km gesteente zou zijn weggeërodeerd. En dit laatste nu is uit geologisch oogpunt onaanvaardbaar; men vindt in de literatuur lagere en zelfs veel lagere schattingen voor de vormingsdiepte. Met behulp van vermindering der belastingdruk door afglijding of uitwalsing van bovenliggende lagen schijnt men er ook nog niet te komen.

Men komt dus tot de overtuiging dat op de een of andere manier de druk aanmerkelijk moet zijn verhoogd, zodat de belastingdruk, en dus de diepte aanmerkelijk geringer is geweest. Voor een dergelijke drukverhoging zijn verscheidene methoden genoemd. Men heeft hierbij in de eerste plaats aan de starheid, aan de weerstand tegen deformatie van gesteenten onder eenzijdige druk en de hierdoor mogelijk gemaakte z.g.

tektonische overdruk gedacht (BIRCH, 1955; DE ROEVER, 1956, p. 127; COLEMAN & LEE, 1962). Een tweede mogelijkheid zou de ,,fluid overpressure'' der Angelsaksische literatuur opleveren, die wel als door snelle produktie van water en/of CO_2 ontstaan is gedacht; een derde mogelijkheid zou thermische overdruk door expansie ten gevolge van plaatselijke verhitting kunnen zijn (RUTLAND, 1965). Zowel bovengenoemde soort van ,,fluid overpressure'' als thermische overdruk schijnen in ons geval niet toepasbaar te zijn. Op een andere variant van de ,,fluid overpressure'' kom ik straks nog terug.

Voor de weerstand tegen deformatie van de meeste gesteenten zijn onder passende temperatuur en alzijdige druk wel waarden gevonden, die groot genoeg zijn om het verschil tussen de in het laboratorium gevonden minimumdrukken en geologisch geschatte belastingdrukken te kunnen overbruggen. Een grote puzzle levert echter de extrapolatie van de factor tijd op. Men vond namelijk bij laboratoriumproeven, dat, naarmate de vormveranderingssnelheid afneemt, de weerstand der gesteenten geringer is. HEARD (1963) deed proeven met een bepaald soort marmer en vond daarvoor bij 500° C bij een tienmiljoenmalige verlangzaming der deformatie een afname der ,,yield strength'' met $1\frac{1}{2}$ kb, nl. van 2 kb tot $\frac{1}{2}$ kb. Zou men deze resultaten mogen extrapoleren, dan zou de ,,yield strength'' bij een volgens hem geologisch redelijke vormveranderingssnelheid die nog eens tienmiljoen maal lager is, in het niet vallen, nl. 10^{-6} kb zijn. Als stevigere gesteenten — marmer is namelijk erg gemakkelijk deformeerbaar — nu een overeenkomstige vermindering te zien zouden geven, zou tektonische overdruk op grond van weerstand tegen deformatie als geologische factor te verwaarlozen zijn.

Er zijn echter m.i. een aantal feiten, die toch voor de bestaanbaarheid van tektonische overdruk van aanmerkelijk belang pleiten, hoewel ,,fluid overpressure'' van tektonische oorsprong een nog belangrijker factor lijkt.

Ten eerste is de temperatuur van 500° C te hoog in vergelijking met ons geval. Bij 300° C haalt de betreffende marmer bij dezelfde vormveranderingssnelheid al een geëxtrapoleerde resistentie van enige tienden kilobar i.p.v. 10^{-6} kb, en bij 200° C zijn deze waarden nog enige malen zo groot. Houden we daarbij in gedachten, dat marmer zeer gemakkelijk deformeerbaar is, dan komen we al tot redelijker bedragen van enige kilobars of meer voor glaukofaanhoudende gesteenten.

Verder is de vormveranderingssnelheid zeer variabel; deze behoeft niet zo laag te zijn geweest. In seismische zones kan men zelfs plaatselijk plotselinge hoge spanningen verwachten, bijv. als bij een aardbeving de beweging der losspringende wanden op een lokaal obstakel stuit. Desalniettemin schijnen we in ons geval niet met sterk wisselende drukken te moeten rekenen, aangezien de leden der amfiboolgroep gevoelige indicatoren van P-T-veranderingen zijn, doch glaukofaan voorzover bekend geen *oscillerende* zonering vertoont. De druk moet dus vrij constant zijn geweest.

De aanwezigheid in glaukofaangesteenten, zoals die van Californië, van oude barsten en spleten opgevuld door glaukofaan, lawsoniet, aragoniet en jadeiet lijkt mij verder een belangrijk argument ten gunste van de mogelijkheid van tektonische overdruk van aanmerkelijk belang te kunnen opleveren. Zulke opgevulde spleten schijnen althans ten dele door open- springen door dilatatie of extensie en niet door schuif te zijn gevormd, en komen volgens mijn ervaring veel in laaggradige regionaalmetamorfe gesteenten voor. MISCH (1962) betoogde dat soortgelijke aders tijdens hun opengaan zijn opgevuld, en wel zo dat er steeds nauwelijks enige ruimte overbleef, m.i. een voorbeeld van het principe dat mineraalsub- stantie elders onder druk werd opgelost en hier op de plaats van de minste druk werd afgezet. De betreffende extensiespleten nu zouden gevormd zijn onder geopetale eenzijdige belastingdruk die volgens de meeste Amerikaanse onderzoekers op dit gebied 8 kb moet hebben gehaald, overeenkomend met een diepte van ongeveer 30 km. Gesteenten, waarin op een dergelijke diepte, onder een dergelijke unilaterale belastingdruk nog extensiebarsten konden worden gevormd, hebben zeker op het moment der barstvorming een grote resistentie gehad, groot genoeg om te mogen concluderen, dat op het moment der spleetvorming aanmerkelijke tek- tonische overdruk kan zijn opgetreden. Het is dus mogelijk, en in het licht van bovengenoemde geringere schattingen van de metamorfose- diepte zelfs waarschijnlijker, dat de belastingdruk nooit 8 kb gehaald heeft, maar aanmerkelijk kleiner was. In beide gevallen zou de ,,fluid pressure" in de barsten 8 kb hebben gehaald, hetgeen zonder meer inhoudt dat de spleten niet op enigerlei wijze met het aardoppervlak waren verbonden. Een ,,fluid pressure" van 8 kb kan natuurlijk veroorzaakt zijn door een belastingdruk van 8 kb; in ons tweede geval daarentegen moet er een flinke ,,fluid overpressure" hebben bestaan. De reden daarvan is in het betreffende milieu niet ver te zoeken; deze moet m.i. van tektonische aard zijn geweest, dus een gevolg van een eenzijdige druk van de orde van grootte van 8 kb. Ook in dit laatste geval zouden extensiespleten zijn ontstaan terwijl er een eenzijdige druk van de orde van grootte van 8 kb bestond, zodat op het moment der barstvorming ook in dit geval de gesteenten grote resistentie moeten hebben vertoond. Dit impliceert dat ook in dit geval tektonische overdruk van aanmerkelijke grootte kan zijn opgetreden. Rekenen we uit, wat in dit laatste geval in het gehele gesteente de hoogste gemiddelde druk oplevert, een ,,fluid overpressure" of de weerstand tegen deformatie, dan blijkt eerstgenoemde effectiever te zijn en dus belangrijker te moeten zijn geweest. Door de aanname van ,,fluid overpressure" van tektonische oorsprong worden ook RUTLAND's (1965) structurele bezwaren tegen het bestaan van belangrijke tektonische over- druk ondervangen.

Aan WINKLER's (1965, p. 151) opmerking, dat het weinig gedeformeerde karakter van vele der betreffende gesteenten een argument tegen het optreden van belangrijke tektonische overdruk vormt, moge ook nog een enkel woord worden gewijd; hij hield hierbij geen rekening met het feit dat blijvende deformatie pas optreedt als de tektonische overdruk een bepaalde grenswaarde heeft overschreden.

Al met al moet de conclusie van dit betoog luiden, dat het er alle schijn van heeft, dat ,,fluid overpressures'' van tektonische oorsprong zowel als gewone tektonische overdruk een rol hebben gespeeld bij de vorming van glaukofaan, etc., zodat de diepte der gesteentemetamorfose in dit geval niet excessief behoeft te zijn geweest.

De herkomst van het water in genoemde spleten kan overigens nog een interessant onderzoekingsobject vormen; men zou kunnen denken aan een verband met de serpentinisatie van peridotieten en aan de lokale Na-toevoer daar in de buurt, en zelfs een oorsprong uit de aardmantel willen overwegen; dit alles zou ons hier echter te ver voeren.

Ik zou hierbij aan het eind van mijn betoog zijn gekomen, ware het niet dat de mogelijkheid van belangrijke overdruk van tektonische oorsprong een nieuw licht werpt op de genese van de hier in hoofdzaak besproken mineralen. Nu overdruk van tektonische oorsprong een belangrijke rol blijkt te kunnen spelen, kan men in de eerste plaats aan de mogelijkheid denken, dat glaukofaan niet echt regionaal optreedt, maar dynamometamorf is gevormd in bepaalde zones van sterke overdruk, zoals door DOBRETSOV (1964) voorgesteld en door NICOLAS (1966) in de West-Alpen gevonden.

Als in een periode van regionale metamorfose glaukofaan en lawsoniet worden gevormd, geschiedt dit altijd in een zeer vroeg stadium (DE ROEVER & NIJHUIS, 1964). Vroeger leek het, dat hiervoor volgens een suggestie van ELLENBERGER de belastingdruk van een aantal dekbladen nodig zou zijn, hetgeen enigszins in strijd lijkt met het gestelde in de voorgaande zin. Nu zou dit probleem zijn opgelost: glaukofaan, etc. zouden onder invloed van overdruk van tektonische oorsprong ook bij afwezigheid van een stapel dekbladen kunnen zijn gevormd, tijdens bewegingen in het geosynklinale sedimentatiebekken. Een dergelijke verklaring zou een zeer goede oplossing opleveren voor het probleem van het voorkomen van glaukofaan, lawsoniet, etc. in gebieden waar volgens de literatuur geen dekbladen zouden voorkomen. En in gebieden met dekbladen kan men aan glaukofaanvorming zowel ná als vóór de dekbladenvorming denken, beginnend in verband met embryonale bewegingen in het geosynklinale bekken, zoals voorgesteld door VIALON (1966). Deze embryonale glaukofaanvorming ten gevolge van overdruk van tektonische oorsprong zou ten dele met het in vele fasen tot stand gekomen tektonische transport van ultramafische aardmantelfragmenten (DE ROEVER, 1957) kunnen samenhangen, zoals verondersteld door COGULU (1965). Het veelvuldig tezamen voorkomen van glaukofaangesteenten met peridotieten en serpentijnen – tevoren nog onverklaard – zou op deze wijze een harmonische verklaring vinden.

LITERATUUR

ALTHAUS, E., Experimentelle Bestimmung des Stabilitätsbereichs von Disthen (Cyanit). — Die Naturwissenschaften **54**, 42–43 (1967).

BIRCH, F., Physics of the crust. — Geol. Soc. America Spec. Paper **62**, 101–117 (1955).

COGULU, E., Remarques sur les schistes à glaucophane et lawsonite de la région de Mihaliççik (Turquie). — Archives des Sciences Soc. de Physique et d'Histoire Naturelle de Genève **18**, 126–131 (1965).

COLEMAN, R. G. & D. E. LEE, Metamorphic aragonite in the glaucophane schists of Cazadero, California. — Amer. Journ. Sci. **260**, 577–595 (1962).

CRAWFORD, W. A. & W. S. FYFE, Lawsonite equilibria. — Amer. Journ. Sci. **263**, 262–270 (1965).

DOBRETSOV, N. L., The jadeite rocks as indicators of high pressure in the Earth's crust. — Int. Geol. Congr., Report 22nd Sess., India, Vol. of Abstr., p. 231–232 (1964).

ERNST, W. G., Polymorphism in alkali amphiboles. — Amer. Min. **48**, 241–260 (1963).

————, Petrochemical study of coexisting minerals from low-grade schists, eastern Shikoku, Japan. — Geochim. Cosmochim. Acta **28**, 1631–1668 (1964).

GREEN, D. H. & A. E. RINGWOOD, An experimental investigation of the gabbro to eclogite transformation and its petrological applications. — Petrology of the Upper Mantle. Publ. **444** Dept. Geophys. and Geoch., Australian Nat. Univ., 1–59 (1966).

HEARD, H. C., Effect of large changes in strain rate in the experimental deformation of Yule marble. — Journ. Geol. **71**, 162–195 (1963).

MISCH, P., New criteria for synkinematic growth of metamorphic minerals. — Geol. Soc. America Spec. Paper **68**, 44–45 (1962).

NEWTON, R. C., Kyanite — andalusite equilibrium from 700° to 800° C. — Science **153** (3732), 170–172 (1966).

NICOLAS, A., I. Etude pétrochimique des roches vertes et de leurs minéraux entre Dora Maïra et Grand Paradis (Alpes piémontaises). II. Le complexe Ophiolites — Schistes Lustrés entre Dora Maïra et Grand Paradis (Alpes piémontaises). Tectonique et métamorphisme. — Fac. Sci. Nantes (1966).

ROEVER, W. P. DE, Some differences between post-Paleozoic and older regional metamorphism. — Geologie en Mijnbouw, nw. ser. **18**, 123–127 (1956).

————, Sind die alpinotypen Peridotitmassen vielleicht tektonisch verfrachtete Bruchstücke der Peridotitschale? Geol. Rundschau **46**, 137–146 (1957).

———— & H. J. NIJHUIS, Plurifacial alpine metamorphism in the eastern Betic Cordilleras (SE Spain), with special reference to the genesis of the glaucophane. — Geol. Rundschau **53**, 324–336 (1964).

RUTLAND, R. W. R., Tectonic overpressures. — Controls of metamorphism (ed. W. S. PITCHER & G. W. FLINN), Edinburgh and London, Oliver & Boyd, 119–139 (1965).

VIALON, P., Etude géologique du massif cristallin Dora-Maira, Alpes Cottiennes internes, Italie. — Trav. Lab. Géol. Fac. Sci. Grenoble, Mém. **4** (1966).

WINKLER, H. G. F., Die Genese der metamorphen Gesteine. — Berlin, Heidelberg, New York, Springer-Verlag (1965).

18

Overpressure of Tectonic Origin or Deep Metamorphism?

W. P. DE ROEVER

This article was translated by W. P. de Roever,
Geological Institute of the University of
Amsterdam, from Koninkl. Nederl.
Akad. Wetensch., Versl. Gew. Vergad.
Afd. Natuurk., **74**(4), 69–74 (1967).

Three large groups of rocks have been considered to have originated especially under high pressures but at comparatively less high temperatures: the eclogites, the kyanite-bearing rocks, and the rocks of the glaucophane-schist facies. In this respect, however, there is still difference of opinion about two of these groups. Green and Ringwood (1966) contended that the characteristic mineral association of the eclogites can be stable in dry rocks of basaltic composition metamorphosed in the almandine-amphibolite facies, the glaucophane-schist facies, or part of the greenschist facies. Therefore, in this lecture eclogites will be left out of further consideration. The second group of rocks subject to difference of opinion is that of the kyanite-bearing rocks. Some years ago a pressure of about 10 kb was thought to be necessary for the formation of these rocks on the basis of well-corresponding direct syntheses and reversed transformations. Later on, much lower values were preferred, mainly on account of indirect determinations and calculations, which, in part, did not correspond with each other. According to several investigators, pressures of a few kb would be sufficient for the formation of kyanite (see, e.g., Newton, 1966). A direct determination that was published three months ago, however, again gave a value of about $6\frac{1}{2}$ kb (Althaus, 1967). These data, similarly, are too contradictory for further use in this lecture.

Finally, one group of rocks is left, that of the glaucophane-schist facies, for which less contradictory results were obtained; for this group, furthermore, the data concerning several different minerals point in the same direction.

For glaucophane—a Na-Mg-Fe-Al-silicate of the amphibole group, in nature virtually always with a considerable content of iron—and, more specifically, for its normal varieties, on the basis of data given by W. G. Ernst (1963, see also 1964), the following approximate minimum pressures of formation can be calculated: $4–4\frac{1}{2}$ kb at 200°C, $7–7\frac{1}{2}$ kb at 300°C, and $10–10\frac{1}{2}$ kb at 400°C; Ernst's data, however, especially for iron-bearing varieties of this mineral, are not very reliable. The values given refer to a glaucophane that is intermediate between $Na_2Mg_3Al_2Si_8O_{22}(OH)_2$ and crossite, with 50% Fe instead of Mg and Al. For this type of crossite itself the corresponding values are 3 kb higher, for glaucophane devoid of iron 3 kb lower. For lawsonite Crawford and Fyfe (1965)—of which the latter obtained comparatively low values for kyanite—estimated the following minimum pressures: $4\frac{1}{2}–5$ kb at 200°C, $5–5\frac{1}{2}$ kb at 300°C, an

5½–6 kb at 400°C. It may be remarked, based on a comparison of these two groups of values, that the introduction of a lawsonite-glaucophane facies or a glaucophane-lawsonite-schist facies instead of part of Eskola's glaucophane-schist facies apparently makes very little sense, unless this introduction is meant to comprise lawsonite associations without glaucophane in addition to glaucophane associations with or without lawsonite. The data for further characteristic minerals of the glaucophane-schist facies, viz., for aragonite and for jadeite occurring together with quartz, are based on thermodynamic data as well as on rather closely corresponding direct observations of reversed transformations, in part with hydrostatic apparatus. For the formation of pure aragonite as a stable phase we may assume as approximate minimum pressures 5½–6 kb at 200°C, 7–7½ kb at 300°C, and 8½–9 kb at 400°C; for pure jadeite together with quartz 10 kb at 200°C, 12 kb at 300°C, and 14 kb at 400°C (see, e.g., Crawford and Fyfe, 1965); however, as jadeite contains some iron, slightly lower values are generally accepted for natural jadeite. If for the temperature of metamorphism, which is not yet adequately known, we follow the most prominent investigators in this field and accept a value of about 200°C, we arrive at minimum pressures of 4–5 kb for the mineral association of the glaucophane-schist facies; it is generally assumed that the pressure reaches at least 8 kb.

The question may be posed of whether this pressure is essentially load pressure, as is presumed by the most prominent American investigators in this field. If this were the case, then the formation of these minerals would have occurred at depths between 15 and 30 km. This would imply that the folded sediments of geosynclinal areas, which during a later stage form mountain ranges, would have been depressed to much greater depths than hitherto assumed. And also that in all localities where these minerals are found, 15–30 km of rock would have been eroded away. This, however, is unacceptable from a geological point of view; in literature on finds lower and even much lower values for their depth of formation. With the aid of a later reduction of the load pressure by gravity tectonics or tectonic thinning by lamination of overlying layers, one does not seem able to reconcile the two different points of view.

So one becomes convinced that in one way or another there must have been an augmentation of the pressure, so that the load pressure, and consequently the depth of formation, has been considerably smaller. Several different mechanisms have been proposed to explain such an augmentation of pressure. In the first place, the rigidity, the resistance against deformation shown by rocks under unilateral pressure has been seized upon, and the so-called tectonic overpressure made possible by this resistance (Birch, 1955; de Roever, 1956, p. 127; Coleman and Lee, 1962). A second possibility would be the "fluid overpressure" of English literature, which was thought to have been originated by rapid production of water and/or CO_2; a third possibility could be thermal overpressure by expansion in consequence of local heating (Rutland, 1965). In our case both the above-mentioned type of "fluid overpressure" and thermal overpressure do not seem to be applicable. I shall return later to another variety of "fluid overpressure."

For the resistance of most rocks against deformation under suitable temperature and confining pressure, values have been found that are large enough to bridge the gap between the minimum pressures found in the laboratory on the one hand, and geologically estimated load pressures on the other hand. A great puzzle, however, is formed by the extrapolation of the factor time. During laboratory investigations it was found that the resistance of rocks shows a proportional lessening when the strain rate is decreased. Heard (1963) made experiments with a certain type of marble; at 500°C and a rate of deformation 10 million times slower he found a lessening of the yield strength of 1½ kb, viz., from 2 to ½ kb. Would it be warranted to extrapolate these results, then, at a strain rate he judged to be geologically acceptable—which again would be 10 million times lower—the yield strength would be insignificantly small, viz., 10^{-6}kb. Now, as a matter of fact, marble is very easily deformed. If more rigid rocks would show a similar lessening of their yield strength, tectonic overpressure due to resistance against deformation would be negligible as a geological factor.

Nevertheless, in my opinion a number of facts favor the existence of tectonic overpressure of considerable importance, although "fluid overpressure" of tectonic origin seems to be still more important.

First, a temperature of 500°C is too high for our case. At 300°C and the same strain rate, the marble under consideration reaches an extrapolated resistance of some tenths of 1 kb instead of 10^{-6}kb, and at 200°C these values are several times greater again. Taking into account that, in comparison with other rocks, marble is very easily deformed, we reach already more acceptable values of some kilobars or more for the yield strength, at the same strain rate, of glaucophane-bearing rocks.

Furthermore, the strain rate is highly variable; it need not have been so low. In seismic zones one may even expect local and sudden high pressures; for instance, if during an earthquake the movement of the loosened walls meets with a local obstacle. Nevertheless, in our case we should not reckon with strongly varying pressures, since the members of the amphibole group are sensitive indicators of changes in P–T conditions, but glaucophane as far as known does not display *oscillatory* zoning. So the pressure must have been fairly constant.

The presence in glaucophane-bearing rocks like those of California of old cracks and fissures filled by glaucophane, lawsonite, aragonite, and jadeite in my opinion seems to provide another important argument in favor of the possibility of tectonic overpressure of considerable importance. Such filled fissures seem to have been formed, at least in part, by opening through dilatation or extension, and not by shear; according to my experience they are of plentiful occurrence in low-grade regionally metamorphosed rocks. Misch (1962) contended that similar veins were filled at the same time they opened, in such a way that hardly any open space was ever present; in my opinion this is an example of the principle that mineral substance went into solution elsewhere under the influence of pressure, and that it was deposited where pressure was least. The extension fissures under consideration would have been formed under geopetal *unilateral load pressure* that, according to most American investigators in this field, must have reached 8 kb, corresponding to a depth of about 30 km. Rocks in which at such a depth, under such unilateral load pressure, extension cracks could still be formed, certainly had a large resistance at the moment of the formation of the cracks, large enough to allow the conclusion that at the moment of fissuring considerable tectonic overpressure could have existed. So it is possible—and in the light of abovementioned smaller estimates of the depth of metamorphism even more probable—that the load pressure never reached 8 kb, but was considerably smaller. In both cases the fluid pressure in the cracks would have reached 8 kb, which implies simply that the fissures were not connected in any way with the earth's surface. A fluid pressure of 8 kb, of course, could have been caused by a load pressure of 8 kb; in our second case, on the other hand, there must have been a considerable "fluid overpressure." In the surroundings under consideration we need not go far afield to seek the cause of this phenomenon; in my opinion, it must have been of tectonic origin, i.e., a consequence of a unilateral pressure of the order of 8 kb. In the latter case, too, extension fissures would have originated while there was a unilateral pressure of the order of 8 kb, so that at the moment of cracking the rocks must have shown a large resistance. This implies that in this case, too, tectonic overpressure of considerable magnitude could have existed. If we calculate for this latter case whether a "fluid overpressure" or the resistance against deformation leads to a greater mean pressure in the whole rock, it appears that the former is more effective and, consequently, of greater importance. By assuming "fluid overpressure" of tectonic origin, Rutland's (1965) structural objections against the existence of important tectonic overpressure are also met.

Some attention may also be paid to Winkler's (1965, p. 151) remark that the little-deformed character of many of the rocks under consideration provides an argument against the occurrence of important tectonic overpressure; he did not take into account that lasting deformation occurs only if the tectonic overpressure exceeds a certain limit.

All in all, *the conclusion of this argument has to be that it looks as if "fluid overpressures" of tectonic origin, as well as normal tectonic overpressure, have played a role during the formation of glaucophane, etc., so that the depth of metamorphism in this case need not have been excessive.*

The origin of the water in the cracks under consideration may be interesting to investigate; one could imagine a connection with the serpentinization of peridotites and with the local Na metasomatism in the neighborhood of these rocks and even be inclined to consider an origin from the earth's mantle; in the present lecture, however, all this would carry us too far.

This would have been the end of my lecture were it not for the possibility that important overpressure of tectonic origin sheds new light on the genesis of the minerals under discussion. As overpressure of tectonic origin apparently can play an important role indeed, one can imagine the possibility that glaucophane is not of real regional occurrence but is of dynamometamorphic formation in certain zones of strong overpressure as has been proposed by Dobretsov (1964) and as had been found by Nicolas (1966) in the Western Alps.

If during a period of regional metamorphism glaucophane and lawsonite are formed, this always occurs in a very early stage of the metamorphic history (de Roever and Nijhuis, 1964). Formerly it seemed that for the formation of these minerals, according to a suggestion by Ellenberger, the load pressure of a number of overthrust sheets would be a necessary condition, and this seems to some extent to be in contradiction with the very early formation of the minerals under consideration. Now this problem would have been solved: glaucophane, etc., could also have been formed under the influence of overpressure of tectonic origin in the absence of a pile of nappes, viz., during movement in the geosynclinal basin of sedimentation. Such an explanation would provide a very good solution to the problem of the occurrence of glaucophane, lawsonite, etc., in areas where according to literature no overthrust sheets would occur. And for regions where overthrust sheets are indeed present, one could think of formation of glaucophane both after and before the formation of the overthrust sheets, beginning in connection with embryonal movements in the geosynclinal basin, as proposed by Vialon (1966). This embryonal formation of glaucophane in consequence of overpressure of tectonic origin could in part be connected with the polyphase tectonic transport of ultramafic fragments of the earth's mantle (de Roever, 1957), as presumed by Cogulu (1965). The often-observed joint occurrence of glaucophane-bearing rocks with peridotites and serpentines, still unexplained, in this way would find a harmonious explanation.

Editor's Note: The list of references can be found at the end of the preceding original article.

19

Reprinted from *Contr. Mineral. Petrol.*, **24**, 93–113 (1969)

Blueschist Alteration during Serpentinization

RANDALL L. GRESENS

Department of Geological Sciences, University of Washington, Seattle, Washington 98105

Received March 23, 1969/Revised July 24, 1969

Abstract. The petrogenesis of Franciscan-type blueschists is controversial or paradoxial in regard to the possible significance of the serpentinite-blueschist association, the isochemical vs. allochemical character of blueschist metamorphism, the significance of "high pressure" mineralogy, and the physical-geologic-tectonic conditions existing during metamorphism. A model for blueschist alteration during serpentinization is presented that departs from conventional treatments of metamorphism because the reaction path rather then the thermodynamically lowest energy state is considered to be a controlling factor.

Alteration of ultramafic rocks to serpentinites requires oxidation of iron in the rock and selective withdrawal of water from saline pore fluids derived from surrounding eugeosynclinal rocks. Pore fluids become concentrated and chemically reducing in the vicinity of serpentinites.

The activated pore fluids may react with surrounding rocks via reactions that include a reduction step. The pore fluids may also affect the reaction path through surface chemical effects existing between mineral surfaces and/or "growth units" and the reducing pore fluid. The degree of polarization of oxygen ions in the silicate structural types, a function of polymerization and aluminum substitution, may control the surface effects and result in the preferential growth of chain silicates, and, more generally, of silicates with low amounts of tetrahedral aluminum.

The ratio Mg^{+2}/H^+ in the pore fluid can change during serpentinization depending on the extent to which magnesium is lost from the original mafic rock. This ratio may be an important control on the growth of jadeite vs. glaucophane in the presence of excess quartz.

The reduction reactions and those involving conventional fluid-solid equilibria cause a change in pore fluid chemistry as the reaction proceeds. Such reactions may explain short-range metasomatic transitions observed in some blueschists. The kinetic controls involving surface chemical effects are catalytic and may explain isochemical phenomena.

The ultimate "drive" for the process is the large negative free energy change of serpentinization that results when ultramafic rocks are emplaced in eugeosynclinal rocks with which they are not in equilibrium. Removal of this overwhelming disequilibrium may induce secondary disequilibrium and activation of pore fluids, producing Franciscan-type blueschists.

Introduction

"Until information is available on the solubility of silicates under the conditions of the glaucophane-lawsonite schist facies, discussion of the relevant solution chemistry is unsatisfactory. But the association of serpentinites and spectacular metamorphism makes it tempting to suggest that fluids derived from serpentinization may be rich in the components necessary for this type of soda-calcium metasomatism" (FYFE and ZARDINI, 1967, p. 827).

"By reference to theoretical treatment and experimental findings, the conclusion is drawn that the metamorphic reaction mechanisms will be found to be specific and to follow detailed paths closely related to the structural peculiarities of individual mineral species" (LACY, 1965, p. 153).

The above quotations reflect the basic themes of this paper, that certain blueschists may be genetically related to the serpentinization of ultramafic rocks and that the specific reaction mechanisms may be an important control on blueschist mineralogy. The hypotheses are presented as alternatives to the more conventional treatment of metamorphism. An excellent summary of other types of chemical effects associated with serpentinites is given by Černý (1968).

Stability-Metastability, Kinetics, and Pore Fluids

A common approach to metamorphism is to (1) note the phases present and their compositions, (2) make a judgement (based on phase rule considerations, textural relationships, or both) as to whether an equilibrium assemblage is present, and (3) interpret the rock in terms of the pressures and temperatures of metamorphism, based largely on experimentally determined stability fields of minerals and aided by geological evidence such as approximate depth of burial. Although this is a valid approach and generally leads to correct interpretations, it may in some cases be misleading. Because of the emphasis on experimentally determined P-T stability fields, the petrologist using this approach becomes conditioned to the concept that minerals form in their stability fields. The possibility of metastable growth is acknowledged, but in practice this possibility is generally ignored.

Metastable processes must involve reaction kinetics, including the path followed by the particular reaction. The strict thermodynamic approach to metamorphic reactions ignores the reaction path and concentrates on the energy states of the beginning and end points. However, the particular reaction path may lead to an end product that does not correspond to the state of lowest free energy and therefore is not the one predicted by thermodynamics. For example, the calcite and aragonite stability fields are thermodynamically independent of anything but pressure and temperature, yet the independent formation of these polymorphs is notoriously sensitive to the chemical environment (McCauley and Roy, 1966). More detailed discussions of the role of kinetics in metamorphism are given by Rast (1965) and Lacy (1965).

Another factor in metamorphism which is acknowledged in theory but sometimes ignored in practice is the pore fluid. Taylor et al. (1963) presented isotopic evidence for pore fluids during metamorphism. Taylor and Coleman (1968) presented similar evidence that at least some blueschists (including some from the Franciscan formation) must have recrystallized in the presence of a pervasive, probably water-rich, pore fluid. Fluid inclusions that may be samples of pore fluid are known from various minerals. Yet equations describing metamorphic reactions are commonly written as reactions between solids, or, at best, with pure water included. Equilibrium between minerals and possible pore fluids of complex composition is not considered. Some notable exceptions to this are the works of Korzhinskii (1959), Helgeson (1967) and, as applies specifically to this paper, Colemans (1967a) treatment of rodingitization.

An example of the importance of pore fluid composition is given by the work of Hemley and Jones (1964), who showed that in the presence of brines and excess quartz, the stability fields of muscovite and feldspar are dependent on the ratio of K^+/H^+ in the fluid. The muscovite-feldspar relationships point out the importance of the pH of the pore fluid as a control on mineral equilibria. The oxidation potential of the pore fluid should also be an important control, particularly for phases containing iron. EUGSTER (1959) discussed some of the implications of oxidation-reduction reactions in metamorphism, and he included the role of water. However, he did not consider the possibility of a pore fluid of more complex composition.

The possible role of mineral reactions that take place in the presence of complex pore fluids will be discussed later in relation to the petrogenesis of certain blueschists.

Franciscan Blueschists

The blueschists of the Franciscan formation of California present a number of paradoxes.

There are several minerals that, taken individually or as a whole, are generally considered to indicate metamorphism at low temperature and, in particular, high pressure. Yet all of the so-called "high pressure" minerals are found in geologic situations suggestive of low pressure. Glaucophane occurs in veins. COLEMAN (1961, p. 236) referred to veins of nearly pure jadeite that cut jadeite-albite schist in the New Idria serpentinite as possibly having formed in low-pressure zones. DAVIS (1960, p. 692) reported the occurrence of a large lawsonite-pumpellyite vein in Franciscan rocks and interpreted it as predominantly an open-space filling. Aragonite in Franciscan rocks commonly occurs in veins (COLEMAN and LEE, 1962). [VANCE (1968) reported the widespread occurrence of aragonite marbles in the San Juan Islands of Northwest Washington. The marbles occur in a metamorphic unit that, according to evidence presented by VANCE, was the result of metamorphism at both low temperature *and* pressure. Although not a part of the Franciscan formation, it is worth noting VANCEs aragonite marbles at this point.] Eclogites, containing the pyroxene omphacite, are traditionally assigned to a metamorphic grade of both high pressure and high temperature. Omphacite in the Franciscan formation occurs not only in veins cutting Franciscan metamorphic rocks, but as an incrustation and filling in vugs (ESSENE and FYFE, 1967). In all of these cases one could argue for the presence of a fluid under high pressure, but it seems unlikely that the fluid pressure could exceed the load pressure in dilational structures.

The emphasis on the need for high pressure to explain blueschist assemblages is largely the result of the direct application of experimental P-T fields of the minerals to the natural occurrence, with the implicit assumption that the minerals formed stably within their P-T fields. Given the general range of temperatures that may have existed during Franciscan metamorphism, it was necessary to find some mechanism that could give rise to high pressures. One of the mechanisms postulated was rapid burial to depths of 20 to 30 kilometers (BAILEY *et al.*, 1964, p. 111; ERNST, 1965, p. 907). On the grounds that stratigraphic thicknesses and possible geothermal gradients make deep burial implausible, COLEMAN (COLEMAN and LEE, 1962, p. 589—594; COLEMAN, 1967 b, p. 495) called on the concept of "tectonic overpressures". However, it is questionable whether the rocks can sustain tectonic overpressures of the necessary magnitude (RUTLAND, 1965).

Another problem of the Franciscan blueschists concerns whether metamorphism is isochemical or metasomatic or both. The debate is a very old one (pre-1909) as summarized by DAVIS (1918, p. 277). CRITTENDEN (1951, p. 26)

7*

believed that blueschists of the San Jose-Mt. Hammilton area were produced metasomatically by fluids derived from ultramafic rocks. In recent times, Fyfe and others presented evidence for metasomatism (Fyfe and Zardini, 1967; Essene *et al.*, 1965). Many investigators, including myself, have seen examples that require introduction of sodium — for example, development of riebeckite in red chert beds. Older accounts report such phenomena as the transition from unmetamorphosed rock to blue amphibole rock over distances of three feet (Davis, 1918, p. 275) or the selective replacement of certain beds (Taliaferro, 1941, p. 123; 1943, p. 165—166).

On the other hand, Coleman and Lee (1963), Ernst (1963b), and Ghent (1965) presented evidence that at least some of the metamorphism is isochemical. The evidence rests on the comparison of rocks that contain recognizable clues to their origin with presumably equivalent unmetamorphosed rocks of the same type.

The interpretation regarding the association between blueschists and serpentines is also controversial. Older reports stressed the relationship as having genetic importance. Taliaferro (1941, p. 123—124; 1942, p. 85; 1943, p. 168) and Switzer (1945) believed in "pneumatolytic" effects of the serpentines on the adjacent rocks, causing blueschist alteration. These beliefs were based on the frequent association of the two rock types and specific field occurrences of "aureoles" of blueschists near serpentinites. Taliaferro (1941, p. 123; 1943, p. 165—1966), for example, reported selective replacement of specific interbeds of sedimentary sequences in contact with serpentinites. Some recent investigators have held to the possibility that a genetic relationship may exist (Essene *et al.*, 1965; Fyfe and Zardini, 1967; Fyfe, 1967).

Also in recent years, certain authorities, although recognizing the association of these two rock types, consider it to be fortuitous (Bailey *et al.*, 1964, p. 105). They occur together, according to this concept, because they are both products of the tectonically active continental border. Deep-seated faults, along which serpentinites are intruded, cut across the deep-seated high-pressure, low-temperature environment that is the locale of blueschist facies metamorphism. Exotic blocks of blueschists carried up along the faults account for the association.

I believe that (1) there is a genetic relationship between serpentinites and Franciscan-type blueschists, (2) the mineralogy of the blueschists may be explained without resorting to the high pressures necessary to place the minerals within their P-T stability fields, (3) the explanation may lie in the consideration of the pore fluids present during serpentinization, the changes in the pore fluid resulting from serpentinization, and the re-equilibration of pore fluids with country rocks, and (4) both metasomatic and isochemical occurrences may be explained by this model.

Effect of Serpentinization on the Pore Fluid

The concept that the parent ultramafic rocks of serpentinites are derived from depth, possibly as part of the earth's mantle, is widely accepted. Olivines and pyroxenes that comprise the bulk of these rocks are stable in a hot, "dry" environment. When they are tectonically emplaced into cooler, wet geosynclinal rocks, a major chemical disequilibrium results. Fyfe (1967, p. 48) called attention to the large ΔH of serpentinization.

It is generally accepted that the fluids necessary for serpentinization come from the enclosing rocks. A common practice is to describe the fluid only as "water", and reactions typically are written this way. In reality, the pore fluids are more likely to be a sodium-rich brine. The geosynclinal rocks must have contained abundant sea water during deposition. It is reasonable to suppose that after equilibration during diagenesis and after some of the fluids have been squeezed out by compaction, the residual pore fluids had compositions perhaps resembling oil-field brines. [Where samples of pore fluids have been directly obtained, e.g., as fluid inclusions in minerals (ROEDDER, 1962) or from direct sampling of active metamorphism (MUFFLER and WHITE, 1969), they are brines.]

During serpentinization, water is removed from the pore fluids. Sodium is not removed since serpentinites are notable for their low sodium content. For crystal chemical reasons, the mineral phases formed during serpentinization are not able to accommodate sodium. One immediate effect of serpentinization is that the concentration and ionic strength of the brine will increase. This alone will increase the chemical activity of the pore fluids.

During serpentinization, the major change is the transformation of anhydrous into hydrous magnesium silicates. Although the original olivines and pyroxenes are able to tolerate considerable substitution of ferrous iron for magnesium, this is not possible to the same degree for the serpentine minerals. As serpentinization proceeds, the chemical potential of iron increases as more and more iron is liberated. This is evident from the work of PAGE (1967a, 1968), who showed that the iron content of serpentine minerals increases from the early to the later minerals. The amount of such iron substitution is not great, but serpentine minerals are forced to accept more and more iron as its chemical potential increases. Nevertheless, for crystal chemical reasons there are limits on the amount of iron can be tolerated (RAMBERG and DeVORE, 1951; RAMBERG, 1952), and, as a result of this, a common process during serpentinization is formation of a stable phase that can accept iron. Magnetite is formed, but this step requires that two-thirds of the ferrous iron be oxidized to ferric iron in each formula unit of magnetite. It is significant that hematite is not formed. The oxidation proceeds only to the state necessary to accommodate the excess iron.

A comparison of published analyses suggests that oxidation of iron is an important part of the serpentinization process. In dunites and peridotites ferrous iron generally predominates, whereas ferric iron generally is dominant in serpentinites. The oxidation step has to be accommodated in some way, and it is reasonable to expect that the pore fluids should become chemically reducing.

There are a number of mechanisms by which the electrochemical transfers could take place[1]. One way is the simple oxidation of iron oxide.

$$4 \text{ FeO (in ultramafic)} + O_2 \text{ (from pore fluid)} \longrightarrow$$
$$2 \text{ Fe}_2O_3 \text{ (in magnetite)} \tag{1}$$

The pore fluid in this case would be affected through the equilibrium,

$$2 \text{ H}_2O \longrightarrow O_2 + 4H^+ + 4e^- \tag{2}$$

_____

[1] Some of the equations that follow are written as half-cell reactions. It is assumed that appropriate chemical species exist in the pore fluid to accept or donate electrons.

Reduction of iron could also be accomplished by means of a cation transfer.

$$2FeO \text{ (in ultramafic)} + (Ca,Mg)O \text{ (in ultramafic)} \longrightarrow$$
$$Fe_2O_3 \text{ (in magnetite)} + (Ca,Mg)^{+2} \text{ (in pore fluid)} + 2e^- \tag{3}$$

This process could be accompanied by a similar, but reverse, process in the country rock as explained below.

Finally, various combinations of (1) and (3) could occur. For example,

$$6FeO \text{ (in ultramafic)} + (Ca,Mg)O \text{ (in ultramafic)} + O_2 \text{ (from pore fluid)}$$
$$\longrightarrow 3Fe_2O_3 \text{ (in magnetite)} + (Ca,Mg)^{+2} \text{ (in pore fluid)} + 2e^- \tag{4}$$

I favour reactions (3) and (4) for the following reasons. Reaction (1) requires only the addition of oxygen. Oxygen contributes most of the volume to a silicate mineral. There is a "volume problem" already because serpentinization solely by means of water addition would require a volume increase (HOSTETLER et al., 1966; THAYER, 1966). Addition of oxygen would require even greater volume increase. Secondly, there is evidence that much of the calcium present in the original ultramafic rock is expelled (PAGE, 1967b; GRESENS, 1967). Others believe that magnesium is sometimes expelled (THAYER, 1966; ČERNÝ, 1968).

In summary, serpentinization of ultramafic bodies should cause chemical activation of the pore fluids in adjacent rocks by making the fluids more concentrated and reducing.

Alteration of Surroundings via Oxidation-Reduction Reactions

If pore fluids in the vicinity of ultramafic bodies undergoing serpentinization are indeed altered in the manner described above, then one obvious means by which they may re-equilibrate with the surrounding rocks is by reactions that involve reduction. If the element being reduced in the country rock is iron, then this stage of the process involves simply the balancing of an oxidation step in the ultramafic body with a similar reduction step in the country rock.

Of the rock types typical of the Franciscan formation the ferruginous cherts and red shales are very rich in ferric iron and low in ferrous iron. Cherts are particularly suitable for demonstrating the metasomatic nature of some blue-schist alteration. The identity of the parent material is commonly recognized, and comparison of metachert compositions with that of unaltered chert commonly demonstrates enrichment in sodium (see analyses in BAILEY et al., 1964, Table 9, p. 63, and Table 14a, p. 108). It also demonstrates a change in the ferric/ferrous ratio such that the metachert is the reduced form.

In general, possible reduction reactions in the country rock are similar to the possible oxidation reactions in the ultramafic bodies.

$$2Fe_2O_3 \text{ (in country rock)} \longrightarrow 4FeO \text{ (in blueschist)} + O_2 \text{ (in pore fluid)} \tag{5}$$

Once again, the pore fluid would change via reaction (2) above. Another possibility is

$$Fe_2O_3 \text{ (in country rock)} + 2Na^+ \text{ (from pore fluid)} + 2e^- \longrightarrow$$
$$2FeO \text{ (in blueschist)} + Na_2O \text{ (in blueschist)} \tag{6}$$

Combinations are possible. For example,

$$3Fe_2O_3 \text{ (in country rock)} + 2Na^+ \text{ (from pore fluid)} + 2e^- \longrightarrow$$
$$6FeO \text{ (in blueschist)} + Na_2O \text{ (in blueschist)} + O_2 \text{ (in pore fluid)} \tag{7}$$

Combining the reactions to show the overall process, and using chert as an example, on can obtain equations such as

$$2FeO \text{ (in ultramafic)} + (Ca,Mg)O \text{ (in ultramafic)} + Fe_2O_3 \text{ (in chert)} + 2Na^+ \text{ (pore fluid)} \longrightarrow$$
$$Fe_2O_3 \text{ (in magnetite)} + 2FeO \text{ (in blue amphibole)} + Na_2O \text{ (in blue amphibole)} \qquad (8)$$
$$+ (Ca,Mg)^{+2} \text{ (pore fluid)}$$

The overall process in this case is one of an oxidation in the ultramafic body balanced by a reduction in the country rock, accompanied by an exchange of $2Na^+$ for $(Ca,Mg)^{+2}$ in the pore fluid.

This reaction is only a model and is subject to considerable variation. For example, it should not be taken to mean that blue amphiboles in metasomatized cherts should contain one ion of ferrous iron for each sodium ion. The equations can be combined in various ways to yield different ratios. Also, the reduction could take place anywhere in the rock and the reduced iron need not always be incorporated in the blue amphibole. Nevertheless, blue amphiboles, including glaucophane, that occur in the entire suite of Franciscan rock types generally contain significant amounts of ferrous iron (5—24 percent FeO). The reactions may also be varied in that sodium in the pore fluids may be directly exchanged for elements from the original rock. FYFE and ZARDINI (1967, p. 827) indicated that this had occurred during blueschist facies metasomatic alteration of conglomerates of the Pacheco Pass area. It is also well known to persons working on Franciscan rocks that the original composition is a key factor in the type of blue amphibole that is formed; glaucophane forms in aluminous rocks, whereas crossite and riebeckite form in Fe^{+3}-rich rocks.

FYFE (1967, p. 50—52) called attention to the loss of minor constituents, including alkalies, from ultramafic bodies during serpentinization as bearing on the metasomatic effects of blueschist metamorphism. As discussed above, it is clear that calcium and magnesium can be released and should affect the equilibrium within the pore fluid and the equilibration of the pore fluid with surrounding rocks. However, I believe that metasomatic effects involving sodium are not the result of sodium lost from the parent ultramafic rock. I believe that sodium comes from without, either as original connate sodium or from breakdown of sodium-rich minerals, and that the sodium, already present in the pore fluid, is activated by the processes described above in the vicinity of serpentinizing ultramafic bodies.

Effect of Pore Fluid on Nucleation and Growth Kinetics

It would be convenient for the hypothesis just proposed if all of the Franciscan blueschists were enriched in sodium and if a change from ferric to ferrous iron could always be demonstrated. However, as discussed above, evidence has been presented by some authorities suggesting that isochemical blueschist metamorphism does occur in the Franciscan, and that the ferric/ferrous ratio is essentially unchanged. Moreover, the processes discussed above do not explain the formation of minerals like pure jadeite and lawsonite. The formation of these minerals does not involve oxidation-reduction steps, and they contain at best only negligible amounts of ferrous iron.

Other explanations must be sought. One explanation lies in the consideration of the role of the reducing brine on the atomic level, during the critical processes of nucleation and growth of the mineral, i.e., the chemical kinetics. The arguments that follow are put forward as possibilities that are hypothetical and that have not been verified by experimentation. The bases for the discussion are the concepts of screening of cations and polarization and mutual deformation of ions that were originally developed by FAJANS (1931) and expanded by WEYL (WEYL, 1953, 1956; WEYL and MARBOE, 1962). WEYL used these concepts to explain phenomena related to the solid state and solid surfaces. This approach also was used by DEVORE (1955, 1956, 1957, 1959, 1963) to explain certain crystal chemical effects associated with silicate minerals.

In a silicate structure, the high field strength of the silicon ions causes deformation or polarization of the "electron clouds" of the oxygen ions inward toward the silicon ions. The "tightening" of the electron cloud produces decreased screening of the oxygen nucleus on the "outward-facing" or apical portion of the oxygen ion. Polarization of the electron cloud toward the silicon ion produces additional screening between the oxygen and silicon nuclei. This results in shorter bond distances and produces a lower energy state. However, polarization of electronic charge away from the apical portion of the oxygen ion results in an energy increase as the oxygen nucleus becomes less well screened. The two conflicting trends must reach a low energy balance that controls the state of polarization of the oxygen ion.

The series of silicate structural types ranging from those of individual silica tetrahedra (nesosilicates) through those of 3-dimensional frameworks (tectosilicates) represents a series of increasing polymerization corresponding to a changing Si:O ratio. As the ratio increases, the screening on the apical portion of the oxygen ion is increasingly diminished as the high field strength of a fixed number of silicon cations is brought to bear on fewer and fewer oxygen ions. The result is that nesosilicates will have the least polarized and tectosilicates the most polarized oxygen ions, based only on the Si:O ratio.

However, the most highly polymerized structures (with the exception of quartz which is a special case) show the greatest substitution of aluminum for silicon in the Si-O structure. The effect of aluminum (which has a lower field strength than silicon) is to relax the inward polarization. Consequently the oxygen ion is less polarized and the apical portion is better screened.

Thus the two factors, polymerization and aluminum substitution, have offsetting effects. Although pyroxenes and amphiboles show some Al substitution, it is in sheet and 3-dimensional structures that Al substitution becomes great. When these factors are taken together, the polarization of oxygen ions should increase as the polymerization of the Si-O structures increases, up to and including chain silicates that are relatively free of tetrahedral aluminum, and then should decrease as aluminum substitution becomes dominating in sheet and 3-dimensional structures[2]. It may therefore be expected that the oxygen ion polarization in Si-O structures reaches a maximum in the chain silicates.

---

2 An analogous discussion of the electronegativity of silicate structural types by RAMBERG
  (1952) was used as a model for this discussion of oxygen ion polarization.

Table. *Idealized O:Si ratios and degree of Al-Si substitution for some common silicates and their blueschist counterparts*[a]

Mineral	Ideal O:Si Ratio	(Tetrahedral Al): (Total tetrahedral Al + Si)		
		High	Low	Mean
Epidote	$11:3 = 3.66$	0.031	0.000	0.009
Pumpellyite	$11:3 = 3.66$	0.000	0.000	0.000
Lawsonite	$7:2 = 3.5$	0.000	0.000	0.000
Augite	$3:1 = 3.0$	0.214	0.016	0.066
Jadeite	$3:1 = 3.0$	0.000	0.000	0.000
Omphacite	$3:1 = 3.0$	0.035	0.010	0.023
Common Hornblende	$11:4 = 2.75$	0.237	0.069	0.142
Glaucophane	$11:4 = 2.75$	0.039	0.000	0.024
Crossite	$11:4 = 2.75$	0.084	0.007	0.049
Riebeckite	$11:4 = 2.75$	0.078	0.000	0.023
Common Muscovite	$5:2 = 2.5$	0.261	0.072	0.206
Biotite	$5:2 = 2.5$	0.363	0.243	0.319
Phengite	$5:2 = 2.5$	0.237	0.125	0.175
Stilpnomelane	$5:2 = 2.5$	0.000	0.000	0.000
Albite	$2:1 = 2.0$			0.250
Anorthite	$2:1 = 2.0$			0.500

[a] Phengites from ERNST (1963a); omphacites from COLEMAN et al. (1965); others from DEER et al. (1962a, b; 1963).

If minute silicate growth units (protonuclei) are forming either in the pore fluid or on mineral surfaces in contact with the pore fluid, then a stable unit in the presence of a reducing fluid should be the one with the highest degree of oxygen polarization, i.e., the chain silicate. The reducing fluid, having a high electron flux[3], should provide a screening effect at the apical (outward) portion of oxygen ions. This may allow increased inward polarization toward the silicon ions and the attendant lowering of energy without causing the energy increase associated with decreased screening of the apical portions of the oxygen ions. By this reasoning, the chain type of silicate (with little tetrahedral aluminum) can best utilize the enhanced screening capacity of the reducing pore fluid to lower the energy relationships.

The Table lists the O:Si ratio of the structural type and the ratio (tetrahedral aluminum: total tetrahedral sites) for a number of silicate minerals. It is

---

3 The term "high electron flux" as used above refers to the relative abundance in the pore fluid of chemical species having the ability to donate electrons in an oxidation-reduction step. This implies that the species carry electrons in excess of that needed for the normal screening demands of their nuclei — in the sense in which this concept is used by WEYL (1953, 1956) — and therefore that their presence in a fluid should enhance the overall screening ability of the fluid. No attempt has been made in this discussion to identify the specific chemical species that may exist in the pore fluid. The argument instead has been directed to the overall effect of the fluid on silicate surfaces, whatever the particular species that control the oxidation-reduction steps in the fluid may be. However, because the various possible chemical species should logically be excepted to have slightly different screening relationships with a given solid surface, the specific identity of the chemical species may be of considerable importance.

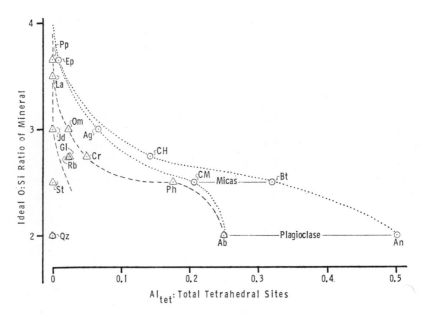

Fig. 1. Plot of the ideal O:Si ratio of some minerals vs. the mean ratio of tetrahedral aluminum (see Table I). Common minerals are shown as circles. The dotted lines show the "main trend" of Al-Si substitution with increasing polymerization. Blueschist minerals are shown as triangles. The dashed lines show the trend of Al-Si substitution with increasing polymerization for blueschist minerals. The trend for blueschist minerals plots off the main trend and is skewed toward the axis of zero tetrahedral aluminum. Code: Ep, epidote; Ag, augite; CH, common hornblende; CM, common muscovite; Bt, biotite; Ab, albite; An, anorthite; Pp, pumpellyite; La, lawsonite; Jd, jadeite; Om, omphacite; Gl, glaucophane; Cr, crossite; Rb, riebeckite; Ph, phengite; St, stilpnomelane; Qz, quartz

immediately clear that minerals typical of blueschists generally contain less tetrahedral aluminum than their more common counterparts. This is also seen in Fig. 1, in which the parameters listed in Table 1 are plotted. The common minerals of each structural type show a regular trend of increasing tetrahedral Al substitution with increasing polymerization. The blueschist minerals plot well off this trend and are skewed toward the axis of low tetrahedral Al[4].

Reducing pore fluids should favor the growth of chain silicates low in tetrahedral Al; thus the common occurrence of jadeitic pyroxenes and blue amphiboles in blueschists may be expected. In every case, the minerals of the blueschist facies contain less tetrahedral aluminum and therefore have Si-O structures with a higher degree of oxygen polarization than their more common counterparts.

Plagioclase contains abundant tetrahedral aluminum. Therefore, even though it is highly polymerized, the oxygen ions are not highly polarized. When plagioclase does occur in blueschists, it is pure albite, as would be expected because albite has more highly polarized oxygen ions as compared with calcic plagioclases. Low temperature is, of course, another control on albite stability. The two

4 The role of high pressure and low temperature as compared with chemical factors as a control on tetrahedral Al are discussed in a later section.

factors — low temperature and reducing pore fluid — both have the effect of making albite the preferred plagioclase.

On the other end of the polymerization scale, neither lawsonite nor pumpellyite contain tetrahedral aluminum. However, lawsonite represents a slightly more polymerized structure because it contains only Si_2O_7 groups, whereas pumpellyite contains both Si_2O_7 and SiO_4 groups.

It is interesting that muscovite and albite contain nearly the same amount of tetrahedral aluminum. Ordinary muscovite is not common in blueschists although albite and phengite are. On the one hand there is a shift toward phengite (the sheet structure is retained). On the other hand there is a shift toward albite (the ratio of tetrahedral Al is roughly retained but in a more polymerized structure).

It is well known that plagioclase breaks down in some blueschists to form jadeite, quartz, and lawsonite. The (albite ———→ jadeite + quartz) transition has been widely cited as an indicator of high pressure. However, in the presence of a reducing pore fluid, the breakdown may be expected. The minerals formed have Si-O structures with a higher degree of oxygen polarization than albite. Similarly, although lawsonite is not highly polymerized, it has more highly polarized oxygen ions than anorthite; in anorthite the Si-O framework is half-filled with aluminum and therefore has an exceptionally low state of oxygen polarization. Thus the breakdown of the plagioclase in general is explained as the attempt of the mineral to adjust to the reducing pore fluid by a process that moves aluminum out of the tetrahedral sites.

This viewpoint suggests that if the energy differences between the stable and metastable states is small, then mechanistic considerations that involve the energy relationships between the solid surface and the fluid may promote metastable growth. For example, the work of HLABSE and KLEPPA (1968) showed that, depending on whether high or low albite is used, the position of the phase boundary for the transition to jadeite plus quartz is shifted 3—4 kilobars at 300° C. This suggests that jadeite existing in the low albite field is metastable only to the extent of the small energy difference between high and low albite.

Speculations on Possible Kinetic Mechanisms

The polarization of oxygen ions of the silicate structural types may exert an important control during the nucleation and growth of the minerals. There are a number of possibilities, and perhaps all of them have some validity. During the breakdown of plagioclase, one possibility is that rearrangements take place at the surface of the mineral and proceed inward. The importance of surface chemistry was stressed by WEYL (1953) and applied to mineral relationships by DeVORE (1956a, 1959). Because plagioclase breakdown involves transfer of aluminum, one could postulate that the reaction proceeds via mechanisms proposed by LACY (1965, Fig. 3), by which aluminum and silicon migrate through rings of three oxygen ions. Because the mineral surface is probably highly distorted (WEYL, 1953) and because of the high concentration of ions in the pore fluid which would aid this process, LACYs mechanism is a distinct possibility.

Another possibility simply has to do with the dynamic equilibrium between minerals and pore fluid. At the mineral surface, statistical fluctuations in the kinetic energies of the ions will cause the ions to temporarily detach from the surface. Similarly, ions in the fluid will sporadically attach to the mineral surface. The dynamic equilibrium is established when the rates of these two processes is equal. If silica is present in the fluid, then there should exist groupings of silicon and oxygen ions that will constantly disaggregate and reform. These should include molecules containing only a single silica tetrahedron and also more polymerized forms. In the reducing environment especially, molecular configurations containing an aluminum tetrahedron should be highly unstable relative to those with silica tetrahedra on the basis of the hypotheses discussed above. The aluminum instead should preferentially form complexes with oxygen and other metal cations, as suggested by DeVore (1957). These transient molecular configurations form the protonuclei or growth units that may transfer as a unit from the fluid to the mineral. As the pore fluid becomes reducing, the rate of production of chain-type protonuclei should increase. Statistically, there should be a higher concentration of these units in the pore fluid during a given interval of time and a corresponding low statistical concentration of units containing tetrahedral aluminum. The possibility of nucleation of minerals such as jadeite and/or lawsonite is increased. After nucleation occurs, the higher statistical concentration of jadeite- und lawsonite-type protonuclei relative to plagioclase-type protonuclei may cause the rate of transfer of growth units to jadeite and lawsonite to exceed the rate of transfer of growth units to plagioclase. Plagioclase might gradually be replaced. Thus the presence of the reducing pore fluid may upset the dynamic equilibria between solid and fluid phases such that jadeite and lawsonite form at the expense of plagioclase.

It was pointed out earlier that when plagioclase does persist in blueschists it occurs as albite. Fyfe (1967, p. 41) observed that the ordering of aluminum and silicon in tetrahedral sites of plagioclase is higher in albite from blueschists than in albite from other natural occurrences. Fyfe suggested that this indicates low temperature. It might also be explained in terms of the influence of a reducing pore fluid. In disordered albite there would exist random domains in which two or more aluminum ions occupy adjacent tetrahedral sites. The relatively low state of oxygen polarization of such a configuration would make it less stable in contact with a reducing pore fluid than a configuration in which aluminum ions are separated. DeVore (1956 b) discussed Al-Si ordering in plagioclase and pointed out (p. 257) that a growing crystal would have some freedom to select favorable bonding relationships and reject unfavorable ones. The reducing pore fluid could increase the sensitivity of the crystal surface to the unfavorable association of Al ions in adjacent tetrahedral sites. Rejection of this configuration during crystal growth could result in highly ordered albite.

The surface tension (surface energy per unit area) of a mineral is the most important factor controlling the nucleation energy barrier (Weyl, 1953). Surface tension is the tension of an *interface*, i.e., the contact between two phases. In the presence of a reducing pore fluid, minerals whose nuclei have the lowest interfacial free energy with the fluid will have the lowest energy barrier to nucleation. It was suggested above that chain silicates with low amounts of tetrahedral Al should provide a mineral surface that is most compatible with a reducing pore fluid, i.e., will have the lowest surface tension (if the type of silicon-oxygen framework is considered to make a major contribution to the surface tension). Thus nucleation of chain silicates whould be favored under reducing conditions. Once nucleation occurs, continued growth should be favored by the relative abundance of suitable pre-structured growth units, as discussed above.

Other Fluid Equilibria

In addition to the kinetic controls and the role of oxidation-reduction reactions, there are numerous other reactions between the pore fluid and solid phases that will be important in determining the local mineralogy in a rock. Of the many possibilities, there are two that have special significance to blueschists. These concern the role of the Mg^{+2}/H^+ ratio as related to serpentinization and jadeite-glaucophane equilibrium.

Serpentinization may be idealized by consideration of the hydration reactions of pure magnesium olivines and pyroxenes (ignoring iron and other constituents). This is commonly expressed by equations such as

$$3Mg_2SiO_4 + SiO_2 + 4H_2O \longrightarrow 2Mg_3Si_2O_5(OH)_4 \tag{9}$$
$$\text{forsterite} \qquad\qquad\qquad\qquad \text{serpentine}$$

$$3MgSiO_3 + 2H_2O \longrightarrow Mg_3Si_2O_5(OH)_4 + SiO_2. \tag{10}$$
$$\text{enstatite} \qquad\qquad \text{serpentine}$$

Serpentinization, viewed this way, requires volume increase, and the only change in the fluid is the extraction of water (and perhaps a change in soluble silica). However, some geologists (THAYER, 1966; ČERNÝ, 1968) consider serpentinization to be a more-or-less constant volume process involving the loss of magnesium. In this case the effects on the pore fluid are more complex. For example, Eq. (9) can be rewritten to hold silica constant.

$$2Mg_2SiO_4 + H_2O + 2H^+(P.F.) \longrightarrow Mg_3Si_2O_5(OH)_4 + Mg^{+2}(P.F.) \tag{11}$$
$$\text{forsterite} \qquad\qquad\qquad\qquad \text{serpentine}$$

The equation can be further rewritten to give a silica loss.

$$3Mg_2SiO_4 + 6H^+(P.F.) \longrightarrow Mg_3Si_2O_5(OH)_4 + SiO_2 + H_2O + 3Mg^{+2}(P.F.) \tag{12}$$
$$\text{forsterite} \qquad\qquad\qquad \text{serpentine}$$

Similar variations of Eq. (10) could be obtained. Thus the equations may be written in a number of ways, depending on the amount of magnesium that is to be lost. Notice that Eq. (12) even represents the production of water during serpentinization. The important thing is that loss of magnesium during serpentinization changes the Mg^{+2}/H^+ ratio of the pore fluid.

The possible significance of this ratio is obvious when one considers the equilibrium between pure jadeite and pure glaucophane in the presence of a pore fluid in which magnesium may exist in solution.

$$2NaAlSi_2O_6 + 4SiO_2 + 4H_2O + 3Mg^{+2}(P.F.) \longrightarrow \tag{13}$$
$$\text{jadeite} \qquad \text{quartz}$$

$$Na_2Mg_3Al_2Si_8O_{22}(OH)_2 + 6H^+(P.F.)$$
$$\text{glaucophane}$$

In the presence of the pore fluid and excess quartz, jadeite-glaucophane equilibrium should be controlled by the Mg^{+2}/H^+ ratio of the pore fluid (more precisely, by the ratio of their activites). Thus, other factors being constant, glaucophane should be favored over jadeite in the vicinity of ultramafic bodies undergoing serpentinization involving loss of magnesium.

Catalytic vs. Consumptive Reactions

The processes involving oxidation-reduction and hydrogen ions are "consumptive" reactions, i.e., the pore fluid is consumed or altered as the reaction progresses. This type of reaction accounts for transitions from altered to unaltered rocks over relatively short distances, such as are described in numerous accounts from the Franciscan. These are usually metasomatic phenomena. For example, as reactive pore fluid penetrates a red chert, iron may be reduced and sodium may be simultaneously introduced to produce a mineral like crossite, riebeckite, or aegirine. However, the reaction changes the oxidation potential of the pore fluid and it becomes unreactive. Fresh pore fluid must be supplied. If the process is interrupted before the alteration can go to completion, the partially altered rocks are preserved, giving the appearance of a rather abrupt change in the physical conditions of metamorphism. Similarly, the description by Fyfe and Zardini (1967, p. 827—830) of conglomerate pebbles that are altered in such a way that they have an outer rind of glaucophane and an inner core of jadeitic pyroxene is explained by the change in the Mg^{+2}/H^+ ratio of the pore fluid as it penetrates the pebbles. Glaucophane is formed on the outside, but the chemical nature of the pore fluid is thereby changed so that jadeite is formed deeper in the pebble.

In contrast to these consumptive, largely metasomatic, reactions that alter the pore fluid, the kinetic control postulated above is essentially a catalytic effect. The breakdown of plagioclase to jadeite, lawsonite, and quartz does not involve oxidation-reduction or proton transfer reactions. The pore fluid would not be consumed or altered. Therefore, this mechanism may be able to affect larger masses of rock more uniformly. This view is consistent with the extensive alteration of large masses of Franciscan greywacke to metagreywacke, without the abrupt short-range changes sometimes observed in metasomatic blueschists. The consumptive reactions and the catalytic reactions may occur together, which may be the case for the rocks of the Pacheco Pass area described by McKee (1962) and Fyfe and Zardini (1967). McKee was the first to recognize the possibility of catalysis when he wrote (p. 609), "I believe solutions were of paramount importance in the metamorphism, and the recrystallization involved some transfer of substances. The possible catalytic effect or the origin of these solutions is unclear."

Rodingites and Eclogites

Fyfe and Zardini (1967, p. 827) called attention to the high mobility of calcium as well as sodium during blueschist metamorphism. It was pointed out previously that most of the calcium that is present in a parent ultramafic body is expelled during serpentinization. Calcium may also be a major component of the original pore fluid, if a typical oil-field brine is used as a model for the pore fluid.

Rodingites are assemblages of hydrous calcium-aluminum silicates believed to result from calcium metasomatism during serpentinization at relatively low temperatures (BILGRAMI and HOWIE, 1960; COLEMAN, 1967a). The aluminous minerals contain virtually no tetrahedral aluminum, and aragonite has been noted in some assemblages. Rodingites are known to occur with jadeitic pyroxene and blue amphibole in greywacke altered at serpentinite contacts (CHESTERMAN, 1960) and as rims around jadeite pods in serpentinites (COLEMAN, 1961). ČERNÝ (1968, p. 1379) ascribed both rodingitization and jadeite formation to chemical effects produced at serpentinite contacts.

Eclogites also occur with blueschist facies rocks. COLEMAN et al., (1965) demonstrated that blueschist eclogites differ from eclogites of other geologic environments. They are less mafic and higher in sodium and calcium. Although they appear to be nearly isochemical, they may have undergone slight sodium and calcium metasomatism. The pyroxenes contain very little tetrahedral aluminum. The garnets are rich in calcium and ferrous iron. If mica is present, it is phengite.

The formation of rodingites and eclogites appears to be related to the same kinds of chemical processes advocated in this paper for the formation of certain blueschists. They possibly represent different conditions of temperature, original fluid composition, water content of activated pore fluid, etc. These possibilities are tentative and require suitable experimental investigation.

Miscellaneous Supporting Evidence for a Reducing Pore Fluid

The hypotheses presented in this paper rely heavily on the presence of concentrated sodium-rich reducing pore fluids during metamorphism. A legitimate question is whether there is any direct evidence for such a pore fluid. There are scattered bits of information that lend support to the hypothesis.

FYFE and ZARDINI (1967) described the "spectacular" metamorphism of the Franciscan metaconglomerates of the Pacheco Pass area, which they regard as having involved extensive sodium and calcium metasomatism. After describing the mineralogy, they observe, "It will be noted that there is a general absence of minerals containing ferric iron unless these are sodic phases". The absence of ferric iron argues for a reducing pore fluid.

COLEMAN and LEE (1963, p. 276) commented on the ferric:ferrous ratio of clinozoisite and pumpellyite in the Ward Creek blueschists and stated, "... the preponderance of pumpellyite over clinozoisite in type III metabasalts may be another manifestation of the relatively low redox potential of the metamorphic environment..."

TAYLOR and COLEMAN (1968, p. 1737) showed that metacherts from the Ward Creek blueschist sequence commonly did not achieve oxygen isotopic equilibrium with metamorphic pore fluids, even though equilibration was nearly complete for other rock types in the area. They attributed this to the more impermeable nature of the original cherts. They further demonstrated (p. 1738) that the greater the deviation from the isotopic composition of the original unmetamorphosed chert, the lower is the ratio of ferric to ferrous iron. If the degree of change of the oxygen isotopic composition is an indication of the degree of alteration, then the cherts have become progressively depleted in ferric iron relative to ferrous iron during their alteration. This supports the contention that the pore fluids responsible for the alteration were chemically reducing.

COLEMAN (1961) described the jadeite pods enclosed in the New Idria serpentinite. He observed (p. 225) that blebs of native copper are present in the rock. The native copper may represent reduction by the pore fluids of trace amounts of copper that originally may have been present in the parent material.

MILTON and EUGSTER (1959) described the geochemical relationships of mineral phases from the Green River formation. There is no question that the minerals formed in a concentrated sodium-rich brine during and after the final evaporation of the prehistoric lake. Preservation of organic material indicates that the environment was reducing. Both temperature and pressure were low. Among the authigenic minerals formed are sodic pyroxenes and blue amphiboles. MILTON and EUGSTER included in their paper a photomicrograph of blue amphibole overgrowths on an older amphibole in an altered tuff of the Green River formation. The relationship is strikingly similar to blue amphibole overgrowths on older amphiboles in some glaucophane-lawsonite rocks.

Supporting Evidence for Chemical Control of Aluminum Substitution

There is general acceptance of the concept that both high pressure and low temperature should favor the formation of silicates low in tetrahedral aluminum. This paper has stressed the possibility that purely chemical effects also may be major factors controlling the structural position of aluminum, and the question arises as to whether there is any direct evidence to support the hypothesis.

Mineralogical relationships reported by SURDAM (1968) for low-grade metamorphic rocks of the Karmutsen Group, Vancouver Island, British Columbia, have a bearing on this question. Two types of amygdaloidal fillings are present — (1) prehnite and native copper and (2) laumontite and chalcopyrite. In one hand specimen adjacent amygdules have different fillings. The difference appears to be due to the presence or absence of altered opaque oxide grains in the vicinity of the amygdule, presumably altering the local chemistry of the pore fluid. The significant point for this discussion is that although prehnite and laumontite are both hydrous silicates of aluminum and calcium, prehnite contains only small amounts of tetrahedral aluminum, whereas in laumontite almost all the aluminum is tetrahedral. SURDAMs evidence supports nearly contemporaneous filling of the amygdules. Thus the P-T conditions were the same for both types of fillings, but the local chemical environment was a factor in controlling the structural position of aluminum.

Aragonite

As discussed earlier, the nucleation and growth of calcite and aragonite is sensitive to the chemical environment. McCAULEY and ROY (1966) showed that pH and concentration of reactants control which polymorph will form. The question as to whether the oxidation potential of the fluid can also influence the stability of calcium carbonate polymorphs must await experimental studies. BUCKLEY (1951, p. 385—386) stated that electrolytic pretreatment of water seems to cause changes in the solution that promotes the removal of boiler scale as aragonite, but the process is not understood.

The low-grade aragonite marbles described by VANCE (1968) demonstrate that high pressure is not necessary for the growth of metamorphic aragonite. KUNZLER (1969, and personal communication) studied the isothermal inversion of aragonite to calcite. Most of KUNZLERs work was on organic aragonite, but the one sample of metamorphic aragonite from the Franciscan formation showed unusual behavior in two respects: (1) The reaction curve was unique with anomalous peaks that were not present for other aragonite specimens. (2) The metamorphic aragonite had the greatest thermal stability. Temperatures of 435—455° C were necessary to obtain measurable rates on metamorphic aragonite, whereas most specimens yielded measurable rates in the range of 350 to 380° C. These observations suggest that the stability of aragonite is still incompletely understood.

If the formation of aragonite is sensitive to reducing conditions, then its formation in blueschists possibly could be attributed to a catalytic mechanism.

Broader Implications for Regional Blueschist Terranes

COLEMAN (1967 b, p. 482) observed that there is a nearly universal association of blueschist and ultramafic rocks in the Circum-Pacific orogenic belts. I am not worried by the fact that certain individual serpentinites are not in direct contact with blueschists, and vice-versa. In the tectonic environment this is to be expected. For example, BLAKE et al. (1967) described the presence of a belt of blueschist facies rocks located along the sole of a regional thrust fault in northwestern California and southwestern Oregon. Although serpentine is not present everywhere along the fault, in general a sheet of partially serpentinized ultramafic rock resides in the plane of the thrust fault. It is quite possible that at any given locality where blueschists are found without serpentine, an ultramafic body may have been present at the time of metamorphism but may have been subsequently sliced out by continuing movement on the fault. DAVIS (1968, p. 915 and map) reported a similar belt of blueschists in the Klamath Mountains of northern California. The rocks lie within a mile of ultramafic rocks emplaced along a thrust fault.

ROUTHIER (1953) and COLEMAN (1967 b) cited New Caledonia as a metamorphic terrane in which a clear regional transition from greenschists to blueschists is exposed. The extreme tectonic intermixing typical of the Franciscan terrane is absent, although deformation has occurred. The description by Routhier of relationships between sericite schists, greenschists, and blueschists (p. 161—162) suggests that the transition to "glaucophanites" is controlled by elevation, with the blueschist rocks at the higher elevation. There is also evidence that a thrust fault along which the New Caledonian ultramafic massif was emplaced had existed above the metamorphic rocks prior to erosion (AVIAS, 1967). Thus the relationships between ultramafic and blueschist rocks may be similar to those described by BLAKE et al. (1967) (see above).

Chemical "Fronts" and the Chemical "Driving Force"

The term "glaucophane front" was introduced by ROUTHIER (1953) for the New Caledonian metamorphic belt. Although he intended it in the sense of

"isograd" (footnote p. 152; p. 161), and although he spoke of sodium metasomatism as a local phenomenon (p. 153), ROUTHIERs use of the term has been
misinterpreted as meaning a chemical metasomatic front. The hypotheses expressed
in this paper would explain the glaucophane front as a chemical front, but not
necessarily requiring large-scale migration of ions. The term "reducing front"
might be better than "glaucophane front". The metamorphism could be largely
isochemical, with local metasomatic effects, as the pore fluid becomes reducing
over an extensive volume of rock. Electrochemical communication via the pore
fluid should be transmitted rather easily over significant distances. A chemical
"front" involving electron transfer is more plausible than other types of
chemical "fronts", involving large-scale migration of ions, that have been postulated from time to time to explain the petrogenesis of various rock types.

The ultimate "drive" for this process is the large negative free energy change
of serpentinization. The initial overwhelming disequilibrium is removed by
converting anhydrous magnesium silicates to hydrous magnesium silicates. In the
process, iron must be partially oxidized so that it can be accommodated in
a stable phase. This step requires the expenditure of work (energy), which is
supplied by the free energy change of serpentinization. The pore fluids are
activated, and they must re-equilibrate with the surrounding rocks.

Summary

1. Saline pore fluids in the country rock become concentrated and reducing
during serpentinization of ultramafic rocks.

2. Because of surface chemical effects existing between the pore fluid and
solids, growth units in the pore fluid are pre-disposed toward silicate structures
with a high degree of polarization of oxygen ions, particularly chain silicates.
These effects express themselves by the avoidance of minerals containing
tetrahedral aluminum.

3. Reducing processes are important, particularly in metasomatic blueschists.

4. Other fluid equilibria are important, particularly the Mg^{+2}/H^+ ratio of the
pore fluid, which can change during serpentinization and which affects the
jadeite-glaucophane equilibrium.

5. The catalytic effects explain the widely and uniformly developed isochemical
phenomena. The oxidation-reduction and proton-transfer reactions explain metasomatic alterations in which local transitions may be abrupt.

By this model, some blueschist facies minerals may form outside of their
P-T stability fields. It is debatable whether this should be referred to as
"metastable formation", or whether the minerals should be considered as stable
under the special chemical conditions that prevail. The fact that the eugeosynclinal
environment is one of low geothermal gradient (relatively high P/T) certainly
would favor the formation of these minerals because they would be close to their
P-T stability fields. Because of this, the possibility of forming the minerals
metastably or the possibility that the P-T stability fields are sufficiently altered
under the special chemical conditions so that the minerals can form stably is
enhanced.

I would not advocate that all blueschists are the result of the processes described in this paper. Many rock types are polygenetic. However, I do believe that Franciscan-type blueschists may be produced by special chemical conditions in the range of temperature and pressure generally typical of lowgrade metamorphism, without the necessity of exceptionally high pressure.

Acknowledgments. My work on the Franciscan blueschists began while I was a postdoctoral fellow. I am grateful to the National Science Foundation and the University of Southern California for providing this opportunity. In particular, discussions and field excursions with Dr. GREGORY DAVIS stimulated my interest in the problem.

The manuscript has been improved by the critical reading and suggestions of GREGORY DAVIS, GEORGE DeVORE, PETER MISCH, BATES McKEE, JOSEPH VANCE, NIKOLAS CHRISTENSEN, EDWIN BROWN, and RANDALL BABCOCK. However, the more radical views expressed herein remain entirely my own.

References

AVIAS, J.: Overthrust structure of the main ultrabasic New Caledonian massives. Tectonophysics **4**, 531—541 (1967).

BAILEY, E. H., W. P. IRWIN, and D. L. JONES: Franciscan and related rocks, and their significance in the geology of western California. Calif. Div. Mines and Geology, Bull. **183** (1964).

BILGRAMI, S. A., and R. A. HOWIE: The mineralogy and petrology of a rodingite dike, Hindubagh, Pakistan. Am. Mineralogist **45**, 791—801 (1960).

BLAKE, M. C., JR., W. P. IRWIN, and R. G. COLEMAN: Upside-down metamorphic zonation, blueschist facies, along a regional thrust in California and Oregon. U. S. Geol. Surv. Profess. Paper **575**-C, p. C1—C9 (1967).

BUCKLEY, H. E.: Crystal growth. New York: John Wiley & Sons, Inc. 1951.

ČERNÝ, P.: Comments on serpentinization and related metasomatism. Am. Mineralogist **53**, 1377—1385 (1968).

CHESTERMAN, C. W.: Intrusive ultrabasic rocks and their metamorphic relationships at Leech Lake Mountain, Mendocino County, California: Internat. Geol. Congr. 21st, Copenhagen 1960, Proc., pt. 13, p. 208—215 (1960).

COLEMAN, R. G.: Jadeite deposits of the Clear Creek Area, New Idria District, San Benito County, California. J. Petrol. **2**, 209—247 (1961).

— Low temperature reaction zones and alpine ultramafic rocks of California, Oregon and Washington. U. S. Geol. Surv. Bull. **1247** (1967a).

— Glaucophane schists from California and New Caledonia. Tectonophysics **4**, 479—498 (1967b).

—, and D. E. LEE: Metamorphic aragonite in the glaucophane schists of Cazadero, California. Am. J. Sci. **260**, 577—595 (1962).

— — Glaucophane-bearing metamorphic rock types of the Cazadero area, California J. Petrol. **4**, 260—301 (1963).

— —, L. B. BEATTY, and W. W. BRANNOCK: Eclogites and eclogites: their differences and similarities. Bull. Geol. Soc. Am. **76**, 483—508 (1965).

CRITTENDEN, M. D., JR.: Geology of the San Jose-Mount Hamilton area, California. Calif. Div. Mines Bull. **157** (1951).

DAVIS, E. F.: The radiolarian cherts of the Franciscan group. Univ. Calif. Pub., Bull. Dept. Geology **11**, 235—432 (1918).

DAVIS, G. A.: Lawsonite and pumpellyite in glaucophane schist, North Berkeley Hills, California, with notes on the x-ray crystallography of lawsonite by A. Pabst. Amer. J. Sci. **258**, 689—704 (1960).

— Westward thrust faulting in the South-Central Klamath Mountains. California. Bull. Geol. Soc. Am. **79**, 911—934 (1968).

8*

Deer, W. A., R. A. Howie, and J. Zussman: Rock-forming minerals, vol. one, Ortho- and ring silicates. London: Longmans, Green & Co., Ltd. 1962 a.
— — — Rock-forming minerals, vol. three. Sheet silicates. London: Longmans, Green & Co., Ltd. 1962 b.
— — — Rock-forming minerals, vol. two, Chain silicates. London: Longmans, Green Co., Ltd. 1963.
DeVore, G. W.: Crystal growth and the distribution of elements. J. Geol. 63, 471—494 (1955).
— Surface chemistry as a chemical control on mineral association. J. Geol. 64, 31—55 (1956 a).
— Al-Si positions in ordered plagioclase feldspars. Z. Krist. 107, 247—264 (1956 b).
— The association of strongly polarizing cations with weakly polarizing cations as a major influence in element distribution, mineral composition, and crystal growth. J. Geol. 65, 178—195 (1957).
— Role of minimum interfacial free energy in determining the macroscopic features of mineral assemblages. I. the model. J. Geol. 67, 211—227 (1959).
— Compositions of silicate surfaces and surface phenomena. Contr. Geology, Univ. Wyoming 2, 21—37 (1963).
Ernst, W. G.: Significance of phengitic micas from low-grade schists: Am. Mineralogist 48, 1357—1373 (1963 a).
— Petrogenesis of glaucophane schists. J. Petrol. 4, 1—30 (1963 b).
— Mineral parageneses in Franciscan metamorphic rocks, Panoche Pass, California. Bull. Geol. Soc. Am. 76, 879—914 (1965).
Essene, E. J., and W. S. Fyfe: Omphacite in Californian metamorphic rocks. Contr. Mineral. and Petrol. 15, 1—23 (1967).
— — and F. J. Turner: Petrogenesis of Franciscan glaucophane schists and associated metamorphic rocks, California. Beitr. Mineral. Petrog. 11, 695—704 (1965).
Eugster, H. P.: Reduction and oxidation in metamorphism. In: P. H. Abelson (ed.), Researches in geochemistry, vol. 1. New York: John Wiley & Sons, Inc., 1959.
Fajans, K.: Radioelements and isotopes: Chemical forces and optical properties of substances. New York: McGraw-Hill Book Co. 1931.
Fyfe, W. S.: Metamorphism in mobile belts: The glaucophane schist problem. Trans. Leicester Lit. and Phil. Soc. 61, 36—54 (1967).
—, and R. Zardini: Metaconglomerate in the Franciscan formation near Pacheco Pass, California. Am. J. Sci. 265, 819—830 (1967).
Ghent, E. D.: Glaucophane schist facies metamorphism in the Black Butte area, northern Coast Ranges, California. Am. J. Sci. 263, 385—400 (1965).
Gresens, R. L.: Composition-volume relationships of metasomatism. Chem. Geology 2, 47—65 (1967).
Helgeson, H. C.: Solution chemistry and metamorphism. In: P. H. Abelson (ed.), Researches in geochemistry, vol. 2. New York: John Wiley & Sons, Inc. 1967.
Hemley, J. J., and W. R. Jones: Chemical aspects of hydrothermal alteration with emphasis on hydrogen metasomatism. Econ. Geol. 59, 538—569 (1964).
Hlabse, T., and O. J. Kleppa: The thermochemistry of jadeite. Am. Mineralogist 53, 1281—1292 (1968).
Hostetler, P. B., R. G. Coleman, F. A. Mumpton, and B. W. Evans: Brucite in Alpine serpentinites. Am. Mineralogist 51, 75—98 (1966).
Korzhinskii, D. S.: Physicochemical basis of the analysis of the paragenesis of minerals (translated from the Russian edition, 1957). New York: Consultants Bureau, Inc. 1959.
Kunzler, R. H.: The aragonite-calcite transformation: Unpub. Ph. D. thesis, Florida St Univ. (Contr. No. 29, Sedimentological Research Laboratory, Florida St. Univ. 1969).
Lacy, E. D.: Factors in the study of metamorphic reaction rates. In: W. S. Pitcher and G. W. Flinn (eds), Controls of metamorphism. New York: John Wiley & Sons, Inc. 1965.
McCauley, J. W., and R. Roy: The effect of pH and concentration of reactants on the crystal growth of $CaCO_3$ (abs). Program, Geol. Soc. Am. Ann. Meet., San Francisco, p. 136 (1966).
McKee, B.: Widespread occurrence of jadeite, lawsonite, and glaucophane in Central California. Am. J. Sci. 260, 596—610 (1962).

MILTON, CH., and H. P. EUGSTER: Mineral assemblages of the Green River formation. In: P. H. Abelson (ed), Researches in geochemistry. New York: John Wiley & Sons, 1959.

MUFFLER, L. J. P., and D. E. WHITE: Active metamorphism of upper Cenozoic sediments in the Salton Sea geothermal field and the Salton trough, southeastern California. Bull. Geol. Soc. Am. **80**, 157—182 (1969).

PAGE, N. J.: Serpentinization at Burro Mountain, California. Contr. Mineral. and Petrol. **14**, 321—342 (1967a).

— Serpentinization considered as a constant volume metasomatic process: a discussion. Am. Mineralogist **52**, 545—549 (1967b).

— Serpentinization in a sheared serpentinite lens, Tiburon Peninsula, California. U. S. Geol. Surv. Profess. Paper **600-B**, B21—B28 (1968).

RAMBERG, H.: Chemical bonds and distribution of cations in silicates. J. Geol. **60**, 331—355 (1952).

—, and G. W. DeVORE: The distribution of Fe^{+2} and Mg^{+2} in coexisting olivines and pyroxenes. J. Geol. **59**, 193—210 (1951).

RAST, N.: Nucleation and growth of metamorphic minerals. In: W. S. Pitcher and G. W. Flinn (eds), Controls of metamorphism. New York: John Wiley & Sons, Inc. 1965.

ROEDDER, E.: Ancient fluids in crystals. Sci. Am. **207**, 38—47 (1962).

ROUTHIER, P.: Étude géologique du versant occidental de la Nouvelle-Calédonie entre le Cola de Boghen et la Pointe d'Arama. Mem. Soc. Geol. France **67**, 1—271 (1953).

RUTLAND, R. W. R.: Tectonic overpressures. In: W. S. Pitcher and G. W. Flinn (eds), Controls of metamorphism. New York: John Wiley & Sons, Inc. 1965.

SURDAM, R. C.: Origin of native copper and hematite in the Karmutsen Group, Vancouver Island, B. C. Econ. Geol. **63**, 961—966 (1968).

SWITZER, G.: Eclogite from the California glaucophane schists. Am. J. Sci. **243**, 1—8 (1945).

TALIAFERRO, N. L.: Geologic history and structure of the central coast ranges of California. Calif. Div. Mines Bull. **118**, 119—164 (1941).

— Geologic history and correlation of the Jurassic of southwestern Oregon and California. Bull. Geol. Soc. Am. **53**, 71—112 (1942).

— Franciscan-Knoxville problem. Bull. Am. Assoc. Petrol. Geologists **27**, 109—219 (1943).

TAYLOR, H. P., JR. A. L. ALBEE, and S. EPSTEIN: O^{18}/O^{16} ratios of coexisting minerals in three assemblages of kyanite-zone pelitic schist. J. Geol. **71**, 513—522 (1963).

—, and R. G. COLEMAN: O^{18}/O^{16} ratios of coexisting minerals in glaucophane-bearing metamorphic rocks. Bull. Geol. Soc. Am. **79**, 1727—1756 (1968).

THAYER, T. P.: Serpentinization considered as a constant-volume metasomatic process. Am. Mineralogist **51**, 685—710 (1966).

VANCE, J. A.: Metamorphic aragonite in the prehnite-pumpellyite facies, Northwest Washington. Am. J. Sci. **266**, 299—315 (1968).

WEYL, W. A.: Wetting of solids as influenced by the polarizability of surface ions. In: R. Gomer, and C. S. Smith (eds), Structure and properties of solid surfaces. Chicago: Chicago Univ. Press 1953.

— Application of the screening theory to chemical reactions involving non-metallic solids. Inst. Internat. Chem. Solvay Conseil Chim, Brussels, p. 401—458 (1956).

—, and E. C. MARBOE: The constitution of glasses. A dynamic interpretation, vol. 1, Fundamentals of the structure of inorganic liquids and solids. New York: Interscience: Wiley 1962.

Dr. R. L. GRESENS
Department of Geology
University of Washington
Seattle, Washington 98105, U.S.A.

20

Reprinted from *Amer. Jour. Sci.*, **270**, 81–108 (Feb. 1971)

DO MINERAL PARAGENESES REFLECT UNUSUALLY HIGH-PRESSURE CONDITIONS OF FRANCISCAN METAMORPHISM?

W. G. ERNST

Department of Geology and Institute of
Geophysics and Planetary Physics,
University of California, Los Angeles, California 90024

ABSTRACT. Metamorphosed graywacke, shale, chert, and mafic volcanic rocks of the Franciscan group of California display low-grade mineral parageneses absent from metamorphic terranes of more continental affinities. Characteristic phases include glaucophane-crossite, jadeitic pyroxene, lawsonite, pumpellyite, and aragonite. Explanations of this distinctive metamorphism fall into two groups: either metasomatism or metastable recrystallization taking place under normal low-grade conditions; or low-temperature production at relatively high pressures resulting from either tectonic overpressures or very deep burial.

Although metasomatic effects are not unknown, existence of metamorphic rock compositions virtually indistinguishable from those of protoliths proves the genetic insignificance of chemical change. Metastable recrystallization is of general importance in metamorphism; however, growth of jadeitic pyroxene + quartz from albite of low structural state argues against a metastable low-pressure origin for Diablo Range metagraywackes. Experimentally determined rock strengths preclude tectonic overpressures exceeding about 1 kb for the "strong" metagraywacke during recrystallization in the presence of an aqueous fluid phase; interbedded jadeitic pyroxene-bearing metashale is even weaker. Phase equilibrium relations obtained in the laboratory are compatible with observed parageneses and suggest lithostatic pressure of 5 to more than 8 kb for the inferred 150° to 300°C temperature range. The required 20 to 30 + km depth of metamorphism is in accord with the hypothesis of Late Mesozoic accumulation and tectonic thickening of the Franciscan in one or a series of oceanic trenches contemporaneously being overridden by the North American lithospheric plate along a Benioff (subduction) zone and with geothermal gradients computed for the descending plate. Imbricate thrusting of more deeply buried, higher pressure metamorphosed rocks on the east over less intensely recrystallized, more westerly sections seems to account for some of the observed structural, temporal, and paragenetic features within the Franciscan terrane.

The subduction zone setting thus accounts for the development of relatively high-pressure, low-temperature blueschist facies mineral assemblages. Postulated alternative origins are not as obviously related to continental margins, and, were one of them to be accepted, it would be necessary to regard the striking world-wide restriction of glaucophane schists and associated rocks to plate junctions as coincidental.

STATEMENT OF THE PROBLEM

Feebly to thoroughly metamorphosed, chiefly clastic sedimentary rocks of the Franciscan group crop out extensively in the California Coast Ranges and more sparsely in Baja California and its off-shore islands as well as in southwestern Oregon. Although some parts of the terrane contain relatively coherent sections of strata, other associated units are lithologic mélanges (Hsü, 1968), due both to sedimentary slumping and tectonic disruption. Because of the paucity of fossils and the lack of widespread marker horizons, the structural relations and

81

both stratigraphic and tectonic thicknesses of the Franciscan are incompletely known. This metamorphosed—and in part chaotically deformed—eugeosynclinal terrane has been thrust beneath the roughly contemporaneous, less deformed, only feebly recrystallized Great Valley sequence (for example, see Irwin, 1960, 1964, 1966; Dickinson, 1966; Blake and others, 1967, 1969; Bailey and Blake, 1969; Page, 1970). The tectonic juxtaposition evidently took place adjacent to and within a Late Mesozoic subduction zone marking the margins of North American and Pacific lithospheric plates (Hamilton, 1969; Ernst, 1970; Page, 1970). For descriptions of the lithologies and the regional geologic relationships, the reader is referred to the detailed study of the California Coast Ranges by Bailey, Irwin, and Jones (1964). The areal extent of these units is shown in figure 1.

In situ metamorphosed Franciscan rocks exhibit a systematic set of characteristic mineral assemblages which, to some extent, are areally distinct and hence are inferred to reflect a progressive metamorphic sequence. Quartz-, phengite-, and chlorite-bearing metaclastic rocks, principally graywackes, micrograywackes, and black shales, display the critical phase compatibilities:

    A. laumontite + albite ± calcite;

    B. pumpellyite and/or lawsonite + albite ± calcite;

    C. lawsonite ± pumpellyite + albite ± aragonite;

    D. lawsonite + jadeitic pyroxene ± aragonite.

The less abundant, interlayered, pod-like masses and somewhat larger piles of metamorphosed pillow lavas concomitantly have developed characteristic zeolitized and pumpellyitized greenstone and lawsonite + blue amphibole and/or omphacite-bearing assemblages. The metamorphic petrology of a number of specific areas has been described by many workers (for example, Bloxam, 1956, 1959, 1960; McKee, 1962a, 1962b; Coleman, 1961, 1965, 1967a; Coleman and Lee, 1962, 1963; Ghent, 1965; Ernst, 1965, in press; Ernst and Seki, 1967; Ernst and others, 1970, chaps. 3 and 4; Blake, Irwin, and Coleman, 1967, 1969; Blake and Cotton, 1969; Bailey and Blake, 1969; Raymond, 1970).

The origin of these phase assemblages of low metamorphic grade has puzzled petrologists since before the turn of the century (a general summary has been presented by Turner, 1968, p. 289-295). Basically, explanations regarding the petrogenesis of mafic blueschists, jadeitic pyroxene-bearing metaclastics, and their mineralogically less exotic associated lithologies can be divided into two groups. (1) The phase assemblages may reflect low-temperature metasomatism accompanying the emplacement and serpentinization of alpine-type peridotites common in most glaucophane schist terranes, or they may represent metastable recrystallization. (2) Alternatively, the dense minerals—such as sodic amphibole, sodic pyroxene, lawsonite, and aragonite, which have been demonstrated in the laboratory to be favored by high pressures—are postulated to have formed during low-temperature recrystallization under high pressures resulting from deep burial or due to the generation of tectonic overpressures substantially exceeding the lithostatic value. Four different schools of thought have thus evolved, two of which favor relatively low-pressure recrystallization, the other two relatively high-pressure recrystallization.

Taliaferro (1943, p. 159-182) ascribed the formation of glaucophane schists and related rocks in the California Coast Ranges to metasomatism accompanying the intrusion of ultramafic plutons. Other workers have also been impressed by the nearly world-wide association of blueschists with alpine-type peridotites and by the peculiar bulk compositions of some glaucophane schists. For these reasons, many investigators of Franciscan mineral parageneses (for example, Brothers, 1954; Chesterman, 1960; Bloxam, 1966; Fyfe and Zardini, 1967; Gresens, 1969) have invoked metasomatism or at least catalytic interaction of an active pore solution with condensed assemblages as the causative agent in the production of glaucophane schists and kindred metamorphics (see also Essene, Fyfe, and Turner, 1965). Most, but not all, have linked the fluid emanations to nearby mafic or ultramafic bodies.

There is no question that the effects of metasomatism are present in metamorphosed Franciscan rocks and that in some cases exchange of material has taken place between serpentinized peridotites and wall rocks or inclusions; examples have been well documented, for instance, by Ransome (1894, p. 222-226), Davis (1918, p. 275-278), Bloxam (1960-1966), Coleman (1961, 1967b, p. 28-31), and Barnes and O'Neil (1969). The fundamental question, however, is not whether local or regional metasomatism has taken place, but whether chemical exchanges between preexisting rocks and a chemically active pore fluid are required to produce the blueschist mineralogic suite. One may further inquire whether the presence of alpine-type serpentinized peridotites is necessary for the formation of such lithologies.

The answer to the second question is most probably negative. On a global basis, it is true that glaucophane schists and ultramafics commonly are spatially associated, as in the Alps, the Caribbean, and New Caledonia, as well as the California Coast Ranges. However, the near absence of coeval serpentinized peridotites in the Shuksan belt of blue amphibole-bearing schists in Washington State (Misch, 1959, 1966, 1969), in the glaucophanitic metabasalt + phyllite complex of Calabria, southern Italy (Hoffman, 1970), and the very limited occurrence of ultramafics in the Sanbagawa glaucophane schist belt of southwestern Japan (Saito and others, 1960, geologic map; see also Miyashiro, 1966, fig. 3) all argue against a genetic link between ultramafic masses and glaucophane schists. Moreover, the extensive emplacement of ultramafic plutons in some orogenic areas, such as the Appalachian belt, evidently has failed to produce blueschist-type phase assemblages. As is discussed below, the general association of blueschists and serpentinized ultramafic rocks may be the

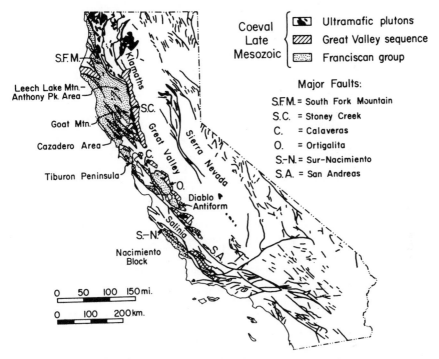

Fig. 1. Regional distribution of Franciscan group and Great Valley sequence rocks in the California Coast Ranges. Note that although the depositional ages of the largely sedimentary and metasedimentary series are reasonably well known, the "age" of the cold ultramafic plutons is considered to be the time of emplacement (California Coast Ranges only). The South Fork Mountain, Stoney Creek, Ortigalita, and Sur-Nacimiento faults are viewed as segments of the Coast Range thrust (Bailey, Blake, and Jones, 1970). The juxtaposition of Salinia, Sierran-type basement, and the Nacimiento Block of Franciscan-type basement with the Diablo Range is thought to be the result of large scale Tertiary strike-slip motion along the San Andreas fault (Hill and Dibblee, 1953).

result, rather, of tectonic interaction between crust and mantle at the suture zone marking convergence of lithospheric plates.

To carry the argument specifically to the Franciscan parageneses, the most remarkably recrystallized portion of this terrane with regard to mineralogy is exposed in the Diablo antiform (Bloxam, 1956; McKee, 1962a, 1962b; Ernst, 1965, in press; Ernst and Seki, 1967; Ernst and others, 1970; Bailey and Blake, 1969); except for the Red Mountain sill[1] (Maddock, 1964, Saad, 1969) this portion of the eugeosynclinal mélange is characterized by a near absence of alpine-type peridotites (for example, see fig. 4; Bailey, Irwin, and Jones, 1964, pl. 1). Those serpentinites that do occur are extremely small bodies and clearly are not commensurate with the areal extent of thoroughly recrystallized metamorphic

_____

[1] Diablo Range localities mentioned in the text are indicated in figure 4, those exclusive of the Diablo Range are located in figure 1.

rocks.[2] Bailey and Blake (1969) have made a similar observation for the northern Coast Ranges.

Furthermore, viewing the Franciscan terrane as a whole, ultramafic rocks are exposed chiefly along the west side of the Great Valley, particularly marking the Stoney Creek fault zone (see fig. 1), and although the underlying eugeosynclinal assemblage has been converted in many places to blueschist-type compatibilities, the tectonically overlying Great Valley strata are either weakly laumontized or are practically unmetamorphosed. According to Bailey, Irwin, and Jones (1970), the serpentinized peridotites are the lowest member of the overthrust plate, hence could not have interacted metsomatically with the Franciscan rocks without similarly affecting the Great Valley section. Moreover, as will be described in the section dealing with tectonic overpressures, the Great Valley (+ basal ultramafic) and Franciscan terranes in general have been juxtaposed too late (60-90 m.y. B.P.) to account for an early period (120-150 m.y. ago) of relatively high-pressure Franciscan recrystallization.

Of the authors who have invoked metasomatism associated with serpentinized peridotites, only Taliaferro (1943, figs. 4-6) provided diagrammatic sketches of the inferred field relations; the metamorphic rocks described as bordering a serpentinite sill on Tiburon Peninsula, San Francisco Bay, are now known to be tectonic blocks and not parts of a distinct aureole (Dudley, ms). No maps purporting to show metamorphic aureoles surrounding similar plutons have even been published.

We may conclude that a *close spatial correlation between ultramafic plutons and Franciscan metamorphic rocks does not exist.*

The more general question of regional metasomatism still remains. At the turn of the century, Washington (1901) and Smith (1907, p. 224-240) clearly demonstrated that among blueschists from various terranes, and the Franciscan in particular, rock bulk compositions need not have changed significantly during the recrystallization. Based on the equivalence of bulk chemistry and on the distinctive phase petrology, Eskola (1939) recognized blueschists and kindred lithologic assemblages as representatives of a separate mineral facies, the glaucophane schist (or blueschist) facies.

With regard to the Franciscan, it is clear that some metavolcanics and metasediments have developed the characteristic glaucophane schist phase assemblages without substantial introduction or removal of material (see chemical analyses presented by Coleman and Lee, 1963, tables 2-4; Coleman, 1965, table 4; Bailey, Irwin, and Jones, 1964, tables 1, 4, 14; Ernst, 1965, tables 10, 12, in press, table 4; Ernst and others, 1970, tables 1, 2, 5-9). The isochemical nature of the metamorphism of certain rocks of metabasaltic composition has been illustrated previously (Ernst, 1959, fig. 26, 1963a, fig. 1; Coleman and Lee, 1963, fig. 19). However, metasedimentary rocks are far more abundant than the mafic meta-

---

[2] In the Panoche Pass area of the Diablo Range (Ernst, 1965, pl. 1), one of three mapped patches of jadeitic metagraywacke does seem to be associated fortuitously with serpentinite (no metasomatic effects were observed in the country rock adjacent to this body), but the other two jadeitic patches definitely are not.

METAGRAYWACKES

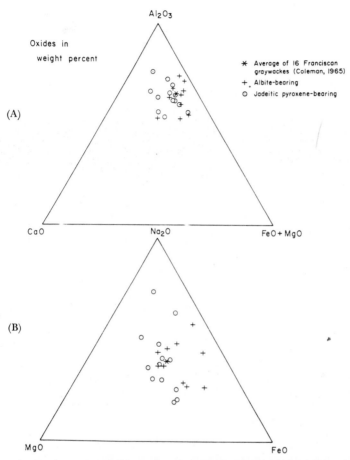

Fig. 2. Chemical range of Franciscan metagraywackes with regard to the oxides: (A) Al₂O₃, CaO, FeO + MgO; and (B), Na₂O, MgO, FeO. Albite-bearing metagraywacke gravimetric analyses are from Ernst (1965, table 12) and Ernst and others (1970, tables 2, 6, and 7); jadeitic pyroxene-bearing metagraywacke gravimetric analyses are from Bloxam (1956, table 2; 1960, table 1), Coleman (1965, table 4), Ernst (1965, table 12), and Ernst and others (1970, tables 6 and 7).

volcanics in the California Coast Ranges, and although the megascopic evidence of recrystallization is not as obvious, a striking metamorphic mineralogy has been produced; thus, because of their volumetric importance, it is critical for the discussion to determine whether or not these rocks exhibit the effects of chemical exchange.

A range of metaclastic bulk compositions as a function of phase petrology is presented in this report as figures 2 and 3; although not complete, a broad overlap in chemical ranges is obvious. For such rocks, physical rather than chemical differences must have produced the con-

METAGRAYWACKES

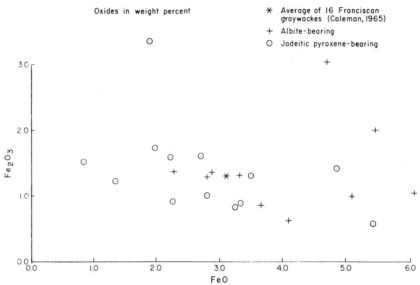

Fig. 3. Proportions of FeO and Fe$_2$O$_3$ for conventionally analyzed Franciscan metagraywackes; samples are the same as those illustrated in figure 2.

trasting assemblages. Even in outcrops exhibiting monomineralic veins and other evidence of local metasomatism, no new phases absent from the host rock have been formed. Thus, although metasomatic effects are apparent in some exposures, *chemical changes cannot be held responsible for generation of the distinctive mineralogy of blueschists and related rocks* but merely account for the observed differences in the phase proportions, which necessarily reflect bulk chemistry.

METASTABLE RECRYSTALLIZATION OF OBSERVED ASSEMBLAGES

The problem of metastability is encountered in the investigation of virtually all metamorphic processes. Undoubtedly this phenomenon is of major significance in considering low-temperature parageneses, for reaction rates under such conditions are of critical importance in determining the exact nature of the assemblage produced from a relatively unstable protolith. The principal constraint operative in such a recrystallization process is that under the metamorphic conditions the observed phase associations must have possessed a lower total Gibbs free energy than the initial reacting material. Several different varieties of metastable recrystallization can be envisioned:

1. low-pressure production of metastable blueschist-type phase compatibilities from initial mineral assemblages of even higher Gibbs free energy;

2. low-pressure stability of what would otherwise be a metastable blueschist assemblage as a result of greatly increased surface

energy (hence molar G) of the low-pressure, inherently stable phase association due to a particularly active pore fluid;

3. high-pressure persistence of metastable low-pressure phase assemblages except where reaction rates were sufficient to allow their conversion to more stable blueschist-type compatibilities.

The first two explanations obviate the necessity of postulating the attendance of high pressures during Franciscan metamorphism, whereas the third ascribes the somewhat irregular development of glaucophane schists and allied rocks to local catalysis under more uniform blueschist facies conditions. Let us consider each of these three hypotheses in turn.

1. Fyfe and Newton (in press) have advocated a proposal, advanced originally by Hlabse and Kleppa (1968), that accounts for the apparently high-pressure production of jadeitic pyroxene in metagraywackes by the breakdown of metastable high albite at moderate pressures. Calorimetric work by Hlabse and Kleppa showed that albite of high structural state (that is, Al and Si disordered among the tetrahedral sites) possesses a molar entropy exceeding that of ordered, low albite by 3.5 entropy units. At modest temperatures, the stable equilibrium jadeite + quartz = low albite lies at pressures 2 to 3 kb higher than the metastable equilibrium jadeite + quartz = high albite (compare Newton and Smith, 1967, fig. 6, with Hlabse and Kleppa, 1968, fig. 1).

In what manner could metastable disordered albite possibly have been present in the Franciscan rocks? Either it might have accumulated as clastic grains of volcanic albite, or it might have represented an intermediate stage in the recrystallization of more calcic detrital plagioclase of various structural states. An implication of the first alternative concerns the possible provenance of high albite: although the Klamath and Sierran terranes easily could have been the source of great amounts of largely plutonic, ordered, Ca-bearing plagioclase as well as volcanic, somewhat disordered calcic plagioclase derived from the superjacent extrusives, occurrences of albite-bearing rhyolites required in vast quantities by the first alternative are practically non-existent in the geologic record of California. X-ray work presented by Seki, Ernst, and Onuki (1969, tables 19 and 20) for 83 samples of Franciscan metagraywackes and metavolcanic rocks indicates that sodic plagioclases from the Diablo Range and its northern extension are uniformly of low structural state. Such feldspars exhibit textural evidence of an early stage of albitization of preexisting detrital grains of intermediate plagioclase, hence the second alternative must be considered.

In general, if intermediate plagioclase recrystallizes to disordered $NaAlSi_3O_8$ as a transitional stage preceding the formation of low albite, it is strange that high albite has not been described from a variety of graywackes and arkoses (for example, see Hawkins, 1967, table 5), nor has jadeitic pyroxene been widely reported as a constituent of metamorphosed feldspathic sandstones. A far commoner situation involves the observed production of ordered albite from more calcic plagioclase, as is true of the more deeply buried portions of the Great Valley sequence, reported by Dickinson, Ojakangas, and Stewart (1969). Inasmuch as the Franciscan and Great Valley strata seem to have had a common source, it is also unlikely that occurrences of jadeitic pyroxene derived from metastable high albite would be confined to the former sequence.

Photomicrographs illustrating the partial or complete replacement of preexisting sodic plagioclase in Franciscan clastic rocks are presented in plate 1. A complex, fine-grained intergrowth of jadeitic pyroxene + minor lawsonite + quartz which faithfully pseudomorphs plutonic—hence ordered—sodic plagioclase subhedra in a granitic pebble from a Panoche Pass metaconglomerate is shown in plate 1-A (Ernst, 1965). A similar metaconglomerate has been studied west of Pacheco Pass (see also Fyfe and Zardini, 1967); here the degree of replacement of plutonic albite in the granitic pebbles is incomplete (see samples X-39, X-30A, X-39B, X-40, X-92, X-93B, X-93C, X-93D, X-207A of Seki, Ernst, and Onuki, 1969, tables 6 and 20; Ernst and others, 1970, fig. 9). In plate 1-B clastic grains of albitized plagioclase exhibit rare acicular sprays of jadeitic pyroxene in a metagraywacke that crops out near the Calaveras Reservoir (Ernst, in press). Like other Diablo Range metasedimentary rocks, these Pacheco Pass and Calaveras Reservoir sodic plagioclases have refractive indices less than that of Canada balsam; whole rock X-ray examination of the illustrated specimens yields $Cu_{K\alpha}$ two θ separations between 131 and 1$\overline{3}$1 peaks of 1.11 and 1.15 respectively, indicating the presence of virtually pure, ordered albite.

Plate 1 also illustrates the general lack of shearing observed in some jadeitic pyroxene-bearing metaclastics, particularly in rocks of the Diablo Range; in many of these samples, recrystallization has taken place with preservation of the original detrital textures. Therefore, accumulated strain cannot be called upon to have promoted mechanically-induced twinning in albite (which thereby would have attained an elevated molar Gibbs free energy relative to strain-free albite).

In summary, *there is no evidence to suggest that disordered albite, or highly strained albite, was an important constituent of Franciscan metaclastic rocks at the time of intense metamorphism. All available data indicate the presence of ordered, low albite as in the Great Valley section; hence for jadeitic pyroxene-bearing metagraywackes of the Franciscan, the appropriate equilibrium to be considered is the one investigated experimentally by Newton and Smith (1967), rather than the lower pressure metastable reaction.*

The occurrence of apparently high-pressure aragonite in Franciscan metamorphic rocks has also been discussed by Ernst and others (1970, chaps. 3, 4, and 13) and by Fyfe and Newton (in press). These authors have pointed out that several processes may lead to metastable formation of the orthorhombic polymorph. These include: (A) precipitation of aragonite from a solution that has become supersaturated with respect to $CaCO_3$ due to the presence of cations such as Mg which inhibit growth of the stable polymorph (see Wray and Daniels, 1957; Bischoff and Fyfe,

PLATE 1

Photomicrographs of albite-jadeitic pyroxene textural relations in Diablo Range

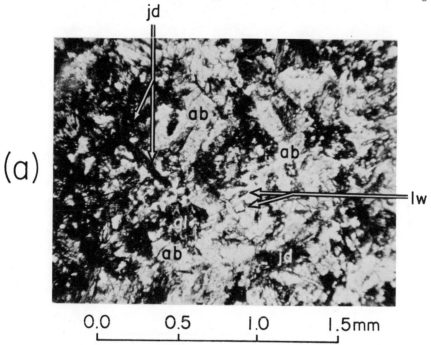

A. Granitic pebble in metaconglomerate from Panoche Pass (sample no. 185). Although the original microscopic examination failed to reveal jadeitic pyroxene, this rock lies within the so-called "jadeitic pyroxene + lawsonite isograd" (Ernst, 1965, table 8 and pl. 1), and the absence of both sodic plagioclase and sodic pyroxene was puzzling (glaucophane is abundant in this pebble, however). Reexamination of the sample during the course of the present study revealed that pseudomorphs after subhedral plutonic plagioclase consist of fine-grained intergrowths of jadeitic pyroxene, lawsonite, and quartz. Plane light.

1968); (B) production of aragonite within the calcite I stability field by metastable inversion of calcite II (Boettcher and Wyllie, 1967, 1968a); and (C) growth of aragonite from strained—hence elevated molar G—calcite within the (strain-free) calcite stability field (Newton, Goldsmith, and Smith, 1969). There is no doubt that each of these mechanisms may be operative in a specific geologic environment. However, Franciscan phase assemblages exhibit the systematic progressive changes albite + quartz ± calcite → albite + quartz ± aragonite → jadeitic pyroxene + quartz ± aragonite, precisely the sequence predictable from experimental phase equilibrium studies. Therefore, it seems unnecessary to call on metastable low-pressure crystallization of aragonite in Franciscan metamorphic rocks, especially where associated sodic pyroxene can be shown to have been produced stably at even higher pressures than required for the formation of the aragonite.

PLATE 1

metaclastics. Abbreviations are as follows: ab = albite; gl = glaucophane; jd = jadeitic pyroxene; lw = lawsonite; ms = white mica; q = quartz.

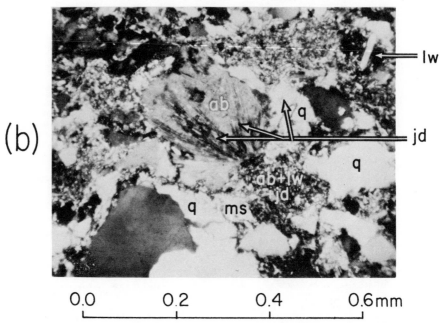

B. Detrital plagioclase grain in metagraywacke (sample no. R-325) from near Calaveras Reservoir (Ernst, in press); the low albite is partly replaced by an acicular spray of jadeitic pyroxene. Crossed nicols.

2. Gresens (1969) has directed attention to a supposed association of *in situ* Franciscan metamorphic rocks with serpentinites and erected a hypothetical solution-recrystallization model based on effects attending hydration of peridotite masses. The process of serpentinization probably results in residual concentration of the saline constituents of an initial country rock pore fluid, because an altering peridotite acts as a desiccating agent in removing H_2O from the surroundings. Oxidation of initially ferrous iron in the preexisting anhydrous ferromagnesian silicates of the ultramafic pluton is thought to have caused a relative lowering of the oxygen fugacity in the aqueous phase. Based on H_2O contents of the involved rock types, crude mass balance calculations indicate that ultramafics and host rocks necessarily would have had to have been present in roughly subequal proportions for the hypothesized effects of serpentinization to have been of more than local importance in altering the composition of the residual interstitial solutions.[3]

[3] Metagraywackes now contain about 3 to 4 wt percent H_2O^+, hydrated ultramafics up to 13 to 14 wt percent H_2O^+. If the original country rocks had lost half their H_2O resulting in only a 2:1 concentration of the residual brines during complete serpentinization of peridotite emplaced in an originally anhydrous condition, then the ultramafic material would have to have been present as an unacceptable 20 to 30 percent of the terrane (see figs. 1 and 4).

Gresens (1969, p. 99-105, and personal commun., 1970) suggests that the resultant saline, reducing hydrothermal fluid might interact with grain surfaces of the originally stable phases such as plagioclase and calcite to produce sodic pyroxene, sodic amphibole, pumpellyite, lawsonite, and aragonite. This process conceivably could take place at low pressures, but only if the surface free energy contribution to the total G of the (formerly) stable assemblage—much of it sand- and silt-sized particles—raised its total Gibbs free energy to a value in excess of that of the metastable blueschist-type compatibility. The hypothesis is as yet untested experimentally. There is no surety that a concentrated, re-

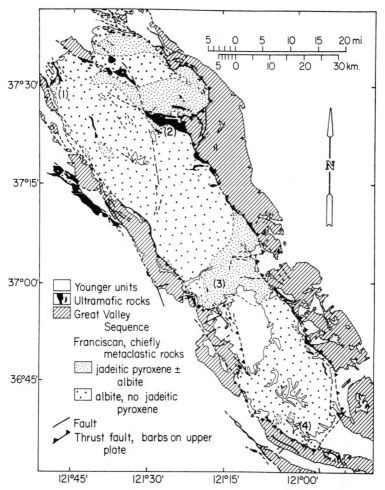

Fig. 4. Distribution of jadeitic pyroxene-bearing and jadeitic pyroxene-free metagraywackes from the Diable Range, California Coast Ranges (simplified from Ernst, in press, fig. 7). Areas mentioned in the text and located on this map include: (1) Calaveras Reservoir; (2) Red Mountain; (3) Pacheco Pass; and (4) Panoche Pass.

ducing brine would in fact significantly raise the total G of albite-bearing phase assemblages, let alone cause it to increase to the extent that such assemblages would attain higher values of total Gibbs free energy than metastable glaucophane schist-type compatibilities for the same bulk composition.

The petrogenetic model is invoked to account for an alleged spatial and genetic association between serpentinized peridotites and the rather unique Franciscan metamorphic rocks, but such a relationship is not evident in any area known to the writer (see discussion in the previous section). As a case in point, the areal distributions of ultramafics and of jadeitic pyroxene-bearing and Na-pyroxene-lacking *in situ* metaclastic rocks of the Diablo Range are illustrated in figure 4. Serpentinized peridotites are volumetrically unimportant in this portion of the Franciscan terrane. A clear correlation between jadeitic metagraywackes and the serpentinites is not obvious.

It may also be noted that, in figures 2B and 3, proportions of the oxides Na_2O, FeO, and MgO, and the Fe_2O_3/FeO ratios of albite-bearing Franciscan metaclastic rocks exhibit *virtually* the same range of values as do those of jadeitic metagraywackes; if the latter assemblage had been produced by reaction of protoliths with a reducing saline pore fluid, one might expect sodic pyroxene-bearing metaclastics to display systematically higher Na_2O contents and lower ferric/ferrous ratios.

In conclusion, although perhaps not impossible, *available field relations, chemical and thermodynamic data do not seem to substantiate the metastable crystallization hypothesis proposed by Gresens.*

3. In the Pacheco Pass area, albite is associated with jadeitic pyroxene in metagraywackes over an extensive region (McKee, 1962a; Ernst and Seki, 1967). Because of the presence of minor Ca, Fe, and Mg in the pyroxene solid solution, this phase must coexist stably with quartz and albite over a P–T zone (Robertson, Birch, and MacDonald, 1957). Calculations by Essene and Fyfe (1967) and experimental studies by Newton and Smith (1967) demonstrate that at low temperatures and in the presence of quartz, $NaAlSi_2O_6$-rich solid solutions complete decomposition to albite-bearing assemblages at a pressure about 1 kb lower than does pure jadeite. Therefore, although widespread association of Na-pyroxene with albite in the vicinity of Pacheco Pass might reflect local domain equilibrium compatibilities produced under physical conditions within this relatively narrow P–T zone, such an explanation requires a somewhat restricted metamorphic geothermal gradient.

Textural relations described by McKee (1962a), Ernst (in press), and Ernst and others (1970) indicate that preexisting albite has been partially or completely replaced by jadeitic pyroxene + quartz in Diablo Range metagraywackes (refer also to pl. 1). Accordingly, the observed associations of these phases in some cases may reflect the metastable persistence of albite. Under this hypothesis, large tracts of the Franciscan terrane now exposed in the Diablo Range may have been subjected more uniformly to the high pressures attending blueschist facies metamorphism, but sluggish reaction rates allowed the local, incomplete preservation of metastable initial phase associations. Recrystallization to more stable mineral compatibilities would be accelerated by the catalytic effect of fluids and by the increased reaction surface produced by granulation accompanying shearing (Ernst, 1963a, p. 17). A similar suggestion to account for an inverted metamorphic sequence described in Franciscan rocks of the South Fork Mountain area, northern Coast Ranges, has been advanced by Bailey and Blake (1969), but it is not the explanation these authors favor (see next section).

To summarize, *petrographic relations indicate that blueschist-type phase associations possessed lower total Gibbs free energies than did the preexisting ones during Franciscan metamorphism. Conceivably other configurations of even lower total G could have existed, but inasmuch as traces of them have not been recognized, the glaucophane schist-type associations are believed to have represented the stable, equilibrium configuration.*

GENERATION OF OBSERVED ASSEMBLAGES BY TECTONIC OVERPRESSURES

Building on the earlier works of Irwin (1960, 1964), Kilmer (1962), and by Blake (ms), Blake, Irwin, and Coleman (1967) have described a sequence of inverted or "upside down" metamorphism in the vicinity of South Fork Mountain, northern California. Here imbricate thrust sheets of the Klamath province overlie Franciscan rocks along a low-angle, east-dipping fault. Within the Franciscan terrane, hydrous calcium-aluminum silicates, which have grown in quartzose, albite-bearing metagraywackes, are thought by these authors to exhibit a progressive west-to-east paragenesis of pumpellyite → lawsonite.

An even lower grade, laumontite-bearing association crops out to the west of the pumpellyitized metagraywacke terrane. Calcite occurs sporadically on the west, whereas aragonite is a critical phase nearer the thrust surface. As described by these investigators, the degree of textural reconstitution also gradually increases, proceeding eastward toward this low-angle fault. Blake, Irwin, and Coleman (1967) have suggested that such relations represent an inverted metamorphic sequence. They invoked the buildup of considerable stress—tectonic overpressures—and concomitantly high aqueous fluid pressures to account for the observed assemblages in the Franciscan rocks lying along the sole of the thrust[4]; the confining pressure ($P_{lithostatic}$) was thought to have been moderate, on the order of 4 kb, at the time of recrystallization. More recently, Blake, Irwin, and Coleman (1969), Blake and Cotton (1969), and Bailey and Blake (1969) have extended this tectonic-metamorphic concept to other portions of the Franciscan terrane to account for the observation that in many areas, more thoroughly recrystallized (for example, jadeitic

---

[4] Other authors (for example, Coleman and Lee, 1962, p. 594; Coleman, 1967a) have also called on large tectonic overpressures to explain apparently high-pressure blueschist facies mineral assemblages in the Franciscan but without specifically relating the strain buildup to structural features.

pyroxene-bearing) metaclastic sections seemingly overlie lower-pressure lithologies.

Blake, Irwin, and Coleman (1969, p. 244) have drawn attention to presumably analogous inverted metamorphic sequences in the outer metamorphic belt of Japan, the Urals, Kamchatka Peninsula, Venezuela, New Caledonia, and New Zealand and have speculated that production of blueschist facies rocks in general may require the attendance of substantial tectonic overpressures. But at least where the present author has some familiarity, namely that part of the outer metamorphic belt of Japan exposed in Shikoku, alternative interpretations are possible. Based on metamorphic and structural studies (Kawachi. 1965, 1968; Ernst and others, 1970, chaps. 7, 8, 15), it has been demonstrated that the metamorphic sequence observed in Shikoku reflects the existence of previously unrecognized recumbent folding. Tectonic emplacement of a higher grade nappe, or series of nappes, over a virtually unmetamorphosed section evidently caused rapid lithostatic pressure increment and an initial blueschist facies recrystallization in the underlying mass, followed by gradual temperature buildup within the latter during the subsequent period of erosive unloading. Structural complexities described in some of the other terranes also hint at the possibility of penecontemporaneous and postmetamorphic tectonic juxtaposition—rather than a simple case of "upside-down" metamorphism.

The generation and persistence during recrystallization of substantial tectonic overpressures require that the rocks so stressed possess considerable strength. Previous experimental results indicate that, as laboratory strain rates decrease, so do dry rock strengths (Heard, 1963). Moreover, in the presence of an aqueous fluid, the strengths of silicates are greatly lessened above moderate temperatures (Griggs and Blacic, 1965); on the base of oxygen isotope studies, Taylor and Coleman (1968, p. 1735-1737) have shown that glaucophane-schist mineral assemblages from the Cazadero area of northern California equilibrated with an ubiquitous hydrous fluid phase. Recent experimental investigation of both strong Franciscan metagraywacke and weak Franciscan metashale by Brace and others (1970) has also shown that, under the appropriate metamorphic conditions (that is, temperatures of about 150°-300°C, moderately high fluid pressures, and geologically reasonable strain rates of approximately 10^{-13} or 10^{-14}), the possibilities of generating tectonic overpressures exceeding about 1 kb on a regional scale are remote for a homogeneous, strong metagraywacke terrane; where such lithologies are intercalated with a subequal proportion of incompetent metashales (as is the case for jadeitic pyroxene-bearing metaclastics exposed in the vicinity of Pacheco Pass), the maintenance of substantial tectonic overpressures seems clearly impossible. The role of solution and recrystallization was not considered by Brace and others, but this process would be expected to cause further weakening of the rocks. Hence, although deformed rocks assuredly have been subjected to at least minor differential stress, laboratory experiments indicate that the magniture of the effect in Franciscan metaclastics must have been quite small.

Furthermore, although the concept of tectonic overpressures has been invoked to account for an inverted, presumably contemporaneous progressive metamorphic sequence along the sole of a major thrust fault (Blake, Irwin, and Coleman, 1967, p. 7, 1969, p. 237-239, 241), the described petrologic and field relationships as yet have not been sufficiently documented either at South Fork Mountain or farther south. The complete absence of mineralogic evidence for stress buildup in the immediately overlying tectonically emplaced Great Valley section has not been explained either. Conceivably, several thrust plates of contrasting phase assemblages (hence P–T histories) could have been juxtaposed in the South Fork Mountain area as is evidently the case in Franciscan rocks of the nearby Leech Lake Mountain–Anthony Peak region (Suppe, 1969a).

In the Diablo Range where the author has been working, the spatial distribution of jadeitic pyroxene-bearing metagraywackes appears to be unrelated to proximity to the bounding Ortigalita thrust fault, as illustrated in figure 4 (see also Ernst, in press). In this area, it is unclear whether the contact between albitic and jadeitic pyroxene ($\pm$ albite)-bearing metaclastic portions of the terrane represents an isograd or an as yet unrecognized tectonic contact; it is possible that in some places this contact may represent an isograd, whereas in other localities it may be a fault.

Several radiometric studies have dealt with the time of metamorphism of thoroughly recrystallized Franciscan rocks located adjacent to the thrust surface. In the Leech Mountain–Anthony Peak region of northern California, Suppe (1969b) presented evidence for a 150 to 151 m.y. metamorphic age for a plate of jadeitic pyroxene-bearing metagraywacke, whereas overlying and underlying blocks of fossiliferous albitic metagraywacke yielded recrystallization dates ranging from 104 to 109 m.y. Reported apparent ages of metamorphism of the nearby South Fork Mountain schist range from 123 to 136 m.y. Also in northern California, coarse-grained mafic blueschists disposed along a thrust fault at Goat Mountain were metamorphosed at least as long ago as 137 to 148 m.y. (Ernst and others, 1970, chap. 14; Suppe, personal commun., 1970). Near Pacheco Pass, recrystallization occurred—or at least argon loss ceased—about 115 to 122 m.y. B.P. (Suppe, personal commun., *in* Ernst, in press). Although not clearly related to a nearby thrust fault, in place metashales associated with blueschists of the Cazadero area (Coleman and Lee, 1963) provide radiometric ages of metamorphism of 130 to 135 m.y., similar to those listed above (Lee and others, 1963). Although a spectrum of apparent ages has been presented, it would appear that metamorphism of the Franciscan rocks took place largely during (or in some areas possibly prior to) the Early Cretaceous. This is significant because Upper Cretaceous Great Valley sequence rocks constitute a large portion of the overlying plate, both in northern California and especially surrounding the Diablo Range, hence the thrust fault relations now recognized probably were produced 30 to 60 m.y. after culmination of the metamorphic event. Conceivably earlier thrusting events occurred, but the structural complexities so generated in the Franciscan itself have not been adequately deciphered to the present time.

In summary, *neither experimental strength-of-materials studies, detailed petrologic mapping, nor metamorphic-stratigraphic temporal relations lend support to the hypothesized production of tectonic overpressures on a regional scale in Franciscan metamorphic rocks.*

PRODUCTION OF OBSERVED ASSEMBLAGES BY DEEP BURIAL

Arguments favoring deep burial, on the order of 20 to 30 km or more, are derived primarily from a comparison of the observed mineral parageneses with experimentally determined phase equilibria. Although the stability relationships of pumpellyite are known only through reconnaissance work (see Hinrichsen and Schürmann, 1969), those involving laumontite and lawsonite have been studied by several investigators (Newton and Kennedy, 1963; Crawford and Fyfe, 1965; Liou, 1968, 1969). Physical conditions of the calcite-aragonite transition have been located by numerous experimentalists employing a variety of techniques (Jamieson, 1953; Clark, 1957; Crawford and Fyfe, 1964; Boettcher and Wyllie, 1967, 1968a; Newton, Goldsmith, and Smith, 1969). The equilibrium between jadeite + quartz and albite and that between jadeitic pyroxene + quartz and albite have been investigated by Birch and Le-Comte (1960), Newton and Smith (1967), and Boettcher and Wyllie (1968b). The P–T order-disorder relations in sodic amphiboles were studied by Ernst (1963b) (see also Papike and Clark, 1968). Equilibrium diagrams are summarized in figure 5.

As previously stated, numerous petrologic investigations of specific portions of the Franciscan terrane have documented systematic mineral parageneses. The progressive metamorphic phase assemblages developed in quartzose clastic rocks are compatible with the experimentally investigated phase equilibria. Thus the change in mineral assemblage, (A) laumonite + albite ± calcite → (B) pumpellyite and/or lawsonite + albite ± calcite → (C) lawsonite + albite ± aragonite → (D) lawsonite + jadeitic pyroxene ± aragonite, represents a sequence of pressure increment at nearly constant temperature as seen from the schematic metamorphic geothermal gradient indicated in figure 5D. Sodic amphibole—the more ordered, glaucophane II polymorph—occurs as a minor constituent in some of the originally more chloritic metaclastics, where it is associated with jadeitic pyroxene and/or aragonite. The volumetrically much less abundant mafic igneous rocks metamorphosed *in situ* display a simpler equilibrium paragenesis ranging from calcitic, albitic greenstones to glaucophane II and/or omphacite + lawsonite + sphene + aragonite (that is, blueschist-type) compatibilities, suggesting an increasing pressure sequence. *The close correspondence of the observed changes in mineralogy with the experimentally determined phase equilibria lends*

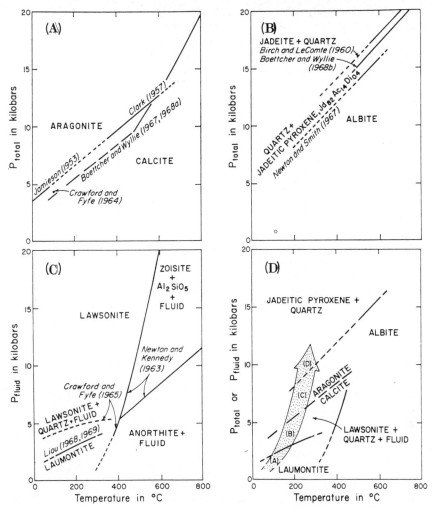

Fig. 5. Some experimentally determined or calculated phase equilibria which pertain to the observed Franciscan mineral parageneses. In (D), the inferred range of physical conditions attending the progressive metamorphism is indicated very diagrammatically.

credence to the hypothesis that the natural occurrences represent a stable sequence.

If so, judging from combination of the oxygen isotopic geothermometry presented by Taylor and Coleman (1968) and the laboratory phase equilibrium data, *the most intensely recrystallized in place Franciscan metagraywackes and associated rocks would seem to have been subjected to physical conditions of approximately 150 to 300°C at from 5 to more than 8 kb total pressure.* Provided the production of such mineral assemblages is not due fundamentally to processes involving metasomatism or metastable recrystallization as argued earlier, high pressures and low temperatures of blueschist metamorphism appear to be indicated for the Franciscan. Moreover, if the criticisms of tectonic overpressures advanced in the last section are valid, it follows that *the 5 to 8+ kb pressure necessitated by the paragenesis must involve deep burial, on the order of 20 to 30 km or more.*

Objections to such extreme depths of burial reflect the fact that nowhere has a Franciscan stratigraphic section of the required magnitude actually been measured in the California Coast Ranges, nor has a downward increase in metamorphic "grade" ever been demonstrated. Inasmuch as the depositional base of the Franciscan has not been recognized, however, and the entire section is considerably disturbed, the tectonic thickness during metamorphism (and at present as well) is unknown. Therefore, as of now, it seems premature if not incorrect to regard this deposit as of insufficient tectonic thickness to provide the required burial depths for sections of rock now exposed. Because the gross fabric of the Franciscan terrane has still to be deciphered, it also may be premature to draw the conclusion that the paragenetic sequence is "right side up" or "upside down"; in any case, the problem seems to be chiefly structural and thus far is unresolved. Nevertheless, the geologic plausibility of such great depths of burial may be appropriately questioned.

DISCUSSION

Franciscan metamorphic mineral assemblages appear to provide compelling evidence for a process that combines high lithostatic pressures with low temperatures. This author has proposed the accumulation of Franciscan, principally clastic, first-cycle debris chiefly within the confines of one or a series of oceanic trenches to account for the seemingly rapid downsinking of large volumes of material to great depths (Ernst, 1965, p. 905-910). It was thought that this mechanism would allow sufficient depression of the isotherms to preserve temperatures on the order of 150 to 300°C at 20 to 30 km below the surface. Nevertheless, lacking additional independent evidence of great depth of burial, various other models for producing the Franciscan metamorphic rocks have been persuasively advocated, namely recrystallization due to metasomatism, metastability, or tectonic overpressure. While admitting the validity and importance of those processes, the arguments presented in previous sections seem to the present author to require the rejection of these as determining factors in production of the Franciscan metamorphic mineral assemblages. *All these other hypotheses, for instance, fail to explain the contrasts in phase petrology between coeval Franciscan and Great Valley rocks of similar provenance; only profound burial of the Franciscan followed by substantial postmetamorphic uplift can account for the remarkable mineralogic differences between these tectonically juxtaposed series.*

Now, with the advent of the concept of new global tectonics (for example, Isacks, Oliver, and Sykes, 1968), the general problem of blueschist facies metamorphism needs to be reexamined. Such low-grade metamorphic belts are confined to the vicinity of present or ancient colliding plate margins—or sutures (Dewey and Bird, 1970; Dewey and Horsfield, 1970). In these zones of convergence, masses of crust and mantle lithosphere appear to have been overridden along the Benioff zone. The relatively rapid spreading rates currently observed, coupled with the inferred superposition of plates at their junctions, provides an overall mechanism for pressure increment at low temperatures: although the increase in lithostatic pressure in the overridden, downgoing slab is instantaneous, the temperature rise depends on the relatively slow process of conductive heat transfer. A dynamic, quasi-steady state is thereby produced in which the isotherms within the descending lithospheric plate exhibit a rather extreme downbowing as illustrated by the computations presented by Oxburgh and Turcotte (1968, 1970), MacKenzie (1969), and Minear and Toksöz (1970). Here, then, is a general process to explain the generation of high lithostatic pressures and low temperatures along suture zones and within the descending slab. Inasmuch as this is the locale of blueschists and related metamorphic rocks, it seems only plausible to relate their formation to an environment characterized by unusually high pressures and low temperatures.

Trenches, too, are confined to suture zones, where they evidently result from the dynamics attending a downgoing slab. But was the Franciscan deposited in a trench—or subduction zone—environment? The local geometry of plate convergence can result in a variety of types of superposition of lithospheric blocks, not all of which necessarily produce trenches. For instance, in eastern New Guinea a slab of oceanic crust + uppermost mantle evidently has been thrust over continental crust (Davies and Milson, 1969), generating relatively high-pressure mineral assemblages in the lower plate. Perhaps a similar tectonic regime gave rise to the metamorphic + ultramafic complex in New Caledonia. (This view is not shared by either Coleman, 1967a, or Brothers, 1970; both of these investigators have mapped in New Caledonia, and both regard the high-pressure mineral assemblages as having been produced by tectonic overpressures.) In Shikoku, as discussed previously, the emplacement of recumbent folds, possibly reflecting convergence between the Japanese arc and the western Pacific plate, is hypothesized to have loaded a section of thin continental crust tectonically, thereby generating relatively high-pressure mineral assemblages in the underlying rocks (Ernst and others, 1970, chap. 15). Such a process conceivably might also account for the observed early stage of Alpine blueschist facies metamorphism (for example, see van der Plas, 1959; Ellenberger, 1960; de Roever and Nijhuis, 1963; Bearth, 1966). The common characteristic of virtually all these glaucophane schist belts is that they appear to be associated with the lower plates (or at least undertucking of material adjacent to the lower plates) in the vicinity of suture zones, where lithostatic pressures would be high and temperatures low.

There are reasons for suggesting that the Franciscan was deposited in an oceanic, trench-type environment among which, however, must be included the following facts. (1) Taking into account the offset produced by Tertiary strike-slip movement on the San Andreas fault (Hill and Dibblee, 1953), the present areal distribution of the Franciscan marks a linear belt of very thick accumulation roughly parallel to but seaward of the more-or-less coeval Klamath, Sierran, and Salinian volcanic-plutonic arcs (for example, see Bailey, Irwin, and Jones, 1964). Present day trenches typically lie approximately 50 to 250 km to the oceanic side of such volcanic arcs (see Menard, 1964, chap. 5; Dickinson and Hatherton, 1967), just as the Franciscan is disposed relative to the Sierran-type terrane. (2) The depositional base of the Franciscan nowhere has been recognized, and the only visible fragments of underlying units consist of basaltic and ultramafic compositions—units typical of oceanic crust and mantle (Ernst, 1965). (3) There is clear evidence for underthrusting of the Franciscan relative to the South Fork Mountain schists and the Great Valley sequence (Irwin, 1960, 1964; Brown, 1964; Page, 1970). (4) The in-part chaotic (mélange) nature of the Franciscan (Hsü, 1968) is compatible with a process involving deposition and both penecontemporaneous and later shearing and tectonic dislocation within a subduction zone. And (5), great quantities of sediment known to have been derived from western North America and shed westward are missing from the geologic record, hence presumably were overridden by the continent and dragged down along a trench-Benioff zone complex (Gilluly, 1969.)

If Pacific sea floor + mantle underflow occurred at the western margin of North America during Late Mesozoic time, the deposition of vast amounts of clastic debris eroded from the continentalward arc would necessarily have resulted in the subduction of Franciscan-type sediments. *Models presented by Hamilton (1969, fig. 5), Page (1970, fig. 9), and Ernst (1970, figs. 3 and 4) all illustrate depths of underthrusting that satisfy the 20 to 30+ km seemingly required by the observed Franciscan metamorphic mineral assemblages. Explanations of these parageneses that call upon metasomatism, metastable crystallization, or tectonic overpressures at best seem to be only incidentally related to such events.*

DYNAMIC MODEL

Let us provisionally accept a trench-subduction zone environment for Franciscan accumulation, tectonic thickening, deformation, and recrystallization. The generalized dynamic model is presented in figure 6A, modified from Ernst (1970, fig. 3); it shows the proposed relationships among downwarped isotherms in the vicinity of the trench deposits, partial melting of the downgoing oceanic crust at much greater depths to provide the Sierran-type island arc volcanics and plutonics, and lithospheric plate convergence. The exposure of what apparently is oceanic crust along the western margin of the Great Valley (Bailey, Blake, and Jones, 1970) is illustrated in figure 6B. According to the model illustrated

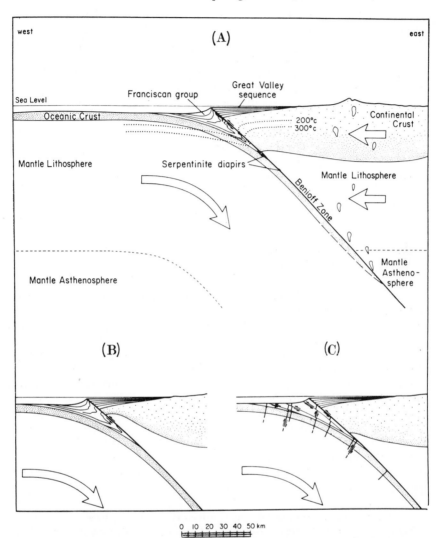

Fig. 6. Tectonic model relating Franciscan group features to a period of active plate consumption, slightly modified from Ernst (1970, fig. 3); no vertical exaggeration implied. (A) presents the overall geometry illustrating downbowing of the isotherms of the trench complex due to sea-floor spreading; basaltic oceanic crust is presumed to have been transformed to eclogite and subsequently to have undergone partial melting at profound depths to yield Sierran, island arc-type igneous rocks (see Ringwood and Green, 1966, p. 420 and fig. 9F). Whereas (A) exhibits a regional setting appropriate to the South Fork Mountain and Nacimiento-Salinia areas, (B) shows relations including a basal contact of overlying Great Valley strata with oceanic crust of the North American plate exposed directly east of the Coast Range thrust in central California as documented by Bailey, Blake, and Jones (1970). (C) presents hypothesized fault sets presumably controlled by the Benioff master shear zone (see Malahoff, 1970, fig. 11) and tensional features in the vicinity of the downturn of the oceanic lithospheric plate (Isacks, Oliver, and Sykes, 1968, fig. 7). Some of the shears subparallel to the Benioff zone may transect oceanic crust and underlying mantle.

in figure 6, Franciscan and Great Valley rocks owe their contrasting lithologies and style of deformation—but essentially common provenance —to the fact that they were laid down on the eastern Pacific and North American plates respectively; their juxtaposition along the South Fork Mountain–Stoney Creek–Ortigalita–Sur-Nacimiento thrust (that is, the Coast Range thrust of Bailey, Blake, and Jones, 1970) is thought to represent the crustal expression of a Late Mesozoic Benioff shear zone (Ernst, 1970).

It still remains to be answered how such a tectonic setting could account for the present structural relations and areal distribution of the Franciscan metamorphic mineral assemblages. Aspects of the model depicted in figure 6C may provide a partial explanation: here, hypothesized differential movements on intersecting fault sets within the subducted prism of tectonically thickened and deformed trench mélange would require the thrusting of deeper, largely older, more thoroughly recrystallized eastern rocks relatively over predominantly shallower, younger, feebly metamorphosed western sections as is observed in some portions of the Diablo Range and its northern extension. Modern day oceanic deeps are typified by what appears to be extensional rifting (normal faulting) as shown for instance by Ludwig and others (1966), Isacks, Oliver, and Sykes (1968, fig. 7), and Malahoff (1970). The last author has interpreted this system of faults in some localities as being subparallel to the Benioff zone, with the sense of movement similar to that along the juncture between the overriding and downgoing slabs (Malahoff, 1970, fig. 11). With regard to the Franciscan, we may conjecture that several sets of faults could have developed at any one time, as shown in figure 6C, leading to successive undertucking of the more oceanward, higher level blocks beneath the more continentalward, more deeply buried sections. Slight changes in the geometry of plate impingence with time would have been reflected in variations in the attitudes of the fault systems. Thus a series of imbricate thrusts—anastomosing with time— would be expected to have shuffled systematically more easterly higher-pressure Franciscan rocks over more feebly recrystallized strata on the west.

The net displacement of the Franciscan suite would have been successively deeper and deeper beneath the more normal, low-pressure Great Valley sequence during active subduction as illustrated in figure 6. However, the relationships observed today are quite different: blueschists and related high-pressure metamorphics which appear to have recrystallized at profound depths evidently have been brought up subsequently to their present level of exposure, principally along eastern portions of the Franciscan terrane. Large-scale structures, mineral parageneses, and the ages of rocks and fossils in different tectonic blocks all indicate that *the Franciscan terrane now spatially associated with the Great Valley sequence was metamorphosed in an environment far removed from the present one.* Because the predominantly metaclastic trench filling possessed a lower aggregate density than the mantle material which it displaced during subduction, diminution in the spreading rate, or cessation of underflow, of the eastern Pacific lithospheric plate would have resulted in the buoyant upward surge of the Franciscan terrane. From the present gross distribution of the lithologic assemblages now observed, it would appear that the transport direction of the ensimatic mélange involved both large western and vertical components. Such movements would be expected further to enhance the imbrication of blueschist facies blocks chiefly on the east over less deeply buried oceanward sections and would result in the asymmetric upturn of the now steeply dipping western limb of the Great Valley synclinorium (see Lachenbruch, 1962; Safonov, 1962).

In the author's opinion, such a model involving deep burial, seemingly required by the observed phase relations, in a Late Mesozoic subduction zone at the western margin of North America, more adequately explains the currently understood character of the Franciscan and its juxtaposition against other rock units than alternative hypotheses thus far advanced.

ACKNOWLEDGMENTS

The somewhat parochial point of view presented in this paper is the result of a continuing association with the Franciscan itself and with scientific colleagues also engaged in its investigation. Diverse portions of the terrane possess contrasting features; hence different opinions are quite justifiably held. The present paper has benefited from the criticisms of E. H. Bailey, M. C. Blake, Jr., and R. G. Coleman, all of the U.S. Geological Survey; R. L. Gresens, University of Washington; R. C. Newton, University of Chicago; and especially John Suppe, University of California, Los Angeles, for which the author is very much obliged. It should be noted, of course, that none of these authorities agrees very completely with the ideas expressed here.

REFERENCES

Bailey, E. H., and Blake, M. C., Jr., 1969, Late Mesozoic sedimentation and deformation in western California: Geotektonika, v. 3, p. 17-34; v. 4, p. 24-34 (in Russian).

Bailey, E. H., Blake, M. C., Jr., and Jones, D. L., 1970, On-land Mesozoic oceanic crust in California Coast Ranges: U.S. Geol. Survey Prof. Paper 700-C, p. 70-81.

Bailey, E. H., Irwin, W. F., and Jones, D. L., 1964, Franciscan and related rocks and their significance in the geology of western California: California Div. Mines and Geology Bull., v. 183, 177 p.

Barnes, I., and O'Neil, J. R., 1969, The relationship between fluids in some fresh alpine-type ultramafics and possible modern serpentinization, Western United States: Geol. Soc. America Bull., v. 80, p. 1947-1960.

Bearth, Peter, 1966, Zur mineralfaziellen Stellung der Glaukophangesteine der Westalpen: Schweizer Mineralog. Petrog. Mitt., v. 46, p. 13-23.

Birch, Francis, and LeComte, Paul, 1960, Temperature-pressure plane for albite composition: Am. Jour. Sci., v. 258, p. 209-217.

Bischoff, J. L., and Fyfe, W. S., 1968, Catalysis, inhibition, and the calcite-aragonite problem. I. The aragonite-calcite transformation: Am. Jour. Sci., v. 266, p. 65-79.

Blake, M. C., Jr., ms, 1965, Structure and petrology of low-grade metamorphic rocks, Blueschist facies. Yolla Bolly area, northern California: Ph.D. thesis, Stanford Univ., 91 p.

Blake, M. C., Jr., and Cotton, W. R., 1969, Inverted metamorphic mineral zones in Franciscan metagraywacke of the Diablo Range, Northern California [abs.]: Geol. Soc. America Abs. with Programs, v. 1, no. 2, p. 6-7.

Blake, M. C., Jr., Irwin, W. P., and Coleman, R. G., 1967, Upside-down metamorphic zonation, blueschist facies, along a regional thrust in California and Oregon: U.S. Geol. Survey Prof. Paper 515-C, p. 1-9.

——————— 1969, Blueschist-facies metamorphism related to regional thrust-faulting: Tectonophysics, v. 8, p. 237-246.

Bloxam, T. W., 1956, Jadeite-bearing metagraywackes in California: Am. Mineralogist, v. 41, p. 488-496.

——————— 1959, Glaucophane-schists and associated rocks near Valley Ford, California: Am. Jour. Sci., v. 257, p. 95-112.

——————— 1960, Jadeite-rocks and glaucophane schists from Angel Island, San Francisco Bay, California: Am. Jour. Sci., v. 258, p. 555-573.

——————— 1966, Jadeite-rocks and blueschists in California: Geol. Soc. America Bull., v. 77, p. 781-786.

Boettcher, A. L., and Wyllie, P. J., 1967, Revision of the calcite-aragonite transition, with the location of a triple point between calcite I, calcite II, and aragonite: Nature, v. 213, p. 792-793.

——————— 1968a, The calcite-aragonite transition measured in the system $CaO-CO_2-H_2O$: Jour. Geology, v. 76, p. 314-330.

——————— 1968b, Jadeite stability measured in the presence of silicate liquids in the system $NaAlSiO_4-SiO_2-H_2O$: Geochim. et Cosmochim. Acta, v. 32, p. 999-1012.

Brace, W. F., Ernst, W. G., and Kallberg, R. W., 1970, An experimental study of tectonic overpressure in Franciscan rocks: Geol. Soc. America Bull., v. 81, p. 1325-1338.

Brothers, R. N., 1954, Glaucophane schists from the North Berkeley Hills, California: Am. Jour. Sci., v. 252, p. 614-626.

——————— 1970, Lawsonite-albite schists from northernmost New Caledonia: Contr. Mineralogy and Petrology, v. 25, p. 185-202.

Brown, R. D., 1964, Thrust-fault relations in the northern Coast Ranges, California: U. S. Geol. Survey Prof. Paper 475-D, p. 7-13.

Chesterman, C. W., 1960, Intrusive ultrabasic rocks and their metamorphic relationships at Leech Lake Mountain, Mendocino County, California: Internat. Geol Cong., 21st, Copenhagen 1960, Rept., Pt. 13, p. 208-215.

Clark, S. P., Jr., 1957, A note on calcite-aragonite equilibrium: Am. Mineralogist, v. 42, p. 564-566.

Coleman, R. G., 1961, Jadeite deposits of the Clear Creek area New Idria district, San Benito, County, California: Jour. Petrology, v. 2, p. 209-247.

——————— 1965, Composition of jadeitic pyroxene from the California graywackes: U.S. Geol. Survey Prof. Paper 525-C, p. 25-34.

——————— 1967a, Glaucophane schists from California and New Caledonia: Tectonophysics, v. 5, p. 479-498.

——————— 1967b, Low-temperature reaction zones and alpine ultramafic rocks of California, Oregon, and Washington: U.S. Geol. Survey Bull. 1247, 49 p.

Coleman, R. G., and Lee, D. E., 1962, Metamorphic aragonite in the glaucophane schists of Cazadero, California: Am. Jour. Sci., v. 260, p. 577-595.

——————— 1963, Glaucophane-bearing metamorphic rock types of the Cazadero area, California: Jour. Petrology, v. 4, p. 260-301.

Crawford, W. A., and Fyfe, W. S., 1964, Calcite-aragonite equilibrium at 100°C: Science, v. 144, p. 1569-1570.

——————— 1965, Lawsonite equilibria: Am. Jour. Sci., v. 263, p. 262-270.

Davies, H. L., and Milsom, J. S., 1969, Eastern Papua geology and gravity [abs.]: Geophys. Union Trans., v. 50, p. 333.

Davis, E. F., 1918, The radiolarian cherts of the Franciscan group: California Univ. Pub. Geol. Sci., v. 11, p. 235-432.

Dewey, J. F., and Bird, J. M., 1970, Mountain belts and the new global tectonics: Jour. Geophys. Research, v. 75, p. 2625-2647.

Dewey, J. F., and Horsfield, B., 1970, Plate tectonics, orogeny and continental growth: Nature, v. 255, p. 521-525.

Dickinson, W. R., 1966, Table Mountain serpentinite extrusion in California Coast Ranges: Geol. Soc. America Bull., v. 77, p. 451-472.

Dickinson, W. R., and Hatherton, T., 1967, Andesitic volcanism and seismicity around the Pacific: Science, v. 157, p. 801-803.

Dickinson, W. R., Ojakangas, R. W., and Stewart, R. J., 1969, Burial metamorphism of the Late Mesozoic Great Valley sequence, Cache Creek, California: Geol. Soc. America Bull., v. 80, p. 519-526.

Dudley, P. P., ms, 1967, Glaucophane schists and associated rocks of the Tiburon Peninsula, Marin County, California: Ph.D. thesis, Univ. of California, Berkeley, 116 p.

Ellenberger, F., 1960, Sur une paragénèse éphémère á lawsonite et glaucophane dans le métamorphisme alpin en Haute-Maurienne (Savoie): Soc. géol. France Bull., v. 7, p. 190-194.

Ernst, W. G., 1959, Alkali amphiboles: Carnegie Inst. Washington Year Book 58, p. 121-126.

——————— 1963a, Petrogenesis of glaucophane schists: Jour. Petrology, v. 4, p. 1-30.

——————— 1963b, Polymorphism in alkali amphiboles: Am. Mineralogist, v. 48, p. 241-260.

——————— 1965, Mineral parageneses in Franciscan metamorphic rocks, Panoche Pass, California: Geol. Soc. America Bull, v. 76, p. 879-914.

——————— 1970, Tectonic contact between the Franciscan mélange and the Great Valley sequence, crustal expression of a Late Mesozoic Benioff zone: Jour. Geophys. Research, v. 75, p. 886-901.

——————— in press, Petrologic reconnaissance of Franciscan metagraywackes from the Diablo Range, Central California Coast Ranges: Jour. Petrology, in press.

Ernst, W. G., and Seki, Y., 1967, Petrologic comparison of the Franciscan and San-bagawa metamorphic terranes: Tectonophysics, v. 4, p. 463-478.

Ernst, W. G., Seki, Y., Onuki, H., and Gilbert, M. C., 1970, Comparative study of low-grade metamorphism in the California Coast Ranges and the Outer Meta-morphic Belt of Japan: Geol. Soc. America Mem. 124, 276 p.

Eskola, Pente, 1939, Die metamorphen Gesteine, *in* Barth, T. F. W., Correns, C. W., and Eskola, P. J., Die Enstehung der Gesteine: Springer, Berlin, p. 263-407.

Essene, E. J., and Fyfe, W. S., 1967, Omphacite in Californian rocks: Contr. Mineralogy and Petrology, v. 15, p. 1-23.

Essene, E. J., Fyfe, W. S., and Turner, F. J., 1965, Petrogenesis of Franciscan glauco-phane schists and associated metamorphic rocks, California: Contr. Mineralogy and Petrology, v. 11, p. 695-704.

Fyfe, W. S., and Newton, R. C., in press, High pressure metamorphism, *in* Bailey, D. K., ed., Experimental Petrology: New York, Academic Press.

Fyfe, W. S., and Zardini, R., 1967, Metaconglomerate in the Franciscan formation near Pacheco Pass, California: Am. Jour. Sci., v. 265, p. 819-830.

Ghent, E. D., 1965, Glaucophane-schist facies metamorphism in the Black Butte area, Northern Coast Ranges, California: Am. Jour. Sci., v. 263, p. 385-400.

Gilluly, James, 1969, Oceanic sediment volumes and continental drift: Science, v. 166, p. 992-994.

Gresens, R. L., 1969, Blueschist alteration during serpentinization: Contr. Mineralogy and Petrology, v. 24, p. 93-113.

Griggs, D. T., and Blacic, J. D., 1965, Quartz, anomalous weakness of synthetic crystals: Science, v. 147, p. 292-295.

Hamilton, Warren, 1969, Mesozoic California and the underflow of Pacific mantle: Geol. Soc. America Bull., v. 80, p. 2409-2430.

Hawkins, J. W., 1967, Prehnite-pumpellyite facies metamorphism of a graywacke-shale series, Mount Olympus, Washington: Am. Jour. Sci., v. 265, p. 798-818.

Heard, H. C., 1963, Effect of large changes in strain rate in the experimental deforma-tion of Yule marble: Jour. Geology, v. 71, p. 162-195.

Hill, M. L., and Dibblee, T. W., Jr., 1953, San Andreas, Garlock and Big Pine faults, California: Geol. Soc. America Bull., v. 64, p. 443-458.

Hinrichsen, von T., and Schürmann, K., 1969, Untersuchungen zur Stabilität von Pumpellyit: Neues Jahrb. Mineralogie Montasch., v. 10, p. 441-445.

Hlabse, T., and Kleppa, O. J., 1968, The thermochemistry of jadeite: Am. Mineralogist, v. 53, p. 1281-1292.

Hoffman, C., 1970, Die Glaukophangesteine, ihre stofflichen Äquivalente und Um-wandlungsprodukte in Nordcalabrien (Süditalien): Beitr. Mineralogie Petrographie, v. 27, p. 283-320.

Hsü, K. J., 1968, Principles of mélanges and their bearing on the Franciscan-Knoxville Paradox: Geol. Soc. America Bull., v. 79, p. 1063-1074.

Irwin, W. P., 1960, Geologic reconnaissance of the Northern Coast Ranges and Klamath Mountains, California, with a summary of the mineral resources: California Div. Mines and Geology Bull. 179, 80 p.

———— 1964, Late Mesozoic orogenies in the ultramafic belts of northwestern California and southwestern Oregon: U.S. Geol. Survey Prof. Paper 501-C, p. 1-9.

———— 1966, Geology of the Klamath Mountains province, *in* Bailey, E. H., ed., Geology of Northern California: California Div. Mines and Geology Bull. 190, p. 19-38.

Isacks, B., Oliver, J., and Sykes, L. R., 1968, Seismology and the new global tectonics: Jour. Geophys. Research, v. 73, p. 5855-5899.

Jamieson, J. C., 1953, Phase equilibria in the system calcite-aragonite: Jour. Chem. Physics, v. 21, p. 1385-1390.

Kawachi, Y., 1965, Finding of overturned graded bedding in spotted crystalline schists of the Sanbagawa metamorphic zone in central Shikoku, Japan: Geol. Soc. Japan Jour., v. 72, p. 311-313.

———— 1968, Large-scale overturned structure in the Sanbagawa metamorphic zone in central Shikoku, Japan: Geol. Soc. Japan Jour., v. 74, p. 607-616.

Kilmer, F. H., 1962, Anomalous relationship between the Franciscan formation and metamorphic rocks, northern Coast Ranges, California [abs.]: Geol. Soc. America Spec. Paper 68, p. 210.

Lachenbruch, M. C., 1962, Geology of the west side of the Sacramento Valley, California, *in* Bowen, O. E., Jr., ed., Geologic guide to the gas and oil fields of Northern California: California Div. Mines and Geology Bull. 181, p. 53-66.

Lee, D. E., Thomas, H. H., Marvin, R. F., and Coleman, R. G., 1963, Isotope ages of glaucophane schists from Cazadero, California: U.S. Geol. Survey Prof. Paper 475-D, p. 105-107.

Liou, J. G., 1968, Zeolite equilibria in the system $CaO \cdot Al_2O_3 \cdot 2SiO_2 - SiO_2 - H_2O - CO_2$, the stabilities of wairakite and laumontite [abs.]: Geol. Soc. America Program Ann. Mtg. 1968, Mexico City, p. 175.

———— 1969, P–T stabilities of laumonite, wairakite and lawsonite [abs]: Am. Geophys. Union Trans., v. 50, p. 352.

Ludwig, W. J., Ewing, J. I., Ewing, M., Murauchi, S., Den, N., Asano, S., Hoffa, H., Hayakawa, M., Asanuma, T., Ichikawa, K., and Noguchi, I., 1966, Sediments and structure of the Japan Trench: Jour. Geophys. Research, v. 71, p. 2121-2137.

MacKenzie, D. P., 1969, Speculations on the consequences and causes of plate motions: Royal Astron. Soc. Geophys. Jour., v. 18, p. 1-32.

Maddock, M. E., 1964, Geology of the Mt. Boardman Quadrangle, Santa Clara and Stanislaus Counties, California: California Div. Mines and Geology Map Sheet 3.

Malahoff, A., 1970, Some possible mechanisms for gravity and thrust faults under oceanic trenches: Jour. Geophys. Research, v. 75, p. 1992-2001.

McKee, Bates, 1962a, Widespread occurrence of jadeite, lawsonite, and glaucophane in central California: Am. Jour. Sci., v. 260, p. 596-610.

———— 1962b, Aragonite in the Franciscan rocks of the Pacheco Pass area, California: Am. Mineralogist, v. 47, p. 379-387.

Menard, H. W., 1964, Marine geology of the Pacific: New York, McGraw-Hill, 271 p.

Minear, J. W., and Toksöz, M. N., 1970, Thermal regime of a downgoing slab and global tectonics: Jour. Geophys. Research, v. 75, p. 1397-1419.

Misch, Peter, 1959, Sodic amphiboles and metamorphic facies in Mount Shuksan belt, Northern Cascades, Washington [abs.]: Geol. Soc. America Bull., v. 70, p. 1736-1737.

———— 1966, Tectonic evolution of the Northern Cascades of Washington State, *in* Gunning, H. C., ed., Tectonic history and mineral deposits of the Western Cordillera: Canadian Inst. Mining Metallurgy Spec., v. 8, p. 108-148.

———— 1969, Paracrystalline microboudinage of zoned grains and other criteria for synkinematic growth of metamorphic minerals: Am. Jour. Sci., v. 267, p. 43-63.

Miyashiro, Akiho, 1966, Some aspects of peridotite and serpentinite in orogenic belts: Japanese Jour. Geology Geography, v. 37, p. 45-61.

Newton, R. C., Goldsmith, R. J., and Smith, J. V., 1969, Aragonite crystallization from strained calcite at reduced pressures and its bearing on aragonite in low-grade metamorphism: Contr. Mineralogy Petrology, v. 22, p. 335-348.

Newton, R. C., and Kennedy, G. C., Some equilibrium reactions in the join $CaAl_2Si_2O_8 - H_2O$: Jour. Geophys. Research, v. 68, p. 2967-2983.

Newton, R. C., and Smith, J. V., 1967, Investigations concerning the breakdown of albite at depth in the earth: Jour. Geology, v. 75, p. 268-286.

Oxburgh, E. R., and Turcotte, D. L., 1968, Problem of high heat flow and volcanism associated with zones of descending mantle convective flow [abs.]: Am. Geophys. Union Trans., v. 49, p. 318.

————— 1970, Thermal structure of island arcs: Geol. Soc. America Bull., v. 81, p. 1665-1668.

Page, B. M., 1970, Sur-Nacimiento fault zone of California: continental margin tectonics: Geol. Soc. America Bull., v. 81, p. 667-690.

Papike, J. J., and Clark, J. R., 1968, The crystal structure and cation distribution of glaucophane: Am. Mineralogist, v. 53, p. 1156-1173.

Peterman, Z. E., Hedge, C. E., Coleman, R. G., and Snavely, P. D., 1967: $^{87}Sr/^{80}Sr$ ratios in some eugeosynclinal sedimentary rocks and their bearing on the origin of granitic magma in orogenic belts: Earth and Planetary Sci. Letters, v. 2, p. 433-439.

Plas, L. van der, 1959: Petrology of the northern Adula region, Switzerland (with particular reference to glaucophane-bearing rocks): Leidse Geol. Meded, v. 24, p. 415-602.

Ransome, F. L., 1894, The geology of Angel Island: California Univ. Pub., Geol. Sci., v. 1, p. 193-240.

Raymond, L. A., 1970, Relationships between blueschists facies metamorphism, folding, and faulting in Franciscan rocks, Seegers Ranch area, northeastern Diablo range, California [abs.]: Geol. Soc. America Abs. with Programs, v. 2, no. 2, p. 133-134.

Ringwood, A. E., and Green, D. H., 1966, An experimental investigation of the gabbro-eclogite transformation and some geophysical implications: Tectonophysics, v. 3, p. 383-427.

Robertson, E. C., Birch, Francis, and MacDonald, G. J. F., 1957, Experimental determination of jadeite stability relations to 25,000 bars: Am. Jour. Sci., v. 255, p. 115-137.

Roever, W. P. de, and Nijhuis, H. J., 1963, Plurifacial alpine metamorphism in the eastern Betic Cordilleras (SE Spain), with special reference to the genesis of the glaucophane: Geol. Rundschau, v. 53, p. 324-336.

Saad, A. H., 1969, Paleomagnetism of Franciscan ultramafic rocks from Red Mountain, California: Jour. Geophys. Research, v. 74, p. 6567-6578.

Safonov, A., 1962, The challenge of the Sacramento Valley, California, in O. E. Brown, Jr., ed., Geologic guide to the gas and oil fields of Northern California: California Div. Mines and Geology Bull. 181, p. 77-98.

Saito, M., Hasimoto, K., Sawata, H., and Shimazaki, Y., 1960, Geology and mineral resources of Japan, 2d ed.: Tokyo, Japan, Geol. Survey, 304 p.

Seki, Y., Ernst, W. G., and Onuki, H., 1969, Phase proportions and physical properties of minerals and rocks from the Franciscan and Sanbagawa metamorphic terranes, a supplement to Geol. Soc. America Mem. 124: Tokyo, Japan, Japan Soc. Promotion Sci., 85 p.

Smith, J. P., 1907, The paragenesis of the minerals in the glaucophane-bearing rocks of California: Am. Philos. Soc. Proc., v. 45, p. 183-242.

Suppe, John, 1969a, Franciscan geology of the Leech Lake Mountain-Anthony Peak region, northern Coast Ranges, California [abs.]: Geol. Soc. America Abs. with Programs, 1969 (v. 1), pt. 2, p. 65-66.

————— 1969b, Times of metamorphism in the Franciscan terrain of the northern Coast Ranges, California: Geol. Soc. America Bull., v. 80, p. 135-142.

Taliaferro, N. L., 1943, Franciscan-Knoxville problem: Am. Assoc. Petroleum Geologists Bull., v. 27, p. 109-219.

Taylor, H. P., and Coleman, R. G., 1968, O^{18}/O^{16} ratios of coexisting minerals in glaucophane-bearing metamorphic rocks: Geol. Soc. America Bull., v. 79, p. 1727-1756.

Turner, F. J., 1968, Metamorphic petrology: New York, McGraw-Hill, 403 p.

Washington, H. S., 1901, A chemical study of the glaucophane schists: Am. Jour. Sci., 4th ser., v. 11, p. 35-59.

Wray, J. L., and Daniels, F., 1957, Precipitation of calcite and aragonite: Am. Chem. Soc. Bull., v. 79, p. 2031-2034.

V
Evolution of Metamorphic Facies Types with Time

Editor's Comments on Papers 21 and 22

21 de Roever: *Some Differences Between Post-Paleozoic and Older Regional Metamorphism*

22 Ernst: *Occurrence and Mineralogic Evolution of Blueschist Belts with Time*

A gradual change in the physical conditions attending metamorphism has been referred to by many authors, including Miyashiro (Papers 1 and 6) and Zwart (Paper 2). These authors recognized that individual metamorphic belts are characterized by a specific range of P–T conditions and moreover that, in general, pre-Mesozoic and especially Precambrian recrystallization took place at relatively higher temperatures and relatively lower pressures than post-Paleozoic events.

The subject is pursued in greater detail in Papers 21 and 22 by de Roever and Ernst, respectively. These authors concerned themselves particularly with glaucophane schists and related rocks, but in so doing, addressed the general problem of the possible secular change in the earth's geothermal gradient. It is recognized that the relatively high pressure, relatively low temperature metamorphic terranes are especially susceptible to obliteration through subsequent recrystallization and tectonism, both by virtue of (1) their extreme departure from normal thermal regimes and mineral equilibrium states during generation, and (2) their spatial location at convergent plate junctions, where later orogenic events such as continental accretion or subcrustal erosion take place. Nevertheless, low-grade metamorphic rocks are preserved in some of the most ancient terrestrial sections. Hence it would be truly remarkable if blueschists had been produced abundantly in the early Precambrian without any being retained as relics.

Moreover, an evolution of glaucophane schist mineralogy appears to have taken place, with higher-pressure (e.g., jadeitic) phase assemblages being confined to post-Paleozoic terranes. Evidently the Phanerozoic era has witnessed a gradual change in P–T conditions of subduction zones. This means that a greater downbowing of the isotherms near the convergent plate junction is characteristic of young sea-floor-spreading regimes. This, in turn, probably is a function chiefly of (a) an increase in thickness of the lithospheric plates; or (b) an increase in the rate of convergence and subduction. Until the mechanism that drives the plates is understood, it is difficult to evaluate hypothesis (b); however, presuming a thermally induced gravitative instability, and considering the earth's limited heat budget, it appears implausible to call upon an increasing rate of plate motion with decreasing age. Hypothesis (a), which suggests a thickening of the lithospheric plates, hence a decreasing geothermal gradient, with the passage of time, is compatible with the ever-decreasing production of radiogenic heat.

A secular cooling of the earth, and an increase in depth to the asthenosphere-lithosphere boundary, is also in harmony with the observed restriction of ultramafic lavas (komatiites) to early Precambrian terranes (see, e.g., Viljoen and Viljoen, 1969; Nesbitt, 1971; Pyke, Naldrett, and Eckstrand, 1973). These magmas, which apparently were extruded at temperatures on the order of 1350–1400°C, most probably reflect the attendance of a relatively higher thermal regime during magma generation which was restricted to early stages of the earth's recorded history. General readers, however, should be cautioned that this secular cooling should not be extrapolated to surface environments, where the evidence of early Precambrian sedimentation, biologic activity, and episodic glaciations recorded throughout geologic time deny any gross thermal changes at the atmosphere/crust and hydrosphere/crust boundaries.

References

Nesbitt, R. W. (1971) Skeletal crystal forms in the ultramafic rocks of the Yilgarn block, Western Australia; evidence for an Archean ultramafic liquid: *Geol. Soc. Australia Spec. Publ.* **3,** 331–347.

Pyke, D. R., Naldrett, A. J., and Eckstrand, O. R. (1973) Archean ultramafic flows in Munro Township, Ontario: *Geol. Soc. America Bull.,* **84,** 955–978.

Viljoen, M. J., and Viljoen, R. P. (1969) Evidence for the existence of a mobile extrusive peridotitic magma from the Komati Formation of the Onverwacht Group: *Geol. Soc. South Africa Spec. Publ.* **2,** Upper Mantle Project, 87–112.

Reprinted from *Geologie en Mijnbouw*, (N.S.)**18e**, 123–127 (Apr. 1956)

SOME DIFFERENCES BETWEEN POST-PALEOZOIC AND OLDER REGIONAL METAMORPHISM

W. P. DE ROEVER[1]

ABSTRACT

Many regionally metamorphosed rocks found in post-Paleozoic orogenic zones show important mineralogical differences from the corresponding rocks in older orogenic belts. Metamorphism in the glaucophane schist facies, for instance, shows a striking preferential distribution in the post-Paleozoic orogenic belts. Furthermore, it seems that all known lawsonite is of post-Paleozoic age. It is suggested that post-Paleozoic regional metamorphism, when compared with pre-Mesozoic regional metamorphism, is characterized by the predominance of less steep geothermal gradients during the main phase of metamorphism. There may have been a general, though possibly oscillating, decrease in the steepness of the geothermal gradients during the main phase of regional metamorphism from the early pre-Cambrian toward the youngest orogenic epochs, involving certain changes in the character of the metamorphic mineral assemblages produced. Seen in this light, it seems by no means impossible, for instance, that lawsonite will indeed appear to be a guide mineral for post-Paleozoic metamorphism. It is hoped that the results of this study will encourage further investigations in this interesting field of historical mineralogy and petrology.

A comparison between post-Paleozoic and pre-Mesozoic regional metamorphism involves many difficulties. In many cases it is still unknown whether a certain type of metamorphism shown by older rocks in a given orogenic belt was produced during the formation of that belt or during a previous orogenic cycle. Moreover in several regions even the age of the original rocks is unknown. Therefore a comparison of post-Paleozoic and older regional metamorphism can only be based on data gathered in comparatively well-known regions, the age relations in adjacent parts of the same orogenic belt being inferred by analogy.

A further difficulty is that the most detailed classification of the different types of metamorphism, the classification according to Eskola's ingenious facies principle, has not yet been generally accepted as one of our most important aids in the study of polymetamorphic rocks[2]. Classification according to the facies principle is particularly useful when the different types of metamorphism shown by polymetamorphic rocks are of an approximately similar grade; this is illustrated by the results of recent investigations of

rocks from Celebes (de Roever, 1947, 1950, 1953) and in Corsica (Brouwer and Egeler, 1952; Egeler, 1956; further references are given in the last-named paper). Particularly in the description of rocks from the Alps, Eskola's classification has only seldom been used. Hence a comparison of post-Paleozoic and older regional metamorphism cannot be adequately based on facies studies. It is to be welcomed, therefore, that the use of the facies classification was recently recommended by the well-known Swiss petrologist Bearth (1952).

A general comparison between post-Paleozoic and pre-Mesozoic regional metamorphism, however, can also be based on the study of those minerals which show a preferential distribution in either the post-Paleozoic or the older orogenic belts. As already recognized by Eskola (1929), one of the most striking differences in this respect is represented by the abundance of glaucophane and crossite in the metamorphic rocks of the post-Paleozoic orogenic zones and their scarcity and local occurrence in the older belts. Glaucophane and crossite are very widely distributed, e.g., in Corsica, along the Franco-Italian border, in Greece, in Celebes and in New Caledonia, where these minerals may occur in extensive regions. On the other hand, the older areas with glaucophane or crossite are small, e.g. those of Ayrshire, Anglesey, Ile de Groix, Queensland (Schürmann, 1951, 1953), and northern Portugal (Cotelo Neiva, 1948, p. 122-125). Of all blue metamorphic amphiboles known, more than 95 percent, perhaps even con-

---

[1] Geological and Mineralogical Institute, University of Leiden, formerly Geological Institute, University of Amsterdam.

[2] The reluctance of many authors to use the facies classification may have been caused partly by the unsatisfactory character of the definitions of a metamorphic facies given in several important textbooks. This question was recently discussed by Ramberg (1952, p. 136), who gave the following more appropriate definition: "Rocks formed or recrystallized within a certain P,T-field, limited by the stability of certain critical minerals of defined composition, belong to the same mineral facies." Again, some authors may have rejected the facies classification for polymetamorphic rocks because the mineral assemblages of such rocks do not closely approximate equilibrium conditions. If, however, the existence of a number of given metamorphic facies has been definitely established, and if the corresponding critical minerals or associations are sufficiently known, it is also warranted to use these data for the disentangling of the history of polymetamorphic rocks.

siderably more, seems to be of post-Paleozoic age.

A still more striking preferential distribution is shown by the mineral lawsonite. This mineral is of plentiful occurrence in many parts of the post-Paleozoic orogenic belts, where it is confined to regions with glaucophane-bearing rocks. On the other hand, no lawsonite seems ever to have been found in Anglesey or Ayrshire (personal communication by Professor Tilley), in Ile de Groix (unpublished investigations by the present author) and in the other older occurrences of glaucophane and crossite. Indeed, it seems that all lawsonite known is of post-Paleozoic age. It may be that some exceptions to this rule will be found, or have already been found, but the fact remains that there are empirical indications that if not all, virtually all lawsonite was produced by post-Paleozoic metamorphism.

On the other hand, there are at least several minerals that seem to be less wide-spread as post-Paleozoic than as older metamorphic products. Staub (1948) assumes that biotite, hornblende, and several other minerals show a preferential distribution in the oldest rocks of the Alps. In several other parts of the post-Paleozoic orogenic belts large quantities of metamorphic biotite [3] seem to be confined to pre-Mesozoic units incorporated in these belts. This is considered to hold true for large parts of Corsica (Egeler, 1956, and references given by this author) and similarly for eastern Celebes and the adjacent island of Kabaena. Biotite may be plentiful, however, in regions showing post-Paleozoic thermal or plutonic metamorphism. Non-fibrous green hornblende seems to show an even more pronounced preferential distribution as a product of older metamorphism; in Corsica and eastern Celebes this mineral was also found to be of pre-Mesozoic age.

The above instances may suffice to illustrate that many regionally metamorphosed rocks found in post-Paleozoic orogenic zones show striking mineralogical differences from the corresponding rocks in older orogenic belts.

The significance of the preferential distribution of the minerals mentioned above can best be discussed in terms of the facies principle. Glaucophane, crossite and lawsonite are all three critical minerals of the glaucophane schist facies, in which biotite and non-fibrous green hornblende are not known as stable constituents [4].

Hence we may conclude that metamorphism in the glaucophane schist facies shows a striking preferential distribution in the post-Paleozoic orogenic belts. Further, those subfacies of this facies which are characterized by the stability of lawsonite, the lawsonite-glaucophanite subfacies and the garnet-lawsonite-glaucophane schist subfacies, seem to be exclusively confined to these belts. In several parts of these post-Paleozoic belts the metamorphism in the glaucophane schist facies is followed by a subordinate late phase of metamorphism in the green schist facies.

Now the question arises to which cause this preferential distribution of rocks of the glaucophane schist facies in younger orogenic zones is to be ascribed. Almost complete erosion of such rocks in the older belts is apparently to be dismissed, since in these belts rocks of the green schist and epidote-amphibolite facies — i.e. of facies that are considered as temperature equivalents of the glaucophane schist facies — are of wide-spread occurrence. Further, almost all glaucophane, crossite and lawsonite that have escaped alteration during later phases of metamorphism, show a remarkably fresh appearance. Hence the scarcity or absence of these minerals in older belts is apparently neither due to their being preferentially altered in the course of the geological history. Evidently we are dealing here with some special characteristics of the orogenic periods in question.

In this respect attention may be drawn to the comparatively high specific gravity of the critical minerals of the glaucophane schist facies (Eskola, 1929) and to the range of temperatures during their formation, which appears to correspond essentially to that of the green schist and epidote-amphibolite facies. As already concluded in a former paper (de Roever, 1955), the conditions giving rise to metamorphism in the glaucophane schist facies are apparently characterized by slightly higher pressures, or, in other words, by a slightly different geothermal gradient during the metamorphism. This gradient is considered to have been lower, i.e. less steep, during metamorphism in the glaucophane schist facies than during metamorphism in the other facies mentioned.

---

[4] It may be remarked here that the existence of a separate glaucophane schist facies has not been generally accepted but, in the opinion of the author, has now been definitely established. Main arguments are (1) the intimate association of lawsonite and glaucophane in numerous metamorphic rocks of very different character, the occurrence of lawsonite being confined to regions with glaucophane-bearing rocks; and (2) the chemical equivalence of many of the natural assemblages with glaucophane and lawsonite and many assemblages containing albite, chlorite, tremolite-actinolite and clinozoisite-epidote (de Roever, 1950, 1955).

---

[3] The biotite-like mineral stilpnomelane, which is often of post-Paleozoic age, should not be mistaken for biotite.

It is suggested, therefore, that post-Paleozoic regional metamorphism, when compared with pre-Mesozoic regional metamorphism, is characterized by the predominance of less steep geothermal gradients during the main phase of metamorphism.

On other grounds some authors have assumed the existence of analogous differences in geothermal gradient between different orogenic epochs. According to Daly (1917), Bucher (1933), and Turner (1948, p. 288) the temperature gradients were apparently steeper in early pre-Cambrian times than afterwards, so that conditions permitting regional metamorphism were reached much closer to the surface than was the case in later geological periods. Bucher goes even farther and assumes a regressive change toward the post-Paleozoic belts.

Therefore there may have been a general, though possibly oscillating, decrease in the steepness of the geothermal gradients during the main phases of regional metamorphism from the early pre-Cambrian toward the youngest orogenic epochs.

Seen in this light, the apparent absence of lawsonite and the scarcity of glaucophane and crossite among the products of pre-Mesozoic regional metamorphism gain in importance. There is apparently not only an evolution of life during the history of the earth, but also some change in the character of the metamorphic mineral assemblages produced during the main phases of regional metamorphism of the various orogenic epochs[5]. It seems by no means impossible that lawsonite will indeed appear to be a guide mineral for post-Paleozoic metamorphism, and that its occurrence in pebbles in non-metamorphic clastic sediments will appear to indicate that the sediments in question are of post-Paleozoic age. Although glaucophane and crossite are of much less value as indicators of the age of a period of metamorphism, even the occurrence of these minerals in extensive regional distribution may similarly indicate a post-Paleozoic age.

---

[5] In this respect it may also be mentioned that eclogites, which are rocks of great density, seem to be absent or rare in pre-Cambrian complexes. The origin of at least part of these rocks may be roughly comparable to that of the rocks of the glaucophane schist facies.

In the opinion of the author, a discussion of the origin of the phenomena described above is beyond the scope of the present paper. It is hoped, however, that the above lines will encourage further investigations in this interesting field of historical mineralogy and petrology.

REFERENCES

Bearth, P. (1952) — Über das Verhältnis von Metamorphose und Tektonik in der penninischen Zone der Alpen. Schweiz. Min. Petr. Mitt., v. 32, p. 338-347.

Brouwer, H. A., and Egeler, C. G. (1952) — The glaucophane facies metamorphism in the schistes lustrés nappe of Corsica. Kon. Nederl. Akad. Wetensch., Verh. Afd. Natuurk., Tweede reeks, v. 48, no. 3.

Bucher, W. H. (1933) — The deformation of the earth's crust. Princeton University Press.

Cotelo Neiva, J. M. (1948) — Rochas e minérios da regiao Bragança — Vinhais. Relatório Serviço de Fomento Mineiro de Portugal, v. 14.

Daly, R. A. (1917) — Metamorphism and its phases. Bull. Geol. Soc. Am., v. 28, p. 375-418.

de Roever, W. P. (1947) — Igneous and metamorphic rocks in eastern central Celebes. Geol. explorations in the island of Celebes under the leadership of H. A. Brouwer, Amsterdam, North-Holland Publishing Co., p. 65-173.

——, (1950) — Preliminary notes on glaucophane-bearing and other crystalline schists from South East Celebes, and on the origin of glaucophane-bearing rocks. Proc. Kon. Nederl. Akad. Wetensch., v. 53, p. 1455-1465.

——, (1953) — Tectonic conclusions from the distribution of the metamorphic facies in the island of Kabaena, near Celebes. Proc. 7th Pacific Science Congr. (New Zealand 1949), v. 2, p. 71-81.

——, (1955) — Some remarks concerning the origin of glaucophane in the North Berkeley Hills, California. Am. Journ. Sci., v. 253, p. 240-244.

Egeler, C. G. (1956) — The alpine metamorphism in Corsica. Geologie en Mijnb., v. 18, p. 115-118.

Eskola, P. (1929) — Om mineralfacies. Geol. För. Stockh. Förh., v. 51, p. 157-172.

Ramberg, H. (1952) — The origin of metamorphic and metasomatic rocks. Univ. of Chicago Press.

Schürmann, H. M. E. 1(951) — Beiträge zur Glaukophanfrage. Neues Jahrb. Min. Monatsh., v. 1951, p. 49-68.

——, (1953) — Beiträge zur Glaukophanfrage (2). Neues Jahrb. Min. Abh., v. 85, p. 303-394.

Staub, R. (1948) — Aktuelle Fragen im alpinen Grundgebirge. Schweiz. Min. Petr. Mitt., v. 28, p. 422-442.

Turner, F. J. (1948) — Mineralogical and structural evolution of the metamorphic rocks. Geol. Soc. Am. Mem. 30.

DISCUSSION

Prof. DE SITTER (Leiden) draws attention to an gradient depends much less on the general geothermal gradient than on the depth of the thermal front, which may be accompanied by intrusive rocks. He cites the conditions near the massif of St. Barthélemy, described by Zwart. Here the succession of the zones of thermal metamorphism is much more rapid where the distance to the contemporaneous earth surface was small, than where this distance was large. Though in the former case the gradient was much steeper than in the latter, there is no difference between the mineral zones developed.

Dr. DE ROEVER points out that the gradients dealt with in his lecture are not world-wide general geothermal gradients but local geothermal gradients in orogenic belts, prevailing during the main phase of regional metamorphism. Such a local geothermal gradient is to be defined as the average thermal gradient above a rock that is being metamorphosed. It is expressed in the ratio between temperature and depth-controlled pressure. A local geothermal gradient thus defined, may indeed be considered as depending not only on a world-wide general geothermal gradient but also on local circumstances, which latter, in fact, may be much more important. Prof. de Sitter's question, however, touches the origin of the phenomena described, for which, at the moment, the speaker is not able to give an adequate explanation, As to the second part of Prof. de Sitter's remark: if there is a thermal front that is distinctly connected with the presence of granitic rocks, as in both St. Barthélemy examples, the local geothermal gradients during metamorphism are steeper than those connected with the formation of glaucophane-bearing rocks; in the large region of rocks of the glaucophane schist facies in central Celebes, for instance, synmetamorphic granitic rocks have not been found. Further, the rather steep gradients of the two St. Barthélemy examples, though greatly different, were apparently of such magnitudes that metamorphism still took place within the P,T-fields of the same mineral assemblages. With regard to the concept of a thermal front, Dr. de Roever judges it by no means impossible that during certain kinds of regional metamorphism there was not a raise but a depression of the isotherms, as contended by Daly. This may have occurred when the effects of downward displacement by folding and geosynclinal subsidence were greater than those of the conduction of heat through the rocks, which is extremely slow.

Dr. KÜNDIG (B.P.M.) remarks that the crust of the earth passes through an evolution, which line is also followed by the orogens. It is difficult, however, to discriminate between factors controlled by the general line and such controlled by phases of orogenesis. Studies of Nantz in the United States and Canada on sediments, hypotheses of changes in the composition of the oceans, etc., lead in the same direction.

Dr. DE ROEVER replies that, in order to avoid the effects of local influences and special phases of orogenesis, he made a comparison in a very general sense, between all exposed parts known of all post-Paleozoic and older orogenic belts. Therefore, the striking preferential distribution of glaucophane, crossite and lawsonite in post-Paleozoic belts, to his opinion, is indeed controlled by the general line of evolution of the earth's crust. Perhaps age-controlled differences shown by some ·other geological objects are also connected with the steepness of the geothermal gradient.

Dr. SCHÜRMANN remarks that the length of time elapsed after the formation of the metamorphic rocks of the older orogenic belts, must be taken into account. Pre-Cambrian and Paleozoic schists have been metamorphosed during so many different orogenic epochs that original glaucophane may have disappeared.

Dr. DE ROEVER: It is not possible to accept the explanation mentioned by Dr. Schürmann, since there are many parts of older orogenic belts that have not been incorporated in younger belts; this is clearly illustrated by the different geographical position of orogenic belts of different ages. This fact is also of some importance in connection with the views of Daly, Bucher und Turner on the steepness of the geothermal gradients in early pre-Cambrian times,

since one of the main arguments of these authors is concerned with the amount of erosion in early pre-Cambrian orogenic belts. In this respect the following may be cited from Bucher (op.cit., p. 296): "It is customary to speak of the structure of the Archean rocks as being the product of processes that have been operative 'at great depths'. But there seems little reason for this assertion. The Epi-Archean peneplain truncates these structures. Why should the structures revealed by Epi-Archean base-levelling differ from those exposed by, say, Epi-Carboniferous erosion to base-level? We are deceived by the subconscious thought of the great sediments that have accumulated on top of the Archeozoic rocks in some parts of the world.

If we are to believe that the Archeozoic structures originated at greater depth than those revealed of later date, we must also assume that greater amounts of rock were removed by erosion during Epi-Archean peneplanation than at any time since. The writer knows of no observation which would support this assumption."

Dr. PANNEKOEK (Geol. Survey) would like to have some additional information on what Dr. de Roever means with a "steep" and what with a "low" geothermal gradient. Do we have to assume that the steep one is many times greater than the low one or is it only slightly greater? Further, Prof. de Sitter mentioned an example of great differences in geothermal gradient without much difference in metamorphism. Other examples show that different minerals may be formed with only small differences in gradient.

Dr. DE ROEVER: This can be illustrated in a P,T-diagram, when we draw a line separating the fields of stability of a group of higher pressure assemblages (e.g. those of the glaucophane schist and eclogite facies) from the fields of stability of a group of lower pressure assemblages (e.g. those of the so-called normal facies series). Within the group of lower pressure fields there may be large differences in P,T-ratio, or, in other words, the group of steeper gradients comprises gradients that may show large differences (e.g. the examples mentioned by Prof. de Sitter). On the other hand, the differences in P,T-ratio between metamorphism on both sides of the separating line may be small, i.e., the differences in gradient between regional metamorphism in the glaucophane schist facies and regional metamorphism in the green schist and albite-epidote-amphibolite facies may be small. The latter differences may be comparable to those between 20° and 25° C per km.

Prof. NIEUWENKAMP (Utrecht) would like to have some information about the importance of the history of a metamorphic rock before metamorphism but after the formation of its original material. Is it important, whether a rock is rapidly or slowly heated, etc.? Further, the mineral assemblage produced need not be an equilibrium assemblage.

Dr. DE ROEVER remarks that, in his opinion, the previous history as meant by Prof. Nieuwenkamp does not influence the character of the mineral assemblages produced as long as there is no previous metamorphism, e.g. by internal movements under conditions favourable for metamorphism. The actual mineral assemblage found often is a non-equilibrium assemblage, as illustrated by the frequent occurrence of unstable relics and hysterogene minerals.

Prof. BROUWER (Amsterdam) asks whether a certain amount of hydrostatic pressure is essential to the formation of glaucophane and lawsonite, and, if so, whether this hydrostatic pressure is only controlled by depth.

Dr. DE ROEVER replies that under the conditions prevailing in nature, i.e., at metamorphic temperatures, a certain amount of pressure is indeed considered to be essential to the formation of the minerals mentioned. This need not hold true, however, for an eventual laboratory synthesis of glaucophane and lawsonite at room temperature. Regional metamorphism in the glaucophane schist facies in considered to be a regional dislocation metamorphism under confining pressures that are slightly higher than those prevailing during regional metamorphism in the green schist and albite-epidote-amphibolite facies. Besides depth-controlled pressure there may be a certain amount of confining pressure originated by the combined effect of unilateral pressure and resistance to deformation. In this respect it may be mentioned that in hardness tests with a Vickers indenter nearly two-thirds of the mean pressure of contact is in the form of a hydrostatic pressure and only one-third remains effective in producing plastic indentation (Tabor, Endeavour, Jan. 1954).

Mr. TOBI (Leiden) asks whether there are any theoretical objections against local genesis of lawsonite and glaucophane by pre-Mesozoic regional metamorphism.

Dr. DE ROEVER: No, there may be local variations in the magnitude of the geothermal gradient during the main phase of regional metamorphism. Since lawsonite, unlike glaucophane, has not been found in pre-Mesozoic orogenic belts, this mineral may require slightly higher pressures for its formation.

Reprinted from *Amer. Jour. Sci.*, **272**, 657–668 (1972)

OCCURRENCE AND MINERALOGIC EVOLUTION OF BLUESCHIST BELTS WITH TIME

W. G. ERNST

Department of Geology and Institute of
Geophysics and Planetary Physics,
University of California, Los Angeles, California 90024

ABSTRACT. Blueschists and related rocks are virtually confined to Phanerozoic metamorphic belts (de Roever, 1956) and are progressively more abundant in younger rocks. Although widespread in younger glaucophane schist terranes, lawsonite is rather uncommon in Paleozoic blueschists; metamorphic aragonite and jadeitic pyroxene + quartz are strictly confined to blueschist belts of Mesozoic and Cenozoic age. These observations are compatible with a suggested systematic decrease with time in the Earth's geothermal gradient (at least adjacent to convergent lithospheric plate junctions, the subduction zone locale where glaucophane schists appear to be generated). Such cooling would have promoted the gradual thickening of lithospheric plates during Phanerozoic time: if so, recorded Precambrian tectonic and petrologic processes may have taken place in the presence of rather thin lithospheric plates.

INTRODUCTION

Blueschist belts occur principally around the margins of the Pacific Ocean, in the Caribbean, and the Alpine-Himalayan region (for examples, see van der Plas, 1959; Dobretsov and others, 1966; Ernst and others, 1970; Coleman, 1971). These low-temperature, high-pressure metamorphic suites appear to be confined largely to Paleozoic and younger orogenic terranes (de Roever, 1956, 1964) where they are associated with ophiolite sequences (for examples, see Bailey, Irwin, and Jones, 1964; Bailey, Blake, and Jones, 1970; Nicolas, 1968, 1969; Coleman, 1971; Hsü, 1971). Glaucophane schist belts are thought to be generated by special physical conditions that characterize recognizable convergent lithospheric plate boundaries (Dewey and Bird, 1970; Ernst, 1970). The generally developed progressive metamorphic facies sequence zeolite → pumpellyite → greenschist + blueschist → eclogite + albite amphibolite has been postulated to reflect the polarity of lithospheric plate descent (Ernst, 1971) in much the same way that the K_2O/SiO_2 ratio indicates depth to the subduction zone—hence direction of dip of the downgoing lithospheric slab (Dickinson and Hatherton, 1967; Dickinson, 1970).

This note summarizes the mineralogic evolution of blueschist belts as a function of time and attempts to relate this phenomenon to a systematic decrease in the Earth's geothermal gradient near consumptive plate junctions through increase in lithospheric plate thickness (or, less likely, increase in plate motion velocities) as the present is approached. Such a situation may also account for the virtual absence of glaucophane schist belts, as well as the rarity of eclogitic rocks, and for the general abundance of high-grade sillimanitic, metamorphic, and migmatitic terranes in Precambrian complexes (Miyashiro, 1961, p. 305). Low-grade greenstone belts are, of course, widely distributed in Precambrian terranes too.

MINERALOGIC EVOLUTION OF BLUESCHIST BELTS

Original and recrystallization ages and aspects of the mineralogy of these low-grade but relatively high-pressure metamorphic belts are listed in table 1. Almost all lithologic suites include rocks that carry blue Na-amphibole (glaucophane or crossite), so this phase is not tabulated. Other critical minerals include: (1) pumpellyite, a distinctive, very low-grade mineral; (2) the successively higher pressure phases lawsonite, aragonite, and jadeitic pyroxene (+ quartz), all characteristic of blueschist facies P–T conditions; and (3) epidote and garnet, typical (in association with Na-amphibole) of physical conditions transitional between the blueschist and greenschist facies.

As de Roever (1956) pointed out long ago, most glaucophane schist belts are of post-Paleozoic metamorphic age. This observation subsequently has been documented through radiometric measurements. Blueschist terranes of undoubted Precambrian age are unknown; the only possible examples that appear to fall in this category are two isolated occurrences, Anglesey, Wales, and the Seward Peninsula of Alaska, where geologic evidence suggests pre-Ordovician (probably latest Precambrian) ages of recrystallization. Unfortunately, no radiometric data are available. Even if these two localities do prove to be late Precambrian in age, it is clear from table 1 that the abundance of glaucophane schist belts increases proceeding toward more recent time. Of course, this phenomenon could be due simply to better preservation of younger metamorphic suites.

However, and perhaps more significantly, table 1 also demonstrates that, although intermediate greenschist-blueschist terranes, characterized by assemblages such as crossite ± actinolite + epidote + albite + chlorite, occur throughout the Phanerozic, lawsonite seems to be rare in schists of Paleozoic metamorphic age, and aragonite and jadeitic pyroxene + quartz are strictly confined to Mesozoic and Cenozoic (including Alpine) metamorphic terranes. These critical index minerals require successively higher pressures for their stabilities, as demonstrated by experimental phase equilibrium studies (for example, see Boettcher and Wyllie, 1968; Liou, 1971; Johannes and Puhan, 1971). Summarized a little more quantitatively: less than a third of the late Precambrian and Paleozoic blueschist parageneses involve lawsonite, whereas none contain either aragonite or jadeitic pyroxene + quartz; in contrast, about two-thirds of the Mesozoic and eight-ninths of the Cenozoic high-pressure, low-temperature suites represented in table 1 contain lawsonite, and approximately half the occurrences also include sodic pyroxene + quartz and/or aragonite. The data of table 1 are displayed graphically in figure 1. Perhaps in passing, it is worth noting that a period of 20 to 30 m.y. or more apparently is required after metamorphism for such blueschist-type belts to become exposed at the Earth's surface.

DISCUSSION

What is the cause of this apparent mineralogic variation with time? Evidently it is a real effect, for blueschists and allied rocks (for example, low-temperature eclogites—Group C of Coleman and others, 1965) are so distinctive that they are unlikely to have been overlooked, even in the course of reconnaissance mapping. Furthermore, such failure to recognize glaucophane schists would have to have been biased in favor of Precambrian terranes and would not account for the increasingly greater restrictions of successively higher pressure phases to more recently metamorphosed terranes.

Blueschist facies mineral assemblages are metastable under surficial and other low-grade metamorphic conditions. For this reason, any subsequent reaction will tend to obliterate the high-pressure assemblages. Aragonite shows clear evidence of back reaction in such terranes (Coleman and Lee, 1962). This phase is absent from old rocks, at least in part because of rapid low-temperature conversion to calcite (Brown, Fyfe, and Turner, 1962; Davis and Adams, 1965). Although rate studies have not been performed on Na-amphibole, lawsonite, and jadeitic pyroxene, the sluggishness with which these phases decompose experimentally is comparable to other silicates; hence the absence of these minerals in old rocks probably cannot be ascribed to near-surface retrograde recrystallization.

Inasmuch as glaucophane schist belts seemingly are formed at plate sutures, continued postmetamorphic underflow and accretion would be expected to result in an oceanward stepping of the subduction zone (for examples, see Elsasser, 1971; Hsü, 1971; Oxburgh and Turcotte, 1971). This process might eventually cause the conversion of preexisting blueschist belts into high-temperature plutonic + metamorphic terranes (for example, see Miyashiro, 1967). Continued, or episodic, reworking of continental margin high-pressure belts would result in their selective removal from the geologic record with time. Nonetheless, great tracts of feebly metamorphosed rocks, including ophiolite- and other greenstone-belts, have been abundantly preserved within Precambrian shields (for instance, see Anhaeusser and others, 1968); blueschists are conspicuously absent from these areas. It is difficult for me to conceive of a process whereby glaucophane schist terranes could have been so efficiently destroyed, meanwhile preserving other equally low-grade rocks, such as ophiolite (greenstone) complexes which evidently have been welded to the continental or island arc margins.

Although older terranes in general are more deeply eroded than younger ones, in sufficiently large tracts all crustal levels appear to be exposed; therefore, it also seems implausible that ancient blueschist belts were once abundant but have been removed entirely from the geologic record by erosion.

The explanation for the total lack of ancient blueschists advanced by de Roever (1956, 1964) is adopted here—namely, that the Earth's geothermal gradient has decreased with time, so that the temperature at

TABLE 1

Mineralogic-temporal relationships in high-pressure, low-temperature metamorphic belts

Location	Sedimentary-volcanic age	Metamorphic age (numerical values determined radiometrically)	Critical minerals	References
South-central Chile	Early-Middle Paleozoic	210-245 m.y.	lawsonite	Saliot (1968); González-Bonorino and Aguirre (1970); González-Bonorino (1971)
Cuba	Mesozoic	Alpine (?)	lawsonite	Schürmann (1936); Hill (1959)
North-central Venezuela	Early Cretaceous (?)	Mid-Cretaceous	epidote	Dengo (1953); Shagam (1960)
Guatemala	Mesozoic (?)	Late Cretaceous-Eocene	lawsonite	McBirney (1963); McBirney, Aoki, and Bass (1967)
Baja California Mexico	Mid-Mesozoic	Mid-Mesozoic	lawsonite	Cohen and others (1963)
Western California	Late Jurassic-Cretaceous	100-150 m.y.	pumpellyite, lawsonite, aragonite, jadeitic pyroxene	Coleman and Lee (1963); Lee and others (1964); Bailey, Irwin, and Jones (1964); Suppe (1969); Ernst and others (1970)
Klamath Mountains, California	Paleozoic-Triassic	Mesozoic	lawsonite	Davis (1966, 1968)
Central Oregon	Late Paleozoic-Mesozoic	> Mid-Cretaceous	lawsonite	Swanson (1969)
North-central Washington	Pre-Mesozoic	latest Permian	epidote	Misch (1966, 1969)
Northwestern Washington	Mid-Paleozoic-Mesozoic	Mid-Cretaceous	pumpellyite, aragonite	Hawkins (1967); Vance (1968)
British Columbia	Permian and older	Permo-Triassic and younger	epidote, pumpellyite, lawsonite	Roddick and others (1967); Monger (1969); Monger and Hutchison (1970)
South-central Alaska	Triassic (?)	Mesozoic	epidote	Forbes and others (1971)
Seward Peninsula, Alaska	Late Precambrian (?)	pre-Ordovician	garnet, epidote, pumpellyite	Sainsbury, Coleman and Kachadoorian (1970); Forbes and others (1971)

414

Locality	Age	Age (m.y.)	Minerals	References
Eastern Siberia-Northwestern Kamchatka	Early-Mid-Paleozoic	325-330 m.y.	pumpellyite, lawsonite	Dobretsov and Kuroda (1969); Firsov and Dobretsov (1969)
Kamchatka	Mesozoic	160 m.y.	epidote	Dobretsov and others (1966); Dobretsov and Kuroda (1969); Lebedev, Tararin, and Lagouskaya (1967)
Hokkaido-Sakhalin	Mesozoic	Cretaceous or younger	lawsonite, jadeitic pyroxene	Seki and Shido (1959); Suzuki and Suzuki (1959); Bogdanov (personal commun., 1968); Dobretsov and Kuroda (1969)
Central Kyushu	Early Paleozoic (?) Paleozoic	429-454 m.y. 205-330 m.y.	epidote pumpellyite, epidote	Hayase and Ishizaka (1967) Banno (1958); Hashimoto (1968); Yamaguchi and Yanagi (1970)
Northwestern Honshu-Kyushu Southwestern Honshu	Late-Paleozoic-Mid-Jurassic	110 m.y.	pumpellyite, lawsonite, jadeitic pyroxene	Matsumoto (1947); Seki (1958, 1960, 1965); Seki and others (1964); Yamaguchi and Yanagi (1970)
Shikoku	Late Paleozoic-Early Mesozoic	> 102 m.y.	pumpellyite, epidote, (lawsonite)	Miller and others (1961); Banno and Miller (1964); Ernst and others (1970)
Taiwan	Late Paleozoic-Mesozoic	Late Mesozoic	epidote	Yen (1966, 1967); Ho (1967); Biq (1971)
Philippines	Late Paleozoic-Mesozoic	Late Mesozoic	lawsonite, aragonite	Coleman and Lee (1962); Gervasio (1967); Seki (personal commun., 1970)
Celebes	Mesozoic	Cretaceous-Tertiary	pumpellyite, lawsonite, jadeitic pyroxene	de Roever (1947, 1955)
New Caledonia	Mesozoic-Eocene	21-38 m.y.	lawsonite, aragonite, jadeitic pyroxene	Coleman (1967); Brothers (1970); Lillie and Brothers (1970); Blake (personal commun. 1971)
New Zealand	Late Permian-Jurassic	108-129 m.y.	pumpellyite, lawsonite, aragonite	Coleman (1966); Harper and Landis (1967), Landis and Coombs (1967)

TABLE 1 (continued)

Location	Sedimentary-volcanic age	Metamorphic age (numerical values determined radiometrically)	Critical minerals	References
New Guinea	Jurassic-Cretaceous	52 m.y.	pumpellyite, lawsonite, aragonite	Verhofstad (1966); Davies (1968); Davies and Smith (1971)
Eastern Australia	Late Paleozoic	Mesozoic (?)	pumpellyite, lawsonite	Joplin (1968)
Urals	Late Precambrian-Early Paleozoic	385-450 m.y.	epidote, lawsonite	Dobretsov and others (1966); Sobolev and others (1967); Hamilton (1970)
Turkey	Early Paleozoic	65-82 m.y.	lawsonite, aragonite, jadeitic pyroxene	van der Kaaden (1966, 1969); Cogulu (1967); Cogulu and Krummenacher (1967)
Southern Italy	Mid-Late Mesozoic	Alpine	pumpellyite, lawsonite, aragonite, jadeitic pyroxene	de Roever and others (1967); Hoffmann (1970)
Corsica	Mesozoic	Alpine	lawsonite, jadeitic pyroxene	Brouwer and Egeler (1952)
Western Alps	Late Paleozoic-Late Mesozoic	> 38 m.y.	pumpellyite, lawsonite, jadeitic pyroxene	Bearth (1962, 1966, 1967); Niggli (1970); Hunziker (1970); Bocquet (1971); Chatterjee (1971)
Southeastern Spain	Triassic and older	Alpine	garnet, epidote	de Roever and Nijhuis (1963); Nijhuis (1964)
Île de Groix, France	Early Paleozoic (?)	Paleozoic	epidote (lawsonite?)	Cogné, Jeannette, and Ruhland (1966); Velde (1970); Nicolas (personal commun., 1971)
Scotland	Early Ordovician	Ordovician	pumpellyite, epidote	Bloxam (1958); Bloxam and Allen (1959)
Wales	Precambrian (?)	pre-Ordovician	epidote (lawsonite ?)	Greenly (1919); Holgate (1951); Shackleton (1969)

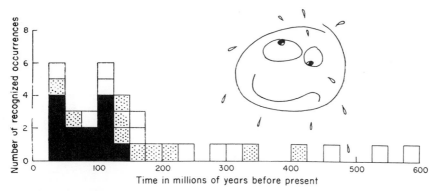

Fig. 1. Histogram showing incidence of blueschists of contrasting mineralogies with time. Open boxes represent epidote-bearing glaucophane schists, stippled pattern lawsonite (± epidote), and black boxes aragonite and/or jadeitic pyroxene + quartz (generally also lawsonite).

any particular depth was greater in the lithologically recorded past than at present (Roering, 1967, and Martin, 1969, arrived at a similar conclusion from a study of South African metamorphic belts). This hypothesized secular decrease in the geothermal gradient could account for the increasing proportions of blueschists in younger metamorphosed rocks as well as for their successively higher-pressure mineralogies.

If P–T conditions required for the generation of these metamorphic rocks reflect the subduction zone environment (Ernst, 1970, 1971), then, provided subduction occurred at all in ancient geologic times, the implications of a temporal change in geothermal gradient should be considered. According to the petrologic-tectonic model previously advanced, the unusually high pressures at low temperatures are attained due to the underflow to considerable depths of a thick, relatively cool lithospheric plate. The faster the underflow and the thicker the plate, the greater should be the downward deflection of the isotherms, hence higher pressure for a given temperature (for example, see MacKenzie, 1969; Oxburgh and Turcotte, 1970, 1971; Minear and Toksöz, 1970; Toksöz, Minear, and Julian, 1971). The ubiquity of only thin lithospheric slabs, or of exclusively slow convergence rates, on the other hand, would fail to provide the requisite environment for the production of blueschists.

Inasmuch as radiogenic heat production is decreasing with time, and considering the thermal-dynamic nature of the general plate tectonic model (for example, see Isacks, Oliver, and Sykes, 1968), it seems unlikely that present-day flow velocities in the upper mantle greatly exceed those of the distant past. However that may be, provided the boundary between lithosphere and asthenosphere represents that region in which the Earth's geothermal gradient intersects the P–T curve for minimum incipient melting of the mantle (Anderson, Sammis, and Jordan, 1971), a higher temperature gradient in the past would be in accord with greater radiogenic heat production and with the existence of thinner lithospheric plates (also see Armstrong, 1970, p. 460; Hart and others, 1970; Wetherill, in press).

If the inferred decrease in lithospheric plate thickness is extrapolated back in time, one may wonder at what stage in the Earth's evolution plates first appeared. Because sialic material partially melts at temperatures below values for the mantle solidus, the presence of lithospheric plates evidently was required before sizable continental masses could have consolidated in early Precambrian time. However, judging from P–T gradients qualitatively deduced from mineral parageneses noted in table 1, the portion of the Precambrian for which rock sections are preserved seemingly was characterized by relatively thin lithospheric plates. If so, ancient plate motions and interactions may have departed somewhat from those observed today.

ACKNOWLEDGMENTS

This rather speculative note has benefited from constructive reviews by R. G. Coleman, U.S. Geological Survey, Menlo Park, Akiho Miyashiro, State University of New York at Albany, and W. P. de Roever, University of Amsterdam. I thank these authorities for pointing out several pertinent literature citations of which I had previously been unaware. The interpretations presented in this paper, of course, are my sole responsibility, although they lean heavily on the earlier works of de Roever.

REFERENCES

Anderson, D. L., Sammis, C., and Jordan, T., 1971, Composition and evolution of the mantle and core: Science, v. 171, p. 1103-1112.

Anhaeusser, C. R., Mason, R., Viljoen, M. J., and Viljoen, R. P., 1968, A reappraisal of some aspects of Precambrian shield geology: Johannesburg, South Africa, Univ. Witwatersrand, Econ. Geology Research Unit, Inf. Circ. 49, p. 1-30.

Armstrong, R. L., 1970, Review of *Age relations in high-grade metamorphic terranes*, edited by H. R. Wynne-Edwards: Am. Jour. Sci., v. 268, p. 459-462.

Bailey, E. H., Blake, M. C., Jr., and Jones, D. L., 1970, On-land Mesozoic oceanic crust in California Coast Ranges: U.S. Geol. Survey Prof. Paper 700-C, p. 70-81.

Bailey, E. H., Irwin, W. P., and Jones, D. L., 1964, Franciscan and related rocks and their significance in the geology of western California: California Div. Mines and Geology Bull., v. 183, 177 p.

Banno, Shohei, 1958, Glaucophane schists and associated rocks in the Omi district, Niigata prefecture, Japan: Japanese Jour. Geology Geography v. 29, p. 29-44.

Banno, Shohei, and Miller, J. A., 1964, Additional data on the age of metamorphism of the Ryoke-Abukuma and Sanbagawa metamorphic belts, Japan: Japanese Jour. Geology Geography, v. 36, p. 17-22.

Bearth, Peter, 1962, Versuch einer Gliederung alpinmetamorphen serien der Westalpen: Schweizer Mineralog. Petrog. Mitt., v. 42, p. 127-137.

————— 1966, Zur mineralfaziellen Stellung der Glaucophan Gesteine der Westalpen: Schweizer Mineralog. Petrog. Mitt., v. 46, p. 13-23.

————— 1967, Die Ophiolithe der Zone von Zermatt- Saas Fee: Beitr. geol. Karte Schweiz, neue folge, v. 132, 130 p.

Biq, C., 1971, A fossil subduction zone in Taiwan: Geol. Soc. China Proc., v. 14, p. 146-154.

Bloxam, T. W., 1958, Pumpellyite from south Ayrshire: Mineralog. Mag., v. 31, p. 811-813.

Bloxam, T. W., and Allen, J. B., 1959, Glaucophane-schist, eclogite, and associated rocks from Knockormal in the Girvan-Ballantrae complex, south Ayrshire: Royal Soc. Edinburgh Trans., v. 64, p. 1-27.

Bocquet, J., 1971, Cartes de répartition de quelques minéraux du métamorphisme alpin dans les Alpes franco-italiennes: Eclogae geol. Helvetiae, v. 64, p. 71-103.

Boettcher, A. L., and Wyllie, P. J., 1968, Jadeite stability measured in the presence of silicate liquids in the system $NaAlSiO_4$–SiO_2–H_2O: Geochim. et Cosmochim. Acta, v. 32, p. 999-1012.

Brothers, R. N., 1970, Lawsonite-albite schists from northernmost New Caledonia: Contr. Mineralogy and Petrology, v. 25, p. 185-202.

Brouwer, H. A., and Egeler, C. G., 1952, The glaucophane facies metamorphism in the schistes lustrés nappe of Corsica: Koninkl. Nederlandse Akad. Wetensch., Afdeeling Natuurk., (Tweede Reeks), pt. 48, no. 3, 71 p.

Brown, W. H., Fyfe, W. S., and Turner, F. J., 1962, Aragonite in California glaucophane schists, and the kinetics of the aragonite-calcite transformation: Jour. Petrology. v. 3, p. 566-582.

Chatterjee, N. D., 1971, Phase equilibria in the alpine metamorphic rocks of the environs of the Dora-Maira Massif, western Italian Alps, Parts I and II: Neues Jahrb. Mineralogie Abh., v. 114, p. 181-245.

Cogné, J., Jeannette, D., and Ruhland, M., 1966, L'île de Groix. Étude structurale d'une série métamorphique à glaucophane en Bretagne méridionale: Services carte géol. Alsace Lorraine Bull., v. 19, p. 41-95.

Cogulu, E., 1967, Etude pétrographique de la région de Mihaliçcik (Turquie): Schweizer. Mineralog. Petrog. Mitt., v. 47, p. 683-824.

Cogulu, E., and Krummenacher, D., 1967, Problèmes géochronométriques dans la partie NW de l'Anatolie centrale (Turquie): Schweizer. Mineralog. Petrog. Mitt., v. 47, p. 825-831.

Cohen, L. H., Condie, K. C., Kuest, L. J., Mackenzie, G. S., Meister, F. H., Pushkar, Paul, and Steuber, A. M., 1963, Geology of the San Benito Islands, Baja California, Mexico: Geol. Soc. America Bull., v. 74, p. 1355-1370.

Coleman, R. G., 1966, New Zealand serpentinites and allied metasomatic rocks: New Zealand Geol. Survey Bull., v. 76, 102 p.

———— 1967, Glaucophane schists from California and New Caledonia: Tectonophysics, v. 4, p. 479-498.

———— 1971, Plate tectonic emplacement of upper mantle peridotites along continental edges: Jour. Geophys. Research, v. 76, p. 1212-1222.

Coleman, R. G., and Lee, D. E., 1962, Metamorphic aragonite in the glaucophane schists of Cazadero, California: Am. Jour. Sci., v. 260, p. 577-595.

———— 1963, Glaucophane-bearing metamorphic rock types of the Cazadero area, California: Jour. Petrology, v. 4, p. 260-301.

Coleman, R. G., Lee, D. E., Beatty, L. B., and Brannock, W. W., 1965, Eclogites and eclogites: their differences and similarities: Geol. Soc. America Bull., v. 76, p. 483-508.

Davies, H. L., 1968, Papuan ultramafic belt: Internat. Geol. Congress, 23d, Prague 1968, Proc., v. 1, p. 209-220.

Davies, H. L., and Smith, I. E., 1971, Geology of eastern Papua: a synthesis: Geol. Soc. America Bull., v. 82, p. 3299-3312.

Davis, B. L., and Adams, L. H., 1965, Kinetics of the calcite $\rightleftharpoons$ aragonite transformation: Jour. Geophys. Research, v. 70, p. 433-441.

Davis, G. A., 1966, Metamorphic and granitic history of the Klamath Mountains, *in* Bailey, E. H., ed., Geology of Northern California: California Div. Mines and Geology Bull., v. 190, p. 39-50.

———— 1968, Westward thrust faulting in the south-central Klamath Mountains, California: Geol. Soc. America Bull., v. 79, p. 911-934.

Dengo, Gabriel, 1953, Geology of the Caracas region, Venezuela: Geol. Soc. America Bull., v. 64, p. 7-40.

Dewey, J. F., and Bird, J. M., 1970, Mountain belts and the new global tectonics: Jour. Geophys. Research, v. 75, p. 2625-2647.

Dickinson, W. R., 1970, Relation of andesites, granites, and derived sandstones to arc-trench tectonics: Rev. Geophysics and Space Physics, v. 8, p. 813-860.

Dickinson, W. R., and Hatherton, T., 1967, Andesitic volcanism and seismicity around the Pacific: Science, v. 157, p. 801-803.

Dobretsov, N. L., and Kuroda, I., 1969, Geologic laws characterizing glaucophane metamorphism in northwestern part of the folded frame of Pacific Ocean: Internat. Geology Rev., v. 12, p. 1389-1407.

Dobretsov, N. L., Reverdatto, V. V., Sobolev, V. S., Sobolev, N. V., Jr., Ushakova, E. N., and Khlestov, V. V. 1966, The map of metamorphic facies of the USSR: Akad. Nauk SSSR, Sibir. Otdel., Inst. Geol. Geof. (Natl. Acad. USSR, Siberian Branch, Inst. Geol. Geophys.), scale 1:7,500,000.

Elsasser, W. M., 1971, Sea-floor spreading as thermal convection: Jour. Geophys. Research, v. 76, p. 1101-1112.

Ernst, W. G., 1970, Tectonic contact between the Franciscan mélange and the Great Valley sequence, crustal expression of a Late Mesozoic Benioff zone: Jour. Geophys. Research v. 75, p. 886-901.

————— 1971, Metamorphic zonations on presumably subducted lithospheric plates from Japan, California and the Alps: Contr. Mineralogy and Petrology, v. 34, p. 43-59.

Ernst, W. G., Seki, Y., Onuki, H., and Gilbert, M. C., 1970, Comparative study of low-grade metamorphism in the California Coast Ranges and the Outer Metamorphic Belt of Japan: Geol. Soc. America Mem. 124, 276 p.

Firsov, L. V., and Dobretsov, N. L., 1969, Age of glaucophane metamorphism at the northwestern fringe of the Pacific Ocean: Acad. Sci. USSR Doklady, Earth Sci. Sec., v. 185, p. 46-48.

Forbes, R. B., Hamilton, T., Tailleur, I. L., Miller, T. P., and Patton, W. W., 1971, Tectonic implications of blueschist facies metamorphism in Alaska: Nature, v. 234, p. 106-108.

Gervasio, F. C., 1967, Age and nature of orogenesis of the Philippines: Tectonophysics, v. 4, p. 379-402.

González-Bonorino, F., 1971, Metamorphism of the crystalline basement of central Chile: Jour. Petrology, v. 12, p. 149-175.

González-Bonorino, F., and Aguirre, L., 1970, Metamorphic facies series of the crystalline basement of Chile: Geol. Rundschau, v. 59, p. 979-994.

Greenly, E., 1919, The geology of Anglesey: Great Britain Geol. Survey Mem., 980 p.

Hamilton, Warren, 1970, The Uralides and the motion of the Russian and Siberian platforms: Geol. Soc. America Bull., v. 81, p. 2553-2576.

Harper, C. T., and Landis, C. A., 1967, K-Ar ages from regionally metamorphosed rocks, South Island, New Zealand, and some tectonic implications: Earth and Planetary Sci. Letters, v. 2, p. 419-429.

Hart, S. R., Brooks, C., Krogh, T. E., Davis, G. L., and Nava, D., 1970, Ancient and modern volcanic rocks: a trace element model: Earth and Planetary Sci. Letters, v. 10, p. 17-28.

Hashimoto, M., 1968, Regional metamorphism of the Katsuyama district, Okayama Prefecture, Japan: Tokyo Univ. Fac. Sci. Jour., sec. II, v. 17, p. 99-162.

Hawkins, J. W., 1967, Prehnite-pumpellyite facies metamorphism of a graywacke-shale series, Mount Olympus, Washington: Am. Jour. Sci., v. 265, p. 798-818.

Hayase, I., and Ishizaka, K., 1967, Rb-Sr dating on the rocks in Japan, I: Japanese Assoc. Mineralogy, Petrology, and Econ. Geology Jour., v. 58, p. 201-211.

Hill, P. A., 1959, Geology and structure of the northwest Trinidad Mountains, Las Villas Province, Cuba: Geol. Soc. America Bull., v. 70, p. 1459-1478.

Ho, C. S., 1967, Structural evolution of Taiwan: Tectonophysics, v. 4, p. 367-378.

Hoffmann, C., 1970, Die Glaukophangesteine, ihre stofflichen Äquivalente und Umwandlungsprodukte in Nordcalabrien (Süditalien): Beitr. Mineralogie Petrographie, v. 27, p. 283-320.

Holgate, Norman, 1951, On crossite from Anglesey: Mineralog Mag., v. 29, p. 792-798.

Hsü, K. J., 1971, Franciscan mélanges as a model for eugeosynclinal sedimentation and underthrusting tectonics: Jour. Geophys. Research, v. 76, p. 1162-1170.

Hunziker, J. C., 1970, Polymetamorphism in the Monte Rosa, Western Alps: Eclogae geol. Helvetiae, v. 63, p. 151-161.

Isacks, B., Oliver, J., and Sykes, L. R., 1968, Seismology and the new global tectonics: Jour. Geophys. Research, v. 73, p. 5855-5899.

Johannes, W., and Puhan, D., 1971, The calcite-aragonite transition, reinvestigated: Contr. Mineralogy Petrology, v. 31, p. 28-38.

Joplin, G. A., 1968, A petrography of Australian metamorphic rocks: New York, Am. Elsevier Pub. Co., Inc., 262 p.

Landis, C. A., and Coombs, D. S., 1967, Metamorphic belts and orogenesis in southern New Zealand: Tectonophysics, v. 4, p. 501-518.

Lebedev, M. M., Tararin, I. A., and Lagouskaya, E. A., 1967, Metamorphic zones of Kamchatka as an example of the metamorphic assemblages of the inner part of the Pacific belt: Tectonophysics, v. 4, p. 445-461.

Lee, D. E., Thomas, H. H., Marvin, R. F., and Coleman, R. G., 1964, Isotope ages of glaucophane schists from Cazadero, California: U.S. Geol. Survey Prof. Paper 475-D, p. 105-107.

Lillie, A. R., and Brothers, R. N., 1970, The geology of New Caledonia: New Zealand Jour. Geology and Geophysics, v. 13, p. 145-183.

Liou, J. G., 1971, P-T stabilities of laumontite, wairakite, lawsonite, and related minerals in the system $CaAl_2Si_2O_8$-SiO_2-H_2O: Jour. Petrology, v. 12, p. 379-411.

MacKenzie, D. P., 1969, Speculations on the consequences and cause of plate motions: Royal Astron. Soc. Geophys. Jour., v. 18, p. 1-32.

Martin, H., 1969, Problems of age relations and structure in some metamorphic belts of southern Africa, *in* Wynne-Edwards, H. R., ed., Age relations in high-grade metamorphic terrains: Geol. Assoc. Canada Spec. Paper 5, p. 17-36.

Matsumoto, T., 1947, The geological research of the Aritagawa Valley, Wakayama Prefecture—a contribution to the tectonic history of the outer zone of southwest Japan: Kyushu Univ. Sci. Rept., v. 2, p. 1-12 (in Japanese).

McBirney, A. R., 1963, Geology of a part of the central Guatemalan cordillera: Calif. Univ. Pub. Geol. Sci., v. 38, p. 177-242.

McBirney, A. R., Aoki, K. I., and Bass, M. N., 1967, Eclogites and jadeite from the Motague fault zone, Guatemala: Am. Mineralogist, v. 52, p. 908-918.

Miller, J. A., Shido, F., Banno, S., and Uyeda, S., 1961, New data on the age of orogeny and metamorphism of Japan: Japanese Jour. Geology and Geography, v. 32, p. 145-151.

Minear, J. W., and Toksöz, M. N., 1970, Thermal regime of a downgoing slab and global tectonics: Jour. Geophys. Research, v. 75, p. 1397-1419.

Misch, Peter, 1966, Tectonic evolution of the Northern Cascades of Washington State, *in* Gunning, H. C., ed., Tectonic history and mineral deposits of the Western Cordillera: Canadian Inst. Mining Metallurgy Spec., v. 8. p. 108-148.

———— 1969, Paracrystalline microboudinage of zoned grains and other criteria for synkinematic growth of metamorphic minerals: Am. Jour. Sci., v. 267, p. 43-63.

Miyashiro, Akiho, 1961, Evolution of metamorphic belts: Jour. Petrology, v. 2, p. 277-311.

———— 1967, Orogeny, regional metamorphism, and magmatism in the Japanese islands: Dansk geol. Fören. Medd. (København), v. 17, p. 390-446.

Monger, J. W. H., 1969, Stratigraphy and structure of Upper Paleozoic rocks, northeast Dease Lake map-area, British Columbia (104J): Canada Geol. Survey Paper 68-48, 41 p.

Monger, J. W. H., and Hutchison, W. W., 1970, Metamorphic map of the Canadian Cordillera: Canada Geol. Survey Paper 70-33, 61 p.

Nicolas, A., 1968, Relations structurales entre le massif ultrabasique de Lanzo, ses satellites et la zone de Sesia Lanzo: Schweizer Mineralog. Petrog. Mitt., v. 48, p. 145-156.

———— 1969, Tectonique et métamorphisme dans les Stura di Lanzo (Alpes Piemontaises): Schweizer Mineralog. Petrog. Mitt., v. 49, p. 359-377.

Niggli, E., 1970, Alpine Metamorphose und alpine Gebirgsbildung: Fortschr. Mineralogie, v. 47, p. 16-26.

Nijhuis, H. J., 1964, Plurifacial alpine metamorphism in the south-eastern Sierra de los Filabres south of Lubrín, SE Spain: Univ. Amsterdam Geol. Inst. Mededeel., no. 297, 151 p.

Oxburgh, E. R., and Turcotte, D. L., 1970, Thermal structure of island arcs: Geol. Soc. America Bull., v. 81, p. 1665-1668.

———— 1971, Origin of paired metamorphic belts and crustal relation in island arc regions: Jour. Geophys. Research, v. 76, p. 1315-1327.

Plas,, L. van der, 1959, Petrology of the northern Adula region, Switzerland (with particular reference to glaucophane-bearing rocks): Leidse geol. Mededeel., v. 24, p. 415-602.

Roddick, J. A., Wheeler, J. O., Gabrielse, H., and Souther, J. G., 1967, Age and nature of the Canadian part of the circum-Pacific orogenic belt: Tectonophysics, v. 4, p. 319-337.

Roering, C., 1967, Non-orogenic granites in the Archean geosyncline of the Barberton Mountain Land: Johannesberg, South Africa, Univ. Witwatersrand, Econ. Geology Research Unit, Inf. Circ. 35, p. 1-13.

Roever, W. P. de, 1947, Igneous and metamorphic rocks in eastern Celebes, *in* Brouwer, H. A., ed., Geological Exploration in the Island of Celebes: Amsterdam, North Holland Publishing Co., p. 65-173.

———— 1955, Genesis of jadeite by low-grade metamorphism: Am. Jour. Sci., v. 253, p. 283-298.

———— 1956, Some differences between post-Paleozoic and older regional metamorphism: Geol. en Mijnb., new ser., v. 18, p. 123-127.

421

Roever, W. P. de, 1964, On the cause of the preferential distribution of certain meta-
    morphic minerals in orogenic belts of different age: Geol. Rundschau, v. 54, p. 933-943.
Roever, W. P. de, and Nijhnis, H. J., 1963, Plurifacial alpine metamorphism in the
    eastern Betic Cordilleras (SE Spain), with special reference to the genesis of the
    glaucophane: Geol. Rundschau, v. 53, p. 324-336.
Roever, W. P. de, Roever, E. W. F. de, Beunk, F. F., and Lahaye, P. H. J., 1967,
    Preliminary note on ferrocarpholite from a glaucophane- and lawsonite-bearing
    part of Calabria, southern Italy: Koninkl. Nederlandse Akad. Wetensch., Afdeeling
    Natuurk., v. 70, p. 543-537.
Sainsbury, C. L., Coleman, R. G., and Kachadoorian, R., 1970, Blueschist and related
    greenschist facies rocks of the Seward Peninsula, Alaska: U.S. Geol. Survey Prof.
    Paper 700-B, p. 33-42.
Saliot, P., 1968, Sur la présence et la signification de la lawsonite dans la cordillère
    côtiére du Chili (île de Chiloé): Acad. Sci. Paris Comptes rendus, v. 267, ser. D,
    p. 1183-1185.
Schürmann, H. M. E., 1936, Lawsonit aus Cuba: Zentralbl. Mineralogie Abh., no. 8,
    p. 245-251.
Seki, Yotaro, 1958, Glaucophanitic regional metamorphism in the Kanto Mountains,
    Central Japan: Japanese Jour. Geology Geography, v. 29, p. 233-258.
———— 1960, Jadeite in Sanbagawa crystalline schists of central Japan: Am. Jour.
    Sci., v. 258, p. 705-715.
———— 1965, Jadeitic pyroxene found as pebbles in lower Cretaceous formation of
    the Kanto Mountains, central Japan: Japan Assoc. Mineralogy, Petrology, and
    Econ. Geology Jour., v. 53, p. 165-168 (in Japanese).
Seki, Yotaro, and Shido, F., 1959, Finding of jadeite from the Sanbagawa and
    Kamuikotan metamorphic belts, Japan: Japan Acad. Proc., v. 35, p. 137-138.
Seki, Yotaro, Oba, T., Mori, R., and Kuriyagawa, S., 1964, Sanbagawa metamorphism
    in the Central part of Kii Peninsula: Japan. Assoc. Mineralogy, Petrology, and
    Econ. Geology Jour., v. 52, p. 73-89.
Shackleton, R. M., 1969, The pre-Cambrian of North Wales, *in* Wood, A., ed., The
    pre-Cambrian and Lower Paleozoic rocks of Wales: Cardiff, Univ. of Wales Press,
    p. 1-22.
Shagam, R., 1960, Geology of central Aragua, Venezuela: Geol. Soc. America Bull.,
    v. 71, p. 249-302.
Sobolev, V. S., Dobretsov, N. L., Reverdatto, V. V., Sobolev, N. V., Ushrakova, E.
    N., and Khlestov, V. V., 1967, Metamorphic facies and series of facies in the
    U.S.S.R.: Dansk Geol. Fören. Medd. (København), v. 17, p. 458-472.
Suppe, John, 1969, Times of metamorphism in the Franciscan terrain of the northern
    Coast Ranges, California: Geol. Soc. America Bull., v. 80, p. 135-142.
Suzuki, J., and Suzuki, Y., 1959, Petrological study of the Kamuikotan metamorphic
    complex in Hokkaido, Japan: Hokkaido Univ. Fac. Sci., Jour., ser. 4, v. 10, p. 349-
    446.
Swanson, D. A., 1969, Lawsonite blueschist from north-central Oregon: U.S. Geol.
    Survey Prof. Paper 650-B, p. 8-11.
Toksöz, M. N., Minear, J. W., and Julian, B. R., 1971, Temperature field and geo-
    physical effects of a downgoing slab: Jour. Geophys. Research, v. 76, p. 1113-1138.
Vance, J. A., 1968, Metamorphic aragonite in the prehnite-pumpellyite facies, north-
    west Washington: Am. Jour. Sci., v. 266, p. 299-315.
van der Kaaden, G., 1966, The significance and distribution of glaucophane rocks
    in Turkey: Mineral Research and Exploration Inst. Turkey Bull., v. 67, p. 36-67.
———— 1969, Zur Entstehung der Glaukophan-Lawsonit- und glaukophanitischen
    Grünschiefer-Fazies, Geländebeobachtungen und Mineralsynthesen: Fortschr. Min-
    eralogie, v. 46, p. 87-136.
Velde, M. B., 1970, Essai d'analyse pétrologique des séries cristallophyliennes anciennes
    de France: mise en évidence de deux épisodes métamorphiques anté-hercyniens:
    Acad. Sci. Paris Comptes rendus, v. 271, p. 2247-2250.
Verhofstad, J., 1966, Glaucophanitic stone implements from West New Guinea (West
    Irian): Geol. en Mijnb., v. 45, p. 291-300.
Wetherill, G. W., in press, The beginning of continental evolution: Tectonophysics.
Yamaguchi, M., and Yanagi, T., 1970, Geochronology of some metamorphic rocks in
    Japan: Eclogae geol. Helvetiae, v. 63, p. 371-388.
Yen, T. P., 1966, Glaucophane schist of Taiwan: Geol. Soc. China Proc., v. 9, p. 70-73.
———— 1967, Structural analysis of the Tananao schist of Taiwan: Taiwan Geol.
    Survey Bull., v. 18, 110 p.

Erratum

On p. 422, in the reference entry under "Sobolov, V. S.," the fifth author's name
should read "Ushakova, E. N."

VI

Summary

The spatial distribution of rocks representing the various metamorphic facies is clearly a function of lithospheric plate dynamics. This is because motions of the several plates influence the local and regional thermal structures, which in turn affect the metamorphic mineralogic assemblages. Figures 1 and 2 provide generalized thermal structures for divergent and convergent plate junctions, respectively (see also Papers 3 and 4). Geologic relationships are also illustrated, as are typical geothermal gradients for subduction zones, magmatic arcs, stable plate interiors, and oceanic ridges. The third variety of plate boundary, a transform fault, would be more-or-less equivalent to one of these other types, inasmuch as only minor vertical motions of material are involved (hence perturbation of the thermal regime would be only local).

Figure 3 presents a schematic petrogenetic grid for crustal rocks. This diagram is based on oxygen isotope geothermometry and experimentally determined phase equilibria, and assumes a high activity of H_2O at low a_{CO_2}. Variation in these parameters, as well as in the activity of oxygen and other mobile species in the fluid phase (if present), and differences in rock bulk compositions markedly influence P–T values for the petrogenetic grid. However, the overall topology appears to be reasonable.

Presuming the appropriateness of the thermal structures presented in Figures 1 and 2, and the petrogenetic grid of Figure 3, the spatial disposition of metamorphic facies in the neighborhood of divergent and convergent lithospheric plate boundaries can be approximated, as shown in Figures 4 and 5, respectively. Chemical equilibrium has been assumed. Local departures from the inferred thermal regime and/or variations in rock bulk compositions and the extent to which phase equilibrium has been attained (a problem of kinetics) undoubtedly account for some of the complicated natural occurrences—and structural intricacies for others—but the general relation-

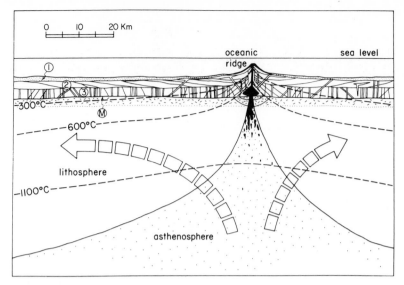

Figure 1. Schematic diagram of an accreting plate margin. The diagram corresponds in general to the East Pacific Rise, which is characterized by relatively rapid sea-floor spreading; for more slowly diverging plate margins, such as the Mid-Atlantic Ridge, areal portions of the rise show evidence of rifting and typically a well-developed medial graben. Increasing degrees of asthenosphere partial melting on ascent (hence depressurization) provide tholeiitic ridge basaltic magmas, shown in black. Oceanic layers are (1) deep-sea pelagic sediments, (2) pillow basalts and breccias, and (3) gabbros + sheeted diabase dike complex. (M) is the Mohorovicic discontinuity. Note the region of cumulate ultramafic occurring as fractional crystallization product of tholeiitic magma (shown as partly solidified). Relatively high temperature, relatively low pressure metamorphism is confined to the vicinity of the rising plume.

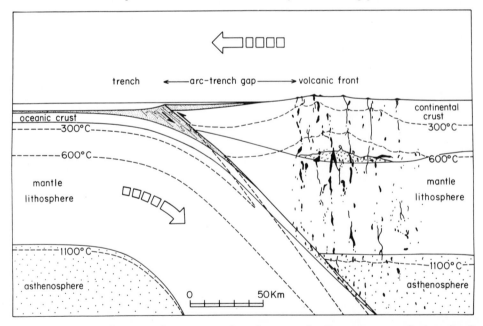

Figure 2. Schematic diagram of a consumptive plate margin. In some cases (Indonesia, for example), imbricate thrusting—or underplating—and uplift at the landward wall of the trench has produced an outer, nonvolcanic arc; for simplicity this has not been illustrated in the figure, which is rather similar to the present margin of northern Chile. Partial melting of preexisting rocks occurs in three regions: (1) the metabasaltic oceanic crust (amphibolite and/or eclogite) at the top of the downgoing plate, giving rise to calc-alkaline melts; (2) the stable, non-subducted asthenosphere (hanging wall) adjacent the descending plate; and (3) the basal, thickened portions of the sialic, H_2O-rich crust. The first two magma types are shown in black, the third with a checked pattern. Relatively high pressure, relatively low temperature metamorphism takes place in the narrow subduction zone mélange, whereas a broad zone of relatively high-temperature, relatively low pressure recrystallization characterizes the magmatic arc.

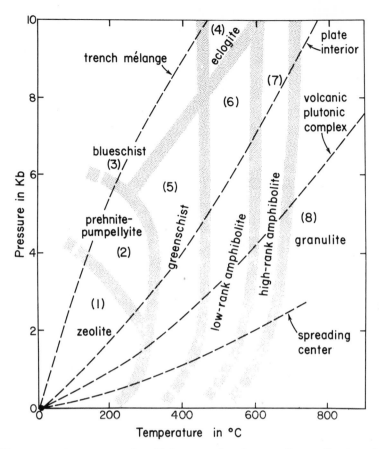

Figure 3. Diagrammatic petrogenetic grid for crustal rock types. Except for the relatively high temperature portions of the diagram, an aqueous-rich fluid phase is presumed to be ubiquitous during prograde metamorphism. Characteristic metamorphic P–T trajectories for contrasting plate tectonic environments are illustrated also.

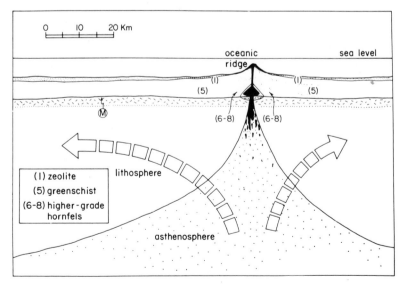

Figure 4. Schematic distribution of metamorphic facies types in the neighborhood of an accreting plate margin, assuming the correctness of Figures 1 and 3.

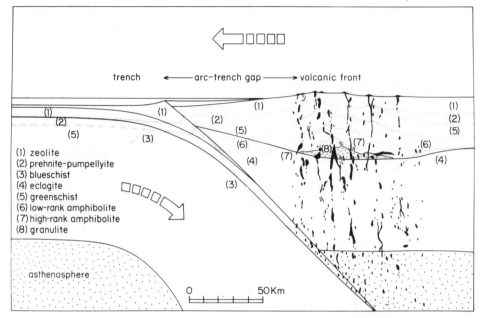

Figure 5. Schematic distribution of metamorphic facies types in the neighborhood of a consumptive plate margin, assuming the correctness of Figures 2 and 3.

ship between recrystallization and sea-floor spreading seems to be reasonably well understood.

It is apparent from Figure 4 that a relatively simple metamorphic zonation is developed in the vicinity of an oceanic ridge. Rocks characteristic of the several high-temperature, low-pressure metamorphic facies are transported laterally during the sea-floor-spreading process; concomitant with declining temperatures, retrograde reactions occur to a greater or lesser degree, depending on the chemical kinetics.

Near a convergent plate junction, such as illustrated in Figure 5, the nature of the thermal structure produces an oceanward, relatively high pressure, relatively low temperature metamorphic belt, and a landward, relatively high temperature, relatively low pressure metamorphic terrane. The former is a narrow, elongate band of rocks that exhibits oceanward vergence of nappes and a pronounced assymmetry in mineral assemblages, with highest grade (most deeply subducted) blueschistic sections lying at the ancient plate junction and successively more feebly recrystallized rocks extending seaward. In contrast, the landward terrane occupies a broad region, shows telescoped, relatively high temperature mineral parageneses, a crude bilateral symmetry to the volcanic-plutonic axis, and in large part vertical tectonics.

Author Citation Index

Subject Index